Data and Methods to Understand Travel 2018

Transportation Research Record: Journal of the Transportation Research Board

Volume 2672, Issue 42

ISBN: 978-1-5108-7837-2

Printed from e-media with permission by:

Curran Associates, Inc.
57 Morehouse Lane
Red Hook, NY 12571

Some format issues inherent in the e-media version may also appear in this print version.

Published by Sage Publications USA

ISBN (Print) 978-1-5108-7837-2
ISBN 2018 Printed Set (All Issues) 978-1-5108-7735-1

Printed by Curran Associates, Inc. (2019)

Additional copies of this publication are available from:

Curran Associates, Inc.
57 Morehouse Lane
Red Hook, NY 12571 USA
Phone: 845-758-0400
Fax: 845-758-2633
Email: curran@proceedings.com
Web: www.proceedings.com

Volume 2672 Number 42 December 2018

TRR
JOURNAL OF THE TRANSPORTATION RESEARCH BOARD

Contents

Data and Methods to Understand Travel

Articles

TRB

TRANSPORTATION RESEARCH BOARD

CORE SPONSORS

Transportation Departments of the 50 States and the District of Columbia

Federal Government

U.S. Department of Transportation
- Federal Aviation Administration (FAA)
- Federal Highway Administration
- Federal Motor Carrier Safety Administration (FMCSA)
- Federal Railroad Administration (FRA)
- Federal Transit Administration
- National Highway Traffic Safety Administration (NHTSA)
- Office of the Assistant Secretary of Transportation for Research and Technology (OST-R)

U.S. Air Force Civil Engineer Center (AFCEC)
U.S. Coast Guard (USCG)
U.S. Department of Energy
U.S. Department of the Interior (DOI)
U.S. Environmental Protection Agency (EPA)

Nongovernmental Organizations

American Public Transportation Association (APTA)
Association of American Railroads (AAR)
California Air Resources Board

The National Academies of
SCIENCES · ENGINEERING · MEDICINE

The **National Academy of Sciences** was established in 1863 by an Act of Congress, signed by President Lincoln, as a private, nongovernmental institution to advise the nation on issues related to science and technology. Members are elected by their peers for outstanding contributions to research. Dr. Marcia McNutt is president.

The **National Academy of Engineering** was established in 1964 under the charter of the National Academy of Sciences to bring the practices of engineering to advising the nation. Members are elected by their peers for extraordinary contributions to engineering. Dr. C. D. Mote, Jr., is president.

The **National Academy of Medicine** (formerly the Institute of Medicine) was established in 1970 under the charter of the National Academy of Sciences to advise the nation on medical and health issues. Members are elected by their peers for distinguished contributions to medicine and health. Dr. Victor J. Dzau is president.

The three Academies work together as the National Academies of Sciences, Engineering, and Medicine to provide independent, objective analysis and advice to the nation and conduct other activities to solve complex problems and inform public policy decisions. The Academies also encourage education and research, recognize outstanding contributions to knowledge, and increase public understanding in matters of science, engineering, and medicine.

Learn more about the National Academies of Sciences, Engineering, and Medicine at **www.national-academies.org.**

The **Transportation Research Board** is one of seven major programs of the National Academies of Sciences, Engineering, and Medicine. The mission of the Transportation Research Board is to increase the benefits that transportation contributes to society by providing leadership in transportation innovation and progress through research and information exchange, conducted within a setting that is objective, interdisciplinary, and multimodal. The Board's varied activities annually engage about 7,000 engineers, scientists, and other transportation researchers and practitioners from the public and private sectors and academia, all of whom contribute their expertise in the public interest. The program is supported by state transportation departments, federal agencies including the component administrations of the U.S. Department of Transportation, and other organizations and individuals interested in the development of transportation.

Learn more about the Transportation Research Board at **www.TRB.org**.

Article

TRR
JOURNAL OF THE TRANSPORTATION RESEARCH BOARD

e-mission: An Open-Source, Smartphone Platform for Collecting Human Travel Data

Transportation Research Record
2018, Vol. 2672(42) 1–12

DOI: 10.1177/0361198118770167
journals.sagepub.com/home/trr
SAGE

K. Shankari[1], Mohamed Amine Bouzaghrane[1], Samuel M. Maurer[1], Paul Waddell[1], David E. Culler[1], and Randy H. Katz[1]

Abstract

GPS-equipped smartphones provide new methods to collect data about travel behavior, including travel survey apps that incorporate automated location sensing. Previous approaches to this have involved proprietary or one-off tools that are inconsistent and difficult to evaluate. In contrast, e-mission is an open-source, extensible software platform that consists of (*a*) an app for survey participants to install on their Android or iOS smartphones and (*b*) cloud-hosted software for managing the collected data. e-mission collects continuous location data, user-initiated annotations, and responses to contextual, platform initiated survey questions. New studies can be set up using the existing University of California, Berkeley, infrastructure with no additional coding, or the platform can be extended for more complex projects. This paper reviews the requirements for smartphone travel data collection, describes the architecture and capabilities of the e-mission platform, and evaluates its performance in a pilot deployment. The results show that the platform is *usable*, with over 150 installations in a month; *stable*, with over 85% of users retaining it for more than 3 days; and *extensible*, with interface and survey customizations accomplished in a little over a week of full-time work by a transportation engineering researcher. We hope that e-mission will be a useful tool for app-based data collection and will serve as a catalyst for related research.

The rapid adoption of GPS-equipped smartphones is transforming data collection for travel behavior research and analysis. As of 2016, in the USA, 77% of all adults, and 92% of adults under 30, owned a smartphone (*1*). Passive records of the approximate locations of smartphones are compiled by service providers (as well as by certain location-based app platforms), but purpose-built travel survey apps remain the best way to obtain detailed data. Travel survey apps allow researchers to collect high-precision GPS traces and pose contextual, on-device survey questions for data validation or to gather supplementary information. Technical challenges include a lack of standardization in smartphone sensing systems, shorter battery life than stand-alone GPS devices, and the complexity of building a comprehensive software platform.

This paper introduces e-mission[1], an open-source platform for collecting prompted, user-reported, and automatically sensed travel data from smartphones. It consists of an app for survey participants to install on their Android or iOS smartphones and cloud-hosted software for managing the collected data. e-mission improves on existing tools by being entirely open, modular, and extensible. This provides two important benefits: (*a*) its algorithms for collecting sensor data, managing power drain, and processing GPS traces can be fully documented, benchmarked, and reproduced; and (*b*) project-specific modifications to the software are easy to implement and re-use. This is consistent with the case for open computing programs for reproducible research outlined in (*2*). At the same time, e-mission already provides an extensive suite of functionality and can be quickly deployed for new studies that follow a standard template.

Travel Survey Data Collection

Transportation planners and researchers use data from travel surveys to build predictive models of travel behavior and infrastructure needs. Typical surveys collect information about trip origin, destination, purpose, timing, travel mode, route, and other related information, using paper- or phone-based or electronic tools. This human-based data collection may contain errors and biases, but is ideal for understanding people's perceptions of their own travel (*3*).

Technologies like GPS can reduce the burden on respondents while providing more precise, accurate, and complete records of survey participants' travel. Travel surveys increasingly supplement self-reported information with automatically sensed location data from stand-alone GPS devices (e.g., [*4, 5*]).

[1]University of California, Berkeley, CA

Corresponding Author:
Address correspondence to K. Shankari: shankari@eecs.berkeley.edu

However, these devices have their own drawbacks, such as expense, that make them difficult to use at scale (*6*).

e-mission is part of a new category of smartphone-based tools that combine the expressiveness of surveys with the detail and precision of location sensing. Most smartphones now have GPS chipsets, as well as other sensors like accelerometers that can facilitate travel mode detection. Smartphone-based data collection can provide better data quality and better ease of use for survey participants, at lower overall cost than stand-alone GPS devices.

User Engagement

Survey participants are recruited to studies that use the e-mission platform in the same manner as a traditional survey, but the platform includes a number of features designed to reduce enrollment friction and keep participants engaged in the study. e-mission facilitates on-boarding through (*a*) direct installation of the app from standard app stores, (*b*) optional study-specific interface customization, and (*c*) a clean user experience.

It also includes features to facilitate long-term user engagement through *information provision* or *gamification*, for studies where this is appropriate. Personal travel analytics may appeal to users who are interested in physical activity or environmental sustainability, or just curious about their own mobility patterns. Gamification through personal targets or social competition can make these apps into tools for behavior modification (e.g., [*7, 8*]), and experiments in this area are the topic of active research (e.g., [*9*]). These features can be disabled in cases where they could interfere with a study.

Related Work

A 2014 TRB report (*3*) provides the most extensive review to date of approaches for collecting and analyzing GPS data to study travel behavior. They identify key challenges for smartphone data collection, including: (*a*) *market fragmentation* (different mobile operating systems and hardware capabilities make it difficult to collect equivalent data from all survey participants); (*b*) *power drainage* (continuous collection of GPS data will rapidly drain a smartphone's battery); (*c*) *costly data plans* (cellular data transmission may not be feasible); and (*d*) *sampling biases* (ownership of smartphones varies by age, income, and education).

Many studies have used smartphones to collect travel data, typically falling into three categories: (*a*) automatically generated travel diaries that avoid the errors and biases of self-reporting (e.g., [*10, 11*]); (*b*) behavior modification based on gamifying travel and providing incentives for particular mode choices (e.g., [*7, 8*]); and (*c*) understanding human perceptions by building route choice models for active transportation modes such as bicycling (e.g., [*12, 13*]).

Technological advancement and increasing market penetration of smartphones are leading to steady progress on each of the four challenges identified in the TRB report. However, we identify a fifth, related challenge: *collection platform robustness*. The complexities of cross-device data collection, power management, and data analysis are best addressed by open, modular, extensible software platforms that encourage widespread adoption. Such platforms can easily be deployed for new projects and reliably benchmarked and adapted as technologies change. Importantly, open-source software can improve reproducibility and provide an opportunity for scholars and practitioners to build a collaborative platform that is controlled by the community.

The remainder of this paper is organized as follows. The next section describes the e-mission platform architecture and data collection capabilities. The subsequent section describes usage and extensibility, and the final sections evaluate pilot deployments, identify future work, and conclude.

System Architecture and Collected Data

The core functionality of the e-mission platform is to *collect* and *assemble* travel data. In this section, we identify key data requirements and briefly describe the architecture of the software platform. Important categories of data are: automatically sensed information, user-initiated reports, and platform-initiated requests such as survey questions. For further use, e-mission assembles the raw data into travel diary components, personalized tour models, and other meaningful outputs.

Categories of Human Travel Data

We divide human travel data into three broad categories, based on the technical requirements and user experience of collecting it from a smartphone (see Figure 1).

Automatically Sensed. This represents data, such as location, accelerometer, or microphone readings, obtained automatically from smartphone sensors without any user intervention. Since this data is obtained automatically, it does not represent a cognitive burden on the user and can be collected in large quantities. However, a naive approach of reading data at high frequency from all possible sensors will lead to significant power drain, and represent its own burden on the user—their smartphone may become unusable during the course of the day. Therefore, the data collection processes need algorithms that can strike a balance between data quality and power drain.

Further, this data is typically not useful in itself; inference algorithms need to be run on top of it to generate useful insights. Multiple inferences can be drawn from the same set of base data—for example, accelerometer data can be used for both road quality (*14*) and for travel mode detection (*15*). However, such inferences are inherently inaccurate, so the inference algorithm needs to be able to quantify its uncertainty, and any action on the inference needs to take this into account.

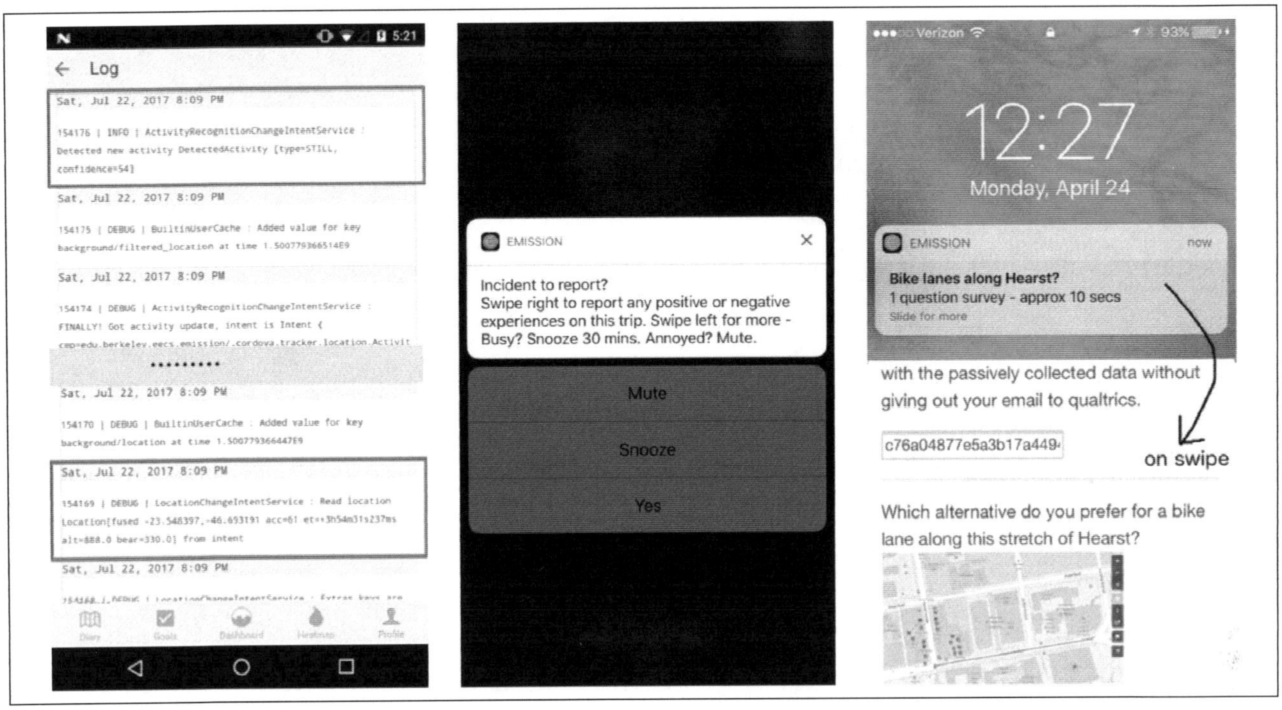

Figure 1. Supported data collection types. Left to right: automatically sensed, user initiated, platform-initiated.

User Initiated. This is data that the user is motivated to report based on his/her surroundings. It is typically perceptual and cannot be inferred by sensor data alone. Examples could be: "The sidewalk here feels empty," "A truck has blocked the bike lane," and so on. Open-ended perceptual data has not typically been integrated into transportation engineering studies but its usage is growing, largely because of data generated on smartphones (e.g., [*16*]).

This type of data is currently collected primarily by commercial projects, such as (*a*) proprietary issue reporting and tracking systems deployed by local public agencies (e.g. SeeClickFix, Comcate, etc.), (*b*) proprietary real-time automobile incident reporting (e.g. Waze), and (*c*) rating systems for points of interest (e.g. Yelp). Including a qualitative component in travel data collection has the potential to provide a richer understanding of human behavior, while supporting new research areas related to data correctness, bias, and heterogeneity of experience (e.g., the four types of cyclists described in [*17*]).

Platform Initiated. This is data that is requested from the user by the platform, such as survey questions. One use of requests is to increase the accuracy of inferred data. Examples of such requests are: (*a*) to obtain *ground truth* for inferences to boost their accuracy; and (*b*) to obtain confirmation of unexpected behavior. However, requesting large amounts of ground truth re-introduces cognitive load on the user. Ground truth acquisition needs to balance accuracy and cognitive load, especially for long- term data collection.

Practitioners can also initiate requests to obtain additional information from a targeted audience. Examples of such requests include: (*a*) obtaining stated preference data about proposed changes from households that will be affected by them; and (*b*) obtain additional demographic information (e.g., bicycle availability) from sub-populations based on their travel patterns (e.g., no recent bicycle trips). These requests are not necessarily tied to sensed or inferred data, and can be fairly complex.

Supported Outputs

The e-mission platform can process collected data into a number of standard outputs (see Figure 2).

Travel Diary. This output is the canonical analysis result. Every one of the prior projects from the literature features a basic travel diary. It is also the building block for the other outputs, so should be considered a core component of a smartphone travel data collection platform.

A travel diary is a linked sequence of *trips* between *places*, each potentially split into *sections*. Each section is associated with a travel *mode* and each trip is associated with a travel *purpose* or *activity*. We suggest using *purpose* to reduce ambiguity because *activity* can have other meanings in a travel context, as in [*18*] and [*19*].

The mobile systems community (since [*20*]) and travel survey community (since [*21*]) have developed several algorithms for automatic mode inference. These typically use

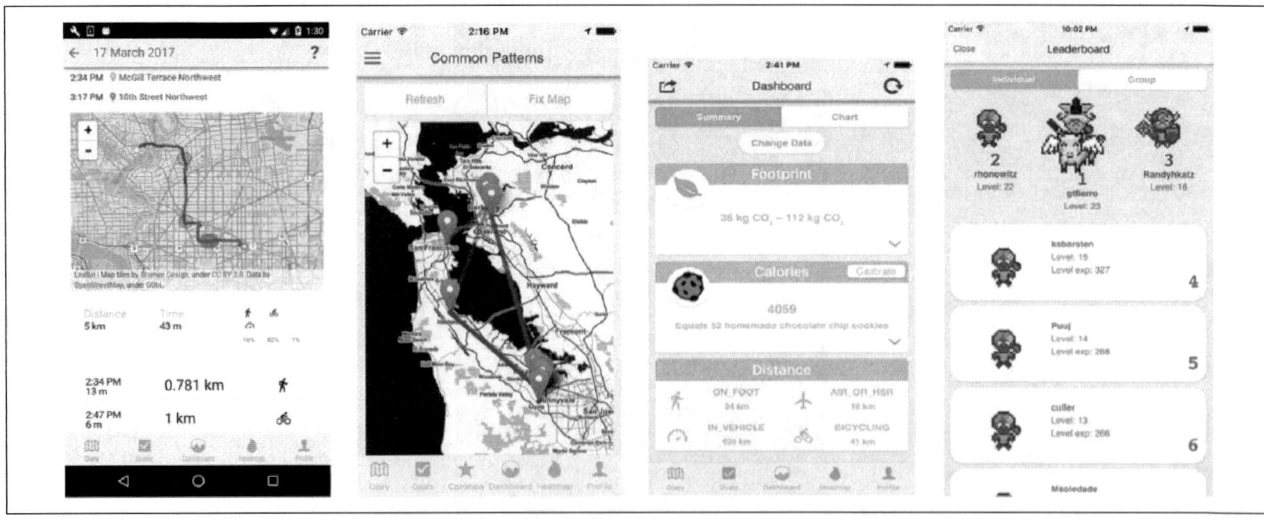

Figure 2. Supported outputs. Left to right: travel diary, tour model, statistics, and game leaderboard.

GPS information alone (*20, 21*) or a combination of GPS and accelerometer data, as in (*22*). They typically support a small number of modes (e.g., walk, bike, car, bus, train) and have accuracies ranging from 70% to 80%. There are fewer algorithms for inferring trip purpose, since it is typically not reflected in sensor readings. Most automated approaches such as (*23*) rely on land use and point-of-interest databases, and the accuracy for locations other than home, work, and school is below 70%.

Personalized Tour Model Graph. This output analyses the trip diary to generate a graph of the common trips taken by the user. This graph is effectively an amalgamation of tour models, similar to the work in (*24*), but with common locations among the tours represented by single nodes in the graph. Collapsing a long sequence of trips into a single graph allows analysis and modeling to focus on a small number of representative trips. For example, detailed semantic data gathering such as stated preference questions can focus on trips in the tour, and reduce user burden. Generating probability distributions over attributes of the common trips (e.g., start and end times) converts it into a Markov model that can be used potentially to predict future behavior, as in (*25*).

Game/Motivation. This output uses the travel diary to generate personalized statistics to motivate travelers to think more carefully about their travel patterns and associated choices. Some projects, such as (*26*), calculate personalized calories burned, carbon footprint, and cost for the traveler. Such projects also typically compare the personalized value to various aggregate statistics to reinforce norm setting, as in (*7*). Other projects, such as (*8*) or (*27*), use gamification techniques such as badges, levels, and challenges to encourage long-term behavior change.

Software Architecture

e-mission follows a sensor–server–client architecture that is standard for Internet of Things (IoT) applications, where everyday devices are used as digital instruments (see Figure 3). In particular, the smartphone app is both the sensor and a client, since personalized information can be viewed on the phone. The server handles communication, long-term storage, data processing, and aggregation. While a detailed description of the architecture and the related work is deferred to a forthcoming paper, we sketch the components and their interaction in this section.

The phone app has a hybrid architecture built using the Apache Cordova mobile app framework. Native plugins for Android and iOS (written in Java and Objective-C) sense *location* and *motion activity* and buffer the data on the phone in a SQLite database. The sensing is automatically turned on at trip start and turned off at trip end to reduce battery consumption. Buffered data is synced to the server after the end of the trip. The trip start and end can also prompt a configurable notification to collect additional information.

The server software is written in Python for ease of extensibility by non-experts. It makes heavy use of Python scientific processing libraries such as scikit-learn, and exposes a REST API for client interaction. It receives data into an input cache, and then saves it into a user-specific section of a shared datastore. The user-specific datastore consists of multiple time series, one for each type of object (e.g., background location, manual incident, etc.). The newly arrived data is then run through a pipeline that generates the applicable outputs. Travel diary information can also be queried for individual or aggregate statistics.

Finally, certain outputs are displayed as Dynamic HTML views. The view in the phone client can display personalized information such as the trip diary, the tour model, and the

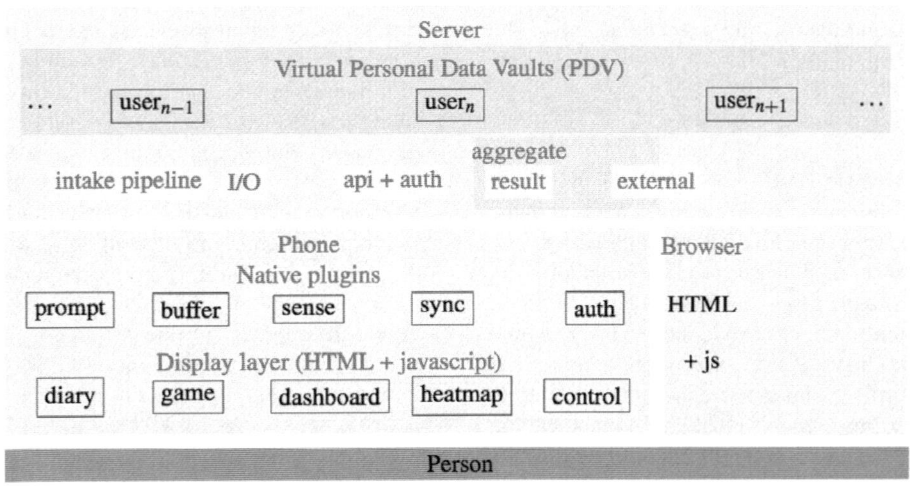

Figure 3. Components of the e-mission architecture.

user's current status in the game. A web app provides a visual display of aggregate, non-personal, usage data.

Usage and Extensibility

In addition to being full-featured, a successful software platform for smartphone data collection must be easy to extend so that it can meet the needs of a variety of projects. Small configuration changes should be easy, and more significant additions to functionality should be achievable using well-defined extension points. Ideally, these changes should be made publicly available for reuse and reproducibility (*12*).

Usage without Customization

If the standard e-mission interface and functionality meet the needs for a study, the practitioner can simply file a research protocol with her institution's review board (IRB) and specify that she will use the e-mission platform for background location data collection. (This is similar to specifying the use of a platform like Qualtrics to collect survey responses.)

The practitioner would then instruct participants to download the e-mission app from the Android or iOS app stores, and obtain separate consent from the participants according to the method specified in the protocol. This consent would need to include the email address that the participant uses to register in e-mission, in order to confirm which users are associated with the study. At the end of the study, the practitioner would show the consent documents to the e-mission lead researcher[2] and receive a copy of the data from those users.[3] (Full in-app consent can be done with simple UI customization; see below.)

Thus, practitioners can collect automatically sensed location and motion activity data without writing any code, simply by directing survey participants to use the app.

Extending the Smartphone App

Easy: Customizing the User Interface (UI). Many practitioners will want to customize the user interface of the app: to add a study logo, to add custom consent, or remove unneeded features. This can generally be done with HTML and CSS changes alone, although functionality related to message prompts involves Javascript.

Because the UI is built using web components, it can be updated without deploying a new app to the stores. The e-mission platform supports multiple UI *channels*, meaning that practitioners can ask survey participants to install the standard e-mission app and then switch to the study-specific channel. A channel can be selected in the UI or by following a special URL or QR code. As soon as a user joins the channel, he/she is presented with study-specific information, consent, and login choices.

Such extensions are shared with the community as new branches on the "e-mission-phone" GitHub repository.[4]

Medium: Extending the Phone App using Existing Plugins. e-mission is built using the Apache Cordova mobile app framework, which allows easy re-use of existing plugins. Functionality like reading a user's calendar or allowing users to take photos can be added in this way. Cordova plugins are controlled using Javascript.

A phone app that has been extended through the addition of new plugins cannot be updated via the UI channels. Instead, a new app would need to be submitted to the stores with a new name and signing key. For iOS, the app must pass the App Store review process. The resulting app would have no obvious connection to the e-mission platform—it could have its own logo, and would be marked as owned by the organization that is submitting it.

Code for such enhancements can be made available to the community by forking the "e-mission-phone" GitHub repository and pushing changes to the fork. Once the project is complete, the enhancement could even be added to the standard e-mission app (in a new UI pane, for example). This would be done by submitting a pull request to the master branch of the "e-mission-phone" repository.

Hard: Writing a New Native Plugin. Some projects may want to use sensing capability that is not currently supported in the Cordova ecosystem, for example, by integrating with a sensor that measures stress from sweat, or using ambient noise to determine whether a car trip is shared or not.

This would require writing native code (in Java and Objective-C or Swift) that reads the appropriate sensors, buffers them, and performs the inference either on the phone or on the server. Such projects can reuse the authentication, buffering, and communication components of the e-mission platform. They can also use the notification component to obtain additional information from the user.

Integration with the e-mission platform would allow the new travel data to be placed in a spatio-temporal context without having to re-write the location tracking and post-processing components. On iOS, restrictions preclude most sensors from being read in the background, but using the e-mission platform would allow plugins to attach themselves to the location tracking callbacks in order to read other sensor data.

Such an extension can be shared with the community by structuring the code as a Cordova plugin and publishing it on GitHub. Projects can then add the plugin like any other.

Extending the Server Functionality

Easy: Adding Queries or Analyses. Aspects of the server software not related to the core outputs are structured as plugins, where new functionality can be added by simply writing a standalone Python script. Some examples are queries to find users who are targets for platform-initiated surveys or notifications to inform users about things related to their travel patterns. New analyses can be added to e-mission by generating a pull request from a fork of the "e-mission-server" repository.[5]

Medium: Modifying Data Pipelines. The existing pipelines for creating travel diaries are open to improvement. Practitioners may want to modify the segmentation, smoothing, or mode inference algorithms used by the core platform. (These pipelines are versioned in GitHub and can be reproduced at any point on a practitioner's own machine. e-mission always retains the original raw data alongside any pipeline outputs.)

These improvements will be more complex to integrate into the core platform, because we need to ensure that they are empirically valid and enough of an improvement to make the default. So while these changes can be contributed using a standard pull request, additional testing will be required before the changes can be merged.

Hardest: Running a Custom Server. Some projects may have data storage and privacy requirements that differ from the core platform and are best achieved by running their own server. Projects that need special external integrations—with an Open Street Maps editor, for example—would also want to run their own server. Projects that modify the core data pipelines could also run a custom server to avoid integrating their changes with the core e-mission platform.

The e-mission server software can run on any Linux, macOS, or other Unix-like system. However, to manage a production backend, you need to be comfortable setting up SSL, obtaining the correct keys for authentication, and monitoring the pipeline logs for errors. Changes to the server software can be shared with the community by publishing the forked code so that it can be used to inform other projects that require similar integrations.

Evaluation

In this section, we evaluate the performance of the platform in two main areas. First, we evaluate its *usability* and *stability* using metrics from a pilot deployment with more than 100 users. Second, we assess the *extensibility* of the platform through a qualitative and quantitative evaluation of the effort required for a non-expert to modify it.

It is also important to note what we do not evaluate: the travel behavior captured by the platform during the pilot. This is because the goal of the pilot was not to generate generalizable results about travel behavior, but to evaluate the use of the platform as a tool to enable others to generate such results. We also do not currently evaluate the accuracy of data collection or trip diary creation, which will be the focus of a future paper. Finally, our goal with this pilot was to assess the installation process and technical stability under varied user interaction patterns. In the upcoming year, we hope to partner with researchers who want to use the platform in their studies. This will allow us to generate usability metrics across more representative populations.

App Usage Metrics

The e-mission platform was launched in a pilot deployment on the U.C. Berkeley campus in fall 2016, covering all the categories of data and outputs described earlier.[6] Participants were not compensated in cash or in kind. The pilot study was linked to an initiative to encourage walking and bicycling to campus. There was no dedicated marketing team or marketing budget associated with the pilot—all publicity was done by researchers associated with the platform. Recruitment was done through email to campus mailing lists, and there was no official endorsement of the pilot as a university initiative.

Installation Rate. This metric captures the rate at which users signed up during the pilot. Ideally, we would use metrics from the app stores to measure this, but the iOS store only reports metrics from users who have opted in to share statistics, so it is not very accurate. Instead, we use calls to the

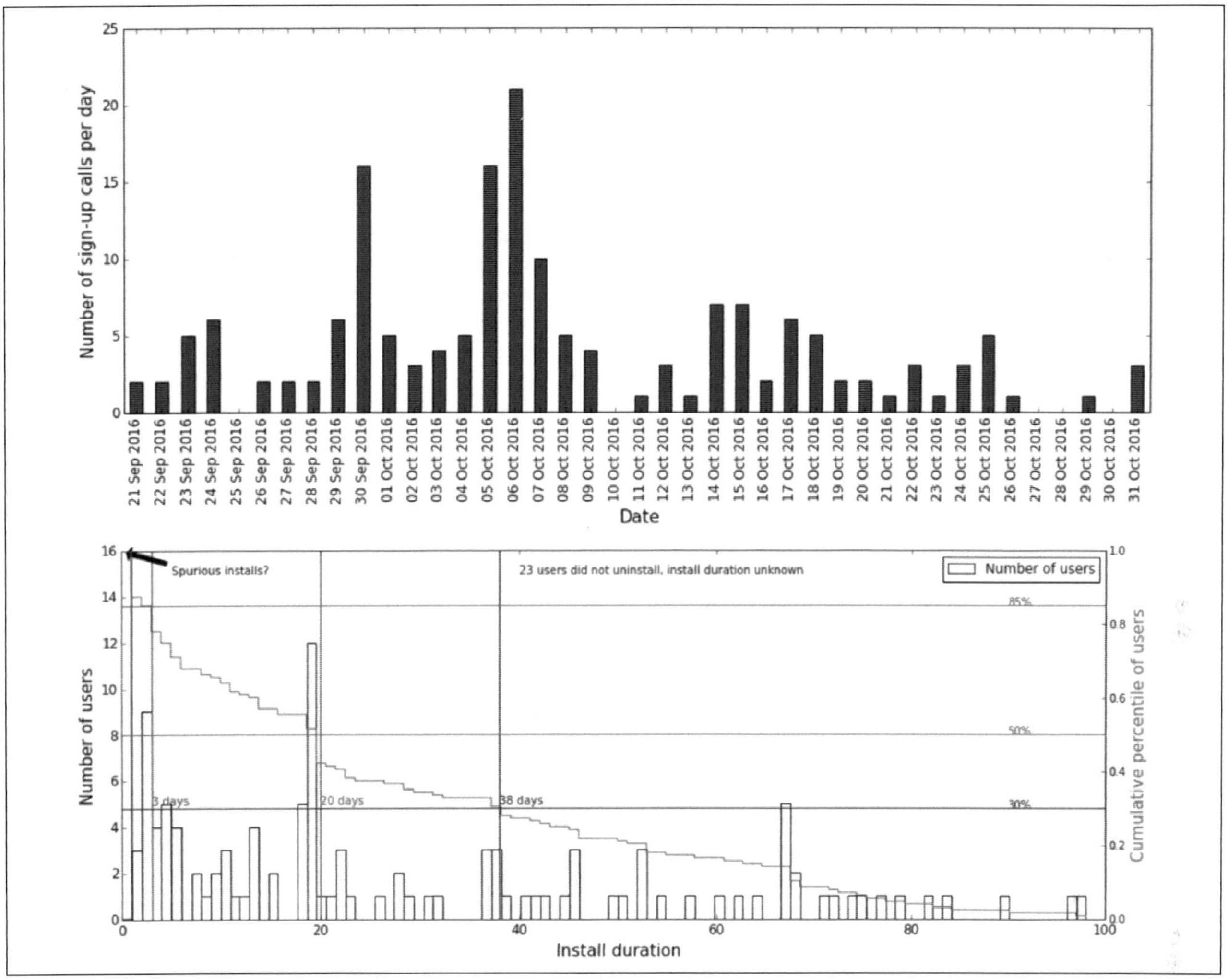

Figure 4. Evaluation of the installation rate and duration. Top: Number of calls to the profile creation API per day. This is a reasonable proxy for the number of installations, since the profile is created when the user signs in. Bottom: Histogram of user install durations. Install duration is represented by the time duration between the first and last data sync for a particular user. Note that 23 users did not uninstall the app, so their data is not included in the histogram.

profile creation API endpoint,[7] which is invoked when a user first launches the app (see Figure 4). This is not a perfect metric because it includes app *re-installs* but it is close (see Table 1). We use calls to the game registration endpoint[8] to detect when a user signed up for the game.

This metric is important because recruitment for traditional travel surveys, and for human- subject research in general, is time-consuming and expensive. While this platform does not claim to solve all problems with recruitment, painless installation ensures that there are no barriers to adoption once participants have been recruited.

These results show that (*a*) the app was installed by more than 150 participants; (*b*) the installations continued for a month after the initial recruitment, presumably through word of mouth; and (*c*) the gamification was interesting to just 50% of users.

Length of Install. This metric evaluates the *stickiness* of the app by measuring the number of days the app was installed. Since the phone app automatically uploads data to the server periodically, we use calls to the data upload API endpoint[9] as a proxy for the app being active. The install duration for a particular user is thus the length of time between the first and last API call. This does not distinguish between a user having suspended tracking, and having no trips for a particular period, so a user who reported exactly two trips 10 days apart would have an install duration of 10 days. If the last call was during the final two days of the analysis period, we assume that the app was not uninstalled.

This metric is important because longer-term data collection enables researchers to capture variability in daily travel patterns. Most GPS-enabled household travel surveys now cover multiple days: for example, the wearable GPS component of

Table 1. Evaluation Metrics for the Phone Application

	Users
Number of calls from unique users	172
Number of sign-ups	170
Number of new clients sending data	151
Number of unique sign-ups for the game	63
Pages	Lines changed
CSS style	+971
Settings	+160
Trip list	−95
Trip detail	+70

the 2012 California Household Travel Survey (CHTS) spanned three days (*5*). Since the recruitment for the pilot was ad-hoc and no compensation was provided to participants, we expect that the duration of data collection would be robust if the app is used for classic travel surveys.

The results are promising. More than 85% of users had the app installed for at least three days. Twenty-three users (not shown in the histogram) still had the app installed at the end of the analysis period. Of the users whose install duration is known, half had the app installed for ≈ 20 days, and a third had it installed for over a month. There were also 16 users who had the app installed for just one day.

App Launches over Time. This metric measures user engagement with the app. It tracks two related metrics: *app launches* and *screen switches* (the latter measures how many times a user switched between screens while using the app). App launches are measured by calls to the server API that populates the dashboard, while screen switches are indicated by client stats.[10]

This metric is important because not all the data is obtained through passive sensors. If we want to interact with the user to capture semantic and perceptual data, we need to have a platform that engages with the user and encourages him or her to provide the information that we seek. Designing for such engagement is challenging, and one goal of the platform is to facilitate it.

However, given the expectations about novelty in user interaction, the results (see Figure 5) are promising. First, they show that although app launches go down after the initial install, they never stop completely, and continue even several months after launch. Second, they show that the distribution of app launches across users is highly skewed—80% of users opened the app fewer than five times, but 10% of users opened it more than 150 times. Finally, they show that in addition to opening the app, users consistently navigate to other screens, even months after the install. However, the screen switching is an order of magnitude lower than the app launches. Finally, we see a marked drop off in both metrics around the end of the study period, which coincided with the winter break.

Extensibility Metrics

In this section, we evaluate the effort required for a transportation engineering student with no prior front-end experience, specifically in HTML/CSS/Javascript, and who has not worked in app or web development before, to build a custom UI for the app. We use quantitative metrics, such as lines of code, in addition to a brief, open-ended qualitative evaluation of the challenges encountered while completing the task. The results show that less than 1500 lines of code and one and a half weeks of full-time work are sufficient to generate a dramatically different user interface. This includes the learning curve for HTML, CSS, and some Javascript, and the platform UI in particular. We estimate that the time can be reduced to little more than a day with better documentation or better examples to draw from.

Lines of Code. Changes to the UI are shown in Figure 6. These changes consisted mainly of eliminating interface panes, keeping only the main ones (profile and diary). Panes were removed by commenting out the relevant sections of code. The contents of the profile pane were modified as well, along with color schemes, element sizes, and certain icons.

The diary tab gives users access to the list of trips they have taken on each date, and to the details of these trips such as speed profiles and travel modes. This tab was also modified to a new theme that changed the flow and amount of information provided to the user. The "details" page in the modified version displays the trip breakdowns in HTML tables. The new page also includes a button that allows users to fill out a survey about that specific trip. Table 1 illustrates the number of lines of code adjusted for each page in addition to the styles page.

Time Required. Another metric to evaluate the amount of effort put into the customization is to measure the amount of time spent understanding the source code, modifying it, and reviewing changes. In this case, getting up to speed with the existing code took approximately 75% of the total time, or 40 hours. Once the source code was understood, implementing modifications was fairly straightforward. Writing new code, debugging it, and testing the results took the remaining 25% of the time, or 15 hours.

Qualitative Comments. The major hurdles in customizing the UI were setting up the development environment and understanding the source code. This may have been due to relative inexperience with this type of software development, but improvements to the documentation would assist newcomers in navigating through the project and reduce the time required to make modifications.

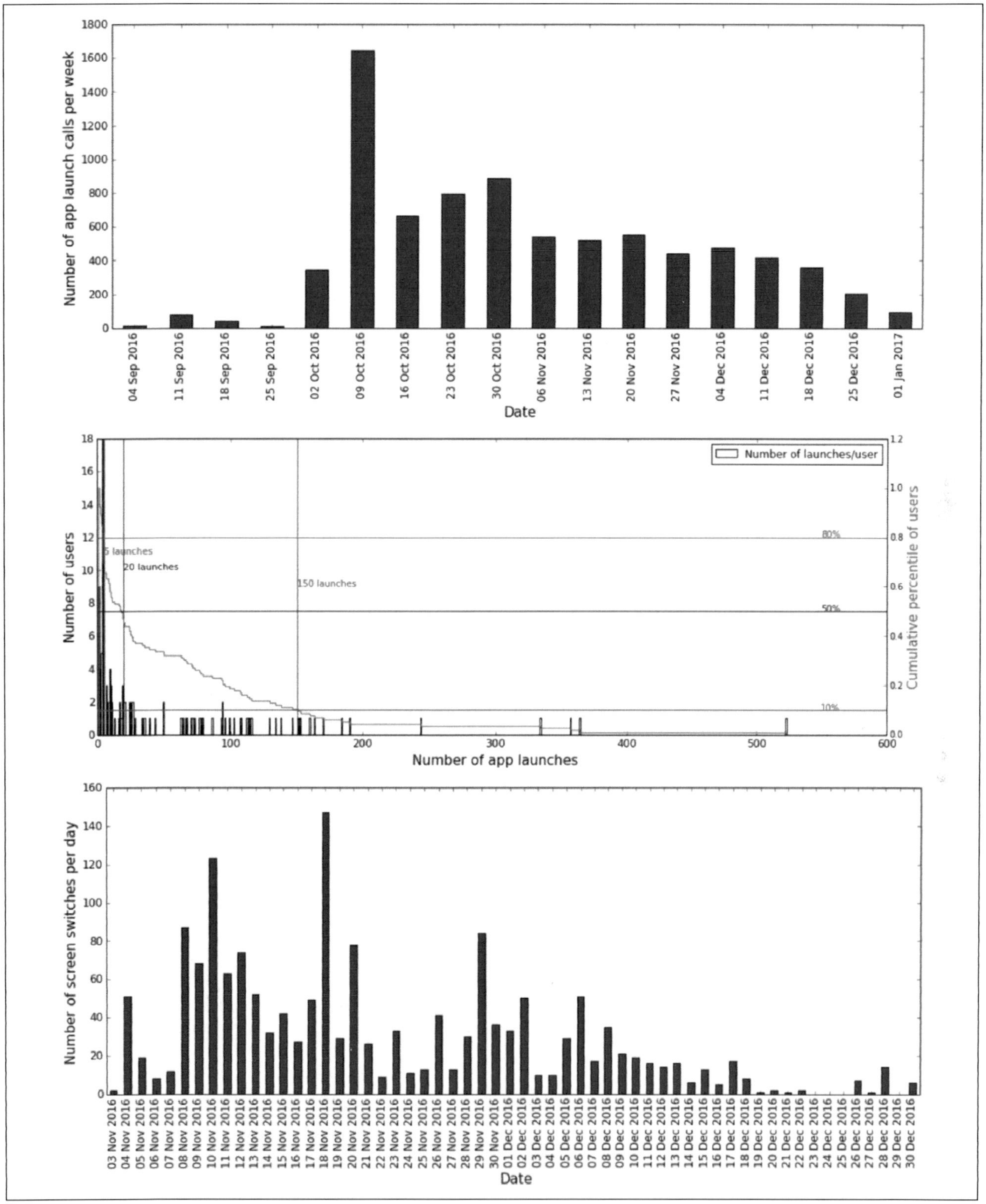

Figure 5. Metrics for user interaction and engagement. Top: Number of app launches per week. Middle: Histogram of the number of app launches per user. Bottom: Screen switches per day, starting in November, when we started tracking that data.

Figure 6. Screenshots of customized UI. Top: Base UI on the master branch of the e-mission-phone repository. Bottom: Customized UI in the joangroup branch of the e-mission-phone repository.

Conclusion

The e-mission platform[11] aims to make state-of-the-art smartphone travel data collection broadly available to researchers and other practitioners. It supports data collection through (*a*) background sensing, (*b*) user-initiated reporting, and (*c*) contextual, platform-initiated survey questions. Its architecture includes native apps for Android and iOS as well as cloud-hosted software for managing the collected data, all of which is modular, extensible, and open-source. e-mission can be used without modification to the interface or functionality simply by instructing participants to download the app (and providing consent documents to the corresponding author, to confirm which users are involved in the study). New UI "channels" can be created with minimal effort, providing study-specific consent forms, branding, and feature combinations.

Future development priorities span the data collection and output categories defined earlier. To streamline collection of user-initiated data, we are looking into a "shake-to-report" feature. This would allow users who want to report open-ended perceptual data to shake the phone to report it immediately. In order to guard against false positives, the system can generate a notification for the user to confirm the report. For travel diary creation, we plan to add support for more travel modes, and to evaluate the collected data against various benchmarks. We also plan to improve aspects of the server architecture, including scalability. Additionally, we would like to explore issues of privacy and data ownership, especially in the context of aggregated results.

As described above, improvements to the platform can be contributed by anyone. Valuable independent projects could include an adaptive sampling routine for longer trips or an option to sync data only over WiFi. Finally, we welcome feedback about the potential to assemble general-purpose travel behavior datasets using e-mission. For example, some study participants may be comfortable with a research protocol stipulating that, after a certain time delay, their data be available to outside researchers who agree to restrictions on its use.

Notes

1. See https://e-mission.eecs.berkeley.edu or https://github.com/e-mission
2. K. Shankari, shankari@eecs.berkeley.edu
3. The standard e-mission consent document is available here: https://e-mission.eecs.berkeley.edu/consent
4. See https://github.com/e-mission/e-mission-phone
5. See https://github.com/e-mission/e-mission-server
6. To aid reproducibility, the Jupyter notebook used to generate these results is available at https://github.com/e-mission/e-mission-eval. The underlying data can be obtained from the corresponding author, subject to restrictions on use.
7. /profile/create
8. /habiticaRegister
9. /usercache/put
10. /results/metrics/timestamp populates the dashboard on app launch. Note that there is also an app_launched client stat, but it does not appear to correspond directly to server calls, so we use the more conservative stat in our analysis. The state_changed client stat, filtered to remove changes to and from the splash state provides the basis for measuring screen switches.
11. See https://e-mission.eecs.berkeley.edu or https://github.com/e-mission

References

1. Pew Research Center. *Mobile Fact Sheet*. Pew Research Center, Washington, D.C., 2017.
2. Ince, D. C., L. Hatton, and J. Graham-Cumming. The Case for Open Computer Programs. *Nature*, Vol. 482, No. 7386, 2012, pp. 485–488.
3. Wolf, J., W. Bachman, M. S. Oliveira, J. Auld, A. K. Mohammadian, and P. Vovsha. *NCHRP 775: Applying GPS Data to Understand Travel Behavior, Volume I: Background, Methods, and Tests*. Transportation Research Board of the National Academies, Washington, D.C., 2014.
4. Wolf, J. *Using GPS Data Loggers to Replace Travel Diaries in the Collection of Travel Data*. PhD thesis. Georgia Institute of Technology, Atlanta, Ga., 2000.
5. Kunzmann, M., and V. Daigler. *2010-2012 California Household Travel Survey Final Report*. California Department of Transportation, Sacramento, CA, 2013.
6. Couper, M. P., D. A. Dillman, L. P. Erhard, P. J. Lavrakas, S. Polzin, G. Rousseau, and C. Tucker. *Expert Panel Review of the 2016 National Household Travel Survey*. Federal Highway Administration, Washington, D.C., 2015.
7. Jariyasunant, J., M. Abou-Zeid, A. Carrel, V. Ekambaram, D. Gaker, R. Sengupta, and J. L. Walker. Quantified Traveler: Travel Feedback Meets the Cloud to Change Behavior. *Journal of Intelligent Transportation Systems*, Vol. 19, No. 2, 2015, pp. 109–124.
8. Jylhä, A., P. Nurmi, M. Sirén, S. Hemminki, and G. Jacucci. MatkaHupi: A Persuasive Mobile Application for Sustainable Mobility. ACM Press, New York, NY, 2013, pp. 227–230.
9. Bucher, D., F. Cellina, F. Mangili, M. Raubal, R. Rudel, A. E. Rizzoli, and O. Elabed. Exploiting Fitness Apps for Sustainable Mobility-Challenges Deploying the GoEco! App. *Proc., 4th International Conference on ICT for Sustainability (ICT4S)*, Atlantis Press, Amsterdam, The Netherlands, 2016, pp. 89–98.
10. Cottrill, C. D., F. Pereira, F. Zhao, I. Dias, H. B. Lim, M. Ben-Akiva, and P. C. Zegras. The Future Mobility Survey: Experiences in Developing a Smartphone-Based Travel Survey in Singapore. Presented at 92nd Annual Meeting of the Transportation Research Board, Washington, D.C., 2013.
11. Winters, P., S. Barbeau, and N. Georggi. *Testing the Impact of Personalized Feedback on Household Travel Behavior (TRAC-IT Phase 2)*. National Center for Transit Research, Tampa, Fla., 2008.
12. Hood, J., E. Sall, and B. Charlton, A GPS-Based Bicycle Route Choice Model for San Francisco, California. *Transportation Letters: The International Journal of Transportation Research*, Vol. 3, No. 1, 2011, pp. 63–75.
13. Broach, J., J. Gliebe, and J. Dill. Bicycle Route Choice Model Developed Using Revealed Preference GPS Data. Presented at 90th Annual Meeting of the Transportation Research Board, Washington, D.C., 2011.
14. Eriksson, J., L. Girod, B. Hull, R. Newton, S. Madden, and H. Balakrishnan. The Pothole Patrol: Using a Mobile Sensor Network for Road Surface Monitoring. *Proc., 6th International Conference on Mobile Systems, Applications, and Services*, Breckenridge, Colo., ACM, New York, 2008, pp. 29–39.
15. Hemminki, S., P. Nurmi, and S. Tarkoma. Accelerometer-Based Transportation Mode Detection on Smartphones. *Proc., 11th ACM Conference on Embedded Networked Sensor Systems*, Roma, Italy, ACM Press, New York, 2013, pp. 1–14.
16. Coffey, C., and A. Pozdnukhov. Temporal Decomposition and Semantic Enrichment of Mobility Flows. *Proc., 6th ACM SIGSPATIAL International Workshop on Location-Based Social Networks*, Orlando, Fla., ACM, New York, 2013, pp. 34–43.
17. Dill, J., and N. McNeil. Four Types of Cyclists? Examination of Typology for Better Understanding of Bicycling Behavior and Potential. *Transportation Research Record: Journal of the Transportation Research Board*, 2014. 2387: 129–138.

18. Zhong, M., J. Wen, P. Hu, and J. Indulska. Advancing Android Activity Recognition Service with Markov Smoother. *Proc., Pervasive Computing and Communication Workshops (PerCom Workshops), 2015 IEEE International Conference on*, St. Louis, Mo., IEEE, New York, 2015, pp. 38–43.
19. Bao, L., and S. S. Intille. Activity Recognition from User-Annotated Acceleration Data. In *Per- Vasive Computing* (A. Ferscha and F. Mattern), Springer, Heidelberg, 2004, pp. 1–17.
20Zheng, Y., Y. Chen, Q. Li, X. Xie, and W.-Y. Ma. Understanding Transportation Modes Based on GPS Data for Web Applications. *ACM Transactions on the Web*, Vol. 4, No. 1, 2010, pp. 1–36.
21. Stopher, P. R., Q. Jiang, and C. FitzGerald. *Processing GPS Data from Travel Surveys*. Presented at 2nd International Colloqium on the Behavioural Foundations of Integrated Land-Use and Trans- Portation Models: Frameworks, Models and Applications, Toronto, Canada, 2005.
22. Reddy, S., M. Mun, J. Burke, D. Estrin, M. Hansen, and M. Srivastava, Using Mobile Phones to Determine Transportation Modes. *ACM Transactions on Sensor Networks*, Vol. 6, No. 2, 2010, pp. 1–27.
23. Krumm, J., and D. Rouhana, Placer: Semantic Place Labels from Diary Data. *Proc., 2013 ACM International Joint Conference on Pervasive and Ubiquitous Computing*, Zurich, Switzerland, ACM, New York, 2013, pp. 163–172.
24. Stopher, P., Y. Zhang, and Q. Jiang. Tour-Based Analysis of Multi-Day GPS Data. *Proc., 12th World Congress on Transport Research*, Lisbon, Portugal, 2010.
25. Dash, M., K. K. Koo, J. B. Gomes, S. P. Krishnaswamy, D. Rugeles, and A. Shi-Nash. Next place prediction by understanding mobility patterns. *Proc., Pervasive Computing and Communication Workshops (PerCom Workshops), 2015 IEEE International Conference on*, St. Louis, Mo., IEEE, New York, 2015, pp. 469–474.
26. Könönen, V., M. Ermes, J. Liikka, A. Lämsä, T. Rantalainen, H. Paloheimo, and J. Mäntyjärvi. Anatomy of Automatic Mobile Carbon Footprint Calculator. *Proceedings of the 6th International Conference on Advances in Grid and Pervasive Computing*, Oulu, Finland, Springer, 2011, pp. 84–93.
27. Broll, G., H. Cao, P. Ebben, P. Holleis, K. Jacobs, J. Koolwaaij, M. Luther, and B. Souville. Tripzoom: An App to Improve Your Mobility Behavior. *Proc., 11th International Conference on Mobile and Ubiquitous Multimedia*, Ulm, Germany, ACM, New York, 2012, p. 57.

The Standing Committee on Travel Survey Methods (ABJ40) peer-reviewed this paper (18-03257).

Article

TRR
JOURNAL OF THE TRANSPORTATION RESEARCH BOARD

Transportation Research Record
2018, Vol. 2672(42) 13–22

DOI: 10.1177/0361198118758297
journals.sagepub.com/home/trr
SAGE

Strict and Deep Comparison of Revealed Transit Trip Structure between Computer-Assisted Telephone Interview Household Travel Survey and Smart Cards

Robert Chapleau[1], Philippe Gaudette[1], and Tim Spurr[2]

Abstract

Large sample household travel surveys (HTSs) are an essential tool for the planning of urban transit systems. The progressive adoption by transit agencies of fare collection systems based on smart cards (SCs) has, for the first time, provided opportunities to compare the survey data with detailed, population-level data collected independently. These comparisons have produced some surprising results. Although the underreporting of non-home-based and off-peak trips was to be expected, the significant overestimation of transit use during peak periods was not anticipated. Using the Greater Montreal Area as a case study, this paper performs a strict and deep comparison of computer-assisted telephone interview (CATI) HTS data and SC data across several dimensions: transit agency usage, departure time from home, number of trips per traveler, and activity durations. The analysis reveals that the HTS constitutes a simplified portrayal of transit usage patterns. Non-home-based trips and trips made for activities of short duration are underrepresented in the survey data, leading to an underestimation of off-peak travel by transit. In addition, the systematic overestimation of peak period transit use appears to be because of the corrective weighting of the 20–29 demographic which is notoriously difficult to reach in a telephone-based household survey.

In the Greater Montreal Area and elsewhere, household travel surveys [HTSs—also referred to as origin-destination (O-D) surveys] have been used for the modeling of travel demand, simulation of transit networks, and the multimodal planning of large urban transportation infrastructure. The survey contacts a spatially stratified random sample of roughly 4% of the region's household by telephone. Each surveyed household and person are weighted based on household structure, age, and gender using population characteristics obtained from the national census. An in-depth understanding of these data, including the socioeconomic characteristics contained within, has permitted the analysis of several population-level behavioral phenomena such as urban sprawl, increasing motorization, and their effects on mode choice. Most of the time, these analyses have focused on proposed interventions in the realm of public transit: transfers at metro stations, capacity of park-and-ride lots, effects of fare policy, accessible transport for elderly citizens, and so forth. Smart cards (SCs) are increasingly used as a data source for transportation planning, but the breadth of information found in a travel survey (household location, demographic information, multimodal behavior etc.) make SCs on their own insufficient for many planning applications (*1*). Therefore, the importance of travel survey data that are accurate and impartial.

Practitioners and researchers who used the HTS were under no illusions as to its limitations—especially the tendency of respondents to underreport their travel. This suspicion has been amply confirmed in the literature, particularly when interviewees are asked to complete a trip diary or when the respondent has only a limited knowledge of the activities of the other members of their household. Until recently, however, no sufficiently detailed and reliable data existed to evaluate the magnitude, or even confirm the existence, of bias in the Montreal HTS.

In the Montreal context, although the HTS has been conducted every 5 years since 1970, it is only since 2010 that a

[1]Department of Civil, Geological and Mining Engineering, École Polytechnique de Montréal, Montréal, QC, Canada
[2]Autorité Régionale de Transport Métropolitain, Montréal, QC, Canada

Corresponding Author:
Address correspondence to Tim Spurr: tspurr@artm.quebec

fare collection system based on SCs (called OPUS) is operational in the metropolitan area. The Montreal travel survey of 2013 was undertaken from September to December 2013 and interviewed some 79,000 households. In parallel, complete SC data were collected for almost 60 days and archived, with approximately 2.2 million validation transactions occurring each day. Whereas the two data sets have differing structures and describe different objects, it is possible to transform each one in such a way as to make them directly comparable.

Under these conditions, the objective of this work is to perform a strict and deep comparison of public transit trips as revealed by the objects represented in a HTS (persons, their age and gender, transit networks, trip purposes, travel times, activity durations, payment type) and the attributes of transactions made by SCs. This paper includes the following elements: a light review of the literature, a presentation of the key concepts and data sets associated with the survey and the SCs, a comparison of temporal distributions of travel, a segmentation of behavioral phenomena according to trip purpose, and finally an analysis of the biases of the survey and possible corrective measures.

Literature Review

Since the work of Richardson, Ampt, and Meyberg, transport planners and modelers have been well advised to account for the potential of travel underreporting in the surveys, whether surveys are of the travel diary, self-completion questionnaires, or computer-assisted telephone interview (CATI) type (*2,3*). More recent observations on the Transportation Tomorrow Survey in Toronto have established that:

> Daily traffic (13 hour counts) was under represented by an average of 24%. TTC [Toronto's largest transit operator] total daily ridership is under represented by 7% but the a.m. peak period is over represented by an average of 17%. Similar patterns occur with a number of the other municipal transit operators. (*4*)

Without knowing by which means this result is obtained, it can be hypothesized that the response rate is higher when transit riders are involved. Similar behavior has been observed for the Montreal case. In a preliminary investigation into the potential contribution of SCs to the enrichment of travel surveys, a comparison of the 2008 survey data with SC transactions from 2010, limited to the subway network, indicated an overestimation of peak period demand in the travel survey (*5*). A subsequent comparison with data from the 2013 survey confirmed this phenomenon (*6*).

The consequences of peak period demand overestimation in the survey are potentially significant as major transit planning projects are usually evaluated and designed based on their peak period load profiles derived from static assignment models. In addition to providing a more accurate estimate of transit demand, the SC data can also yield more detailed forecasts produced by microsimulation, notably the volume of passengers within each vehicle between each stop [see, for example, (*7*)].

These studies have demonstrated the importance of a rigorous investigation of the overestimation problem using a strict comparison of public transit travel data, as declared in the telephone-based travel survey with the complete information recorded by the SC fare collection system.

Concepts and Methodology

The following paragraphs describe the main concepts and objects described in the HTS and SC data sets. For the sake of convenience and simplicity, numbers are rounded to the nearest thousand.

The Montreal CATI HTS focuses on the measurement of travel by all modes during an average weekday (AWD). In 2013, the survey contacted 79,000 households containing 188,000 persons making 411,000 trips. Every household responding to the survey is assigned a weighting factor (the household adjustment factor) designed to reproduce the spatial distribution of the population portrayed by the Canadian census data. A second weighting factor (the age adjustment factor) is applied to each person recorded in the survey based on the demographic distribution. The travel survey expansion method is explained in Chapleau et al., and its application is uniform across all households and persons, regardless of the declared mode of transport (*8*).

In the Montreal HTS, the following objects are examined. Persons (with attributes age, gender, fare type) who make transit trips. Trips are defined by a single person traveling between an origin and a destination for a single purpose. Trip attributes include transit itineraries, boarding locations, purpose, and activity duration (the time difference between the start of the current trip and the start of the next trip). Transit itineraries are derived from the transit line sequence provided by respondents and are validated by a routing algorithm. Declared fare types permit the distinction between regular riders (full or reduced monthly pass) and occasional riders (full or reduced fare).

The SC data do not represent an AWD, but rather measure travel every day. Comparison with the HTS requires the selection of a specific sufficiently representative day. No weighting of observations is necessary as, in principle, the SC system captures not a sample but rather the entire population of transit users. The primary object represented by the OPUS system is the fare validation transaction, which has the following attributes: fare type, transaction type (first boarding or transfer), transaction time, and the transaction location (represented by the unique identifier of the validating machine). The transaction chain of a specific SC is decomposed into a sequence of trips (trip chain) for the day, according to a set of rules that conforms to the definition of a person trip in the travel survey, assuming that each card is used by a single person.

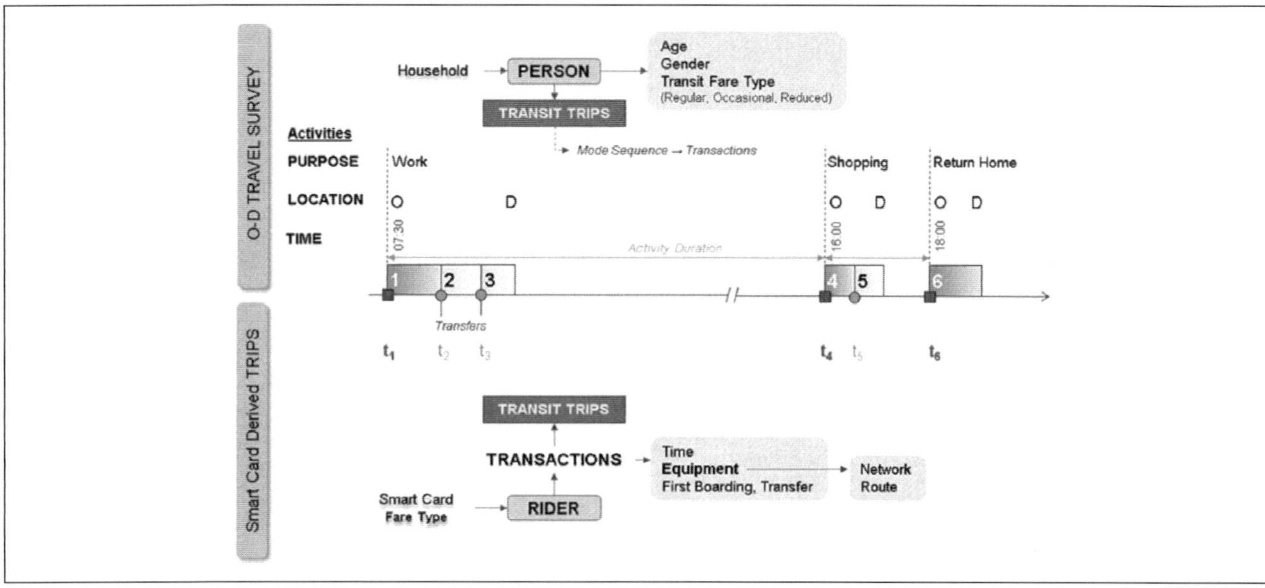

Figure 1. Processing of transit trips (sequence of routes taken) in the O-D survey and the SC environments.

Figure 1 explains how the objects pertaining to each data set are related. The analysis described in this paper presumes a certain equivalence between objects in survey and objects in the SC data. To begin with, each SC is assumed to be used by a single person. Also, an SC validation is considered to be a "boarding" event—either on a bus or at the turnstile in a metro station. A transit trip as defined in the survey is approximated in the SC data by one or more transactions according to some simple rules: consecutive transactions must be <120 min apart; no consecutive metro boardings; no consecutive boardings on the same line. Finally, activity duration is defined in the travel survey as the difference between the start times of two consecutive trips. In the SC data, activity duration is the difference between the times of consecutive "first boardings" (boardings that mark the beginning of a trip).

Starting from the assumption that the population-weighted AWD transit demand measured by the HTS is representative of a typical weekday in fall 2013, distinct comparative analyses are performed for the Montreal case in 2013 across several dimensions. First, data sets are normalized to permit a strict comparison. For example, responses including the commuter rail mode in the O-D survey are excluded because of the lack of SC control. In any case, the AWD rail ridership for the fall of 2013 is estimated at 75,000, whereas the O-D survey value is 110,000, an overestimate of 47% (*9*). This result alone strongly suggests that transit users feel more directly implicated and therefore more inclined to respond to the Montreal HTS. This serious bias merits more scrutiny, and the current project proposes a complete comparison of the objects defined in each data set.

- For the Greater Montreal Area, the HTS of fall 2013 reveals around 2.1 million transactions per day for 1.42 million trips, numbers similar to those obtained from SCs. To ensure the conservation of the analysis objects, a specific day, rather than average day, has to be chosen to obtain comparable SC data. November 21, 2013 was determined to be typical of a weekday in fall 2013 based on ridership trends during that period.
- Figure 2 illustrates the distribution of transit trips over the day for the last three HTSs conducted in Montreal (2003, 2008, and 2013) in contrast with the SC data of November 21, 2013; many other SC days yield the same distribution (*10*). The average distribution for the week of November 21 is also included and is nearly identical to the distribution for the single day. Consequently, even considering some incidental limitations, this SC data set will be considered as the reference and as the true behavior of the transit system.

Based on Figure 2, it is clear that the a.m. and p.m. peaks are overestimated whereas off-peak trips are underreported. To understand more deeply the perception of the travel survey interviewees of the transit network, some explanations of the Greater Montreal Area transit system are necessary.

The Greater Montreal Area Transit Organization

The metropolitan region—with a population of over 4 million people—is now served by four transit operators. Excluding the commuter rail system, one may distinguish

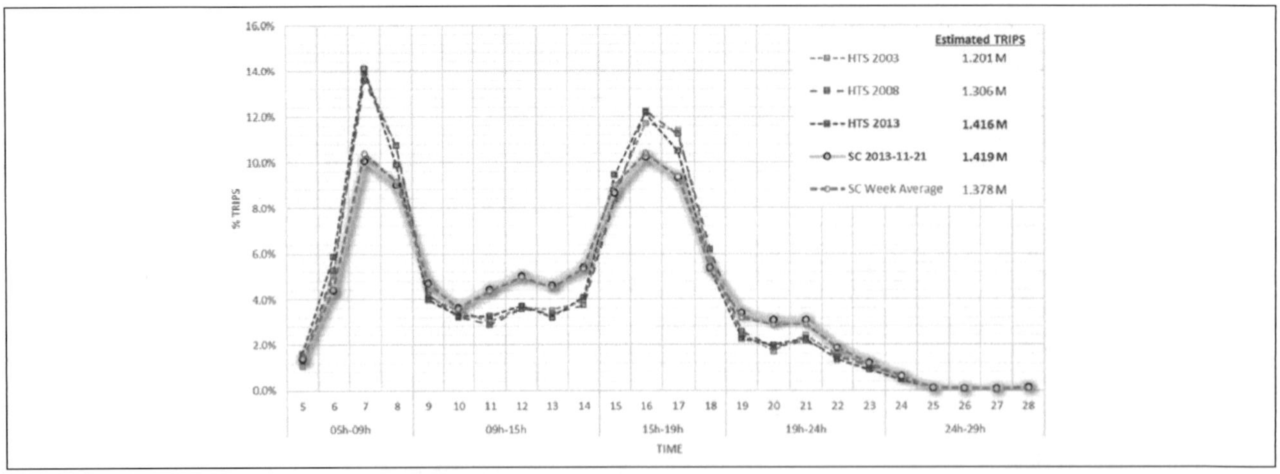

Figure 2. Comparison of temporal distribution of trips over the day for three household surveys, 1 day of SC data and 1 week of SC data.

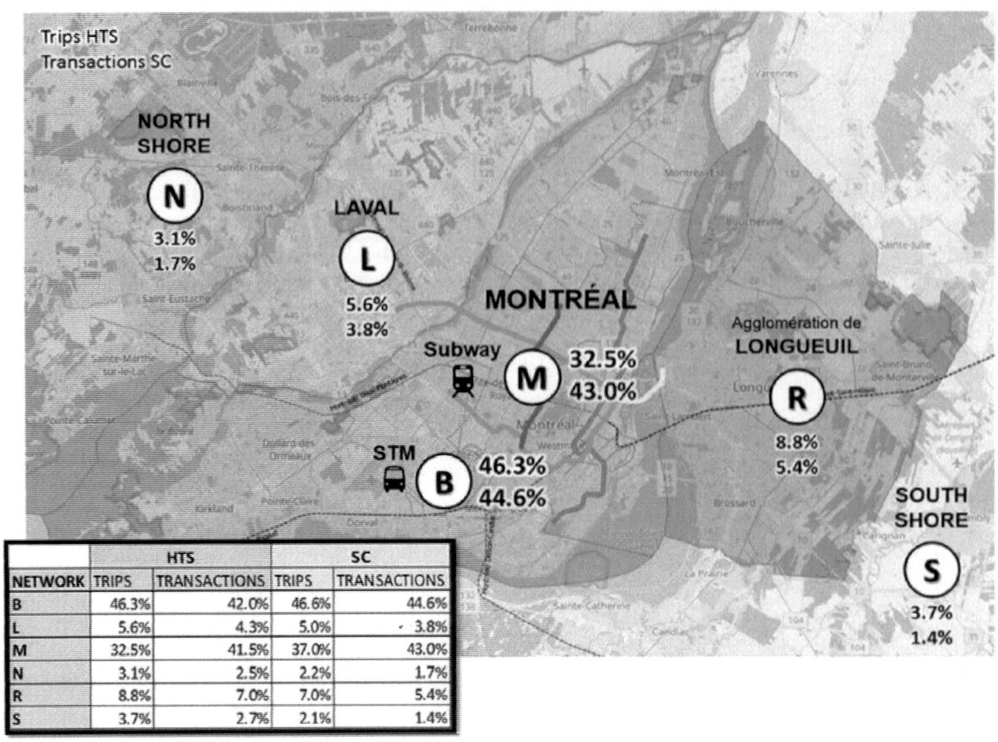

	HTS		SC	
NETWORK	TRIPS	TRANSACTIONS	TRIPS	TRANSACTIONS
B	46.3%	42.0%	46.6%	44.6%
L	5.6%	4.3%	5.0%	3.8%
M	32.5%	41.5%	37.0%	43.0%
N	3.1%	2.5%	2.2%	1.7%
R	8.8%	7.0%	7.0%	5.4%
S	3.7%	2.7%	2.1%	1.4%

Figure 3. Greater Montreal Area Transit Organization: percentage of origin trips in the HTS and percentage of transactions as measured by the SC system.

between six subnetworks, as shown in Figure 3. The Société de Transport de Montréal (STM) has two components: M for metro (subway) and B for the bus system. The other authorities provide surface bus transport only: L (STL-Laval), R (RTL-Longueuil), N (North shore) and S (South shore). The table inset in Figure 3 shows the distribution of trips and transactions across these subnetworks, as measured by the SC and by the HTS. It reveals that the number of trips and transactions declared in the HTS on suburban networks (N, L, R, S) is much higher than the volume observed with the SC. Although all the transactions in the central part of the region are recorded by the OPUS SC

Table 1. Statistics from Decomposition of Persons, Trips and Transactions

				Smart Cards								
	Riders from HTS 2013			2013-11-21			Weekday average			SD		
(×1000)	a.m.	9:00+	Total	a.m.	9:00+	Total	a.m.	9:00+	Total	a.m.	9:00+	Total
Persons	444	253	697	345	361	706	342	347	689	24	19	20
	Transit Trips											
05:00–09:00	446	0	446	353	0	353	350	0	350	24	0	24
09:00–15:00	72	233	305	96	298	395	94	285	378	5	15	19
15:00–19:00	359	172	530	258	219	477	257	214	471	22	8	20
19:00–24:00	48	76	124	59	121	180	53	113	166	6	15	20
24:00–29:00	2	9	11	3	11	14	3	10	13	2	4	6
Total	926	490	1,416	770	649	1,419	756	622	1,378	44	40	46
	Transactions											
05:00–09:00	672	0	672	540	0	540	536	0	536	35	0	35
09:00–15:00	106	336	443	157	434	590	153	415	568	7	20	26
15:00–19:00	533	242	775	402	319	720	400	311	711	35	10	33
19:00–24:00	70	109	179	92	174	266	82	163	245	9	19	27
24:00–29:00	2	13	15	4	16	21	4	15	20	2	6	8
Total	1,383	700	2,084	1,194	943	2137	1,175	904	2,079	68	52	68

system, in the peripheral regions it is possible, though rare in occurrence, for riders to pay cash without obtaining a proof of payment. This phenomenon alone, however, is too marginal to explain the enormous overestimation of trips in these zones. The travel survey omits, to a significant degree, the number of trips initiated on the subway (32.5% versus 37% of all origin transactions). Most of these trips occur after the morning peak period in the central city, far from the homes of most regular transit users.

Decomposition of the Demand by Departure Time

For an in-depth analysis of the peak overestimation/off-peak underestimation phenomena illustrated in Figure 2, it is useful to decompose the demand based on a finding from the travel survey, which is that 97% of trips undertaken during the morning peak period originate at home. Intuitively, these trips are the least likely to be underreported in a HTS because of the strong link between the "person" object and "home-based" trips. The hypothesis can be tested by analyzing separately trips occurring before 9 a.m. and trips occurring later. The idea is to follow each person (or SC) throughout the day and examine their behavior with regard to activities and their durations.

Summary statistics for the persons, trips, and transactions made before and after 9 a.m. (Table 1) show a consistent behavior within each group. The survey records 29% more people traveling before 9 a.m. (dubbed "early birds") relative to the SCs. However, the survey also omits 30% of people traveling after 9 a.m. ("night owls"). As a natural consequence, trips recorded by the travel survey during the peak periods are overestimated and off-peak trips are underestimated. In addition, these two deviations cancel each other out, leading to a negligible difference of persons, trips, and transactions over a 24-h period. The averages and standard deviations (SDs) for the week of SC data are included to illustrate the representativeness of the single day.

The two HTS and SC data sets are decomposed to permit a strict comparison between the early birds and night owls. Figure 4 shows the temporal distribution of the trips of both subsets. The overestimation of trips made by early birds in the HTS is considerable (926,000 versus 770,000—a discrepancy of 20%), whereas the trips of night owls are underreported by 24%. It would seem that trips made by household members who first leave their home later in the day are less likely to be detected by the survey. This suggests that night owls are less likely to be interviewees and are more likely to have their trips described by an early bird who, having left home before the night owl, can't accurately report the night owl's trips. As most survey interviews occur in the evening, the night owl often isn't home to answer on their own behalf. Here, as well, the inclusion of the average distributions for the week of November 21 shows that they are almost indistinguishable from the data from that specific day.

Analysis of Departure Times and Activity Durations

In the context of a travel survey, a trip is associated with a purpose and, by extension, an activity. For SCs, the construction of trips from transactions is based on rules that respect the purpose-activity logic: transfers are possible within a 2-h time window, and consecutive transactions within a single

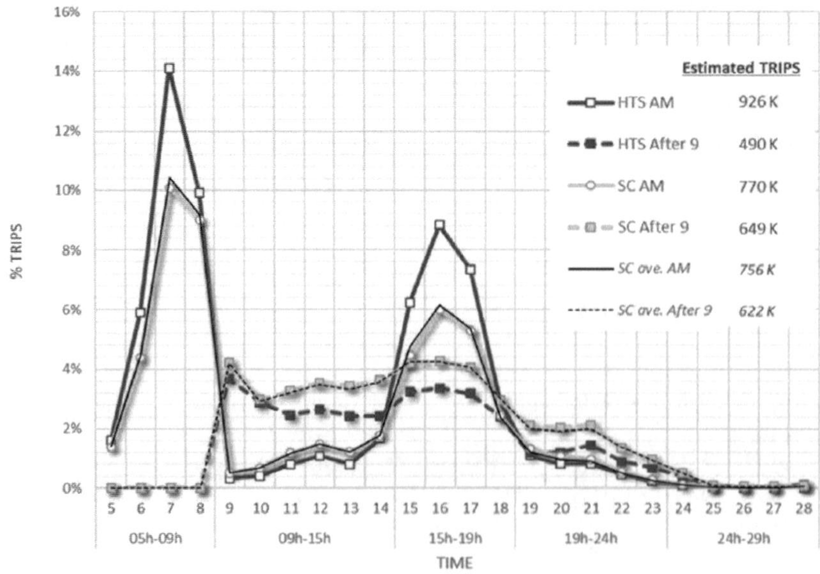

Figure 4. Distribution of the transit trips decomposed by the subsets given by travelers before and after 9 a.m.

trip cannot be made on the subway or on the same bus line. Subsequently, as shown in Figure 1, it is possible to analyze activities according to their duration, which is defined as the time difference between the beginnings of two consecutive trips. It goes without saying that the last trip—usually the return home—does not entail an activity.

To properly illustrate the activity patterns revealed by both the travel survey and the SC data, the Circos graphic technology is used. For the present analysis, the Circos graphic is composed of two circular axes (Figure 5): on the left, the 24-h day is decomposed into five periods (morning peak from 5 to 9 a.m., mid-day from 8 a.m. to 3 p.m., afternoon peak from 3 to 7 p.m., evening from 7 p.m. to midnight and night from midnight to 5 a.m.); the axis on the right represents activity durations divided into 1-h blocks. A link between the two axes represents a volume of trips initiated during the specified time period for the purpose of an activity having the specified duration. For example, link 1 in Figure 5 shows 130,000 travel survey trips undertaken during the morning peak period for activities lasting 9 h. The corresponding data track (the outside ring containing point 3) is a histogram showing the distribution of fare types (regular fare monthly pass in gray, reduced fare monthly pass in blue, occasional fare ticket in pale green, and reduced fare ticket in dark green) used to make these trips. Link 1 contains mostly regular fare monthly passes whereas link 2 (trips with activity durations of 7 h) is composed mostly of reduced fare monthly passes, suggesting a concentration of student travelers.

To demonstrate the importance of the trip-activity structure of travel demand, Figures 5 and 6 illustrate the activity durations by time of departure using the HTS data and the SC data, respectively. Clearly, relative to the SC data, the HTS reports a much greater volume of trip departures during the morning peak period for which the activity durations are very long. In addition, the SC data reveal the existence of numerous short-duration activities not captured in the HTS, although the total volume of activities observed in each data set are almost identical (719,000 versus 713,000). In both figures, it is important to note, at the top of the left-hand axis, the high proportion of pass holders (the blue histogram bands). Later in the day, toward the bottom of the left axis, emerges a significant proportion of ticket users, shown in green.

Figure 7 summarizes the differences in behavior as observed in the travel survey relative to the actual activities measured by the SCs. Morning peak period trips are not only overestimated but also correspond to an overly simplified activity structure. Long-duration activities (>7 h) are more present in the HTS than in the SCs. Activities of short duration are considerably more frequent in the SC data.

Analysis of Transit Usage Patterns

In addition to the characterization of trips by their associated activities, it is possible to examine the daily trips of public transit users. Figure 8 compares the number of trips per day per transit user as measured by the HTS and by the SC. In both images, the left-hand axis represents the departure time period (as in Figures 5 and 6) and the right-hand axis is the number of daily trips increasing from one at the bottom to five or more at the top. The comparison is not exact because in the SC data a single traveler can use multiple tickets. Figure 8*a* (based on the HTS), shows a large proportion of

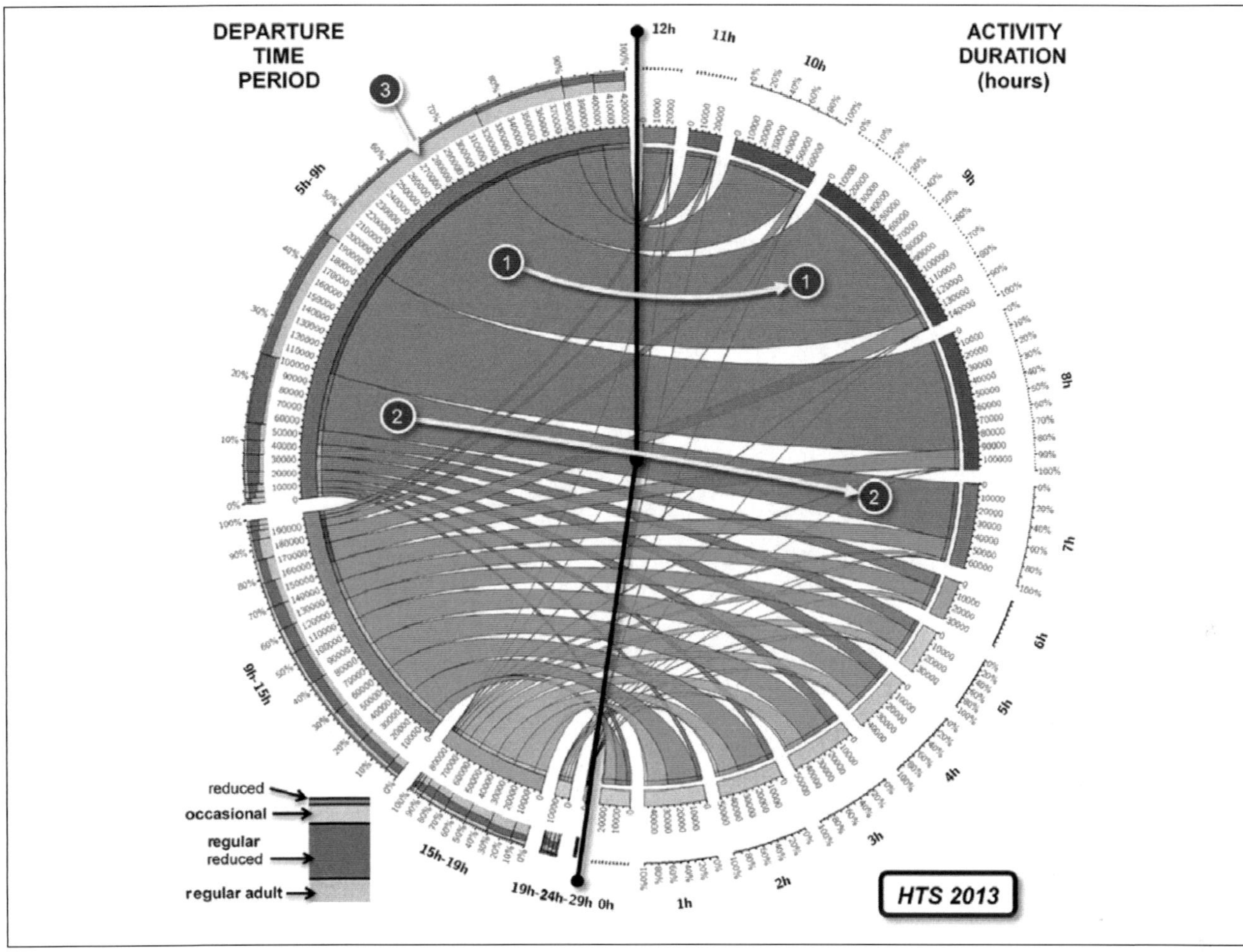

Figure 5. From the HTS 2013: activity duration for transit trips having a departure time at specific time periods, over an average fall weekday (719,000 activities).

monthly pass users represented by the predominantly blue data track on the left axis. The corresponding data track in Figure 8*b* (based on SC data) indicates a much higher proportion of single tickets, leading to an overestimation of the number of individual travelers. If it is assumed that most single ticket users make two trips during the day, then the categorization would of course be altered. Nevertheless, the travel survey image shows that in addition to the overestimation of peak period travelers, a preponderance of travelers making exactly two trips per day. The SC data suggest that real traveler behavior often includes longer trip chains.

A possible explanation for the overly simplified travel patterns portrayed in the HTS can be found in an examination of the transit usage patterns of survey respondents themselves. In Table 2, survey respondents are classified according to their age and gender, and whether they belong to a "transit household" in which at least one member makes at least one trip by transit. The table shows that nearly 60% of survey respondents were women, a rate comparable to their share of transit ridership. However, there is a large discrepancy in age distribution between people answering the phone in transit households (49% aged 40–60 years old) and people who are declared as transit riders (47% under 30). Furthermore, the 20-year gap between peak transit usage and peak survey response suggests that the responding households are more likely to include college-age students residing with their parents.

This finding draws attention to how survey records are weighted. Normally, each household in the survey sample is weighted to reflect the distribution of household sizes found in the corresponding analysis zone (*8*). As a result, every trip is weighted according to a so-called household (HH) factor. At this stage, an examination of the age distribution of the population reveals the necessity of an age adjustment factor to compensate for the low representation of young people, typically in the 20–29 cohort. Table 3 is a compilation of these factors for each of the transit riding groups defined in Figure 3. Similar to the case of the Toronto travel survey in

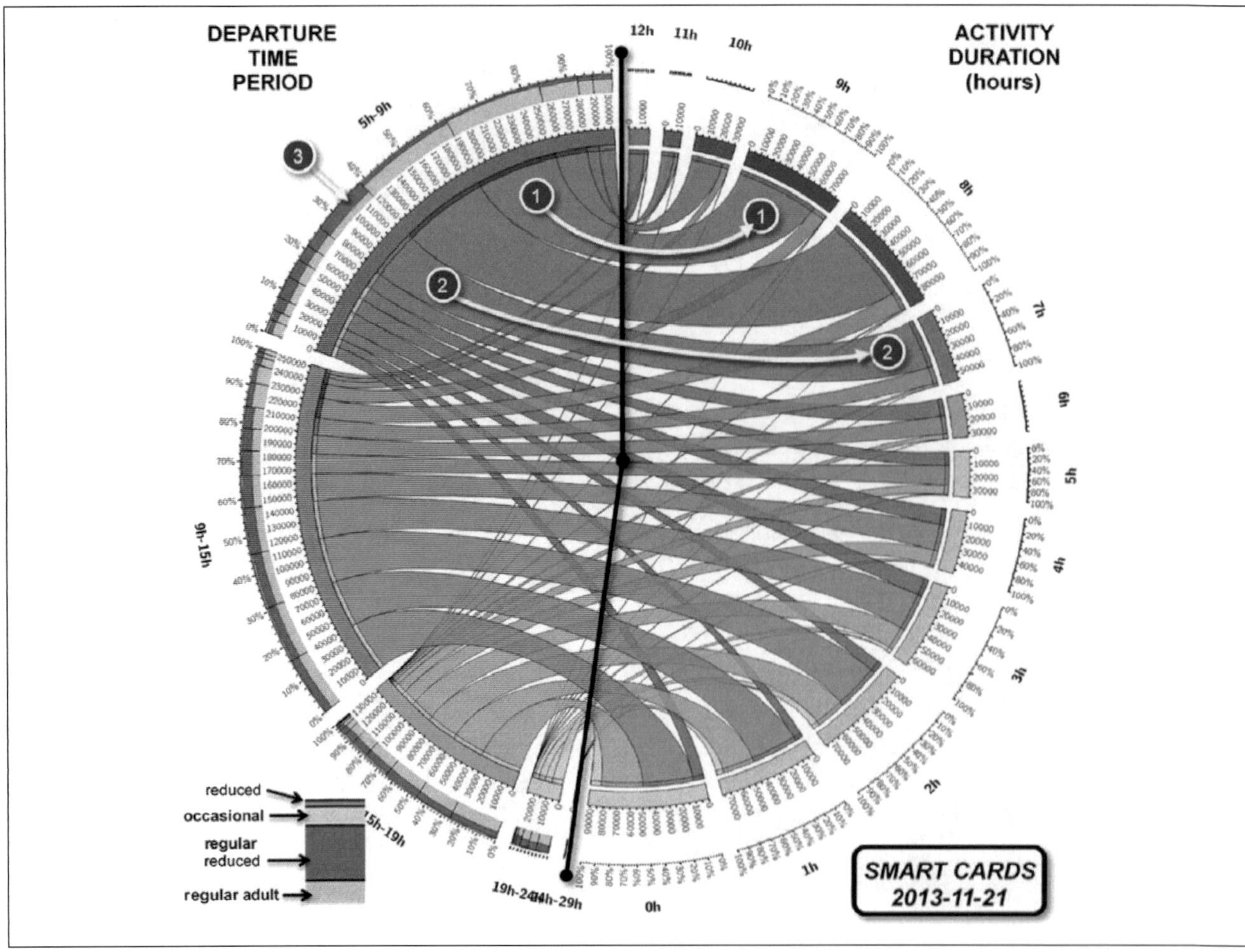

Figure 6. From the SC data on November 21, 2013: departure time period versus activity duration (713, 000 activities).

Revealed Activities	Departure Time Period					
	05h-09h	09h-15h	15h-19h	19h-24h	24h-29h	Total
HTS2013	421 577	199 644	84 516	11 362	1 891	718 990
Smart Cards	302 389	251 942	132 264	25 793	495	712 883

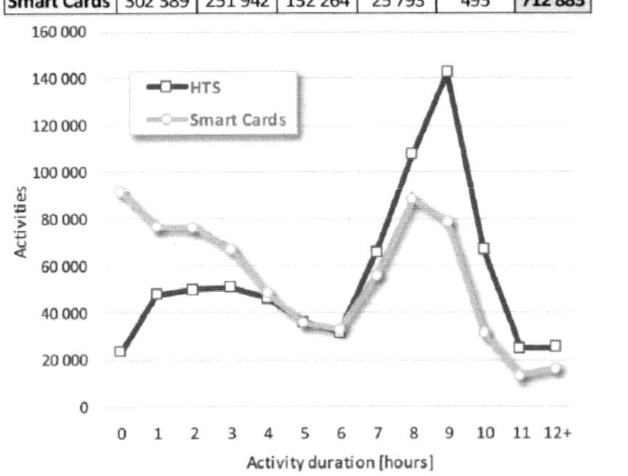

Figure 7. Comparison of the distribution of activity durations (HTS versus SC).

which the adjustment factor for the 18–32 group is 1.535, the 20–29 age group is multiplied by 1.40 for those using the metro (*4*). This clientele consists primarily of students attending university downtown. It could be hypothesized that the travel behavior of these students is not representative of the 20–29 cohort as a whole. The members of this cohort who do not live with their parents are difficult to reach using the telephone survey.

Conclusion

The goal of this project was to perform in an-depth characterization of the users of an urban public transit system using both a large sample travel survey and an archived SC transaction database. A multidimensional characterization included the adoption of the Circos visual representation that effectively illustrates the differences in travel demand structure revealed by the two independent data sources. The analysis shows that the HTS tends to capture simplified travel patterns, especially trips made from home and then toward the home. The absence of intermediate trips causes an underestimation of off-peak

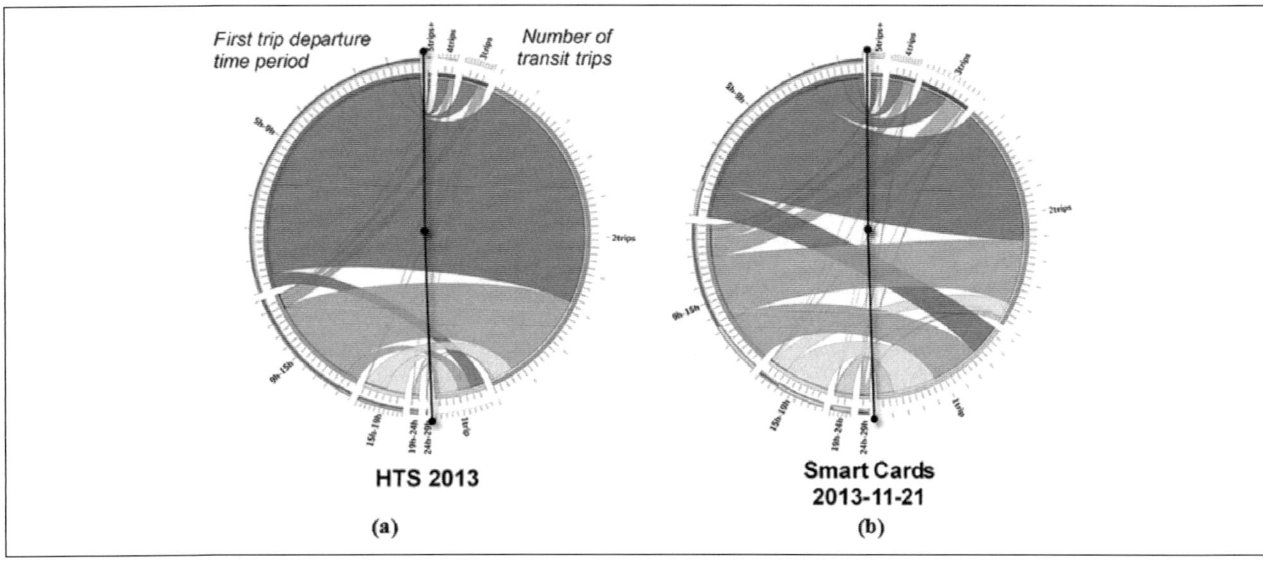

Figure 8. Comparison of number of daily transit trips according to (*a*) the HTS 2013, and (*b*) the SC data set of November 21, 2013.

Table 2. Demographic Comparison of HTS Respondents and Surveyed Transit Riders

	Survey respondents		
	HTS 2013 (%)	Transit HH (%)	Riders (%)
Gender			
Male	40.6	41.4	45.3
Female	59.4	58.6	54.7
Age			
<20	1.5	4.1	21.5
20–29	7.4	15.6	25.5
30–39	16.7	19.1	17.3
40–49	19.2	26.0	14.3
50–59	20.5	23.0	12.0
60–69	18.0	7.9	5.5
70–79	10.9	3.2	2.8
80–100	5.8	1.2	1.0

Note: HH = household.

Table 3. Age Adjustment Factor for Correcting the Household Expansion Factor of the Travel Survey, by Transit System Used

	Ratio person factor/HH factor						
Age	B	L	M	N	R	S	Total
<20	1.08	1.02	1.08	1.14	1.08	1.06	1.08
20–29	1.31	1.13	1.40	1.19	1.17	1.13	1.31
30–39	1.06	1.19	1.06	1.02	1.11	1.05	1.07
40–49	1.00	1.02	1.00	1.00	1.01	1.02	1.00
50–59	0.92	0.93	0.93	0.94	0.92	0.96	0.93
60–69	0.81	0.89	0.79	0.89	0.90	0.88	0.82
70+	0.98	0.97	0.92	1.00	0.94	1.02	0.96
Total	1.06	1.04	1.09	1.09	1.05	1.04	1.07

transit use in the HTS. In addition, the application of weighting factors to correct for the underrepresentation of the student cohort seems to be a significant contributing factor to the HTS's overestimation of transit use during peak periods.

Based on these results, it is clear that, for the purposes of analyzing public transit at least, particular care must be taken in the derivation of weighting factors that correct for population demographics. The overrepresentation of home-based trips as well as the underrepresentation of off-peak trips could both be rectified by using SC data differentiating between the two groups of users (early birds and night owls) illustrated in Table 1 and Figure 4. An even better approach would be to determine the destinations of SC-derived trips based on information contained in the travel survey. Such an exercise will be part of future research.

Acknowledgments

The authors gratefully acknowledge the resources provided by the Autorité Régionale de Transport Métropolitain (ARTM) as well as the collaboration of the Société de Transport de Montréal (STM).

References

1. Kurauchi, F., and J. D. Schmöcker (eds.). *Public Transport Planning with Smart Card Data*. CRC Press, Boca Raton, FL, 2016.
2. Richardson, A. J., E. S. Ampt, and A. H. Meyburg. *Survey Methods for Transport Planning*. Eucalyptus Press, Parkville, VIC, 1995.
3. Ashley, D., T. Richardson, and D. Young. *Recent Information on the Under-Reporting of Trips in Household Travel Surveys*. Presented at the Australasian Transport Research Forum, Auckland, New Zealand, 2009.

4. Data Management Group. *Transportation Tomorrow 2011 TTS Version 1.0 Data Expansion & Validation*. Department of Civil Engineering, University of Toronto, Toronto, 2013.
5. Spurr, T., R. Chapleau, and D. Piché. Discovery and Partial Correction of Travel Survey Bias Using Subway Smart Card Transactions. *Transportation Research Record: Journal of the Transportation Research Board*, 2014. 2405: 57–67.
6. Spurr, T., A. Chu, R. Chapleau, and D. Piché. A Smart Card Transaction "Travel Diary" to Assess the Accuracy of the Montréal Household Travel Survey. *Transportation Research Procedia*, Vol. 11, 2015, pp. 350–364. https://doi.org/10.1016/j.trpro.2015.12.030.
7. Spurr, T., R. Chapleau, K. K. A. Chu, and W. Doucet-Koussaya. *Smart Cards, GIS, GTFS and TRANSIMS: A Plot for the Dynamic Microsimulation of the Montreal Subway*. Presented at the 13th International Conference on Travel Behaviour Research, Toronto, 2012.
8. Chapleau, R., and D. Piché. *Weighting Method of Household Travel Survey to Alleviate Discrepancies in Transport Planning Models*. Presented at the 12th World Conference on Transport Research, Lisbon, Portugal, 2010.
9. Agence métropolitaine de transport. *Annual Report*, Montréal, 2013.
10. Spurr, T., A. Leroux, and R. Chapleau. *Comparative Structural Evaluation of Transit Travel Demand Using Travel Survey and Smart Card Data for Metropolitan Transit Financing*. Presented at 97th Annual Meeting of the Transportation Research Board, Washington, D.C., 2018.
11. Krzywinksi, M., J. Shein, I. Birol, J. Connors, R. Cascoyne, D. Horsman, S. J. Jones, and M. A. Marra. Circos: An Information Aesthetic for Comparative Genomics. *Genome Research*, Vol. 19, No. 9, 2009, pp. 1639–1645. http://doi.org/10.1101/gr.092759.109.

The Standing Committee on Travel Survey Methods (ABJ40) peer-reviewed this paper (18-01596).

The authors assume full responsibility for all results presented, as well as their interpretation.

Article

TRR
JOURNAL OF THE TRANSPORTATION RESEARCH BOARD

Transportation Research Record
2018, Vol. 2672(42) 23–32

DOI: 10.1177/0361198118758337
journals.sagepub.com/home/trr
SAGE

Application of the "Foot-in-the-Door" Compliance Technique for Traveler Intercept Surveys

Benjamin R. Sperry[1], Emily Siler[1], and Tristan Mull[1]

Abstract

Traveler intercept surveys are used to obtain essential data for transportation planning purposes in a variety of contexts and situations. This paper describes an application of the foot-in-the-door (FITD) compliance technique for traveler intercept surveys as a way to increase participation and improve the overall experience for both the traveler and the organization conducting the survey. Outcomes related to survey participation and response characteristics for three such surveys are presented. The two-stage FITD method described in this paper was successful at collecting a small amount of data from a majority of travelers at the locations being studied while minimizing the inconvenience to the traveler. A follow-up survey conducted via the Internet had a response rate that varied between the different case studies and also among different groups of travelers. Approximately one-third of follow-up survey responses were provided on the same day as the initial interaction between the traveler and the researcher, whereas one-quarter of respondents utilized a mobile device to complete the follow-up survey. The results of these case studies are informative for the development of traveler intercept surveys. Transportation planners are encouraged to consider the FITD technique described in this paper when developing traveler intercept surveys within their jurisdictions.

Traveler intercept surveys are used to obtain data from travelers in a wide variety of contexts in which it is necessary to obtain feedback from individual travelers on trip details, behaviors, and preferences at the time that the trips of interest are being made. Intercept surveys have traditionally been the only reliable method for identifying essential transportation planning data such as the location of the start and end of a trip, the purpose of travel, and other details on the characteristics and preferences of travelers, particularly for non-home-based trips or travel on specific modes. The traveler intercept survey method is used extensively for external station surveys, workplace and establishment surveys, transit on-board surveys, and other contexts in which it may be difficult to obtain travel behavior data through traditional household survey methods (e.g., *1–4*). Although some technological advancements have been made to allow approximations of certain data (e.g., estimating origin/destination and purpose using "big data" from probes), obtaining data on the personal characteristics of travelers and other details of their behavior and preferences can only be obtained through personal surveys. Collecting such data has proved difficult in recent years as the public is increasingly hesitant to provide personal information and cannot be bothered with the time burden of providing data, even for legitimate transportation planning purposes. This trend is juxtaposed with the increasing demand on transportation planning agencies to adopt a data-driven planning and decision-making process, which necessarily requires extensive amounts of observational data on system performance as well as personal data on the current and future preferences of the traveling public.

The use of the Internet for conducting surveys has been adopted in a wide range of fields, including transportation planning (e.g., *4–10*). The use of the Internet for personal surveys offers numerous benefits to the organization conducting the survey as well as the respondents themselves (e.g., *10–12*). For example, responses received from Internet surveys are usually compiled automatically into a computer database, drastically reducing the resources required to translate responses into a usable format for analysis. Internet surveys can also automate certain aspects of the survey question flow, such as skips, branches, or randomization, allowing for these features to be "invisible" to the respondent. Finally, more advanced features, such as question customization, visual aids or animations, and the ability to obtain information about respondent behavior (e.g., response

[1]Department of Civil Engineering, Russ College of Engineering and Technology, Ohio University, Ohio University, Athens, OH

Corresponding Author:
Address correspondence to Benjamin R. Sperry: sperry@ohio.edu

times), are also possible using Internet surveys. With reliable estimates placing Internet use at nearly 90% among the U.S. adult population (*13*), the use of the Internet for personal surveys has become the norm in many facets of modern life in which personal surveys are needed. Though the benefits of using the Internet for personal surveys are numerous, the primary issue remains how to recruit or attract respondents from the population of interest to take the survey. That is to say, the main challenge lies in linking individuals from the target population to a device connected to the Internet that would allow them to complete the survey. With so many demands on the time of respondents and (potentially) numerous requests for feedback, a secondary challenge lies in getting individuals from the target population to actually complete the survey even if they can be easily reached.

Data obtained from traveler intercept surveys are an essential part of a broader data-driven process for transportation planning and decision-making. Given that the challenge of implementing traveler intercept surveys via the Internet lies primarily in the recruitment of potential respondents, any opportunities for innovation will benefit both transportation planning agencies and the traveling public. This paper presents two case studies that demonstrate the application of an innovative survey recruitment technique known as the "foot-in-the-door" (FITD) method for traveler intercept surveys that are to be conducted primarily using the Internet. The remainder of this paper is organized as follows. Following this introduction, the next section provides an overview of the process, history, and application of the FITD compliance technique for both general population surveys and transportation planning surveys. Next, a detailed description of the two case studies and the research methods used in each study is provided. Analysis of the results from the two case studies examines trends and patterns in the participation and response rates as well as the total response time among different traveler groups. The final section presents the conclusions of the paper and a discussion of future research opportunities in this area of study.

Foot-in-the-Door Compliance Technique

Overview of Technique

The FITD compliance technique is a two-stage method in which researchers attempt to gain a research subject's compliance with the request of interest by first gaining his or her compliance with a smaller request. The theory is that if the research subject is willing to comply with the smaller first request, he or she is more likely to comply with the larger request (which is actually the request of interest to the researcher). The FITD method first originated with Freedman and Fraser's 1966 study of the household products of a sample of housewives in Palo Alto, California (*14*). In that study, Freedman and Fraser found that the subjects that had agreed to participate in a short telephone-based survey of consumer preferences for household soaps were more likely to allow researchers to spend 2 h in their homes conducting an inventory of household cleaning and cooking products at a later date than subjects who were not initially contacted. They theorized that compliance with the initial phone call request activated positive feelings of being helpful among the research subjects, leading to a greater likelihood of complying with the second, larger request. More than five decades after the publication of the first study on the method, the FITD compliance technique has been deployed extensively in a wide range of contexts; several meta-analyses have consistently shown the two-stage method of first seeking compliance with a smaller task to be effective at achieving compliance with the larger task of interest to the researcher (*15–18*).

FITD Applications for Transportation-Related Surveys

Traveler intercept surveys are used in a diverse range of contexts and situations in which feedback is needed at the time that the trips of interest are being made; however, the application of the FITD method for recruiting participants for such surveys, or in transportation-related surveys more broadly, has been fairly limited. Sheskin and Stopher described the design and implementation of a two-stage on-board passenger survey for bus passengers in three urban areas consisting of a shorter portion to be completed during the ride and a detachable longer portion that could be returned via mail (*2*). Taylor and Booth-Butterfield implemented an FITD-based study designed to encourage alcohol-impaired bar patrons to comply with the bartender's request to call a taxi, in an effort to reduce drunk-driving crashes (*19*). Their analysis found that bar patrons that had signed a petition against drunk driving at the bar were more likely to comply with the bartender's request than patrons that had not. Seethaler and Rose (*20*) utilized an FITD-type experiment as part of a broader study of persuasion techniques to encourage community-based travel behavior changes. Specifically, they found that a pre-intervention gift of a "green" reusable shopping bag before requests to participate in a voluntary travel behavior modification program was effective at raising awareness of the broader travel demand management campaign. Delhomme et al. utilized an FITD-based approach to study the rehabilitation of the most serious speeding violators in 12 locations in France; the researchers developed a short questionnaire to be completed at the end of a driver training course in which the respondents would commit to comply with posted speed limits each time they drove (*21*). They found that the course participants who agreed to comply with posted speed limits were more likely to have reported compliance with the speed limits than those participants who did not complete the commitment questionnaire at the training. Sperry et al. conducted a two-stage FITD survey of travelers utilizing a shuttle connecting a rail station to the airport terminal at the Milwaukee General Mitchell International Airport in Wisconsin (*4*). Shuttle passengers were asked to participate in an online survey about their use of the air-rail link at Milwaukee Airport. The response rate to the online survey was nearly 20% with no significant difference in response rate observed among key traveler groups, with the one exception that residents of the Milwaukee-Chicago region had greater participation than nonresidents. Additional details of this survey are discussed later in this paper. Gueguen et al. reported on the results of an experiment in which drivers at a green traffic signal were blocked by an experimental vehicle with a fake mechanical failure (*22*). The experiment showed that blocked drivers who had provided directions to a nearby store to a passer-by prior to being blocked (the FITD condition) were less likely to demonstrate aggressive behavior than those drivers who were not asked for directions.

Research Methods

General Approach

The primary objective of the research design was to develop a traveler intercept survey procedure that would apply the basic concepts of the FITD compliance without pressure technique to the recruitment of participants (*14*). The general research approach applied in each of the case studies discussed in this paper consisted of the following process.

1. Researchers, strategically positioned at the location where travelers are to be intercepted for the survey, ask travelers if they wish to participate in a brief, two-question survey about their travel patterns at the survey location. For each traveler approached, the researcher records basic observational data such as the size of the travel party, the gender of each traveler, and the estimated age group of each traveler.
2. The brief initial survey, hereafter referred to as the "initial interview" portion of the survey, includes two brief questions of significant interest to the researcher. Although the two questions can be on any topic, for the case studies described in this paper, the two questions were the purpose of the respondent's travel that day (broadly defined) and the respondent's home ZIP code. The initial interview responses are recorded alongside the observational data obtained in the previous step.
3. The respondent is then asked if they wish to participate in a longer, web-based survey related to their travel patterns at the survey location. This survey, hereafter referred to as the "follow-up survey" portion, includes all questions that are of primary interest to the researcher. If the respondent agrees to participate, the researcher provides the respondent with a brightly colored invitation postcard, which includes more details on how to access the online survey, a unique numeric access code to be entered at the start of the survey, and a requested "please complete by" date. The respondent's unique access code is recorded alongside the observational and initial interview data.
4. Respondent completes the follow-up survey at their convenience.

The two-question initial interview forms the FITD between the researcher and the participant. Following the basic theory of the FITD process, it is assumed that compliance with the initial smaller request (the two-question initial interview) will result in greater compliance with the actual request of interest to the researcher (completion of the follow-up survey). Assuming that the primary objective is to obtain responses to an intercept survey of travelers at a particular location, the process described above has the following benefits:

- Observational data is obtained for every traveler passing by the location of interest, whether they choose to participate or not; data on the purpose and home residential location of travelers that participate in the initial interview are also obtained. As a result, researchers have at least some information about a majority of travelers.
- The two-stage process requires less time on the part of the respondent while they are making their trips (a short two-question survey while traveling, the longer survey to be completed later), reducing delays and inconvenience to the traveler.
- The short interaction between the traveler and the researcher increases the legitimacy of the entire survey process in the view of the traveler.
- All the benefits of web-based surveys (e.g., automated response tabulation, automated survey question flow, and ability to use advanced graphics) are available for the follow-up survey, allowing for a broader range of questions and topics to be included.
- The unique access code allows certain aspects of the follow-up survey to be personalized to the respondent's experience. For example, the researcher will know what day the initial interview took place; this detail can be incorporated into the survey to customize the questions and aid in respondent recall of the date of the interaction with the researcher.
- The use of the unique access code, linked to the observational and initial interview data, allows for tracking of response rate and response time, and for such measures to be analyzed for different segments of the travelers being studied. Additionally, responses to the online survey can be weighted to account for nonparticipation if necessary.

Table 1. Summary of Traveler Intercept Survey Case Studies

	U.S. 33 Rest Area Visitor Survey	Dayton Bicycle Facility User Survey	Milwaukee Airport (MKE) Rail Station Shuttle Survey (*4*)
Study location	Logan, Ohio	Dayton, Ohio Xenia, Ohio	Milwaukee, Wisconsin
Target population	Visitors to safety rest area located on U.S. Highway 33	Users of two bicycle facilities in Dayton and Xenia, Ohio	Riders of shuttle connecting Milwaukee Airport Rail Station with MKE airport terminal
Study time frame	June 2015 (6 weekdays)	August 2015 (7 days)	May-June 2011 (15 days)
Initial interview	Trip purpose and home ZIP code	Trip purpose and home ZIP code	Trip purpose and home ZIP code
Follow-up survey content	Online survey consisting of 10 questions related to use of rest area facilities and regional travel patterns	Online survey consisting of 16 questions related to use of bicycle facilities in Dayton region	Online survey consisting of 22 questions related to travel patterns and use of rail station to access Milwaukee Airport
Participation incentive	State of Ohio highway map provided at time of initial interview	$100 gift card via random drawing for completion of follow-up survey	$250 gift card via random drawing for completion of follow-up survey

Description of Case Studies

A summary of the traveler intercept survey case studies analyzed in this paper is provided in Table 1, which includes the details of the target population, location and time frame, initial interview and follow-up request, and incentives used in each study. Additional discussion of the details of each case study is provided in this section. Two case studies, the U.S. 33 Rest Area Visitor Survey and the Dayton Bicycle Facility User Survey, are discussed in detail. Details of the third study, the Milwaukee Airport Rail Station Shuttle Survey, are presented elsewhere (*4*), but are summarized in Table 1 for the convenience of the reader.

U.S. 33 Rest Area Visitor Survey. The purpose of the U.S. 33 Rest Area Visitor Survey was to examine the characteristics, use of facilities, and general travel patterns of visitors to a safety rest area located on U.S. Highway 33 near Logan, Ohio. The study was conducted in June of 2015 and included field data collection during 3 weekdays (Tuesday through Thursday) between the hours of 9:00 a.m. and 4:00 p.m. at the rest area on each side of the highway (a total of 6 days of surveys). The initial interview obtained trip purpose and home ZIP data for rest area visitors, and the follow-up survey consisted of 10 questions. Initial interview participants who agreed to complete the follow-up survey received the invitation postcard as well as a State of Ohio highway map as an incentive to participate. The rationale for including the highway map alongside the invitation postcard was to replicate the survey process with the resource constraints of a transportation agency that may not be interested in offering monetary incentives but do have other tokens of appreciation, such as highway maps, that can be offered. Consequently, the FITD condition for the U.S. 33 Rest Area Visitor Survey consisted of the two-question initial interview plus the highway map incentive.

Dayton Bicycle Facility User Survey. The purpose of the Dayton Bicycle Facility User Survey was to examine the travel patterns, attitudes, and preferences for bicycle travel among users of two bicycle facilities in the Dayton, Ohio region. Surveys were collected from cyclists at two locations in the Dayton region to obtain a cross-section of regular (i.e., frequent) and recreational (i.e., infrequent or leisure) facility users. Surveys were collected in August 2015 and included a total of 7 days of fieldwork between 10:00 a.m. and 6:00 p.m. each day. For this study, participants who completed the follow-up survey were able to provide their contact information on a separate survey for the opportunity to be randomly selected for a $100 gift card as an incentive for completion. The FITD condition for the Dayton Bicycle Facility User Survey consisted of the two-question initial interview plus the gift card lottery incentive.

Both surveys utilized the Qualtrics web-based survey tool and were designed such that the respondent could complete the surveys using a desktop computer or mobile device. Figure 1 shows an example of the follow-up survey recruitment postcard (1*a*) and a screenshot of the follow-up survey as viewed by a desktop computer (1*b*) and a mobile device (1*c*) for the Dayton Bicycle Facility User Survey. The unique survey access code (noted by a circle in Figure 1*a*) provided a link between the date of the initial interview and customization of the follow-up survey questions (noted by a circle in Figure 1*c*).

Case Study Results

The results of the FITD case studies are presented in this section. Responses to the surveys themselves are not the focus of this analysis; rather, this analysis focuses on the following survey outcomes: (1) analysis of participation in the surveys as measured by various acceptance and response rates; (2)

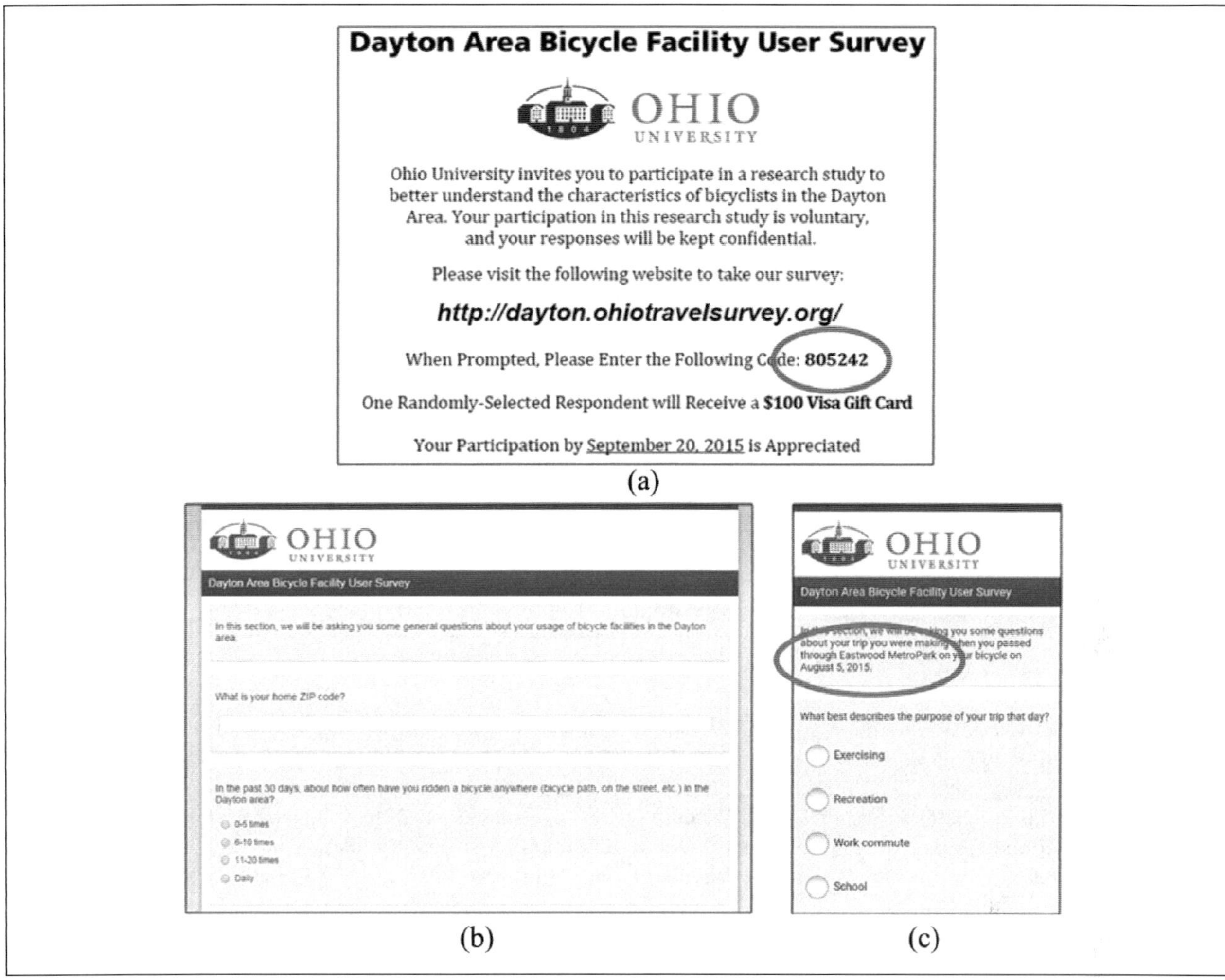

Figure 1. Recruitment postcard and survey views for desktop and mobile device for Dayton Bicycle Facility User Survey follow-up survey: (*a*) follow-up survey recruitment postcard; (*b*) desktop view of follow-up survey; (*c*) mobile device view of follow-up survey.

analysis of response time to the follow-up survey; and (3) participation and response time patterns that relate to various characteristics of the participants, including whether a mobile device was used to complete the follow-up survey.

Participation Analysis

For the FITD traveler intercept surveys analyzed in this paper, participation in the survey is measured three different ways, as follows:

- Initial Interview Acceptance Rate: Percentage of travelers observed using the location of interest that stopped and participated in the initial interview;
- Follow-Up Invitation Acceptance Rate: Percentage of initial interview participants that agreed to accept the follow-up survey invitation and took a recruitment postcard;
- Follow-Up Survey Response Rate: Percentage of participants who agreed to take a follow-up survey recruitment postcard that actually completed the follow-up survey.

Table 2 reports each of the three metrics of participation for each of the two case studies plus the corresponding data from the Milwaukee Airport Rail Station Shuttle Survey (*4*). For clarity, the calculations corresponding to each participation metric are also shown in Table 2.

Each of the metrics in Table 2 provides insight into the level of "success" at each stage of the process used in the FITD traveler intercept survey method. Approximately 80% of travelers that were observed during the survey period agreed to provide responses for the two-question initial interview (based on the two current case studies, as such data was not available for the Milwaukee survey). Among those individuals, the willingness to accept the follow-up survey

Table 2. Summary of Participation Metrics for FITD Traveler Intercept Surveys

	U.S. 33 Rest Area Visitor Survey	Dayton Bicycle Facility User Survey	Milwaukee Airport (MKE) Rail Station Shuttle Survey (*4*)
Travelers approached or observed	801	966	NA
Accepted initial interview request	84.0% (673/801)	80.7% (780/966)	961
Accepted follow-up survey invitation	86.9% (585/673)	96.2% (750/780)	88.2% (848/961)
Completed follow-up Internet survey	12.5% (73/585)	36.4% (273/750)	18.3% (155/848)

Note: NA = data not obtained during survey.

Table 3. Summary of Response Times for FITD Traveler Intercept Surveys

	U.S. 33 Rest Area Visitor Survey	Dayton Bicycle Facility User Survey	Milwaukee Airport (MKE) Rail Station Shuttle Survey (*4*)
Average time to complete follow-up survey	5.53 days	2.99 days	6.75 days
Completed on day of initial interview	31.5% (23/73)	39.2% (107/273)	20.0% (31/155)
Completed within 1 day of initial interview	49.3% (36/73)	64.1% (175/273)	32.3% (50/155)
Completed within 1 week of initial interview	68.5% (50/73)	87.9% (240/273)	64.5% (100/155)
Completed within 2 weeks of initial interview	90.4% (66/73)	94.1% (257/273)	81.9% (127/155)
Completed within 3 weeks of initial interview	93.2% (68/73)	97.1% (265/273)	94.2% (146/155)

recruitment postcard was greater, with approximately 90% of those interviewed accepting the postcard. The response rate to the follow-up survey was 12.5% for the rest area survey and 36.4% for the bicycle facility survey, with the corresponding response rate for the Milwaukee survey being 18.3%. The results presented in Table 2 suggest that the Dayton bicycle survey respondents appeared to be more enthusiastic about participating, as evidenced by higher follow-up invitation acceptance and follow-up survey response rates; however, the initial interview acceptance rate was lower for this study due to some potential respondents cycling at high speeds and being unable to stop at the survey location.

Response Time Analysis

Each follow-up survey recruitment postcard included a numeric access code that the respondent would enter on the survey "splash page" prior to starting the survey. This access code was unique to each individual that participated in the initial interview and was associated with the traveler's responses to the initial interview, as well as the observational data. This arrangement allowed for tracking of a variety of metrics associated with survey participation, including an analysis of the "response time" for the follow-up survey, defined here as the number of days between when the respondent received the follow-up survey invitation and when that individual completed the follow-up survey (this information is provided by the web survey software).

Table 3 displays a summary of the average response time and cumulative proportion of responses obtained at different points in time for the case studies. The average response time for the Dayton survey was approximately 3 days, much lower than the rest area survey (5.5 days) and the shuttle survey (6.8 days). Similar to the participation analysis, the Dayton bicycle survey participants appeared to be more enthusiastic, as evidenced by a relatively greater proportion of responses obtained for that survey at the different time points noted in Table 3.

Figure 2 displays a visualization of the cumulative response rate to the follow-up survey as a function of response time for each of the three FITD traveler intercept surveys. Also shown in Figure 2 is a composite cumulative distribution curve fitted to the three-study average cumulative response rate. The prediction equation for the three-study average fitted curve is provided below Figure 2. The prediction equation indicates that, on average, approximately one-third of follow-up surveys were completed on the same day that the initial interview took place. Additionally, the cumulative distribution of response time suggests that approximately 75% of surveys were completed within 1 week of the initial contact, and 90% were completed within 2 weeks. After 3 weeks, approximately 95% of the follow-up surveys had been received for all three surveys.

Respondent Characteristics Analysis

In addition to analyzing the response time, linking the unique survey access code to the observed and initial interview data also allows for a more substantial analysis of study participation by various segments of travelers being studied. Table 4 shows survey participation (using the three participation

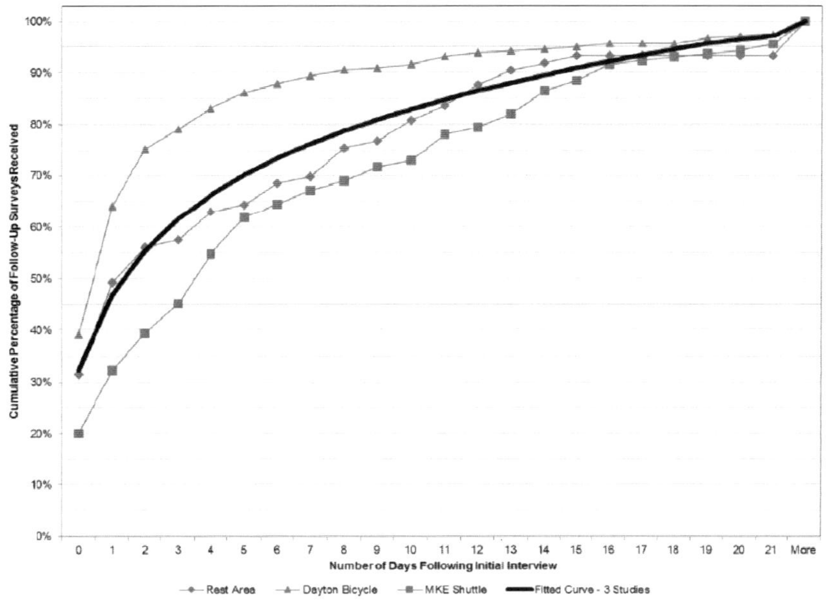

Figure 2. Cumulative distribution of follow-up survey response time. Cumulative response (%) = 0.2109 * ln(Number of days) + 0.3232.

metrics as previously discussed) for four traveler segments (gender, age group, trip purpose, and home residence) for the U.S. 33 Rest Area Visitor Survey and the Dayton Bicycle Facility User Survey. The distribution of travelers in each segmentation is also provided in Table 4 for reference (e.g., 71% of rest area visitors during the survey period were male).

The data presented in Table 4 provide valuable insights into the participation and response patterns of the various traveler segments analyzed. For example, travelers that appeared to be age 65 or older tended to have a higher rate of accepting the initial interview request but had a lower rate of accepting the follow-up survey invitation. Anecdotal evidence from the rest area survey suggests that older travelers declined the follow-up survey invitation because they could not access the Internet; however, even with those concerns, the follow-up survey response rate was not significantly lower among these individuals.

Analysis of participation as related to trip purpose and home residential location was only possible for the follow-up survey invitation acceptance rate, and the follow-up survey response rate, because purpose and residence data were only available if the traveler had accepted the initial interview. For the U.S. 33 Rest Area Visitor Survey, no significant difference in follow-up survey response rate was detected across the four traveler segments. However, several significant differences in follow-up survey response rate were identified among traveler groups in the Dayton bicycle survey. Male travelers, travelers between the ages of 35 and 64, travelers with a trip purpose of business or work commute, and local residents all had a higher than average follow-up survey response rate. One possible explanation for these results is that these travelers may have been regular users of bicycle facilities in the Dayton region and perceived that their responses to the follow-up survey would improve cycling conditions. Additionally, there were no significant differences in follow-up survey response time among the four traveler segments across the two surveys (thus, these data are not shown in Table 4).

Mobile Device Usage

The follow-up surveys were designed to be completed on a desktop computer or on a mobile device, as detected automatically by the survey software. The operating system of the device used by the respondent to complete the survey was recorded along with the other survey data; this information was used to determine if a mobile device was utilized on the survey. The proportion of responses submitted using a mobile device is shown in the far right column of Table 4. For the U.S. 33 Rest Area Survey, 20 out of 73 follow-up surveys (27.4%) were completed using a mobile device. For the Dayton Bicycle Facility User Survey, 53 out of 273 surveys (19.4%) were completed on a mobile device. Analysis of mobile device usage among the four traveler segments indicated that female travelers and travelers aged 18–34 years were more likely to complete the follow-up survey on a mobile device.

The response time characteristics of mobile device users (not shown in Table 4) can also be analyzed using the available survey behavior data, providing insight into the question of whether making traveler surveys available on such devices

Table 4. Participation and Mobile Device Use Analysis by Respondent Characteristics

	Accepted interview (%)	Accepted invitation (%)	Completed survey (%)	Used mobile device (%)
U.S. 33 Rest Area Visitor Survey				
Overall	84.0	86.9	12.5	27.4
Male (71%)	85.1	85.6	12.3	19.6*
Female (29%)	81.3	90.4	13.0	45.5*
Age 18–34 (12%)	80.9	90.8	8.7	83.3*
Age 35–64 (58%)	82.8	91.2*	13.1	26.1
Age 65 and Older (30%)	87.6	77.7*	12.8	14.3
Purpose: Business (16%)	NA	84.5	12.9	50.0
Purpose: Personal (51%)	NA	84.8	10.3	30.0
Purpose: Vacation (33%)	NA	91.4*	15.3	16.1
Home ZIP within 25 miles (15%)	NA	85.6	9.6	37.5
Home ZIP 25–75 miles (50%)	NA	88.8	12.2	25.0
Home ZIP 75–250 miles (24%)	NA	86.5	14.8	20.0
Home ZIP beyond 250 miles (12%)	NA	87.2	13.2	44.4
Dayton Bicycle Facility User Survey				
Overall	80.7	96.2	36.4	19.4
Male (58%)	78.1*	96.0	39.1*	16.1*
Female (42%)	89.7*	96.5	28.5*	32.7*
Age 18–34 (13%)	83.3	98.1	24.3*	20.0
Age 35–64 (55%)	76.8*	97.6*	39.9*	21.3
Age 65 and older (32%)	86.6*	93.2*	35.8	15.9
Purpose: Business/Work (11%)	NA	96.5	64.6*	22.6
Purpose: Personal/Recreation (89%)	NA	96.1	32.9*	18.6
Home Residence: Local (75%)	NA	96.1	38.8*	19.8
Home Residence: Out of Town (25%)	NA	96.4	29.5*	21.7

Note: Number in parenthesis indicates percentage of observed or initial interview participants corresponding to each segment of participants.
NA = accepted interview percentage could only be calculated based on observational data (gender and age group) and could not be calculated for trip purpose and home residence because these data could only be obtained from the initial interview.
*Indicates percentage significantly different than average based on the standardized residual for the Pearson's Chi-Squared Test of Independence being >±2.0 (*23*).
"Home Residence" for Dayton survey defined as follows: Local = home residence within the two counties in which surveys were conducted (Greene and Montgomery); Out of Town = all other participants.

results in a prompter response. The average response time among the Dayton Bicycle Facility User Survey participants that completed the follow-up survey using a mobile device was 1.71 days, which was significantly less than the response time among non-mobile-device participants (3.30 days, $t = 2.17$, $p = .016$). Participants using a mobile device to complete the follow-up survey in the U.S. 33 Rest Area Visitor Survey also completed the survey sooner than the non-mobile-device participants (4.45 days versus 5.94 days); however, the difference was not statistically significant.

Discussion and Recommendations

Summary of Case Studies

This paper presented an analysis of the outcomes of several traveler intercept surveys that incorporated elements of the FITD compliance technique as a way to increase participation and improve the overall experience for both the traveler and the organization conducting the survey. The FITD survey technique described in this paper was implemented at a safety rest area in Southeast Ohio and on two bicycle facilities in the Dayton, Ohio region. The results of a third study that applied this method, a survey of airport travelers in Milwaukee, Wisconsin, have been reported elsewhere (*4*) but were discussed in this paper due to the similarity of the methods utilized. The main feature of the FITD survey technique was the initial interview portion, in which researchers asked travelers to provide answers to two short questions at the intercept location. Travelers that participated in the initial interview were asked to complete an Internet-based follow-up survey at a later time. A unique survey access code was used to link the follow-up survey data to the data obtained during the initial interview; the use of the access code allowed for greater insight into the participation and response characteristics among different traveler groups.

Results Discussion

Analysis of survey participation revealed several notable results. Approximately 80% of travelers accepted the initial interview request and provided details of their trip purpose and home ZIP code. The follow-up survey response rate was 12.5% for the rest area survey and 36.4% for the Dayton survey. The proportion of travelers that accepted the invitation to complete the follow-up survey was also higher in the Dayton survey. The Dayton survey travelers also completed the follow-up surveys more promptly than the rest area visitors (2.99 days versus 5.53 days). It is unclear from the available data whether these differences are due to the difference in incentive ($100 gift card lottery for follow-up survey completion compared with a free highway map for all participants), differences in the survey setting, or differences in un-measured characteristics of bicycle facility users compared with general travelers. Approximately one-third of follow-up surveys were completed on the day of the initial interview, and 95% were completed within 3 weeks. Some differences in participation among different traveler groups were noted, primarily among elderly travelers, travelers that are local to the facility being studied, and those travelers with a business or work-related trip purpose. Approximately one-quarter of follow-up surveys were completed using a mobile device.

Recommendations and Future Research

The FITD-based traveler intercept survey method described in this paper offers a wide range of benefits to both travelers and the survey sponsor. The results of this analysis are limited in that they are only based on three studies that have used the two-stage FITD technique. Nevertheless, the results are informative and provide valuable insight for the design of traveler intercept surveys based on FITD concepts. Based on the results of this analysis, the following recommendations are provided. First, the follow-up survey response rate for the rest area survey (12.5%) could be considered a minimum response rate for other intercept surveys; this information could be used to develop sample size requirements for future surveys. Second, it is recommended that web-based surveys remain open and able to receive responses up to 3 weeks following the final distribution of recruitment materials. Third, it is recommended that web-based surveys be available and functional for both desktop computers and mobile device web browsers. Finally, the use of a unique access code to link web-based survey responses with observational data is recommended to allow for a more detailed analysis of survey participation.

Future research on this topic should examine the use of the FITD method for other intercept survey contexts, such as external station surveys. Additionally, surveys should be conducted at locations where a sufficiently large number of a relatively homogeneous population of travelers can be surveyed such that the FITD portion can be removed to determine the true effect of including the initial interview and related elements in the intercept survey process and survey outcomes. Surveys of travelers will always be needed as part of a comprehensive and metric-driven approach to transportation systems planning and project development. Any improvements to the process that will allow for the necessary data to be obtained with minimal inconvenience to the traveler while maximizing the use of agency resources will result in better data at lower costs. Transportation planners are encouraged to consider the FITD technique described in this paper when developing traveler intercept surveys within their jurisdictions.

Acknowledgments

The authors gratefully acknowledge the Ohio Department of Transportation, Five Rivers MetroParks, and the Greene County Parks & Trails department supporting the research studies from which the content of this paper was drawn. Financial support provided by the Department of Civil Engineering, the Russ College of Engineering and Technology, and the Honors Tutorial College, all of Ohio University, is also gratefully acknowledged.

References

1. Pratt, J., M. Lee-Gosselin, and S. Burbridge. Chapter 3.0: Options for Travel Surveys. In *The Online Travel Survey Manual: A Dynamic Document for Transportation Professionals*. http://www.travelsurveymanual.org/Chapter-3-1.html. Accessed July 18, 2017.
2. Sheskin, I. M., and P. R. Stopher. Surveillance and Monitoring of a Bus System. *Transportation Research Record: Journal of the Transportation Research Board*, 1982. 862: 9–15.
3. Sperry, B. R., B. T. Chigoy, L. K. Green, and E. Hard. Development of Improved Trip Attraction Rates for Small and Medium-Sized Travel Demand Models. *Transportation Research Record: Journal of the Transportation Research Board*, 2016. 2568: 48–54.
4. Sperry, B. R., S. Larson, D. Leucinger, S. Janowiak, and C. A. Morgan. Design and Implementation of Internet-Based Traveler Intercept Survey. *Transportation Research Record: Journal of the Transportation Research Board*, 2012. 2285: 84–90.
5. Son, S., A. Khattak, X. Wang, and J. Y. Chen. Transferring Telephone-Based National Household Travel Survey to the Internet: Application to University Students. *Transportation Research Record: Journal of the Transportation Research Board*, 2012. 2285: 91–99.
6. Smith, C., and G. Spitz. Internet Access: Is Everyone Online Yet and Can We Survey Them There? *Transportation Research Record: Journal of the Transportation Research Board*, 2010. 2176: 35–41.
7. Stinson, M. A., and C. R. Bhat. Frequency of Bicycle Commuting: Internet-Based Survey Analysis. *Transportation Research Record: Journal of the Transportation Research Board*, 2004. 1878: 122–130.
8. Adler, T., L. Rimmer, and D. Carpenter. Use of Internet-Based Household Travel Diary Survey Instrument. *Transportation*

Research Record: Journal of the Transportation Research Board, 2002. 1804: 134–143.
9. Abdel-Aty, M. A., and H. T. Abdelwahab. Toll Road Origin-Destination Travel Survey: Internet and Mailback Response Analysis and Implications for Future Surveys. *Transportation Research Record: Journal of the Transportation Research Board*, 2001. 1768: 61–70.
10Alsnih, R. Characteristics of Web Based Surveys and Applications in Travel Research. In *Travel Survey Methods: Quality and Future Directions* (P. Stopher and C. Stecher, eds.), Elsevier, Oxford, UK, 2006.
11. Dillman, D. A. *Mail and Internet Surveys: The Tailored Design Method*. John Wiley and Sons, Hoboken, New Jersey, 2007.
12. Couper, M. *Designing Effective Web Surveys*. Cambridge University Press, New York, 2008.
13Pew Research Center. Pew Research Center Internet & Technology: Internet/Broadband Fact Sheet. Jan. 2017. http://www.pewinternet.org/fact-sheet/internet-broadband/. Accessed Jul. 25, 2017.
14. Freedman, J. L., and S. C. Fraser. Compliance Without Pressure: The Foot-in-the-Door Technique. *Journal of Personality and Social Psychology*, Vol. 4, No. 2, 1966, pp.195–202.
15. Beanman, A. L., C. M. Cole, M. Preston, B. Klentz, and N. M. Steblay. Fifteen Years of Foot-in-the Door Research: A Meta-Analysis. *Personality and Social Psychology Bulletin*, Vol. 9, No. 2, 1983, pp. 181–196. https://doi.org/10.1177/0146167283092002.
16.Dillard, J. P., J. E. Hunter, and M. Burgoon. Sequential-Request Persuasive Strategies: Meta-Analysis of Foot-in-the-Door and Door-in-the-Face. *Human Communication Research*, Vol. 10, No. 4, 1984, pp. 461–488. https://doi.org/10.1111/j.1468-2958.1984.tb00028.x.
17. Burger, J. M. The Foot-in-the-Door Compliance Procedure: A Multiple-Process Analysis and Review. *Personality and Social Psychology Review*, Vol. 3, No. 4, 1999, pp. 303–325. https://doi.org/10.1207/s15327957pspr0304_2.
18. Pascaul, A., and N. Gueguen. Foot-In-The-Door and Door-In-The-Face: A Comparative Meta-Analytic Study. *Psychological Reports*, Vol. 96, No. 1, 2005, pp. 122–128. https://doi.org/10.2466/pr0.96.1.122-128.
19. Taylor, T., and S. Booth-Butterfield. Getting a Foot in the Door with Drinking and Driving: A Field Study of Healthy Influence. *Communication Research Reports*, Vol. 10, No. 1, 1993, pp. 95–101. http://dx.doi.org/10.1080/08824099309359921.
20. Seethaler, R. K., and G. Rose. Six Principles of Persuasion to Promote Community-Based Travel Behavior Change. *Transportation Research Record: Journal of the Transportation Research Board*, 2006. 1956: 42–51.
21. Delhomme, P., K. Grenier, and V. Kreel. Replication and Extension: The Effect of the Commitment to Comply with Speed Limits in Rehabilitation Training Courses for Traffic Regulation Offenders in France. *Transportation Research Part F*, Vol. 11, No. 3, 2007, pp. 192–206. https://doi.org/10.1016/j.trf.2007.10.002.
22.Guéguen, N., A. Martin, F. Silone, and M. David. Foot-in-the-Door Technique and Reduction of Driver's Aggressiveness: A Field Study. *Transportation Research Part F*, Vol. 36, 2016, pp. 1–5. http://dx.doi.org/10.1016/j.trf.2015.10.006.
23. Agresti, A. *Categorical Data Analysis*, 2nd ed. John Wiley and Sons, Hoboken, New Jersey, 2003.

The Standing Committee on Travel Survey Methods (ABJ40) peer-reviewed this paper (18-00939).

The views expressed in this paper are solely the opinions of the authors and do not represent the official views of Ohio University or any other organization. Any errors, inaccuracies, or omissions are the sole responsibility of the authors.

Article

TRR
JOURNAL OF THE TRANSPORTATION RESEARCH BOARD

Transportation Research Record
2018, Vol. 2672(42) 33–44

DOI: 10.1177/0361198118790376
journals.sagepub.com/home/trr
SAGE

Factors Affecting Interview Duration in Web-Based Travel Surveys

Pierre-Léo Bourbonnais[1] and Catherine Morency[1]

Abstract

Historically, travel surveys have been conducted face-to-face, by mail, or by phone. With the increasing share of households having access to the Internet, other survey modes have been deployed. This paper focuses on web surveys. Among other advantages, using the web to conduct surveys reduces costs and helps mitigate poor response rates among young households. Very few studies have been conducted on interview duration and its determinant using paradata from web travel surveys. Such knowledge is necessary to validate the context in which travel data are gathered and can be used to understand sample and data quality. Interview duration modeling is also essential for allocating survey servers and monitoring interviews during the data collection phase. This paper models interview duration using paradata from nine web surveys conducted in the Quebec province from 2010 to 2014. The main objectives of the model are to assist the monitoring of interviews by detecting outliers, provide a better estimate of the interview duration to respondents and survey managers during the interview, and allow a more precise evaluation of the server performance needed before conducting web travel surveys. Using a multiple regression model, we observed that the most important variables in explaining interview duration were number of car and transit trips as well as number of unique places visited during a day. Conducting the interview on a small-screen device also increased interview duration. The model also provides a baseline estimate of interview duration on the basis of demographic features and questionnaire design.

Three main media were traditionally used to conduct origin–destination travel surveys: telephone, mail, and face-to-face. With the increasing sample sizes involved in large-scale and regional surveys, computer-assisted telephone interviews (CATI) were considered the best tool available at the time to both reduce costs and increase data quality. Since the beginning of the last decade, however, web questionnaires have been suggested as a way to expand the reach of conventional CATI surveys. Web-based travel surveys were a welcome addition to the transportation analyst's toolbox, mostly to cope with problems associated with human-resource-intensive CATI surveys, but also to provide a tool based on a medium that an increasing share of the population uses in its daily activities. Declining response rates and difficulty constructing a representative sampling frame using only landline phone numbers are now the main drawbacks of conventional telephone-based surveys. According to the Residential Telephone Service Survey (*1*), 60% of young households – 18 to 34 years old – use only cell phones, the numbers for which are not usually found in public directories, and do not have a landline phone at home. Although the web does not help in itself to get good and representative samples, it may attract respondents who are not willing to participate in telephone surveys, or that are not reachable using conventional survey modes or during typical periods in which phone interviews are conducted. Furthermore, for specific types of trip generators, web-based questionnaires are easier to administer and less expensive in the long run.

In the last 6 years, web-based surveys have been deployed in the Quebec province to gather travel data from people and households. This paper uses paradata from nine web-based surveys to analyze and model interview duration. The paper proposes various descriptive statistics regarding outcome rates and interview duration as well as two interview duration models, one for people-based surveys and the other one for household-based surveys.

[1]Department of Civil, Geological and Mining Engineering, Polytechnique Montreal, Montreal, Canada

Corresponding Author:
Address correspondence to Pierre-Léo Bourbonnais: cmorency@polymtl.ca

The paper is organized as follows. The Background section (Part 2) introduces the survey process data collection by defining the term "paradata." It also summarizes previous research focused on modeling the interview duration. Context and data follow in Part 3 and provide context for the nine web travel surveys conducted by the authors and their partners. It also gives an overview of the paradata collected during web interviews. Part 4 presents the methodology used to undertake the respondent behavior analysis and summarizes the steps involved in the process of obtaining models for interview durations. Results and Findings (Part 5) consolidate different key variables and indicators by comparing all nine web surveys. It also reveals the results related to the interview duration models. Part 6 discusses limitations, as well as perspectives that will allow the reader to appreciate what should be done to enhance paradata collection and level up the overall quality of future travel surveys conducted on the web.

Background

This section proposes various background elements relevant to the research, namely discussion on paradata and their use in travel survey data analysis, as well as previous researches on interview duration.

Paradata: Measuring the Interview Process Itself

"Paradata," first proposed by Couper and explained in Kreuter et al. is used as a specialized term to represent automatically collected data associated with computer-assisted survey interviews for monitoring, management, and data quality assessment (*2*, *3*). Respondent behavior is therefore evaluated after analyzing paradata. Although interview durations (response times) and breakoffs (incomplete interviews, also called drop-offs) are almost always kept for record, more precise information could also be collected during web interviews. In particular, technical information including the browser, the device, and the operating system used by the respondent are usually recorded. However, even if the amount of paradata collected during interviews is increasing due to a larger array of available data collection tools, the resources required to integrate, validate, and analyze paradata are overwhelming. In this context, paradata collection and management need more structure and would benefit from better standards (*4*).

Paradata Analysis in CATI Travel Surveys

Regional household travel surveys are held once or twice per decade in major urban areas of the Quebec province. Ever since the beginning of these surveys, there has been a strong relationship between partners, both on the academic side (Polytechnique Montreal) and on the stakeholders' side (metropolitan transportation agencies, Ministry of Transportation, transit service providers, and other local partners). This unique long-term partnership has ensured the greatest possible population coverage and enhanced data quality from the hundreds of thousands of telephone interviews carried out over the years (between 50,000 and 70,000 completed household interviews per survey). The first full-scale CATI survey was conducted in 1993 in the Montreal region and since then, there has been continuous methodological and technical enhancements made to the survey process (*5*). Over the course of these surveys, a great amount of paradata has been collected and has allowed Morency, and Chapleau to establish a structured way to evaluate travel survey interview quality (*6*, *7*). By measuring call status, interview start/end time and by recording every interviewer's actions, it was even possible to emulate the interview later on for training or for data quality assessment.

Modeling the Interview Duration in CATI Travel Surveys

Using paradata from the 1998 and 2003 Montreal regional CATI surveys, Morency, and Chapleau proposed a similar interview duration model for both years (*6*, *7*). As expected, interview duration was strongly related to the number of household members (a little more than 30 s per person on average) and the number of trips declared (approximately 1.5 min for transit trips, down to an average of 1 min for non-transit trips). Additionally, they found that factors including the interviewer's experience, that the interview was completed after 5 p.m., that the respondent was female, or that the respondent was at least 65 years old all had a tendency to reduce interview duration. For both years, the model explained a little more than 40% of the interview duration variability observed (R^2 = 0.45 in 2003 compared with 0.43 in 1998).

Modeling the Interview Duration in Web Surveys

Interview durations are difficult to compare between surveys of different kinds due to their strong dependence on the amount of questions asked and their complexity. Nonetheless, Gummer and Rossmann recently proposed an interview duration model by conflating 21 web surveys (23,779 respondents) (*8*). Although all surveys were about political attitudes and behaviors, valuable hypotheses were nevertheless tested. These were related to question complexity and format, device used, and respondent characteristics like age group, educational level, and interest in the subject. Results showed that older respondents, respondents with lower education levels, and respondents with a greater interest in the subject took longer to respond. They also found that the use of smaller screen devices like smartphones demanded significantly more time to complete a survey. However, to date, it seems that interview duration has never been modeled for web travel surveys specifically. Consequently, models by Morency, Chapleau, and Gummer and Rossmann will serve as a starting point for our own modeling of interview duration (*6–8*).

Context and Data

To support the development of a sustainable transportation plan for a university setting, a web-based travel survey questionnaire was developed by Bourbonnais and Morency (*9*). It was first used in the fall of 2010 to gather data among polytechnic students and employees with a response rate of 18%. It was again administered the next year along with the rest of University of Montreal campus, giving a target population of more than 65,000 people, with a response rate of almost 20% for the polytechnic and 11% for the rest of the campus. Following the success of these initiatives, and considering the low cost of the process, other organizations and agencies in the Quebec province were eager to get involved. Between 2010 and 2013, The Quebec Ministry of Transportation, the Agence Métropolitaine de Transport, the University of Sherbrooke and Sherbrooke Cegep (local college), the carsharing company Communauto, and the bikesharing company Bixi, all either asked or agreed to conduct adapted versions of the travel web survey.

Two Types of Questionnaire: Person-Based and Household-Based

The first version of the questionnaire was built to collect data on the trips and travel behaviors of a single respondent at a time and, in that sense, was person-based. However, the next logical step was to upgrade the web survey application to allow households to declare trips and socio-demographic characteristics for every household member older than 4 years old, which is the minimum age threshold established for the CATI regional surveys. To date, seven person-based and two household-based surveys have been conducted using the same custom-built survey software. For an appreciation of the questionnaire interface and design, see Bourbonnais and Morency (*10*).

Context

For each survey, Table 1 includes the period of the year in which it was conducted, the recruiting method, the sampling frame, and the main objective.

Paradata Collected

For every survey, a timestamp was recorded at the beginning and at the end of the questionnaire. Excluding the Polytechnique Montreal 2010 survey (PY10), the system recorded each time the respondent entered or left a specific section. Each new location (visited place during the last working day), each transportation mode, and every geocoding action were also recorded. In the case of drop-off interviews, completed sections were preserved as a means to identify during which section respondents abandoned the survey.

Proxy Respondents

In most CATI travel surveys, one person usually responds for all other household members and is then called a proxy when answering for others. Verreault and Morency (*11*) showed that proxy interviews affect data quality and precision and can create a non-negligible bias for travel surveys. When conducting a household-based web survey, it is impossible to find the exact household member who responded using paradata alone. For this reason, the 2013 Montreal regional survey included a question that asked for the name of the specific household member who provided information for each person.

Sample and Interview Durations' Descriptive Statistics

Sample and descriptive statistics related to interview durations are summarized in Table 2 (household-based surveys) and Table 3 (person-based surveys).

Methodology

Calculation of Samples and Outcome Rates

The calculation of outcome rates (response and breakoff rates) follows standard definitions from the American Association for Public Opinion Research (*12*). For PY10, PY11, UM11 and SH13 (see Table 1 for full survey names), a list-based sampling frame of email addresses was used. The BX13 list of email addresses had no information on the status of the membership, meaning it was not possible to confirm recipients were still active members of the service or not. The response rate could not be calculated for the CM13 survey as invitations were made through Communauto's website and were not monitored, however, it can be assumed that only active members received the invitation to participate. TR11 used two different samples: a local sample of cell phone numbers and a sample of 1,000 random postal addresses drawn from the regional household survey sampling frame. Unfortunately, there was a disruption of postal services during the mailing due to a union strike. Consequently, many letters were lost or greatly

Table 1. Context for Seven Person-Based and Two Household-Based Web Travel Surveys Conducted between 2010 and 2013

Survey	Acronym	Period	Recruiting method	Sampling frame	Main objective
Polytechnique Montreal *Person-based*	PY10	Fall 2010	Email (official list from the institution)	Students and employees	Sustainable mobility plan
Polytechnique Montreal *Person-based*	PY11	Fall 2011	Email (official list from the institution)	Students and employees	Sustainable mobility plan
University of Montreal *Person-based*	UM11	Fall 2011	Email (official list from the institution)	Students and employees	Sustainable mobility plan
Trois-Rivières regional survey *Person-based*	TR11	Spring 2011	Cellphone and postal	Cellphone users and households in phonebooks	Test the survey application for a regional survey
Sherbrooke regional survey *Person-based*	SH13	Spring 2013	Email (official list from the institution)	Student population (main universities and colleges)	Data fusion with CATI for students not in phonebooks
Communauto (carsharing) *Person-based*	CM13	Fall 2013	Email (official list of members)	Carsharing members	Study the travel behaviors of carsharing users
BIXI (bikesharing) *Person-based*	BX13	Fall 2013	Email (official list of members)	Bikesharing members	Study the travel behaviors of bikesharing users
Quebec regional survey *Household-based*	QC11	Fall 2011	Postal, email (student residences) and social networks (invitation made by the local transit agency)	Households in phonebooks, students living in residence and social network users	First test of the household-based version of the questionnaire
Montreal regional survey *Household-based*	MTL13	Fall 2013	Phone refusals and partners' websites	Households in phonebooks	Data fusion with CATI for phone refusals and respondents who prefer the online version

Table 2. Sample and Descriptive Statistics for Two Household-Based Web Travel Surveys Conducted in 2011 and 2013

					Interview duration							
Survey	Population of concern	# invitations (# reminders)	# started interviews	# completed interviews	Mean	SD	IQR	Min	Q1 25%	Median	Q3 75%	Max
QC11	Postal: NA	1,000 letters sent, 749 received (no reminder)	138	98	23.7	13.4	14.2	5.6	14.7	20.1	29.0	63.0
	Students in residences: NA	Unknown (no reminder)	60	41	11.4	7.2	11.3	2.2	5.6	9.7	16.9	32.3
	Social networks: NA	Unknown (no reminder)	36	26	12.7	6.4	8.8	4.0	7.6	12.2	16.4	28.1
MTL13	Phone refusals: NA	533 accepted to participate 456 valid email received (1 reminder)	70	46	23.0	13.1	19.3	3.9	12.6	22.0	31.9	56.8
	Partners' websites: NA	Unknown (no reminder)	599	443	21.4	12.5	13.8	1.4	12.8	19.0	26.6	92.0
Total *Household-based*		Unknown	903	654	20.9	12.6	14.2	1.4	12.1	18.1	26.3	92.0

Note: IQR = interquartile range; Q1 25% = first quartile; Q3 75% = third quartile; NA = not available.

Table 3. Sample and Descriptive Statistics for Seven Person-Based Web Travel Surveys Conducted between 2010 and 2013

					Interview duration							
Survey	Population of concern	# invitations (# reminders)	# started interviews	# completed interviews	Mean	SD	IQR	Min	Q1 25%	Median	Q3 75%	Max
PY10	8,618	8,618 (2 reminders)	1,972	1,530	13.2	6.5	7.4	2.5	8.7	11.6	16.1	19.8
PY11	8,576	8,576 (2 reminders)	1,929	1,673	12.6	5.9	7.0	1.7	8.5	11.4	15.4	75.9
UM11	58,483	58,483 (2 reminders)	7,948	6,501	14.1	6.9	8.0	1.8	9.4	12.8	17.4	74.3
TR11	Cellphones: NA	1,644 (no reminder)	333 accepted 109 started interview	81	11.9	7.5	7.5	2.5	7.3	10.0	14.8	47.6
	Postal: NA	<1,000 (no reminder)	54	35	13.0	7.2	9.9	2.7	7.9	11.8	17.8	33.0
SH13	University: 16,194	16,194 (1 reminder)	2,399	1,838	13,0	5,5	6,7	2,5	9,1	12,1	15,7	48,1
	Colleges: 6,820	6,820 (no reminder)	683	467	13.7	6.5	7.3	9.4	9.4	12.3	16.7	53.7
CM13	~30,000	Unknown (no reminder)	3,143	2,527	13.8	7.1	8.2	1.9	8.9	12.3	17.1	63.2
BX13	78,068	74,905 (1 reminder)	6,191	4,423	13.6	7.3	8.2	2.1	8.7	12.1	16.9	69.6
Total *Person-based*		>180,000	24,428	19,075	13.6	6.8	7.8	1.7	9.0	12.2	16.7	75.9

Note: PY10 sample is not included in the interview duration models because the timestamp paradata for this survey were not precise enough to obtain genuine validated interview durations. IQR = interquartile range; Q1 25% = first quartile; Q3 75% = third quartile; NA = not available.

delayed in the process, meaning that the exact number of households that did receive the letter remains unknown. Regarding the QC11 household-based survey, the response rate was calculated only for the mailing sample as the number of invitations received for the other two samples is unknown. The response rate for the MTL13 survey will be given for households that refused to complete the interview on the phone but then agreed to participate online. Response rates are not provided for respondents contacted through survey partners' websites due to the absence of information about the number of clicks on each website.

Partial Interviews

In most web surveys, interviews in which some questions are left unanswered (partial interviews) are common and, in that sense, are not always considered equivalent to completed interviews. However, for the surveys covered in this study, it was not possible for a respondent to complete the questionnaire with partial answers, except for some facultative questions like personal and household income levels or qualitative questions asked at the end of the survey that were not mandatory for the analysis of travel behavior per se. Even if this design choice influences breakoff rates and could induce some bias, the great complexity of the relations between questions appearing in a travel survey and the amount of human resources needed for post-validation justified the decision.

Interview Duration Models

The paradata used to measure interview duration incorporated several timestamps from the interview logs. First, the duration was calculated as the time between the first click on the "Start" button and the last click on the "Complete interview" button at the end. Moreover, to make sure the duration did not include long periods of time during which the respondent was doing something else, or was disconnected from the survey software, every duration of more than 5 min between two sequential timestamps was converted to the average duration of the other actions associated with the same respondent. A timestamp was recorded every time a respondent opened a section, closed a section, used the map to select or modify a location, selected or modified a transportation mode, or selected an arrival or departure time at a specific location. As there was almost no single action in the surveys studied for which the duration was even near 5 min, this threshold did not modify genuine interview durations and allowed a better analysis of interview duration. Moreover, it was not possible to detect periods during which the respondent was doing something else but came back for the next action before the 5-min threshold. Consequently, some interview durations were overestimated, and will therefore increase the number and the magnitude of model residuals. As a last note, the first survey conducted (PY10) was excluded from the models because its interview durations could not be measured correctly with the paradata available at the time.

Using results from Morency, Chapleau, and Gummer and Rossmann, several assumptions were proposed (*6–8*). From the beginning, it was decided to study household-based and person-based interviews separately. Their mean interview durations are significantly different, and proxy respondents involved in the household-based interview make it more difficult to include attributes related to the respondent. In fact, for the household-based model, only the average age group of adults in the household was included in the model (≥ 18 years old). Table 4 presents the hypotheses related to household-based (HB) and person-based (PB) web travel surveys.

To account for the complexity of pinpointing locations on the map, the number of declared visited places was also included as a predictor, although places with the same latitude and longitude were counted only once because the respondent could then use shortcut buttons.

According to Gummer and Rossmann's model for web surveys, elderly persons need more time to respond (*8*); however, when Morency, and Chapleau modeled CATI interview durations, they found that respondents older than 65 years old tended to present reduced interview durations (*6, 7*). As a consequence, hypotheses HB3 and PB3 were formulated into two opposite views (HB3a/b and PB3a/b). To obtain a measure of the respondent age for the household-based surveys, the average age of all non-proxy respondents should have been used, however as the QC11 survey did not collect information about proxies, it was decided to use the average age of all household members older than 4 years old as a way to keep comparability between the QC11 and MTL13 surveys. For better comparability, the age groups are the same as in Gummer and Rossmann (*8*): 18–29, 30–39, 40–49, 50–59 and 60 + .

Hypothesis PB8 was proposed to verify whether respondents with higher interest in the subject would spend more time responding because of their motivation to be more precise and thorough when answering the survey – as was found in Gummer and Rossmann (*8*). To measure the interest of the respondent, the model relied on the answer given to the question "Would you like to participate to other surveys on travel behavior?" As this question was asked only in PY11, SH13, CM13, and BX13 interviews, a dummy variable was added to include "unknown" answers. Hypothesis PB10 tried to determine whether the number of custom-specific questions added to the last section of the questionnaire would indeed significantly increase interview durations.

Tables 5 and 6 comprise descriptive statistics for every variable used as a predictor in the models, for both household-based and person-based surveys.

A hierarchical regression was applied in which the variables were added in order of predicted effect size. As in Gummer and Rossmann (*8*), the natural logarithm of

Table 4. Hypotheses Used for Modeling Interview Duration in Household-Based (HB) and Person-Based (PB) Web Travel Surveys

Survey	Source	Hypothesis
HB1	CATI	The number of trips declared is proportional to the interview duration. (predicted effect: large)
HB2	CATI	The household size proportionally increases interview duration. (predicted effect: large)
HB3a	(Gummer & Rossmann 2014)	Households with elderly persons need more time to complete the survey. (predicted effect: medium)
HB3b	CATI	Households with elderly persons need less time to complete the survey. (predicted effect: medium)
HB4	CATI	Interviews started in the evening (after 5 p.m.) take longer time to complete. (predicted effect: small)
HB5	New proposition	The presence of a high-speed Internet connection at home decreases the interview duration. (predicted effect: small)
HB6	(Gummer & Rossmann 2014)	The use of a small-screen mobile device increases interview duration. (predicted effect: small)
HB7	New proposition	The fact that the respondent has declared no household trip for the last working day will decrease interview duration even more (even if the number of trips is a predictor in the model). In fact, there should be an overhead duration associated with reading explanations about trips and modes when respondents declare at least one trip for the household. (predicted effect: small)
PB1	CATI	The number of trips declared is proportional to the interview duration. (predicted effect: large)
PB2	CATI	The household size proportionally increases interview duration, though the effect should be smaller than for household-based surveys. (predicted effect: large)
PB3a	(Gummer & Rossmann 2014)	Older respondents need more time to complete the survey. (predicted effect: medium)
PB3b	CATI	Older respondents need less time to complete the survey. (predicted effect: medium)
PB4	CATI	Interviews started in the evening (after 5 p.m.) take longer time to complete. (predicted effect: small)
PB5	New proposition	The presence of a high-speed Internet connection at home decreases the interview duration. (predicted effect: small)
PB6	(Gummer & Rossmann 2014)	The use of a small-screen mobile device increases interview duration. (predicted effect: small)
PB7	New proposition	The fact that the respondent declared no trip for the last working day will decrease interview duration even more (even if the number of trips is a predictor in the model). In fact, there should be an overhead duration associated with reading explanations about trips and modes when respondents declare at least one trip. (predicted effect: small)
PB8 PY11, SH13, CM13, and BX13 only	(Gummer & Rossmann 2014)	Respondents more interested in the subject have longer interview durations because of their motivation to be more precise (available only with surveys asking the question, "Would you like to participate in other surveys on travel behavior?"). (predicted effect: small)
PB9	CATI	The fact that the respondent is a woman decreases interview duration. (predicted effect: small)
PB10	New proposition	The number of specific questions asked at the end of the survey will affect interview duration, that is, more questions will increase duration. (predicted effect: medium)
PB11	New proposition	A large number of unique transportation modes declared will increase interview duration due to complexity (different questions are asked for each mode). (predicted effect: small)

the duration was preferred for the dependent variable. The important skewness of the untransformed variable (skewness = 1.477) could be made less substantial by using the natural logarithmic transformation, skewness of *ln*(interview duration) = −0.414. Most importantly, the correlation between the duration and its natural logarithm was very high (r = 0.920). For the person-based surveys, using the same transformation made the skewness even less important, from 1.683 to −0.092. As for the type of regression chosen, as the questionnaires were very similar, it was decided to test a standard multiple regression instead of a multilevel regression that would have included a survey level and a respondent level. The intraclass correlation coefficient between the interview duration and the survey class was very low (0.01 for household-based and 0.006 for person-based).

Table 5. Descriptive Statistics of Variables to Be Tested as Predictors of Interview Duration in a Household-Based Web Travel Survey Model

Variable	Mean $\bar{X}$	SD	Min	Max
Number of distinct visited places	5.041	2.905	0	17
Number of car segments	3.199	3.818	0	25
Number of transit segments	2.760	3.479	0	28
Number of non-car, non-transit segments	1.687	2.587	0	15
Household size	2.291	1.123	1	7
Average age 18–29 (reference)	0.226	0.419	0	1
Average age 30–39	0.378	0.485	0	1
Average age 40–49	0.190	0.392	0	1
Average age 50–59	0.113	0.317	0	1
Average age 60 +	0.093	0.291	0	1
Interview started ≥5 p.m.	0.327	0.470	0	1
High-speed Internet access at home	0.861	0.346	0	1
Small-screen mobile device used	0.017	0.129	0	1

Table 6. Descriptive Statistics of Variables to Be Tested as Predictors of Interview Duration in a Person-Based Web Travel Survey Model

Variable	Mean $\bar{X}$	SD	Min	Max
Number of distinct visited places	3.611	1.956	0	19
Number of car segments	1.145	1.876	0	16
Number of transit segments	1.679	1.950	0	15
Number of non-car, non-transit segments	1.036	1.614	0	14
Number of unique modes	1.635	1.018	0	6
Household size	2.533	1.289	1	20
Number of specific questions	19.226	18.532	0	42
Average age 18–29 (reference)	0.507	0.497	0	1
Average age 30–39	0.234	0.424	0	1
Average age 40–49	0.121	0.326	0	1
Average age 50–59	0.099	0.299	0	1
Average age 60 +	0.039	0.195	0	1
No trip declared (non-mobile)	0.092	0.289	0	1
Interview started ≥5 p.m.	0.328	0.469	0	1
High-speed Internet access at home	0.862	0.345	0	1
Small-screen mobile device used	0.042	0.202	0	1
Gender: Female	0.592	0.492	0	1
Would like to participate in other surveys = yes	0.317	0.465	0	1
Would like to participate in other surveys = no	0.222	0.416	0	1
Would like to participate in other surveys = unknown	0.461	0.498	0	1

Results and Findings

Outcome Rates

Table 7 integrates response, breakoff, refusal, and contact rates for each sample for which the required data was available.

Household-Based Interview Duration Model

When conducting the multiple regression including both household-based samples, the Durbin–Watson (D-W) test of independent residuals gave a value very close to 2 with a *p*-value a lot greater than 0.05 (D-W = 1.957, *p*-value = 0.604), thus the assumption of independence is certainly met.

Figure 1 summarizes the results and statistics for the household-based interview duration model. All predictors were significant except for HB4, for this reason it was removed from the model. The HB3a hypothesis was supported instead of HB3b, and, in that sense, older respondents seemed to have significantly longer interview durations. The assumption of no multicollinearity was met as no predictor had a VIF (variance inflation factor) larger than 5. Residuals were randomly distributed, except for fitted interview durations of more than 55 min (natural logarithmic >4.0). These residuals were clustered below the zero line, but very long interview durations are rare and could have been caused by small periods of time not corrected by the system in which the respondent was doing something else. However, one

Table 7. Outcome Rates for Household-Based and Person-Based Web Travel Surveys Conducted between 2011 and 2013

Survey	Sample	# contacts	# breakoff interviews	# completed interviews	Response rate	Breakoff rate	Refusal rate	Contact rate
QC11	*Postal* 1,000 letters sent	749 letters received	40	98	9.8%	29.0%	65.1%	74.9%
MTL13	*Phone refusals* 553 agreed to participate	456 valid emails received the invitation	24	46	8.6%	34.3%	73.2%	81.8%
	Partners' websites	Unknown	156	443	NA	26.0%	NA	NA
PY10	Students and employees	8,618	442	1,530	17.8%	22.4%	82.2%	100%
PY11	Students and employees	8,576	256	1,673	19.5%	13.3%	80.5%	100%
UM11	Students and employees	58,483	1,447	6,501	11.1%	18.2%	88.9%	100%
TR11	Cellphones	1,644 333 agreed to participate	28	81	4.9%	25.7%	95.1%	100%
	Postal	<1,000	19	35	NA	35.2%	NA	NA
SH13	University	16,194	561	1,838	11.3%	23.4%	88.7%	100%
	Colleges	6,820	216	467	6.8%	31.6%	93.2%	100%
CM13	Members: ~ 30,000	Unknown	616	2,527	NA	19.6%	NA	NA
BX13	Members: 78,068	74,905	1,768	4,423	5.7%	28.6%	90.3%	95.9%

Note: Calculations follows standard definitions and guidelines from American Association for Public Opinion Research (*12*). NA = not available.

should be careful if using the model to predict very long interview durations. When evaluating influential cases, there appeared to be no cases for which Cook's distance even approached 1, thus confirming that no case had too much influence on the predictor estimates. The histogram and the quantile-quantile plot of residuals both attest that the normality and linearity of residuals were acceptable.

Person-Based Interview Duration Model

Figure 2 includes the same information for the person-based interview duration model. In this case, all predictors were significant and again, the HB3a hypothesis was supported rather than HB3b. However, contrary to what was predicted by the CATI surveys models, the respondent being female did not tend to reduce interview duration; instead, it revealed a small increase in interview duration, but with a small effect. The assumption of no multicollinearity was again met (no VIF larger than 5). No case in the sample had an excessive influence on the results (Cook's distances were always much smaller than 1). Residuals were still mostly randomly distributed, except for fitted interview durations longer than 40 min (natural logarithmic >3.7), which revealed some heteroscedasticity. For the same reason as given in the household-based model analysis, long interview durations did not fit as well in the model as shorter durations. The histogram and the quantile-quantile plot of residuals both exhibited a small deviation on normality and linearity for the longest interview durations (heavy right tail on the quantile-quantile plot), but the residuals' distribution was nonetheless considered acceptable for the analysis proposed. In fact, when including only interview durations with a percentile of 95 or less in the model, the quantile-quantile plot did not display significant deviation from normality. To verify the effect of the different survey samples on the coefficients, a multilevel regression with fixed effects based on the survey level was also carried out. The average difference between the interview durations obtained with the standard regression and the multilevel regression was less than 5 s. Furthermore, the standard error of the model was the same (0.32 for both the multilevel regression and the standard regression). Accordingly, our choice of standard multiple regression instead of a multilevel regression was thus confirmed.

Discussion and Conclusion

Relying on paradata from nine web-based surveys conducted between 2011 and 2013, this paper has provided some insight into the variables that can explain interview durations. This section describes the key findings and the main limitations of the approach. It also identifies some possible enhancements and provides some research perspectives.

Key Findings

Using the standard coefficients β in both figures 1 and 2, the relative effect of each variable on the model can be estimated. As predicted, variables related to the number of trips and segments explain a large amount of the variation in the interview duration. Car segments seemed to take longer to declare than transit segments, but the difference was almost negligible (less than 1 s) and was greatly influenced by the number of questions related to

Sample Size *n*	Regression SS	Residuals SS	*R*	R^2	Adjusted R^2	Residual Std. Error	Durbin-Watson Test
654	173.321 (10 df)	77.128 (643 df)	0.832	0.692	0.687	0.346	• 1.957

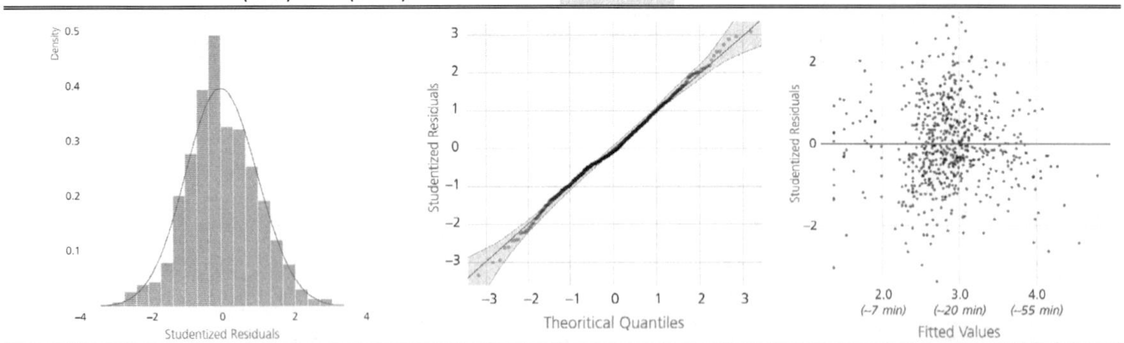

Dependent: *ln*(interview duration)	Coef. B	B Confidence Intervals 2.5%	B Confidence Intervals 97.5%	B Std. Error	Std. coef. β	t	*q*-value*	Collinearity Statistics Tolerance
Constant	2.124	2.015	2.232	0.055		38.597	< .0001	
HB1 Number of car segments	0.060	0.047	0.073	0.006	• 0.370	9.359	< .0001	• 0.306
HB1 Number of transit segments	0.046	0.029	0.065	0.009	• 0.261	9.296	< .0001	• 0.608
HB1 Number of unique visited places	0.047	0.037	0.056	0.005	• 0.219	5.085	< .0001	• 0.258
HB1 Number of non-car/non-transit segments	0.046	0.031	0.061	0.008	• 0.191	6.033	< .0001	• 0.478
HB2 Household size	0.082	0.053	0.111	0.015	• 0.148	5.546	.0013	• 0.669
Dummy variables:								
HB7 Number of trips = 0	-0.701	-0.826	-0.576	0.064	•-0.275	-11.001	< .0001	• 0.768
HB3a Age: 60+	0.401	0.300	0.503	0.052	• 0.189	7.750	< .0001	• 0.809
HB3a Age: 50-59	0.258	0.169	0.346	0.045	• 0.132	5.723	< .0001	• 0.901
HB6 Small screen device used	0.504	0.296	0.712	0.106	• 0.105	4.766	< .0001	• 0.991
HB5 High speed internet at home	-0.133	-0.214	-0.053	0.041	•-0.075	-3.243	.0032	• 0.904

* the *q*-value is the standardized *p*-value to a sample size of 100: $q = p\sqrt{\frac{n}{100}}$ (Woolley, 2003)

Figure 1. Interview duration model for household-based web travel surveys.

the specific mode declared. Household size, age, and device used were also significant in both types of interview, but respondents younger than 40 did not present a significant variability in interview duration when age alone was compared.

Model Purpose

The model developed will be quite useful for survey administration and monitoring. Before conducting the interviews, the model will help establish the server capabilities needed. In a cloud-server environment, the model could be of great help to set up load balancing and to choose the number of server instances required in real-time. Server administrators will be able to detect outliers in the interviews and act accordingly by providing live help to respondents or fixing server problems when technical deficiencies occur. This precise management of resources could reduce the cost of conducting surveys and reduce the probability of having to delete or correct several invalid interviews.

Limitations

Difficulty in Measuring Real Interview Durations. The greatest challenge when analyzing interview durations of web surveys is to get accurate active durations while being able to reject inactive durations, for instance, the time during which the respondent is doing something other than filling in the survey. The solution proposed in this paper was to detect all specific actions that took longer than 5 min and convert these to the average duration of all the actions made by the same respondent. However, the 5-min threshold was, in the case of the surveys conducted for this research, very conservative, and the effect of inactive durations was still noticeable as a right heavy tail exposed in the quantile-quantile plot for the person-based model (Figure 2). Knowing that, more precise timestamps to separate active and inactive durations are

Sample Size n	Regression SS	Residuals SS	R	R^2	Adjusted R^2	Residual Std. Error	Durbin-Watson Test
17545	2178.68 (16 df)	1783.67 (17528 df)	0.741	0.550	0.549	0.319	• 1.990

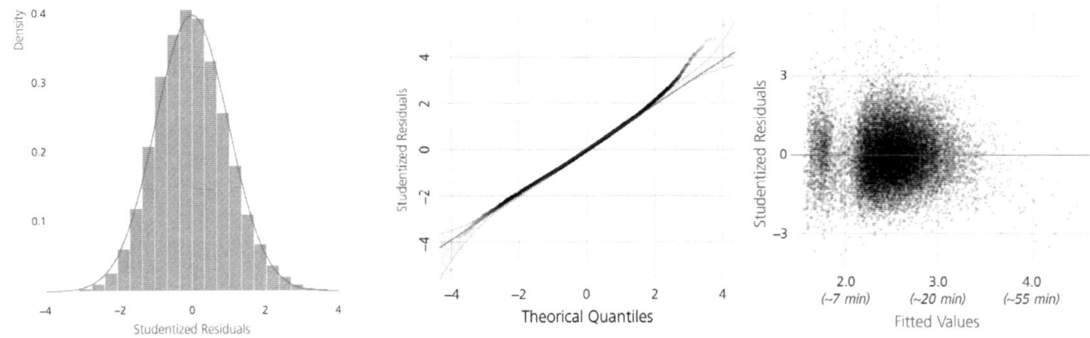

Dependent: *ln*(interview duration)	Coef. B	B Confidence Intervals 2.5%	B Confidence Intervals 97.5%	B Std. Error	Std. coef. β	t	q-value*	Collinearity Statistics Tolerance	Collinearity Statistics VIF
Constant	1.799	1.776	1.821	0.012		156.369	< .0001		
PB1 Number of car segments	0.071	0.067	0.076	0.002	• 0.282	32.067	< .0001	• 0.332	• 3.012
PB1 Number of unique visited places	0.065	0.061	0.070	0.003	• 0.269	26.054	< .0001	• 0.240	• 4.165
PB1 Number of transit segments	0.061	0.057	0.066	0.002	• 0.252	29.838	< .0001	• 0.359	• 2.784
PB1 Number of non-car/non-transit segments	0.054	0.049	0.059	0.002	• 0.184	21.867	< .0001	• 0.363	• 2.752
PB10 Number of specific questions	0.003	0.002	0.003	< 0.001	• 0.101	15.890	< .0001	• 0.636	• 1.573
PB2 Household size	0.026	0.023	0.030	0.002	• 0.072	13.464	< .0001	• 0.908	• 1.101
PB11 Number of unique modes	0.027	0.020	0.034	0.004	• 0.058	7.167	< .0001	• 0.397	• 2.521
Dummy variables:									
PB6 Small screen device used	0.401	0.378	0.425	0.012	• 0.170	33.188	< .0001	• 0.977	• 1.023
PB3a Age: 60+	0.349	0.324	0.374	0.013	• 0.143	27.472	< .0001	• 0.949	• 1.054
PB3a Age: 50-59	0.215	0.198	0.231	0.008	• 0.135	26.030	< .0001	• 0.953	• 1.049
PB7 Number of trips = 0	-0.215	-0.236	-0.194	0.011	•-0.131	-19.776	< .0001	• 0.587	• 1.704
PB3a Age: 40-49	0.117	0.102	0.132	0.008	• 0.080	15.513	< .0001	• 0.958	• 1.044
PB8 Would like to participate to other surveys	0.046	0.034	0.058	0.006	• 0.045	7.430	< .0001	• 0.707	• 1.414
PB4 Interview start time >= 5 p.m.	0.039	0.029	0.049	0.005	• 0.038	7.502	< .0001	• 0.982	• 1.018
PB9 Gender: female	0.034	0.024	0.044	0.005	• 0.035	6.834	< .0001	• 0.952	• 1.051
PB5 High speed internet at home	-0.031	-0.045	-0.017	0.007	•-0.023	-4.422	.0001	• 0.975	• 1.025

* the q-value is the standardized p-value to a sample size of 100: $q = p\sqrt{\frac{n}{100}}$ (Woolley, 2003)

Figure 2. Interview duration model for person-based web travel surveys.

needed. In fact, the new version of the survey software includes a new timestamp on each question answered.

Transferability. The results are partially transferable to other travel surveys in the sense that origin–destination questionnaires are well documented and have been in the field for quite some time. Having objective measures of how long it takes to provide answers on some typical elements could help in evaluating the expected duration of an online survey. It could also be useful in providing more accurate information to administrators and to the respondent during their participation.

Proposed Enhancements

Although the tool was continuously upgraded and improved throughout the conduct of the surveys, there are still some elements that could be added to enhance respondent experience when filling in the survey, to improve data quality (both survey data and paradata), and research possibilities. Among them, let us note:

- Getting paradata for window size, resolution, window in focus and complete record of mouse movements, clicks and keystrokes (recent faster browsers can record this information with almost no lag)

- Randomizing question order (when possible) to assess whether order can affect drop-off or data quality
- Creating a screen-size-responsive version of the questionnaire to make sure the interface adapts itself to the screen resolution and size of the device used to complete the interview

Perspectives

Challenges of the Mobile Web. The mobile web presents a challenge for designing complex surveys: hardware with a smaller screen requires a different design approach. As confirmed by the duration model, using a small-screen device has a significant and positive impact on duration. Numerous survey experiences are being conducted using the smartphone and good integration of both approaches will provide a set of relevant tools for the conduct of transport surveys.

What about the Quality of the Interface Design Itself? More researches about interface design are welcome, even though a handful of web design guidelines and examples can easily be found and applied to both the email invitation and the questionnaire. That some respondents were grateful enough to explicitly comment about the time and effort that was spent on the interface design should not be forgotten. In that sense, the involvement of web graphic designers and interface usability experts should be promoted. Future researches on the ergonomics of web travel surveys should promote the use of panels in which the respondents are asked to complete the questionnaire using distinct designs.

Improving the Survey Experience for the Respondent. Research and development needs to be pursued in the domain of interface design to improve usability. The potential contribution of fun theory (changing people's behavior by having them do fun and novel activities) or of gaming approaches to collecting transport survey data should be assessed. Furthermore, historical paradata could probably contribute to adapt the questionnaire in real-time based on typical patterns of response.

Acknowledgments

The authors thank Julien Faucher of the Réseau de transport de Longueuil, Hubert Verreault of Polytechnique Montreal, Louiselle Sioui and Pierre Tremblay of the ministère des Transports, de la Mobilité durable et de l'Électrification des transports, and Assumpta Cerda of the Autorité régionale de transport métropolitain.

Author Contributions

The authors confirm contribution to the paper as follows: study conception and design: P-LB, CM; data collection: P-LB; analysis and interpretation of results: P-LB; draft manuscript preparation: P-LB. All authors reviewed the results and approved the final version of the manuscript.

References

1. Statistics Canada. *Residential Telephone Service Survey.* Statistics Canada, Ottawa, ON, Canada, 2013.
2. Couper, M. P. *Measuring Survey Quality in a CASIC Environment. Proc., Survey Research Methods Section,* American Statistical Association, Alexandria, 1998.
3. Kreuter, F., M. P. Couper, and L. E. Lyberg. The Use of Paradata to Monitor and Manage Survey Data Collection. *Proc., Joint Statistical Meetings,* American Statistical Association, Alexandria, 2010, pp. 282–296.
4. Kreuter, F. *Improving Surveys with Paradata.* John Wiley & Sons, Hoboken, N.J., 2013.
5. Trépanier, M., R. Chapleau, and C. Morency. Tools and Methods for a Transportation Household Survey. *Urban and Regional Information Systems Association Journal,* Vol. 20, No. 1, 2008, pp. 35–43.
6. Morency, C. Enhancing the Travel Survey Process and Data Using the CATI System. *Transportation Planning and Technology,* Vol. 31, No. 2, 2008, pp. 229–248.
7. Chapleau, R. Measuring the Internal Quality of the Montreal CATI Household Travel Survey. In P. Jones and P. R. Stopher (eds.), *Transport Survey Quality and Innovation,* Emerald Group Publishing, Bingley, UK, 2003, pp. 69–87.
8. Gummer, T., and J. Rossmann. Explaining Interview Duration in Web Surveys: A Multilevel Approach. *Social Science Computer Review,* Vol. 33, No. 2., 2015, pp. 217–234.
9. Bourbonnais, P. L., and C. Morency. Web-Based Origin-Destination Surveys: An Analysis of Respondent Behaviour. Presented at the 13th World Conference on Transport Research, Rio de Janeiro, Brazil, 2013.
10. Bourbonnais, P. L., and C. Morency. Web-Based Travel Survey: A Demo. In *the Travel Survey Toolkit: Where to from Here?* In Munizaga, M., J. A. Carrasco, J. Zmud, and M. Lee-Gosselin (eds.), *Proc., 9th International Conference on Transport Survey Methods,* Emerald Group Publishing, Bingley, UK, 2013.
11. Verreault, H., and C. Morency. The Chronic Issue of Proxy Respondent Bias. Presented at the 10th International Conference on Transport Survey Methods, Sydney, Australia, 2014.
12. American Association for Public Opinion Research. *Standard Definitions: Final Dispositions of Case Codes and Outcome Rates for Surveys.* AAPOR, Chicago, IL, 2011.

The Standing Committee on Travel Survey Methods (ABJ40) peer-reviewed this paper (18-02951).

Article

TRR
JOURNAL OF THE TRANSPORTATION RESEARCH BOARD

Transportation Research Record
2018, Vol. 2672(42) 45–56

DOI: 10.1177/0361198118758687
journals.sagepub.com/home/trr
SAGE

Network Level Clearance Assessment using LiDAR to Improve the Reliability and Efficiency of Issuing Over-Height Permits on Highways

Suliman Gargoum[1], Lloyd Karsten[1], and Karim El-Basyouny[1]

Abstract

Commercial vehicles on highway networks are only permitted to use routes that are designed to accommodate their sizes and weights. When issuing an over-height permit, agencies consider vertical clearances on all route options on a highway network before directing a vehicle to the optimal route. Inefficient routing of commercial vehicles could cause excessive delays, which would result in undesirable economic impacts. Moreover, inaccurate assignment of vehicles to routes where the road infrastructure cannot handle the vehicle's size could cause significant damage to the roadway, resulting in potential safety risks. Although routing programs and permit-issuing agencies try to avoid inaccurate assignments, this is not always possible since they rely on a database of information prone to human error and one which is not always up to date. To create a more reliable database of information, Departments of Transport need more efficient methods to collect information on highways. This paper aims to increase the efficiency of collecting data required to issue over-height permits by utilizing LiDAR data to automatically assess vertical clearance on highways. The method used involves detecting all overhead objects on a highway corridor and estimating the clearance at each object before mapping the data on a GIS map. The method was tested on three different highway corridors in Alberta, Canada ranging in length from 130 to 400 km. Testing revealed that the proposed method is effective in performing network-level assessment of vertical clearance, which has significant impacts on the efficiency of routing over-height vehicles on a network.

Alberta's economy is highly reliant on the efficient movement of commercial vehicles on its highway network. The province is home to 605,335 registered commercial vehicles (*1*) and 694,222 registered commercial drivers (*2*). Due to their sizes and weights, commercial vehicles are only permitted to use specific routes on a highway network. For instance, some routes have underpasses with low clearances and cannot accommodate trucks that exceed those clearance limitations. To manage the presence of commercial vehicles efficiently on a network, oversized vehicles are often required to obtain oversize and/or over-height permits before undertaking a journey. In 2013 alone, 208,382 oversize and over-height permits were issued in Alberta (*3*).

Permits are typically issued by agencies who work under the provincial Departments of Transportation (DOT). The permit-issuing authority assesses the size of the commercial vehicle and, depending on clearance information stored within their database, optimal route options are proposed while also highlighting potential restrictions on each route. In many parts of Canada over-height permits are mandatory when the height of a vehicle or load exceeds 4.15 m (*4, 5*). Processing times to issue permits could reach 15 business days depending on the permit type and the issuing agency (*4*). For example, permits in Ontario can be of Annual, Project, Single Trip or Special Vehicle Configuration type with each permit having a different processing time (*4*).

To improve the efficiency of the permit issuing process and, consequently, improve the efficiency of managing commercial vehicles on highway networks, several researchers have worked on developing applications which would provide up-to-date routing information to permit-issuing agencies and commercial driving companies on a GIS platform. Adams et al. (*6*) presented a conceptual framework through which cross-platform permit information can be unified using GIS. Similarly, Garcia-Diaz et al. (*7*) proposed a decision

[1]Department of Civil and Environmental Engineering, University of Alberta, Edmonton, AB, Canada

Corresponding Author:
Address correspondence to Suliman Gargoum: gargoum@ualberta.ca

support algorithm that could be used by permit-issuing offices to route oversized vehicles throughout a highway network. All possible routes between an origin and destination pair would be identified and the ideal route chosen based on potential restrictions. Other studies went even further by analyzing permit data issued on a network level with the aim of rerouting traffic and improving the level of service on the network. In that context, Coley et al. (*8*) used GIS to visualize historical permits to determine trip origin/destination and identify heavily used commercial corridors. Dayan et al. (*9*) processed archived permits within West Virginia for GIS visualization and routed vehicles through the West Virginia DOT Linear Referencing System.

While these studies help present clearance data in a more efficient manner, clearance information displayed on those GIS maps is based on historical records which may be inaccurate, not up to date, or only available for portions of a highway corridor. In fact, part of the clearance information used to issue over-height permits is often collected in bridge inspection programs, which occur once every two years at best (*10*). In summary, although there has been work on how to present and communicate clearance information to improve the efficiency of issuing over-height permits and, consequently, the level of service of commercial vehicles on a highway network, there is a lack of work on how such a robust clearance database can be created. To create such a database, this paper proposes using an efficient tool to obtain clearance data on highway networks.

Clearance Assessment Tools

Unfortunately, assessing vertical clearances on a highway network using conventional surveying tools is an error-prone and burdensome process. This is prevalent when assessment is required in places like Alberta where approximately 4,500 bridge structures exist on the highway network (*11*). To improve the efficiency and accuracy of clearance assessments a variety of alternative methods have been considered. Many DOTs use digital measuring rods and electronic measuring devices to obtain more accurate clearance measures (*12*). Clearance assessment using photolog data has also been considered (*13*). Another innovative method which has recently gained popularity is clearance assessment using Light Detection and Ranging (LiDAR) data. Watson et al. (*14*) found that LiDAR was an effective alternative to traditional bridge clearance measurement techniques since, just like other digitized techniques, it minimizes the human error associated with assessing clearance using conventional surveying tools.

LiDAR is a remote sensing technology which uses light rays to collect positional information about surrounding objects. LiDAR integrates laser scanners, sensors, Inertial Measurement Unit and Global Navigation Satellite System (IMU/GNSS) into one system known as a laser scanning system. The laser scanner in the system constantly emits light beams at objects in the surrounding environment. The beams hit target objects and then reflect back to the scanning system. Using information on the time, wave length and energy of the reflected beam, the intensity and positional information of points on specific objects can be computed. LiDAR scans can be aerial (scanned from the air) or terrestrial (scanned from the ground). Further, terrestrial scans can be either static, where the scanner is mounted on a tripod, or mobile, where scanning equipment is mounted to a vehicle traveling down the highway. The highly dense 3D point cloud that mobile laser scanning (MLS) creates makes extracting road features and measuring the geometric attributes of the data extremely accurate as illustrated in previous studies (*15–26*).

Riveiro et al. (*27*) combined photogrammetry techniques and terrestrial LiDAR scans to conduct a clearance assessment, with results showing an accurate measurement of vertical clearance for both methods. Liu et al. (*28*) proposed a method whereby static terrestrial LiDAR was used to assess bridge clearance. The authors developed an algorithm with which scanned ground points are automatically matched to a bridge's deck points to estimate clearance across the bridge's deck. In recent work, Puente et al. (*29*), used mobile LiDAR data to assess vertical clearance in tunnels. The authors propose a semi-automated algorithm which involves extracting cross sections along the tunnel, defining the edges of the travel lanes using lane markings, and connecting the different elements of the tunnel's cross section using a convex hull before measuring the clearance.

While there have been a few attempts to extract clearance information from LiDAR data, to the best of our knowledge, no study has attempted the extraction on a network level, which is essential to commercial vehicle routing applications and for permit-issuing purposes. Moreover, the majority of studies that exist focus on the extraction of bridge clearances only, with no attention to other objects such as overhead signs and powerlines as shown in Figure 1. In an attempt to address the aforementioned limitations and help facilitate more efficient clearance assessments, this paper employs a method which can detect, classify and measure vertical clearance at all overhead objects on LiDAR scanned highways. The method involves fully automated detection of overhead objects on a highway segment and the identification of clearance information for each of the objects. The method was tested on three major highway corridors connecting major commercial hubs in Alberta and was successful in detecting all overhead structures on each corridor while also providing clearance estimates. The overhead objects were then mapped onto the network on a GIS platform.

Clearance Assessment Procedure

The methodology employed in this paper detects all overhead structures on a highway before measuring their clearances. This is essential when developing a comprehensive

(a)

(b)

Figure 1. LiDAR point cloud: (a) highway corridor; (b) LiDAR point cloud (overhead sign).

clearance assessment tool used in routing commercial transportation, since, for routing applications, information is typically required on a network level. In that sense, the proposed method provides end-users with a tool where they can import a LiDAR data set for a highway segment, automatically detect locations where overhead structures exist, and return clearance information for those objects. The method is divided into four steps (trajectory definition, non-ground point extraction, object detection, and clearance estimation). The following paragraphs provide a description of each of the steps.

Trajectory Definition

The first step of the detection process involves defining points parallel to the road's centerline which trace the road's trajectory and run along the entire road segment. These can be points obtained in separate GPS surveys or points filtered out of the LiDAR point cloud. The only condition here is that the density of those points must be high. In this study, the set of points tracing the path of the data collection truck were used, the point density was around 10 points/m. These points were obtained by filtering the LiDAR point cloud based on the scanner angle in MATLAB.

Non-Ground Point Extraction

Once the trajectory points are defined, the next step is to filter out the non-ground points from the LiDAR road segment. For this step, the LiDAR point cloud data is imported into ArcScene, which is a software that allows visualizing 3D layers in a GIS platform. The "Classify LAS Ground" tool is then used to classify the point cloud into ground and non-ground points. The ground points are then omitted in MATLAB, leaving the user with non-ground data which contains the overhead structures, vegetation, roadside objects, and other features of the road's environment.

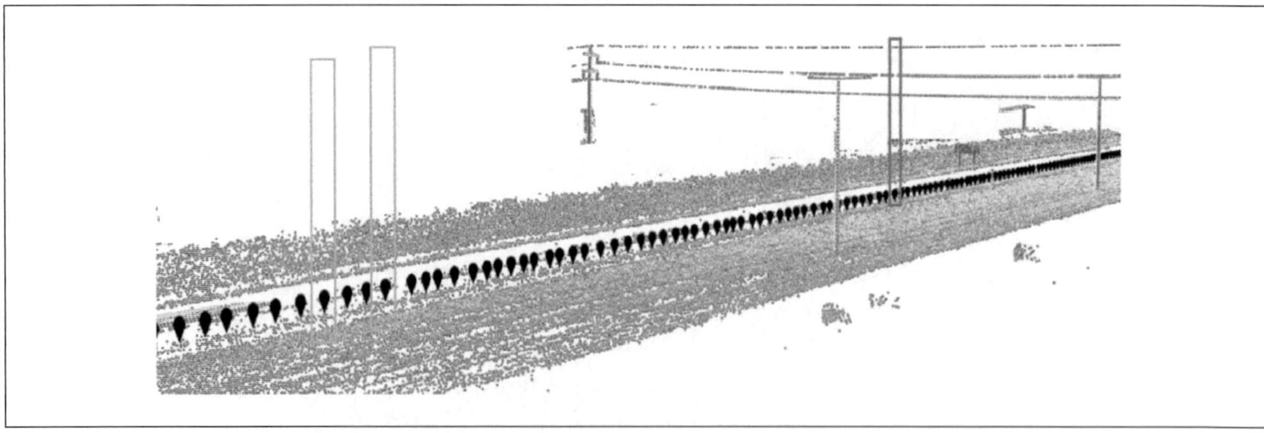

Figure 2. Overhead object detection.

Object Detection

The third step of the detection process involves locating overhead objects, clustering points representing overhead objects into groups, and classifying the clusters into bridges and non-bridges.

Locating Overhead Objects. This step involves matching the trajectory data with the non-ground points to locate the overhead structures. At every point along the defined trajectory, the k-Nearest Neighbor (kNN) search algorithm (*30*) was employed to search for overhead structures above the trajectory point. As displayed in Figure 2, the objective is to find the nearest overhead point to each trajectory point on the ground. The MATLAB algorithm loops through all trajectory points (black pins shown in Figure 2) and returns a list of points for which an overhead match was found. For instance, for the two blue frames in Figure 2, no overhead points exist above the trajectory points, therefore, no overhead object is detected. In contrast, overhead points do exist for the third frame (red) and, therefore, the code records this as a location where an overhead structure exists.

Clustering and Classification. To identify the number of overhead structures determined on a particular segment, trajectory points along the road's surface for which a match was found are clustered. In this study, the Density-Based Spatial Clustering of Applications with Noise (DBSCAN) clustering algorithm was used (*31*). The DBSCAN algorithm is a density-based algorithm that works by regrouping points based on proximity (ε) and hit count. The proximity measure defines how close multiple points within a cluster are to one another. Hit count is a measure of the number of points within a certain cluster. If points are further away from one another or if only a few points are within close proximity, it is likely that these points do not represent a cluster. Clustering identifies groups of points which are potential candidates for being a cluster or, in this case, an overhead object. In this paper, a hit count of 3 and an ε of 10 m is used to cluster overhead objects. This ensures that all clusters contain at least three points and that points are no farther than 10 m away from one another.

Clusters were also filtered through the following processes: standard deviation of vertical clearance, cluster size, and legal vertical clearance. These filters were applied to remove any noise present in the non-ground points. Clusters with a standard deviation of vertical clearance greater than 0.2 m were filtered to retain only those points within one standard deviation of the mean. To account for noise which may exist due to objects on the road surface a filter of 4.15 m was added. This is the legal vertical clearance as set by Alberta Transportation (AT) (*32*) and ensures that false positive clusters such as vehicles on the road are eliminated. A final filter is used to eliminate noise that exists within reasonable elevation range to be considered an overhead object. This was done by applying a kNN (*30*) search from the extracted minimum clearance point. Unlike false positives which are represented by a set of scattered points, a true overhead object will have closely spaced points regardless of object type. As a result, the mean of the distance between the minimum point and the 10 nearest neighbors is used to distinguish objects from noise. If the mean distance is >2.5 m the detected cluster is removed.

After clustering points into groups representing different overhead objects, the clustered were classified into bridges or other objects based on the following if statement.

$$Cluster: \begin{cases} p_{ci} \le \overline{p}_c & \rightarrow \quad \text{Not bridge} \\ p_{ci} > \overline{p}_c & \rightarrow \quad \text{Bridge} \end{cases}$$

where, p_{ci} denotes the number of points in cluster i and $\overline{p}_c$ is determined through an analysis of the histogram of the cluster sizes along a road segment. A $\overline{p}_c$ of 15 points was used, where bridge clusters were expected to have at least 15 points.

Figure 3. Point cloud of test segments: (a) Highway 2; (b) Highway 63; (c) Highway 881.

Table 1. Overhead Object Detection Results

Road segment	HW2	HW63	HW881
Length (km)	242	346	134
Actual number of overhead objects	120	30	68
Detected number of overhead objects	120	31	71
False positives	0	1	3
Minimum clearance (meters)	5.506	5.85	5.577
Maximum clearance (meters)	28.522	28.248	13.288

Clearance Estimation

For points where a match is detected, the MATLAB algorithm returns information about the exact position of the ground point, the exact position of the non-ground point, the difference in lateral placement between the two points, and the difference in elevation between the two points. The difference in elevation represents the clearance at that specific

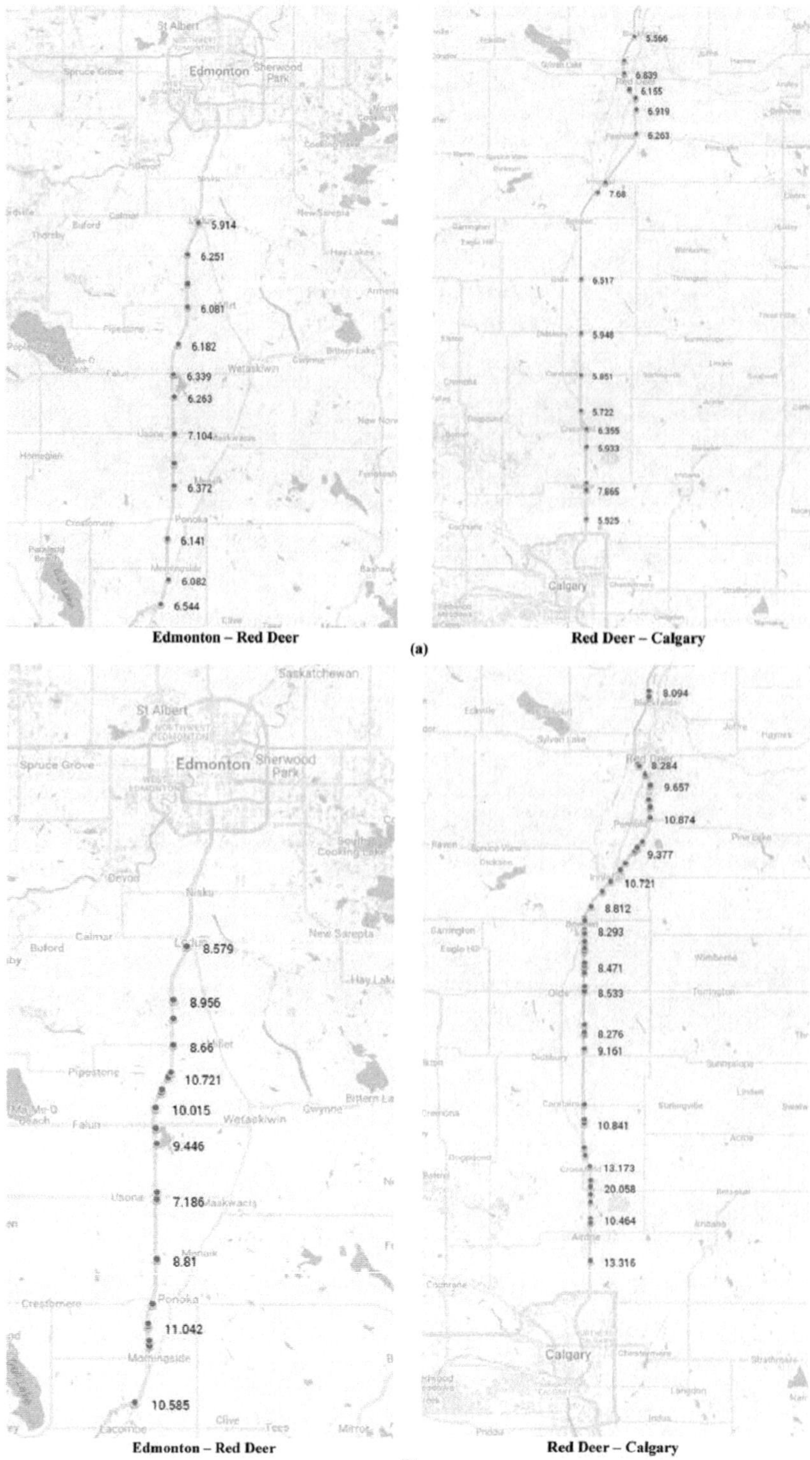

Figure 4. Overhead objects detected (Highway 2): (a) bridge structures; (b) powerline cables.

point beneath the overhead structures. This information is reported for all points in every cluster facilitating the estimation of minimum clearance for each object.

Case Study

The procedure proposed in this paper was tested on three different highway corridors in Alberta, Canada. Two of the three corridors were selected because they connected commercially attractive hubs. This was done to illustrate how important such a method would be in efficient routing of vehicles on strategic highways. The following paragraphs provide information about the laser scanning system used to collect the LiDAR data and the three highway segments used in the Case Study.

LiDAR Data

The LiDAR data used in this study was collected by AT using third party services. The data was collected using a RIEGL VMX-450 scanning system which is mounted on a pavement surface profiling vehicle (PSP-7000). The RIEGL VMX-450 system is equipped with two VQ-450 scanners and an IMU/GNSS unit. The laser scanners are symmetrically configured on the left and right sides, pointing toward the rear of the vehicle at a heading angle of approximately 145°. The VQ-450 scanner has a scan rate of up to 1.1 million points per second and a scan speed of 400 lines per second (*33*). The scanning system collects 3D point cloud data as the data collection truck travels through the highway corridor being surveyed. Surveys are conducted within normal traffic flow while travelling at the posted speeds of up to 100 km/h. Provincial surveys conducted at 90 km/h result in LiDAR point densities on the pavement surface of 150–1,000 points/m^2 (*34*).

Test Segments

Alberta Highway 2. The section of Highway 2 (also known as the Queen Elizabeth II Highway) considered in the analysis extends 242 km and lies in the central part of the Province of Alberta. The segment is part of a 4-lane divided highway located between the cities of Calgary and Edmonton—two major economic regions in Alberta. The speed limit on the segment is 110 km/h and it is well used as it is the main transportation route connecting Northern and Southern Alberta. The segment is primarily bordered by low lying vegetation as seen in Figure 3*a*. In addition, there is physical separation of the two travel approaches. The type of median varies along the segment (depressed versus raised) as does the horizontal alignment of the segment. The 242-km segment consists of over 1 billion points resulting in 41 GB of raw LiDAR data.

Alberta Highway 63. Highway 63 extends 346 km and lies in the western part of the Province of Alberta. The corridor connects Edmonton to the Northern Alberta region, which plays a significant role in the development of the national petroleum industry. The speed limit on the segment is either 100 or 110 km/h, depending on the location. The segment also has varied density of trees and vegetation on either side of the road as seen in Figure 3*b*. The point cloud files for the total 346-km segment contain around 1.4 billion points resulting in 45 GB of raw LiDAR data. It is worth noting that this segment played a pivotal role in the evacuation of 90,000 residents from Fort McMurray during the 2016 wildfire.

Alberta Highway 881. The section of Highway 881 considered in the analysis extended a length of 134 km and lies in the western part of Alberta. As seen in Figure 3*c*, the segment is part of a two-lane undivided highway. The speed limit on the highway is 100 km/h and it is located east of the city of Edmonton. Unlike the other two highway corridors analyzed, this segment does not connect two major commercial hubs. However, it does provide an alternative route from central Alberta to the northern part of the province. The point cloud files for the 134-km corridor contains approximately 500 million points resulting in 23 GB of raw LiDAR data.

Results and Discussion

Manual inspections of all overhead structures on Highways 2, 63, and 881 was conducted to verify the accuracy of the detection process. The number of overhead structures was used as ground truth and was compared with the number of structures detected using the LiDAR assessment.

Table 1 provides a summary of the comparison between ground truth and the detection results. The method was able to detect all 120 overhead objects on Highway 2 including bridges, cables, and overhead signs. No false positives (i.e., a cluster that was detected but did not represent an actual overhead object) were detected on Highway 2. Unlike Highway 2, one false positive was detected on Highway 63, this false positive could be a cluster of dust or the result of the laser beam being refracted by a surface. On Highway 881, three false positives were detected. In this case, the false positives were due to tree branches extending over the highway.

Figures 4 and 5 show plots of the overhead structures that were detected after running the procedure on Highways 2 and 63. Figure 4*a* shows a map with all the bridge structures that were detected on Highway 2 between the City of Edmonton and the City of Calgary (split for better visualization of clearance information). Similarly, Figure 5*a* shows a map of bridges that lie on Highway 63 between the City of Edmonton and Fort McMurray. Both maps also show the minimum clearance at each of those structures. Figures 4*b* and 5*b* show all powerline cables detected on Highways 2 and 63, respectively, while Figure 4*c* shows the overhead signs that were detected on Highway 63. As expected, the bridge clearance on the highways ranges between 5.5 m and 8.7 m while the

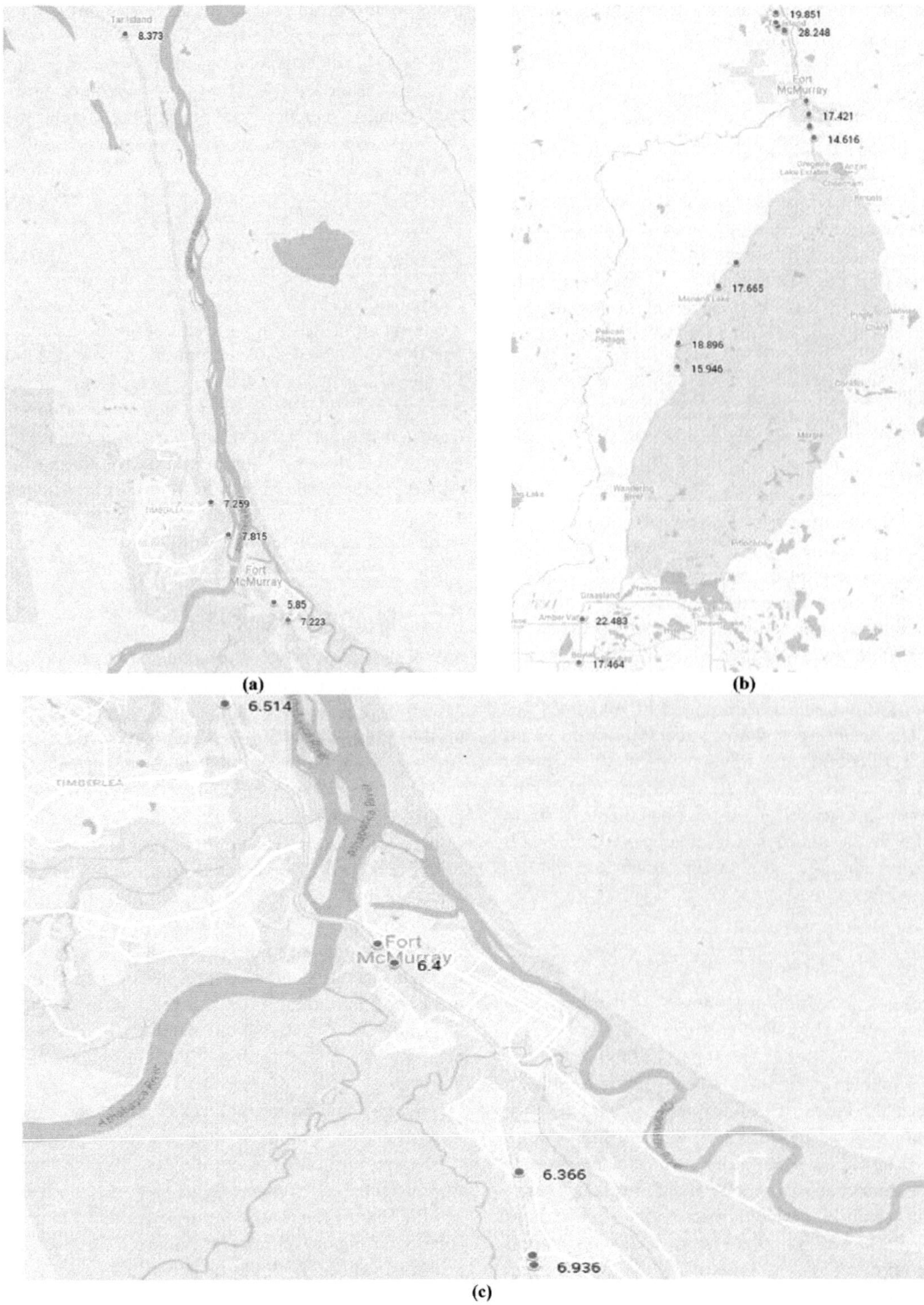

Figure 5. Overhead objects detected (Highway 63) : (a) bridge structures; (b) powerline cables; (c) overhead signs.

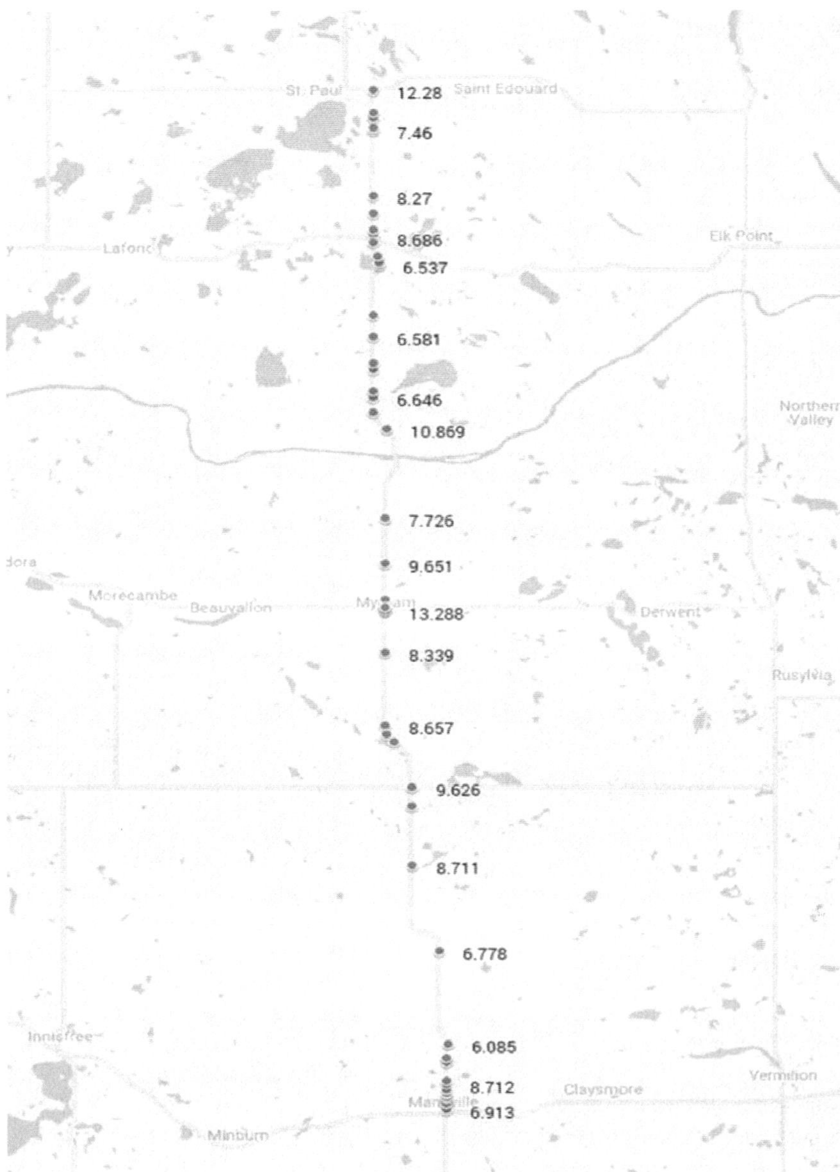

Figure 6. Overhead objects detected (portion of Highway 881).

clearance values for powerline cables are greater than 8 m and reach heights of 28.5 m for high voltage cables.

When considering all objects along the 242-km Highway 2 corridor, clearance varies between 5.51 m and 28.5 m. The efficacy in which the clearance values are estimated for an entire corridor and the ease with which this information can be displayed on a GIS platform is essential to increasing the efficiency of routing commercial vehicles on the highway network. This can be done by allowing vehicles to use parts of a highway corridor where they meet clearance requirements before being diverted to other highways when they do not satisfy the clearance requirements anymore. For instance, if a truck travelling from Edmonton to Calgary is 5.8 m high, based on the information extracted in this paper and presented in Figure 4, this truck would meet clearance requirements on the Highway corridor as far as a bridge just north of the City of Red Deer (around 150 km south of Edmonton) where the clearance is 5.51 m. Therefore, it may be possible for the truck to use part of Highway 2 before being

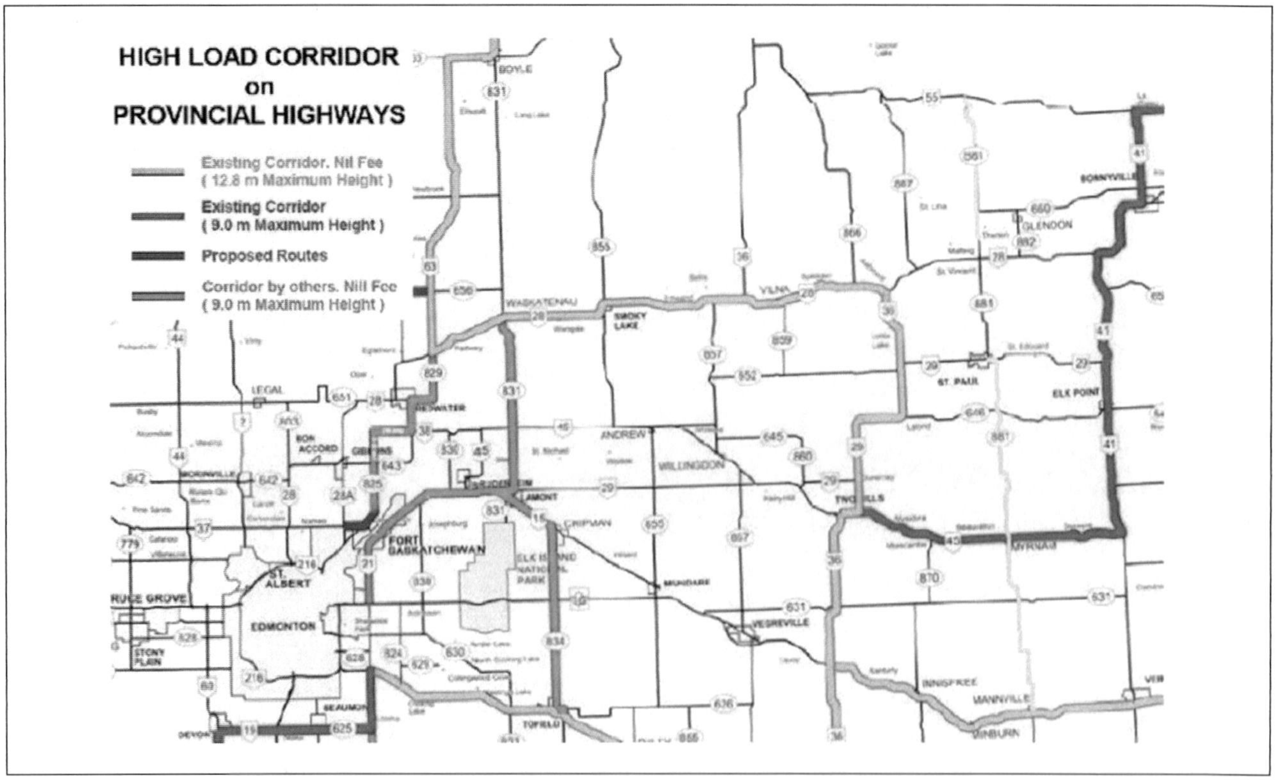

Figure 7. High load corridor map in Alberta (Highway 881).

rerouted. This helps permit issuing and routing agencies maximize link utilization on their highway network, which consequently improves the level of service for commercial vehicles on the entire network. The reason this is not always possible in current practice is that some agencies only have information about the absolute minimum clearance at specific locations on a highway corridor.

Network level assessment of clearances presented in this paper could also help agencies efficiently determine the shortest path for over-height vehicles that are diverted off a specific highway corridor due to clearance limitations. For instance, if a truck travelling from Edmonton to Calgary with a height of 5.8 m is diverted before reaching Red Deer, clearance information on all adjacent corridors may help minimize the amount of travel required by the truck before it is allowed to merge back onto the Highway 2.

Furthermore, the availability of information for an entire corridor also helps decrease the likelihood of inaccurate route assignments when issuing over-height permits. In current practice, even though DOTs issue permits, permit holders are still asked to accept responsibility for any damage that may be caused to overhead structures. This indicates that clearance information used to issue permits might not be perfectly reliable (*4*).

The assessment of HW881 was also conducted to show the possibility of High Load Corridor (HLC) assignment using a network-level clearance assessment. Highway 881 is a minor rural highway which lies in the eastern part of Alberta. Travel volumes and population in that area are, therefore, relatively low. Unlike the other two highways, no bridge structures were detected on the analyzed portion of Highway 881. However, there were several powerlines as shown in Figure 6. Highway 881 is strategically located when considering the locations of future HLCs on the Alberta highway network. In the Alberta Highway Design Guide, HLCs are defined as corridors that can accommodate vehicles of 9 by 9 m dimensions throughout (*32*). As displayed on Figure 7, the analyzed potion of Highway 881 runs parallel to the corridor on Highway 41, which is designated as an HLC. The fact that no bridge structures exist on Highway 881 means that it can potentially be upgraded to an HLC for oversized vehicles to access the northeastern part of the province. The absence of bridges from the segment means that upgrading the segment would only require lifting powerlines with low clearance to meet the dimensional specifications of an HLC. Figure 7 highlights segments which AT has proposed for upgrade to meet HLC specifications. The proposed method would make accurate clearance information readily available to DOTs, assisting them in their decision-making process.

Conclusions and Future Research

Factors such as the minimum allowable clearance and the maximum pavement capacity must be considered when routing commercial vehicles along a highway network. Routing vehicles where vertical clearances are insufficient can have major economic and safety consequences. The damage to persons and infrastructure as a result of unreliable clearance information is avoidable with a more robust clearance assessment procedure is employed. Moreover, a more robust tool that can assess clearances on a network level provides DOT with invaluable information that can have significant impacts on the efficiency of routing commercial vehicles on the network. This paper illustrates the efficiency of assessing vertical clearance on a highway corridor using LiDAR data. The method used works by detecting all overhead structures on a highway and estimating the clearance at each of those structures. After clearance information is obtained for all overhead structures (bridges and non-bridges), clusters are mapped on a GIS platform to help visualize clearance data along a highway corridor. The method was tested on three different highway corridors in Alberta, Canada and its value in network-level clearance assessment is illustrated. The method introduces a robust network-level clearance assessment technique which improves the reliability of clearance information available to DOT and the efficiency in which this information is obtained. This allows agencies to issue future permits with greater confidence, ensuring the safety of all drivers on the network, avoiding network delays, and reducing the chance of economic setbacks. Moreover, the paper also demonstrates the value of the efficient network level assessment when decisions regarding upgrading corridors into High Load Corridors are required.

Acknowledgments

The authors would like to thank Alberta Transportation for sponsoring and providing data for this study. Acknowledgments are also extended to Alberta Innovates and Alberta Advanced Education.

References

1. Alberta Transportation. *Alberta Operator Statistics: By Licence Class and Sex as of March 31, 2015*. Alberta Transportation, Office of Traffic Safety, Edmonton, AB, 2015.
2. Alberta Transportation. *Alberta Vehicle Statistics: Number of Active Registrations By Vehicle Registration Classes as of March 31 2015*. Alberta Transportation, Office of Traffic Safety, Edmonton, AB, 2015.
3. Transportation Stats and Facts. http://www.transportation.alberta.ca/Content/docType437/Production/StatsandFacts.pdf, Accessed May 2017.
4. Ontario Ministry of Transportation. *A Guide to Oversize/Overweight Vehicles and Loads in Ontario*, 2009. http://www.mto.gov.on.ca/english/trucks/oversize-overweight-manual.shtml.
5. Manitoba Infrastructure. *Over-height Permits*. Manitoba Government, Manitoba, 2017. http://www.gov.mb.ca/mit/mcd/mcpd/ohp.html. Accessed April 2017.
6. Adams, T. M., S. Malaikrisanachalee, C. Blazquez, and A. Vonderohe. GIS-Based Automated Oversize/Overweight Permit Processing. In *Computing in Civil and Building Engineering*, 2000, pp. 209–216.
7. Garcia-Diaz, A., P. P. Balestrassi, and K. Dhanabalan. Optimal Procedures to Route Over-Dimensioned and Overweight Trucks in a Highway-Bridge Network. *Proc., IIE Annual Conference*, Institute of Industrial and Systems Engineers (IISE), 2011, p. 1.
8. Coley, N., H. H. Titi, and V. Latifi. Mapping Overweight Vehicle Permits for Pavement Engineering Applications. *Journal of Transportation Engineering*, Vol. 142, No. 10, 2016, p. 04016044.
9. Dayan, S., A. Nichols, C. S. Chou, S. Yoo, T. Nguyen, and E. Pennington. Methodology to Map Routes from Truck Permit Database, Using a Linear Reference System and Network Analysis. *Transportation Research Record: Journal of the Transportation Research Board*, 2014. 2460: 47–57.
10. FHWA. *National Bridge Inspection Standards (NBIS)*. FHWA, 1988.
11. Alberta Transportation. *Budget 2016 Highlights*. Alberta, 2016. http://www.transportation.alberta.ca/5141.htm. Accessed November 2016.
12. Alberta Transportation. *Vertical Clearance Measurements (VCL2)*. Alberta Transportation, 2014.
13. Lauzon, R. G. *Automated Vertical Clearance Measurement During Photolog Operations*. Report no. 2220-F-2000-4, 2000.
14. Watson, C., S. E. Chen, H. Bian, and E. Hauser. Three-Dimensional Terrestrial LiDAR for Operational Bridge Clearance Measurements. *Journal of Performance of Constructed Facilities*, Vol. 26, No. 6, 2011, pp. 803–811.
15. Soilán, M., B. Riveiro, J. Martínez-Sánchez, and P. Arias. Traffic Sign Detection in MLS Acquired Point Clouds for Geometric and Image-Based Semantic Inventory. *ISPRS Journal of Photogrammetry and Remote Sensing*, Vol. 114, 2016, pp. 92–101.
16. Gargoum, S. A., K. El-Basyouny, J. Sabbagh, and K. Froese. Automated Highway Sign Extraction Using LiDAR Data. *Transportation Research Record: Journal of the Transportation Research Board*, 2017. 2643: 1–8.
17. Guan, H., J. Li, Y. Yu, C. Wang, M. Chapman, and B. Yang. Using Mobile Laser Scanning Data for Automated Extraction of Road Markings. *ISPRS Journal of Photogrammetry and Remote Sensing*, Vol. 87, 2014, pp. 93–107.
18. Ai, C., and Y. J. Tsai. An Automated Sign Retroreflectivity Condition Evaluation Methodology Using Mobile LiDAR and Computer Vision. *Transportation Research Part C: Emerging Technologies*, Vol. 63, 2016, pp. 96–113.
19. Truong-Hong, L., D. F. Laefer, T. Hinks, and H. Carr. Combining an Angle Criterion with Voxelization and the Flying Voxel Method in Reconstructing Building Models from LiDAR Data. *Computer-Aided Civil and Infrastructure Engineering*, Vol. 28, No. 2, 2013, pp. 112–129.

20. Holgado-Barco, A., B. Riveiro, D. González-Aguilera, and P. Arias. Automatic Inventory of Road Cross-Sections from Mobile Laser Scanning System. *Computer-Aided Civil and Infrastructure Engineering*, Vol. 32, No. 1, 2017, pp. 3–17.
21. Park, H., H. Lee, H. Adeli, and I. Lee. A New Approach for Health Monitoring of Structures: Terrestrial Laser Scanning. *Computer-Aided Civil and Infrastructure Engineering*, Vol. 22, No. 1, 2007, pp. 19–30.
22. Holgado-Barco, A., D. González-Aguilera, P. Arias-Sanchez, and J. Martinez-Sanchez. Semiautomatic Extraction of Road Horizontal Alignment from a Mobile LiDAR System. *Computer-Aided Civil and Infrastructure Engineering*, Vol. 30, No. 3, 2015, pp. 217–228.
23. Kumar, P., C. P. McElhinney, P. Lewis, and T. McCarthy. An Automated Algorithm for Extracting Road Edges from Terrestrial Mobile LiDAR Data. *ISPRS Journal of Photogrammetry and Remote Sensing*, Vol. 85, 2013, pp. 44–55.
24. Riveiro, B., P. B. Lourenço, D. V. Oliveira, H. González-Jorge, and P. Arias. Automatic Morphologic Analysis of Quasi-Periodic Masonry Walls from LiDAR. *Computer-Aided Civil and Infrastructure Engineering*, Vol. 31, No. 4, 2016, pp. 305–319.
25. Jiménez, F., and J. E. Naranjo. Improving the Obstacle Detection and Identification Algorithms of a Laserscanner-Based Collision Avoidance System. *Transportation Research Part C: Emerging Technologies*, Vol. 19, No. 4, 2011, pp. 658–672.
26. Castro, M., L. Iglesias, J. A. Sánchez, and L. Ambrosio. Sight Distance Analysis of Highways Using GIS Tools. *Transportation Research Part C: Emerging Technologies*, Vol. 19, No. 6, 2011, pp. 997–1005.
27. Riveiro, B., D. Jauregui, P. Arias, J. Armesto, and R. Jiang. An Innovative Method for Remote Measurement of Minimum Vertical Underclearance in Routine Bridge Inspection. *Automation in Construction*, Vol. 25, 2012, pp. 34–40.
28. Liu, W., S. E. Chen, and E. Hasuer. Bridge Clearance Evaluation Based on Terrestrial LiDAR Scan. *Journal of Performance of Constructed Facilities*, Vol. 26, No. 4, 2011, pp. 469–477.
29. Puente, I., B. Akinci, H. González-Jorge, L. Díaz-Vilariño, and P. Arias. A Semi-Automated Method for Extracting Vertical Clearance and Cross Sections in Tunnels Using Mobile LiDAR Data. *Tunnelling and Underground Space Technology*, Vol. 59, 2016, pp. 48–54.
30. Altman, N. S. An Introduction to Kernel and Nearest-Neighbor Nonparametric Regression. *The American Statistician*, Vol. 46, No. 3, 1992, pp. 175–185.
31. Ester, M., H. P. Kriegel, J. Sander, and X. Xu. A Density-Based Algorithm for Discovering Clusters in Large Spatial Databases with Noise. *Kdd*, Vol. 96, No. 34, 1996. pp. 226–231.
32. Alberta Transportation. *Highway Geometric Design Guide*. Edmonton, Alberta, 1999.
33. RIEGL. *RIEGL VMX-450 Data Sheet*. RIEGL Laser Measurment Systems, 2017. http://www.riegl.com/uploads/tx_pxpriegldownloads/DataSheet_VMX-450_2015-03-19.pdf. Accessed April 2017.
34. Steel, P., D. Mesher, and J. Adamson. Development of a Road Safety Audit Network Screening Tool. In *Twenty-Fourth Canadian Multidisciplinary Road Safety Conference*, Vancouver, 1-4 June 2014.

The Standing Committee on Statewide Transportation Data and Information Systems (ABJ20) peer-reviewed this paper (18-04576).

Article

TRR
JOURNAL OF THE TRANSPORTATION RESEARCH BOARD

Transportation Research Record
2018, Vol. 2672(42) 57–70

DOI: 10.1177/0361198118755713
journals.sagepub.com/home/trr
SAGE

Interstate Speed Profiles

Wenjing Pu[1]

This paper draws the first set of high-level, national speed profiles for the entire Dwight D. Eisenhower National System of Interstate and Defense Highways (Interstate system) in the United States based on the 2016 year-long National Performance Management Research Data Set (NPMRDS) and a conflated NPMRDS-HPMS (Highway Performance Monitoring System) geospatial network. This set of quantitative profiles include: (*a*) national average speeds of 2016, (*b*) national average speed time of day variations, (*c*) national average speed day of week variations, (*d*) national average speed seasonal variations, and (e) state average speed and travel time distributions in peak hours. This work demonstrates that the integration of the private sector's emerging big travel-time data and the public sector's HPMS has provided a powerful resource to monitor travel-time-related performance of the nation's highways. As the United States is transforming the Federal-aid Highway Program into a performance-based program with enhanced accountability and transparency, this integrated resource will help states and metropolitan planning organizations (MPOs) to monitor their performance and progress towards achieving targets, and enable the Federal Highway Administration (FHWA) not only to draw high-level national highway performance profiles but also to pinpoint the exact where, when, and how much the challenges are.

The Dwight D. Eisenhower National System of Interstate and Defense Highways (Interstate system) are the backbone of the United States' surface transportation systems. Safe and efficient operations of the Interstate system is vital to the nation's economy and quality of life. The Federal Highway Administration (FHWA) closely monitors the performance of the Interstate system. Since 2013, the FHWA has acquired the National Performance Management Research Data Set (NPMRDS), a third-party dataset providing segment-level travel times in 5-minute increments for the National Highway System (NHS), including the Interstate system. With unprecedented coverage and granularity, this data has enabled the FHWA to draw high-level speed profiles of the entire Interstate system for the first time.

In July 2013, the FHWA acquired the first version of the NPMRDS from HERE that provided data through January 31, 2017 (*1*). In April 2017, the FHWA purchased the second version of the NPMRDS from a team consisting of the University of Maryland, INRIX, Texas A&M Transportation Institute, and others. This new contract provides data covering from February 1, 2017 to up to five years later and the data became available on July 10, 2017. State departments of transportation (DOTs), metropolitan planning organizations (MPOs), and authorized contractors and researchers can have gratis access to the NPMRDS upon signing a data sharing agreement with the vendor. The NPMRDS procurements are pioneering and innovative in the FHWA's use of private sector passively collected, crowdsourced data.

The NPMRDS is a unique package of the private-sector, vehicle-probe data that satisfies FHWA's requirements. The "probe" could include mobile phones (including apps with location-based services), portable navigation devices, and vehicles. Data are collected anonymously by vendors, and no private information can be identified from the NPMRDS. The NPMRDS provides travel times in 5-minute increments for road segments geocoded by a private sector–driven industry standard called traffic message channel (TMC). A typical TMC path is from ramp to ramp on freeways and intersection to intersection on arterials for each direction of travel. The 2016 HERE NPMRDS has a total of 385,000 TMC paths with an average length of 1.74 miles per TMC path.

The FHWA Office of Highway Policy Information has conflated the 2013 Highway Performance Monitoring System (HPMS) geospatial network with the 2015 NPMRDS TMC network and the results, called Transportation Performance Network, are released online (*2*). Through this conflation, travel times and speeds can be examined by HPMS data attributes (*3*) such as urban code, functional class, and facility type.

Existing national or international congestion or mobility reports, such as the FHWA's Urban Congestion Report (*4*), Texas A&M Transportation Institute's Urban Mobility Scorecard (*5*), INRIX Global Traffic Scorecard (*6*), and TomTom Traffic Index (*7*) usually focus on urban areas,

[1]Transportation Specialist, Federal Highway Administration, United States Department of Transportation, Washington, D.C.

Corresponding Author:
Address correspondence to Wenjing Pu: wenjing.pu@dot.gov

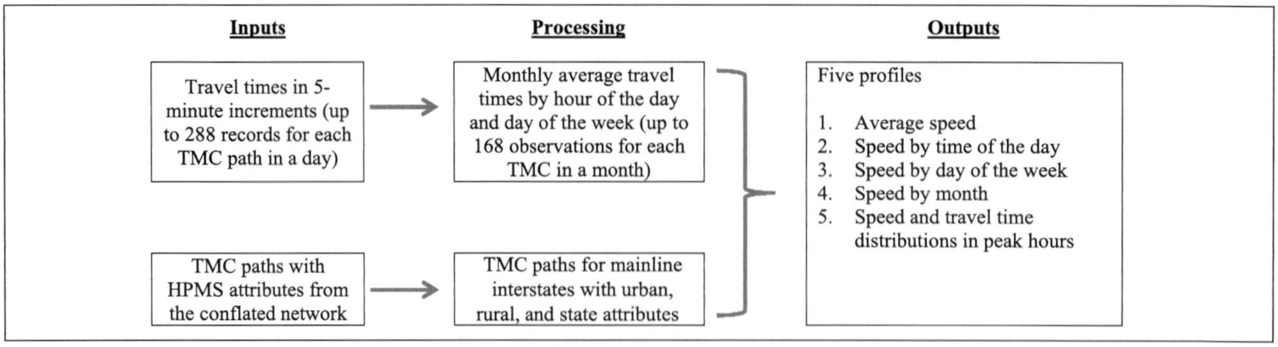

Figure 1. Scheme of data processing.

include both interstate and other highways, and provide annual or monthly summaries. This paper's interstate speed profiles include only mainline Interstate highways in the entire nation, explore the differences in urban and rural areas, and present both average conditions and temporal variations.

Methodology

This analysis uses the 2016 HERE NPMRDS to calculate the average speeds of the Interstate system at different geographic and temporal scales. The results are five national and state speed profiles that represent the average of observed conditions in 2016. Figure 1 illustrates the scheme of data processing.

First, the raw travel-time data are aggregated to monthly average travel times by hour of the day and day of the week for each TMC path. This step significantly reduces data sizes while keeping necessary temporal resolutions for further analyses. Meanwhile, the TMC paths for the mainline Interstate system are selected from the conflated HPMS-NPMRDS geospatial network.

Based on the monthly average data and interstate TMC paths, this analysis then generates five interstate speed profiles. The following explains the calculation steps for each profile.

1. Average Speed

a. All Interstate Highways, All the Time
   i. Calculate annual average travel time by TMC path
   ii. Divide the total length by the total travel time of all TMC paths
b. Rural/Urban Interstate Highways, All the Time
   i. Calculate annual average travel time by TMC path
   ii. Divide the total length by the total travel time of all rural/urban TMC paths, respectively
c. Rural/Urban Interstate Highways, AM/PM Peak Hour
   i. Calculate annual average travel time by TMC path and hour of the day
   ii. Divide the total length by the total travel time of all rural/urban TMC paths in the AM/PM peak hour, respectively

2. Speed by Hour of the Day

i. Calculate annual average travel time by TMC path, weekday/weekend, and hour of the day
ii. Divide the total length by the total travel time of all rural/urban TMC paths for weekday/weekend in each hour of the day, respectively

3. Speed by Day of the Week

i. Calculate annual average travel time by TMC path, day of the week, and hour of the day
ii. Divide the total length by the total travel time of all rural/urban TMC paths for each day of the week in each hour of the day, respectively

4. Speed by Month

i. Calculate annual average travel time by TMC path, month, and AM/PM peak hour
ii. Divide the total length by the total travel time of all rural/urban TMC paths for each month of the year in the AM/PM peak hour, respectively

5. Speed and Travel Time Distributions in Peak Hours

i. Calculate annual average travel time by TMC path and AM/PM peak hour for all urban interstate
ii. Label each TMC path with the following six speed ranges: (0,15], (15,35], (35,45], (45,50], (50, 55], and (55, +∞)
iii. Distribution by length: summarize the total length of TMC paths by the six speed ranges,

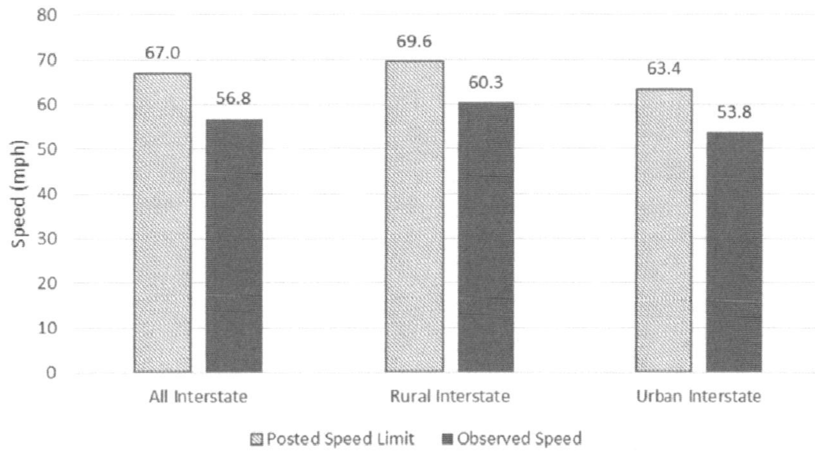

Figure 2. Average of posted speed limit and average of observed speed on interstate highways in 2016.

state, and AM/PM peak hour, and then divide each speed range's total TMC length by the length of all urban interstate TMC paths in each state to get the speed distribution

iv. Distribution by travel time: summarize the total travel time of TMC paths by the six speed ranges, state, and AM/PM peak hour, and then divide each speed range's total travel time by the total travel time on all urban interstate TMC paths in each state to get the travel-time distribution

This analysis did not impute missing values—the NPMRDS only reports travel times if there are observed probes; if none, no data is reported. The TMCs used in the analysis come from the conflated 2013 HPMS and 2015 NPMRDS network, a small portion of which could be outdated for the 2016 travel-time data. Because of the above two reasons, the calculated interstate mileages do not necessarily match the HPMS inventory mileages at national or state levels. This analysis did not identify or remove any possible outliers, as the focus is on annual averages and national trends.

Results

Profile 1: Average Speed

The average speed driven on the entire Interstate system in 2016 was 56.8 mph (Figure 2). The average speed of cars driven on rural interstate highways was 60.3 mph, which was 6.5 mph higher than that of speed driven on urban interstates, 53.8 mph. The average observed speeds were generally 10 mph slower than the average posted speed limits. The observed speeds included almost all conditions that could slow down the traffic, such as adverse weather conditions, work zones, incidents, special events, and traffic congestions. The average speeds also covered all directions of travel, including non-peak directions. More analysis will be presented below on where, when, and how much interstate travel speed slowed down.

Profile 2: Speed by Time of Day

Traffic conditions generally vary by time of day, especially in urban areas where demands could exceed supply during peak travel times and along major commute routes. Figure 3 is the annual average speed by hour of day on the Interstate system by urban/rural areas and weekday/weekend.

Urban interstate highways on weekdays had the largest time-of-day speed variations among all four categories compared. The NPMRDS data clearly identified that (*a*) the AM peak hour is approximately between 7:00 and 8:00 a.m., (*b*) the PM peak hour is 5:00–6:00 p.m., and (*c*) the PM peak hour was more congested than the AM peak hour.

Urban interstate highways on weekends had, on average, slightly higher speeds in the morning than in the afternoon, with no significant slowdowns throughout the day.

Rural interstate highways operated at almost consistent speeds (around 60 mph) at all times. There were no significant differences between weekdays and weekends, and the speeds were 3–14 mph higher than those on urban interstate highways.

Profile 3: Speed by Day of the Week

For urban interstate highways, average speed driven varied not only by hour of the day, but also by day of the week (Figure 4). For rural interstate highways, the national average speed driven was almost constant throughout the day and the week (Figure 5). Several notable phenomena in the data

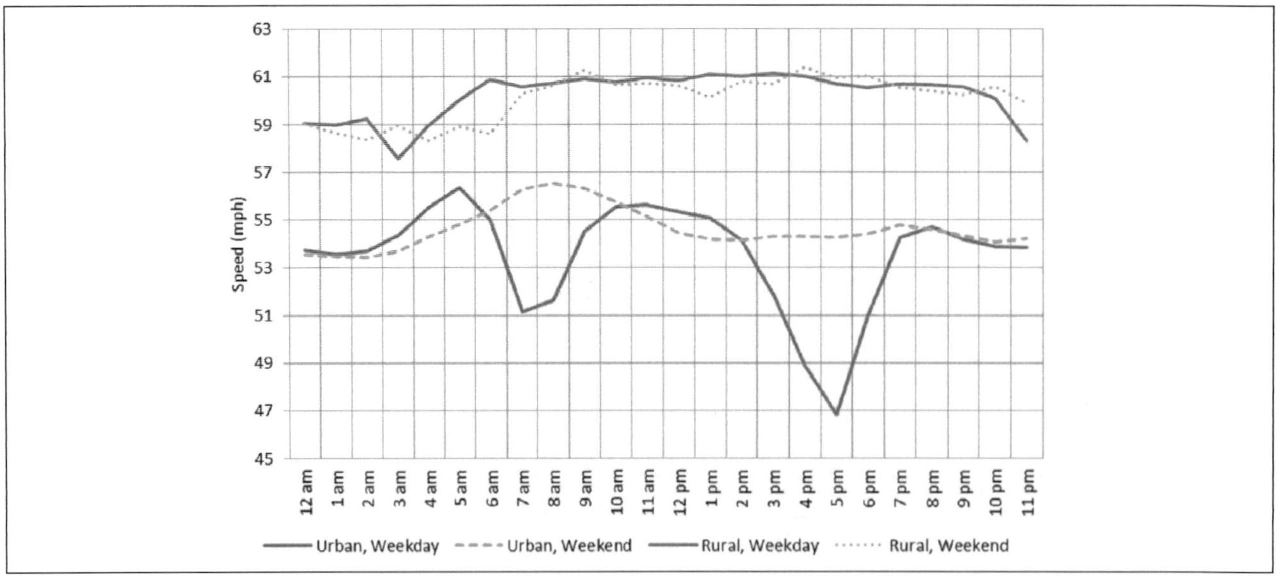

Figure 3. Annual average speed driven by hour of day on interstate highways in 2016.

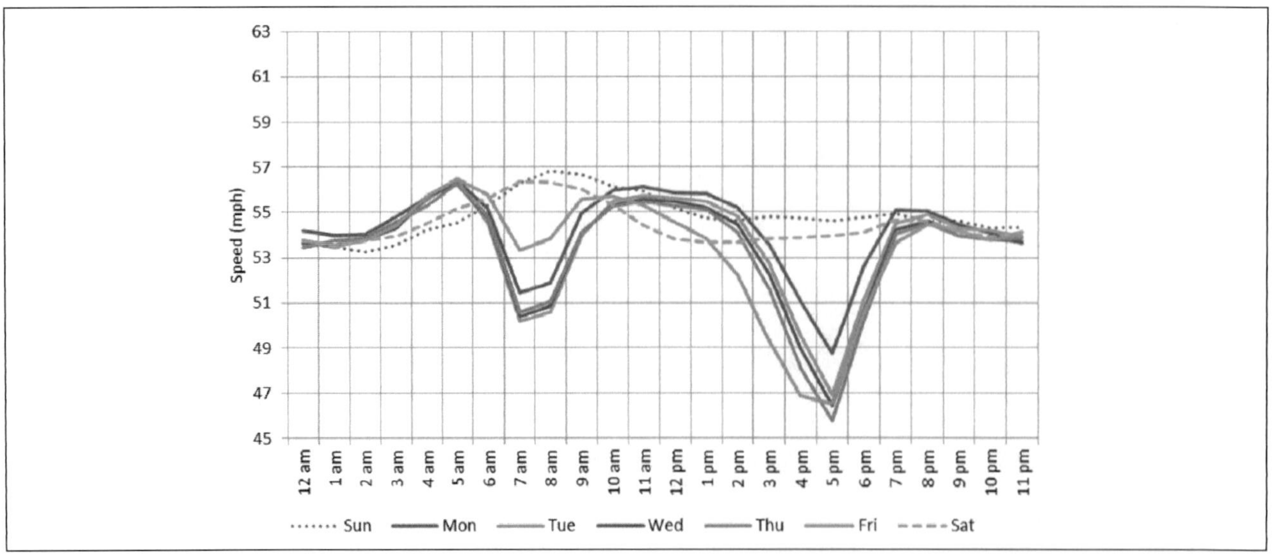

Figure 4. Average speed driven on urban interstate highways by hour of the day and day of the week in 2016.

from the urban interstate highway speeds driven (Figure 4) are listed below.

1) The average speed driven on the weekend was very different from that driven on the weekdays. In addition, the weekend average hourly speed profile lacked the typical diurnal peaks.
2) While all weekdays had similar speed trends, speed profiles by various weekdays were different:
   a. Monday was the least congested weekday, except for the Friday morning peak period.
   b. The three middle weekdays had very similar traffic conditions and the most congested morning peak and afternoon peak of the weekdays.
   c. On Friday, the afternoon peak period was about one hour earlier than the middle weekdays.
3) The most congested afternoon peak period occurs on Thursday, although Friday's afternoon peak period tends to be longer than on the other weekdays.
4) For all weekdays, the afternoon peak period is always more congested than the morning peak period.

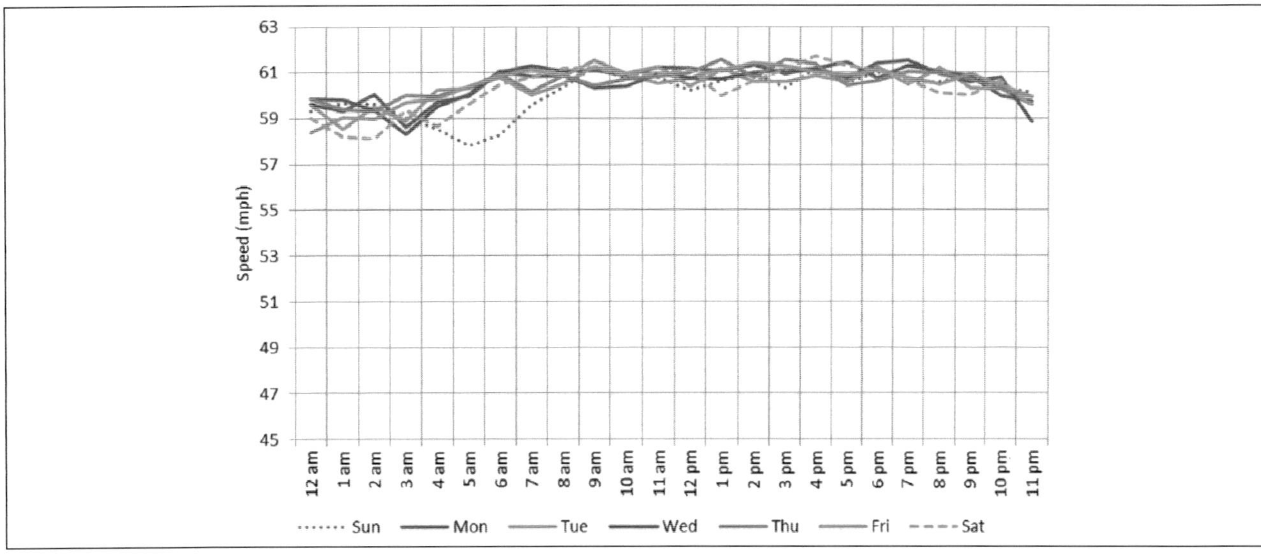

Figure 5. Average speed driven on rural interstate highways by hour of the day and day of the week in 2016.

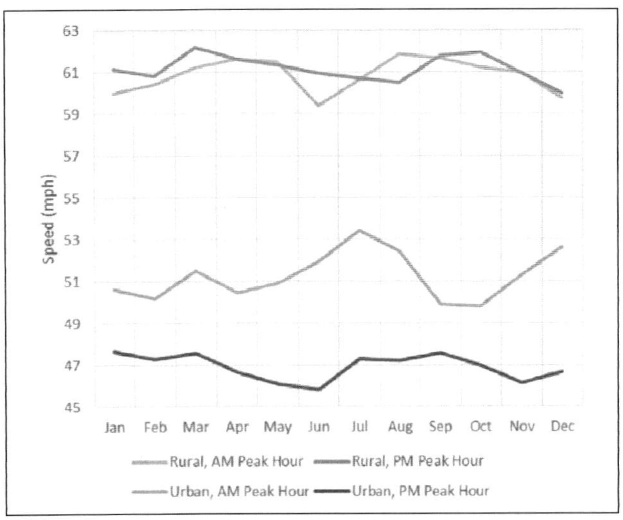

Figure 6. Peak hour speed by month on the Interstate in 2016.

Profile 4: Speed by Month

There were notable differences with regard to seasonal average speeds with more announced variations observed on urban Interstate (Figure 6).

On urban interstate highways, the AM peak hour average speeds driven in summer months from June to August were higher than speeds driven in other months. December, a holiday season, had high speeds similar to the summer months. April, September, and October were among the months with the lowest speeds. In the PM peak hour, May and June had the lowest speeds of the year. September was a special month as it had higher speeds than most time of the year in the PM peak hour, but the lowest speeds of the year in the AM peak hour.

On rural interstate highways, both AM and PM peak hour speeds were slightly lower during summer months and the holiday season compared to that in spring and fall.

Profile 5: Urban Interstate Speed Driven and Travel-Time Distributions in Peak Hours

To further explore the travel speeds on interstate highways in each of the states and the District of Columbia, two new parameters are developed. The first parameter is the percentage of interstate highways operating under six speeds bins. In this parameter, segments of interstate are grouped based on the six speed bins as listed below. Length of segments are summarized accordingly and percentages are computed from the summarized length data.

1) speed ≤ 15 mph
2) 15 < speed ≤ 35 mph
3) 35 < speed ≤ 45 mph
4) 45 < speed ≤ 50 mph
5) 50 < Speed ≤ 55 mph
6) speed > 55 mph

The percentages of roadways as measured by directional length under the six speed bins for each state are listed in Tables 1 and 2 and illustrated in Figures 7 and 8. The definition of regions in the tables are consistent with that of the FHWA Traffic Volume Trends report (*8*). The states in the charts are sorted by the total percentage of directional miles with speed ≤ 35 mph. Obviously, the higher the percentage of lower speed bins, the more significant the challenges are.

Table 1. Urban Interstate Speed Distribution in 2016 AM Peak Hour 7:00–8:00 a.m.

	Percentage of Distance in Speed (S) Range						
Region and State	S ≤ 15 mph	15< S ≤35 mph	35< S ≤45 mph	45< S ≤50 mph	50< S ≤55 mph	S >55 mph	Total
Northeast							
Connecticut	0.0%	3.3%	4.5%	3.2%	10.0%	78.9%	100.0%
Maine	0.3%	4.6%	2.8%	3.9%	4.5%	83.8%	100.0%
Massachusetts	1.3%	8.6%	6.2%	4.6%	6.2%	73.1%	100.0%
New Hampshire	0.4%	3.8%	0.5%	2.2%	12.2%	80.9%	100.0%
New Jersey	1.5%	6.7%	9.1%	8.4%	10.6%	63.6%	100.0%
New York	1.9%	6.5%	5.4%	3.6%	6.9%	75.8%	100.0%
Pennsylvania	1.0%	6.0%	4.7%	3.0%	6.1%	79.2%	100.0%
Rhode Island	0.0%	5.9%	8.9%	3.9%	11.6%	69.6%	100.0%
Vermont	0.0%	3.3%	1.7%	11.3%	2.7%	81.1%	100.0%
Subtotal	**1.3%**	**6.3%**	**5.6%**	**4.3%**	**7.3%**	**75.3%**	**100.0%**
South Atlantic							
Delaware	0.0%	0.5%	5.3%	6.6%	11.7%	75.9%	100.0%
Dist. of Columbia	9.7%	29.0%	12.4%	19.2%	17.1%	12.5%	100.0%
Florida	0.3%	4.1%	5.4%	3.0%	4.2%	83.0%	100.0%
Georgia	0.4%	6.4%	2.8%	2.6%	3.5%	84.3%	100.0%
Maryland	0.1%	5.5%	10.9%	11.1%	13.4%	59.0%	100.0%
North Carolina	1.6%	7.1%	5.6%	3.4%	5.4%	76.9%	100.0%
South Carolina	0.0%	4.0%	4.5%	3.5%	5.4%	82.6%	100.0%
Virginia	1.2%	9.3%	6.3%	5.7%	7.2%	70.4%	100.0%
West Virginia	0.0%	0.1%	2.0%	2.2%	4.0%	91.7%	100.0%
Subtotal	**0.6%**	**5.7%**	**5.1%**	**4.1%**	**5.6%**	**78.8%**	**100.0%**
North Central							
Illinois	1.2%	3.3%	3.9%	2.9%	5.9%	82.7%	100.0%
Indiana	0.0%	1.3%	1.8%	2.3%	4.9%	89.6%	100.0%
Iowa	0.2%	1.4%	2.0%	3.9%	7.2%	85.3%	100.0%
Kansas	0.1%	3.0%	3.0%	2.5%	4.8%	86.6%	100.0%
Michigan	0.1%	2.9%	4.0%	2.0%	5.0%	85.9%	100.0%
Minnesota	0.9%	9.9%	10.0%	7.8%	7.4%	64.1%	100.0%
Missouri	0.3%	3.7%	4.9%	4.5%	6.1%	80.5%	100.0%
Nebraska	0.0%	1.5%	2.4%	2.2%	7.8%	86.1%	100.0%
North Dakota	0.0%	0.4%	1.0%	1.2%	9.1%	88.4%	100.0%
Ohio	0.2%	4.2%	4.7%	4.2%	7.3%	79.4%	100.0%
South Dakota	0.9%	1.1%	0.2%	0.9%	2.0%	94.9%	100.0%
Wisconsin	0.2%	4.4%	3.3%	1.2%	3.1%	87.7%	100.0%
Subtotal	**0.4%**	**3.7%**	**4.1%**	**3.3%**	**5.9%**	**82.5%**	**100.0%**
South Gulf							
Alabama	0.0%	2.2%	3.1%	2.7%	2.8%	89.2%	100.0%
Arkansas	0.0%	1.4%	3.0%	2.0%	3.4%	90.2%	100.0%
Kentucky	0.4%	4.9%	6.8%	3.4%	9.2%	75.4%	100.0%
Louisiana	0.6%	3.2%	4.2%	3.2%	4.7%	84.1%	100.0%
Mississippi	0.0%	0.6%	1.0%	0.7%	3.4%	94.3%	100.0%
Oklahoma	0.2%	3.3%	5.6%	2.2%	4.4%	84.3%	100.0%
Tennessee	0.5%	3.8%	3.4%	3.5%	5.2%	83.7%	100.0%
Texas	5.6%	16.2%	7.9%	4.8%	6.9%	58.7%	100.0%
Subtotal	**2.2%**	**7.7%**	**5.2%**	**3.5%**	**5.4%**	**76.0%**	**100.0%**
West							
Alaska	3.0%	15.4%	22.0%	9.2%	24.4%	26.0%	100.0%
Arizona	3.8%	6.6%	2.6%	1.9%	3.7%	81.3%	100.0%

(continued)

Table 1. (continued)

Region and State	Percentage of Distance in Speed (S) Range: S ≤ 15 mph	15< S ≤35 mph	35< S ≤45 mph	45< S ≤50 mph	50< S ≤55 mph	S >55 mph	Total
California	8.5%	17.6%	9.1%	6.4%	9.9%	48.5%	100.0%
Colorado	4.0%	13.4%	9.0%	5.0%	9.0%	59.6%	100.0%
Hawaii	4.2%	23.9%	14.8%	11.5%	15.6%	30.0%	100.0%
Idaho	0.2%	0.6%	1.6%	2.8%	3.3%	91.5%	100.0%
Montana	1.4%	1.2%	1.8%	2.8%	2.1%	90.7%	100.0%
Nevada	1.3%	4.3%	4.8%	2.0%	5.2%	82.4%	100.0%
New Mexico	0.0%	0.7%	1.8%	3.2%	3.2%	91.0%	100.0%
Oregon	0.5%	6.7%	6.3%	5.6%	11.8%	69.2%	100.0%
Utah	0.0%	1.2%	1.5%	2.5%	4.9%	89.9%	100.0%
Washington	1.6%	8.7%	8.0%	3.8%	5.0%	72.9%	100.0%
Wyoming	0.1%	3.4%	3.4%	1.3%	5.1%	86.6%	100.0%
Subtotal	**4.4%**	**10.9%**	**6.9%**	**4.8%**	**7.7%**	**65.3%**	**100.0%**
Total	**1.7%**	**6.6%**	**5.3%**	**3.9%**	**6.3%**	**76.2%**	**100.0%**

Table 2. Urban Interstate Speed Distribution in 2016 PM Peak Hour 5:00–6:00 p.m.

Region and State	Percentage of Distance in Speed (S) Range: S ≤ 15 mph	15< S ≤35 mph	35< S ≤45 mph	45< S ≤50 mph	50< S ≤55 mph	S >55 mph	Total
Northeast							
Connecticut	1.2%	14.7%	6.6%	8.5%	11.6%	57.4%	100.0%
Maine	0.8%	5.8%	5.0%	4.7%	5.4%	78.2%	100.0%
Massachusetts	2.3%	13.6%	8.8%	4.1%	6.2%	65.0%	100.0%
New Hampshire	3.1%	0.8%	7.4%	7.9%	15.4%	65.4%	100.0%
New Jersey	1.2%	15.7%	10.7%	7.8%	10.1%	54.4%	100.0%
New York	3.1%	8.9%	6.3%	3.6%	5.5%	72.6%	100.0%
Pennsylvania	1.7%	7.6%	6.3%	5.5%	8.3%	70.5%	100.0%
Rhode Island	0.0%	14.7%	19.0%	6.7%	7.4%	52.2%	100.0%
Vermont	0.0%	4.6%	2.8%	5.7%	1.7%	85.2%	100.0%
Subtotal	**2.1%**	**10.1%**	**7.4%**	**5.2%**	**7.5%**	**67.7%**	**100.0%**
South Atlantic							
Delaware	0.0%	13.3%	7.3%	11.9%	12.4%	55.1%	100.0%
Dist. of Columbia	33.0%	54.5%	7.7%	0.0%	0.0%	4.8%	100.0%
Florida	0.8%	8.2%	5.3%	4.8%	5.2%	75.7%	100.0%
Georgia	1.6%	8.0%	5.2%	5.4%	4.9%	74.8%	100.0%
Maryland	1.6%	19.1%	14.6%	12.5%	10.1%	42.1%	100.0%
North Carolina	2.2%	11.3%	8.7%	4.3%	4.4%	69.1%	100.0%
South Carolina	0.5%	7.3%	6.9%	3.8%	7.2%	74.2%	100.0%
Virginia	1.9%	11.9%	9.1%	8.7%	9.2%	59.2%	100.0%
West Virginia	0.0%	0.8%	2.2%	4.3%	7.2%	85.5%	100.0%
Subtotal	**1.4%**	**9.6%**	**7.1%**	**6.1%**	**6.5%**	**69.4%**	**100.0%**
North Central							
Illinois	1.7%	6.1%	4.2%	3.2%	6.7%	78.1%	100.0%
Indiana	0.1%	2.0%	4.0%	4.6%	9.9%	79.5%	100.0%
Iowa	0.2%	1.7%	3.3%	4.5%	9.1%	81.2%	100.0%
Kansas	0.2%	4.4%	3.8%	3.3%	5.8%	82.6%	100.0%
Michigan	0.5%	5.8%	5.7%	3.6%	4.8%	79.7%	100.0%
Minnesota	2.1%	14.9%	9.8%	5.4%	7.5%	60.3%	100.0%
Missouri	0.5%	5.7%	7.6%	5.1%	7.0%	74.2%	100.0%

(continued)

Table 2. (continued)

Region and State	Percentage of Distance in Speed (S) Range						Total
	S ≤ 15 mph	15< S ≤35 mph	35< S ≤45 mph	45< S ≤50 mph	50< S ≤55 mph	S >55 mph	
Nebraska	0.0%	1.3%	6.4%	3.1%	6.6%	82.7%	100.0%
North Dakota	0.0%	0.7%	1.4%	2.6%	5.3%	90.0%	100.0%
Ohio	0.5%	6.5%	7.2%	4.8%	7.1%	73.9%	100.0%
South Dakota	0.9%	1.2%	0.1%	1.1%	3.9%	92.8%	100.0%
Wisconsin	0.3%	5.3%	3.1%	2.1%	3.8%	85.4%	100.0%
Subtotal	**0.8%**	**5.7%**	**5.4%**	**3.9%**	**6.7%**	**77.4%**	**100.0%**
South Gulf							
Alabama	0.2%	3.2%	3.8%	3.5%	5.4%	83.9%	100.0%
Arkansas	0.0%	2.9%	2.6%	3.8%	6.8%	83.9%	100.0%
Kentucky	1.1%	11.5%	11.5%	6.5%	8.1%	61.3%	100.0%
Louisiana	1.2%	5.8%	4.2%	5.3%	7.8%	75.8%	100.0%
Mississippi	0.0%	1.3%	1.5%	2.0%	3.8%	91.5%	100.0%
Oklahoma	0.4%	2.9%	5.2%	3.0%	5.8%	82.6%	100.0%
Tennessee	0.8%	7.4%	5.0%	4.4%	7.7%	74.7%	100.0%
Texas	7.1%	20.0%	10.2%	6.1%	6.1%	50.5%	100.0%
Subtotal	**3.0%**	**10.6%**	**6.6%**	**4.8%**	**6.4%**	**68.6%**	**100.0%**
West							
Alaska	5.7%	28.6%	16.1%	13.7%	16.4%	19.6%	100.0%
Arizona	4.6%	6.8%	4.1%	2.9%	5.3%	76.4%	100.0%
California	13.1%	20.7%	10.4%	8.1%	10.5%	37.3%	100.0%
Colorado	7.1%	11.2%	12.7%	7.4%	8.6%	52.9%	100.0%
Hawaii	7.0%	23.3%	17.9%	4.9%	6.4%	40.5%	100.0%
Idaho	0.2%	1.1%	4.4%	1.8%	2.0%	90.5%	100.0%
Montana	0.7%	1.9%	1.7%	3.4%	2.0%	90.3%	100.0%
Nevada	1.9%	8.2%	5.4%	6.0%	8.4%	70.1%	100.0%
New Mexico	0.0%	1.9%	2.2%	3.1%	5.1%	87.7%	100.0%
Oregon	7.3%	12.4%	3.2%	2.7%	11.3%	63.2%	100.0%
Utah	0.1%	6.1%	6.1%	3.1%	6.8%	77.7%	100.0%
Washington	2.8%	14.3%	8.4%	6.1%	8.5%	59.9%	100.0%
Wyoming	2.2%	1.4%	3.4%	0.5%	3.8%	88.7%	100.0%
Subtotal	**7.1%**	**13.5%**	**8.3%**	**6.0%**	**8.3%**	**56.8%**	**100.0%**
Total	**2.7%**	**9.5%**	**6.8%**	**5.1%**	**7.0%**	**68.9%**	**100.0%**

Nationally, cars on 8.3% of urban interstate highways were operating under the speed of 35 mph in the AM peak hour, and cars on 12.2% of urban interstate highways were operating under the speed of 35 mph in the PM peak hour in 2016.

The second parameter is the percentage of travel time experienced by travelers under each of the above prescribed six speed bins. In this parameter (and like the first parameter), segments of the Interstate system are grouped based on the six speed bins. Travel time of segments (unlike the first parameter in which length of segments was used) are summarized accordingly, and percentages are computed from the summarized travel time data.

The percentages of roadways as measured by travel time under the six speed bins for each state are listed in Tables 3 and 4 and illustrated in Figures 9 and 10. The states in the charts are sorted by the total percentage of travel time experienced with speed ≤ 35 mph.

Nationally, 22.6% of total travel time of cars on urban interstate highways was spent under speed 35 mph in the AM peak hour, and 30.9% under speed 35 mph in the PM peak hour in 2016.

In summary, although only a relatively small portion of urban interstate highways—8.3% of total directional miles, operated at speed ≤ 35 mph during the AM peak hour—the time spent on these low-speed roads accounted for 22.6% of

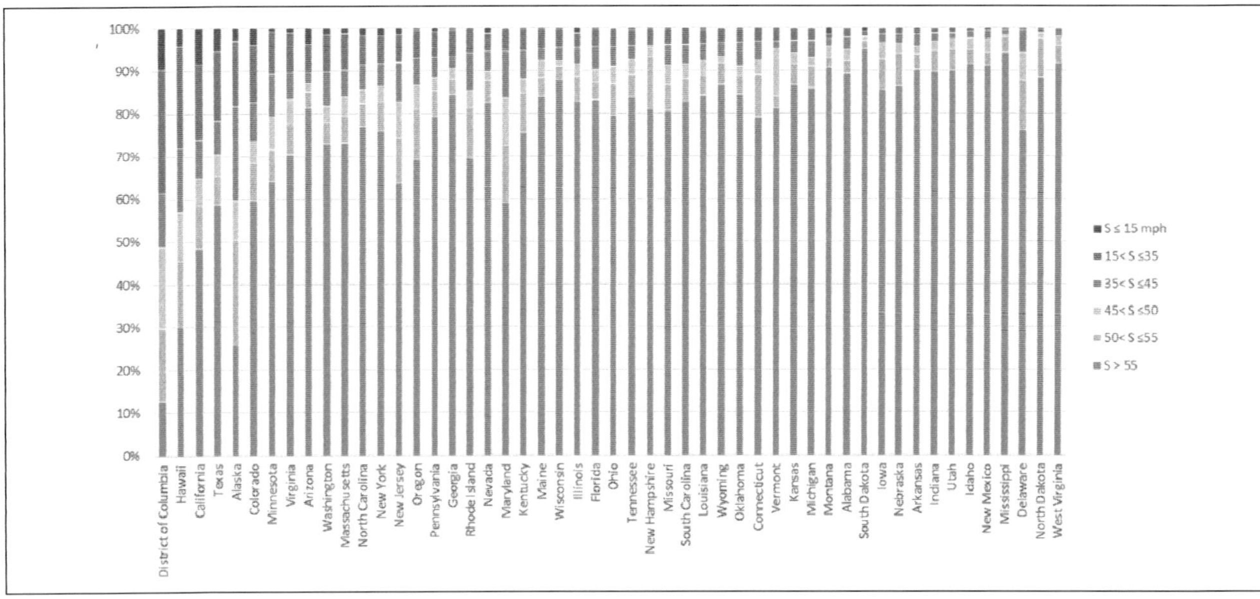

Figure 7. Percent of directional miles on urban interstate highways in each speed range during AM peak hour 7:00–8:00 a.m. in 2016*.
* States are sorted by the total percentage of directional miles with speed ≤ 35 mph.

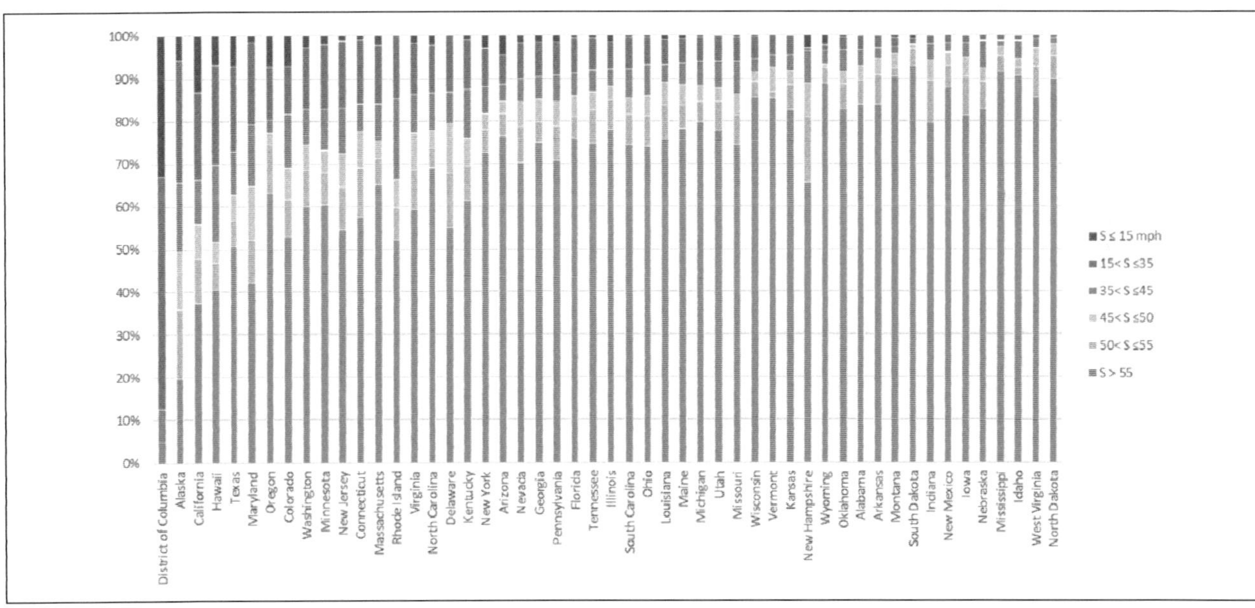

Figure 8. Percent of directional miles on urban interstate highways in each speed range during PM peak hour 5:00–6:00 p.m. in 2016*.
* States are sorted by the total percentage of directional miles with speed ≤ 35 mph.

total travel time of the entire urban Interstate system. During the PM peak hour, 12.2% of total directional miles with speed ≤ 35 mph accounted for 30.9% of total travel time on urban interstate highways.

Conclusions

This paper has drawn the first set of high-level national speed profiles for the Interstate system. Highlights of the findings include:

Table 3. Urban Interstate Travel-Time Distribution in 2016 AM Peak Hour 7:00–8:00 a.m.

	Percentage of Travel Time in Speed (S) Range						
Region and State	S ≤ 15 mph	15< S ≤35 mph	35< S ≤45 mph	45< S ≤50 mph	50< S ≤55 mph	S >55 mph	Total
Northeast							
Connecticut	0.0%	6.8%	6.2%	3.8%	10.8%	72.4%	100.0%
Maine	1.6%	9.6%	4.1%	4.8%	5.0%	75.0%	100.0%
Massachusetts	7.3%	16.8%	7.7%	4.9%	5.9%	57.6%	100.0%
New Hampshire	1.6%	10.9%	0.6%	2.6%	12.8%	71.5%	100.0%
New Jersey	7.8%	12.2%	11.3%	8.9%	10.1%	49.7%	100.0%
New York	9.4%	13.5%	6.7%	3.8%	6.6%	60.1%	100.0%
Pennsylvania	7.3%	14.1%	7.1%	3.8%	6.9%	60.7%	100.0%
Rhode Island	0.0%	10.7%	12.2%	4.4%	11.9%	60.8%	100.0%
Vermont	0.0%	5.4%	2.5%	13.0%	2.8%	76.4%	100.0%
Subtotal	**7.3%**	**13.3%**	**7.4%**	**4.8%**	**7.4%**	**59.8%**	**100.0%**
South Atlantic							
Delaware	0.0%	1.4%	7.5%	8.2%	13.7%	69.3%	100.0%
Dist. of Columbia	50.1%	25.7%	6.5%	8.2%	6.7%	2.8%	100.0%
Florida	2.0%	9.2%	7.9%	3.7%	4.7%	72.6%	100.0%
Georgia	1.9%	14.5%	3.9%	3.1%	3.8%	72.8%	100.0%
Maryland	0.5%	10.5%	14.3%	12.4%	13.4%	48.9%	100.0%
North Carolina	7.3%	15.0%	7.2%	3.7%	5.3%	61.4%	100.0%
South Carolina	0.1%	8.7%	6.5%	4.2%	5.8%	74.7%	100.0%
Virginia	6.6%	18.0%	7.7%	5.9%	6.7%	55.1%	100.0%
West Virginia	0.0%	0.2%	3.0%	2.8%	4.8%	89.2%	100.0%
Subtotal	**3.4%**	**12.3%**	**7.0%**	**4.8%**	**5.9%**	**66.7%**	**100.0%**
North Central							
Illinois	7.6%	7.7%	5.3%	3.4%	6.1%	69.9%	100.0%
Indiana	0.2%	2.8%	2.6%	2.8%	5.6%	85.9%	100.0%
Iowa	0.7%	2.7%	3.1%	5.0%	8.1%	80.5%	100.0%
Kansas	0.5%	7.1%	4.3%	3.0%	5.4%	79.6%	100.0%
Michigan	1.2%	6.2%	5.9%	2.5%	5.6%	78.6%	100.0%
Minnesota	4.0%	18.8%	12.3%	8.2%	7.0%	49.8%	100.0%
Missouri	1.7%	7.4%	6.9%	5.5%	6.7%	71.9%	100.0%
Nebraska	0.0%	3.3%	3.5%	2.8%	8.8%	81.7%	100.0%
North Dakota	0.0%	0.9%	1.5%	1.5%	10.6%	85.6%	100.0%
Ohio	0.9%	8.4%	6.6%	5.0%	7.9%	71.2%	100.0%
South Dakota	5.4%	2.6%	0.3%	1.1%	2.3%	88.4%	100.0%
Wisconsin	1.5%	10.4%	4.9%	1.5%	3.5%	78.3%	100.0%
Subtotal	**2.7%**	**7.8%**	**5.8%**	**4.0%**	**6.4%**	**73.3%**	**100.0%**
South Gulf							
Alabama	0.1%	4.6%	4.6%	3.5%	3.2%	83.9%	100.0%
Arkansas	0.0%	3.4%	4.5%	2.6%	4.0%	85.5%	100.0%
Kentucky	3.7%	10.6%	8.9%	3.8%	9.2%	63.8%	100.0%
Louisiana	3.5%	7.2%	6.0%	3.8%	5.1%	74.4%	100.0%
Mississippi	0.1%	1.3%	1.6%	0.9%	4.1%	91.9%	100.0%
Oklahoma	1.3%	6.6%	8.0%	2.7%	5.0%	76.5%	100.0%
Tennessee	2.3%	8.3%	4.8%	4.2%	5.6%	74.8%	100.0%
Texas	20.8%	27.2%	7.6%	4.0%	5.0%	35.4%	100.0%
Subtotal	**10.7%**	**16.0%**	**6.4%**	**3.6%**	**5.0%**	**58.3%**	**100.0%**

(continued)

Table 3. (continued)

Region and State	S ≤ 15 mph	15< S ≤35 mph	35< S ≤45 mph	45< S ≤50 mph	50< S ≤55 mph	S >55 mph	Total
	Percentage of Travel Time in Speed (S) Range						
West							
Alaska	11.2%	24.3%	21.8%	7.5%	18.2%	17.0%	100.0%
Arizona	15.5%	13.9%	3.3%	2.0%	3.5%	61.7%	100.0%
California	28.7%	25.9%	7.8%	4.6%	6.4%	26.7%	100.0%
Colorado	15.5%	24.8%	9.4%	4.5%	7.1%	38.6%	100.0%
Hawaii	13.7%	38.5%	13.6%	8.9%	11.1%	14.3%	100.0%
Idaho	1.4%	1.4%	2.4%	3.7%	3.8%	87.3%	100.0%
Montana	8.1%	3.5%	2.3%	3.3%	2.2%	80.5%	100.0%
Nevada	7.3%	8.9%	6.4%	2.3%	5.2%	69.9%	100.0%
New Mexico	0.0%	1.6%	2.8%	4.2%	3.8%	87.5%	100.0%
Oregon	5.8%	13.2%	7.8%	5.7%	11.0%	56.6%	100.0%
Utah	0.0%	3.1%	2.2%	3.3%	5.7%	85.7%	100.0%
Washington	8.3%	17.0%	9.5%	3.8%	4.5%	56.9%	100.0%
Wyoming	0.6%	10.5%	4.4%	1.5%	5.4%	77.5%	100.0%
Subtotal	**18.4%**	**19.9%**	**7.4%**	**4.2%**	**6.2%**	**43.9%**	**100.0%**
Total	**8.8%**	**13.8%**	**6.7%**	**4.2%**	**6.1%**	**60.3%**	**100.0%**

Table 4. Urban Interstate Travel-Time Distribution in 2016 PM Peak Hour 5:00–6:00 p.m.

Region and State	S ≤ 15 mph	15< S ≤35 mph	35< S ≤45 mph	45< S ≤50 mph	50< S ≤55 mph	S >55 mph	Total
	Percentage of Travel Time in Speed (S) Range						
Northeast							
Connecticut	4.8%	27.2%	7.6%	8.2%	10.0%	42.3%	100.0%
Maine	4.2%	11.9%	6.8%	5.4%	5.6%	66.1%	100.0%
Massachusetts	11.7%	23.4%	9.8%	3.8%	5.3%	45.9%	100.0%
New Hampshire	12.1%	1.3%	9.2%	8.3%	15.4%	53.8%	100.0%
New Jersey	5.1%	27.7%	12.3%	7.5%	8.7%	38.6%	100.0%
New York	15.2%	16.6%	7.3%	3.5%	4.8%	52.6%	100.0%
Pennsylvania	10.5%	17.0%	8.6%	6.5%	8.8%	48.6%	100.0%
Rhode Island	0.0%	26.8%	21.9%	6.4%	6.5%	38.5%	100.0%
Vermont	0.0%	8.0%	4.0%	6.8%	2.0%	79.3%	100.0%
Subtotal	**10.7%**	**19.4%**	**9.0%**	**5.4%**	**7.0%**	**48.6%**	**100.0%**
South Atlantic							
Delaware	0.0%	24.2%	9.4%	12.7%	12.0%	41.7%	100.0%
Dist. of Columbia	60.5%	36.7%	2.6%	0.0%	0.0%	0.2%	100.0%
Florida	3.7%	17.8%	7.1%	5.4%	5.3%	60.7%	100.0%
Georgia	8.1%	16.9%	6.6%	5.7%	4.7%	58.0%	100.0%
Maryland	5.4%	31.3%	15.6%	11.3%	8.3%	28.1%	100.0%
North Carolina	9.7%	22.6%	10.2%	4.2%	3.9%	49.5%	100.0%
South Carolina	2.0%	15.3%	9.0%	4.3%	7.2%	62.3%	100.0%
Virginia	8.0%	21.9%	10.3%	8.4%	8.0%	43.4%	100.0%
West Virginia	0.0%	1.5%	3.2%	5.5%	8.2%	81.6%	100.0%
Subtotal	**6.4%**	**19.3%**	**8.7%**	**6.3%**	**6.1%**	**53.1%**	**100.0%**

(continued)

Table 4. (continued)

Region and State	Percentage of Travel Time in Speed (S) Range						Total
	S ≤ 15 mph	15< S ≤35 mph	35< S ≤45 mph	45< S ≤50 mph	50< S ≤55 mph	S >55 mph	
North Central							
Illinois	9.5%	12.6%	5.3%	3.4%	6.6%	62.4%	100.0%
Indiana	0.3%	4.9%	5.8%	5.5%	10.9%	72.7%	100.0%
Iowa	0.8%	3.3%	4.8%	5.5%	10.2%	75.4%	100.0%
Kansas	1.0%	10.1%	5.4%	3.9%	6.4%	73.3%	100.0%
Michigan	2.9%	12.3%	7.8%	4.1%	5.0%	67.9%	100.0%
Minnesota	9.3%	26.7%	10.6%	5.0%	6.3%	42.0%	100.0%
Missouri	2.5%	12.2%	10.2%	5.7%	7.1%	62.3%	100.0%
Nebraska	0.0%	2.5%	9.5%	3.8%	7.4%	76.8%	100.0%
North Dakota	0.0%	1.5%	2.0%	3.3%	6.1%	87.0%	100.0%
Ohio	2.2%	13.5%	9.5%	5.3%	7.2%	62.2%	100.0%
South Dakota	5.7%	2.8%	0.1%	1.5%	4.4%	85.6%	100.0%
Wisconsin	1.6%	12.2%	4.5%	2.5%	4.1%	75.2%	100.0%
Subtotal	**4.1%**	**12.2%**	**7.3%**	**4.4%**	**6.8%**	**65.1%**	**100.0%**
South Gulf							
Alabama	0.8%	6.6%	5.6%	4.4%	6.1%	76.5%	100.0%
Arkansas	0.1%	5.8%	3.8%	4.8%	7.8%	77.6%	100.0%
Kentucky	5.7%	21.2%	13.6%	6.5%	7.3%	45.7%	100.0%
Louisiana	5.2%	13.3%	5.5%	5.9%	7.8%	62.3%	100.0%
Mississippi	0.1%	2.7%	2.2%	2.6%	4.4%	88.0%	100.0%
Oklahoma	2.3%	6.9%	7.3%	3.6%	6.3%	73.6%	100.0%
Tennessee	3.6%	16.3%	6.4%	4.8%	7.7%	61.1%	100.0%
Texas	23.9%	30.6%	9.1%	4.5%	4.1%	27.8%	100.0%
Subtotal	**12.9%**	**20.4%**	**7.6%**	**4.7%**	**5.6%**	**48.9%**	**100.0%**
West							
Alaska	17.0%	37.5%	13.3%	10.0%	11.0%	11.1%	100.0%
Arizona	19.9%	13.6%	4.8%	2.8%	4.6%	54.4%	100.0%
California	39.2%	25.4%	7.4%	4.9%	5.7%	17.3%	100.0%
Colorado	23.9%	18.6%	12.4%	6.2%	6.4%	32.5%	100.0%
Hawaii	20.9%	35.4%	15.8%	3.7%	4.2%	20.0%	100.0%
Idaho	1.5%	2.7%	6.3%	2.4%	2.4%	84.8%	100.0%
Montana	3.2%	5.8%	2.3%	4.0%	2.2%	82.5%	100.0%
Nevada	9.3%	16.9%	6.6%	6.2%	7.8%	53.2%	100.0%
New Mexico	0.0%	4.3%	3.2%	3.9%	5.9%	82.6%	100.0%
Oregon	27.4%	20.0%	3.0%	2.1%	8.0%	39.5%	100.0%
Utah	0.4%	13.1%	8.3%	3.7%	7.2%	67.3%	100.0%
Washington	11.9%	25.2%	8.8%	5.4%	6.8%	41.8%	100.0%
Wyoming	9.4%	3.5%	4.4%	0.6%	4.1%	78.0%	100.0%
Subtotal	**26.6%**	**21.3%**	**7.7%**	**4.7%**	**5.9%**	**33.7%**	**100.0%**
Total	**12.6%**	**18.3%**	**7.9%**	**5.0%**	**6.3%**	**49.9%**	**100.0%**

1) The average speed of the entire Interstate system in 2016 was 57 mph. The speed on rural interstate highways (60 mph) was faster than on urban interstate highways (54 mph). The observed average speeds were about 10 mph below the posted speed limits.
2) The time of day variation of speed was pronounced on urban interstate highways during weekdays. It clearly identified that the AM peak hour is 7:00–8:00 a.m. (51 mph), the PM peak hour is 5:00–6:00 p.m. (47 mph), and the PM peak hour was more congested than the AM peak hour. The average speed driven on rural interstate highways kept almost unchanged at 60 mph throughout the year, and the night speed was slighter slower than the daytime speed.

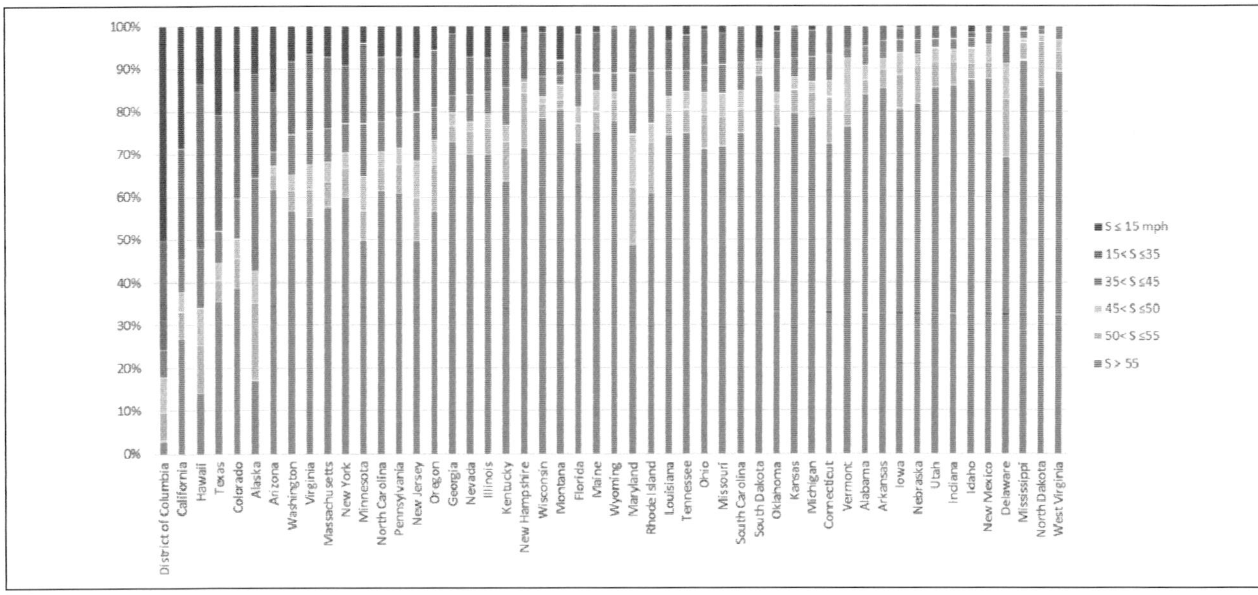

Figure 9. Percent of urban Interstate travel time in each speed range during AM peak hour 7:00–8:00 a.m. in 2016*.
* States are sorted by the total percentage of travel time experienced with speed ≤ 35 mph.

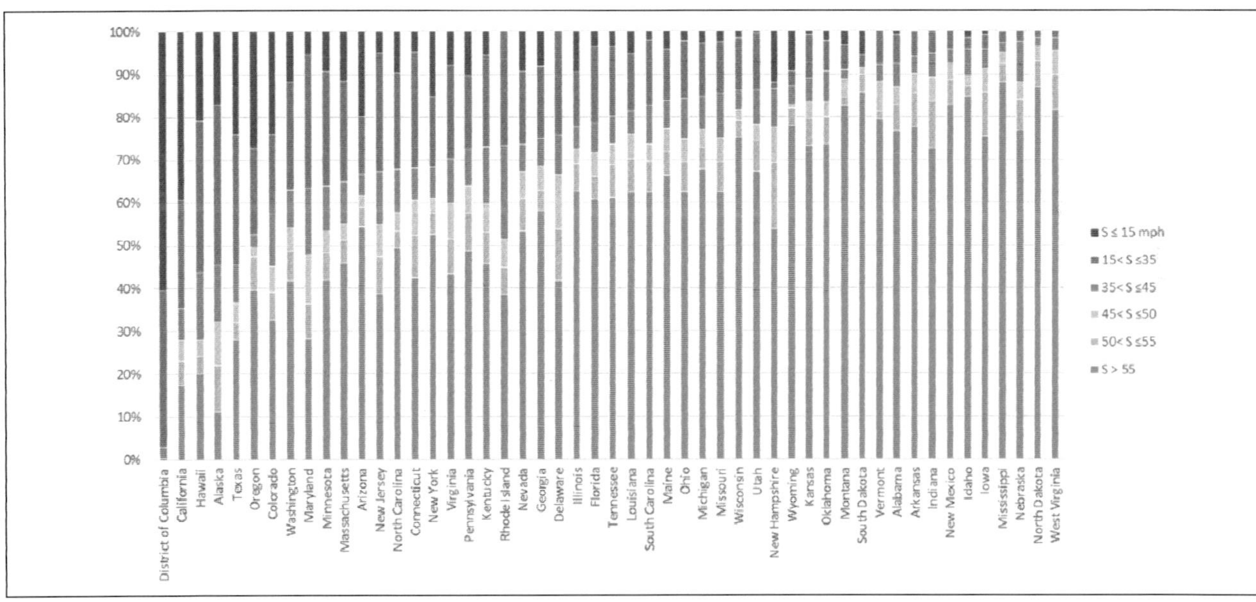

Figure 10. Percent of urban interstate travel time in each speed range during PM peak hour 5:00–6:00 p.m. in 2016*.
* States are sorted by the total percentage of travel time experienced with speed ≤ 35 mph.

3) Urban interstate highways also showed noteworthy day-to-day variations of speed. Monday was the least congested weekday, except in the morning peak period in which Friday had the least congestion. The three middle weekdays had very similar traffic conditions and the most congested morning and afternoon peaks among all five weekdays. On Friday, the afternoon peak period was about one hour earlier than the middle weekdays but ended at almost the same time as other weekdays.
4) Seasonal variations in travel speeds existed on urban interstates highways, and the AM peak and PM peak behaved differently. During the AM peak hour, interstate highways were less congested in

the summer months from June to August than most time of the year. During the PM peak hour, the highways were most congested in the months of May and June. Cars on the interstate highways had higher speeds than most time of the year in the PM peak hour, but the lowest speeds of the year in the AM peak hour.

5) Although only a relatively small portion of urban interstate highways—8.3% of total directional miles, operated at speed ≤ 35 mph during the AM peak hour—the time spent on these low-speed roads accounted for 22.6% of total travel time of the entire urban Interstate system. During the PM peak hour, 12.2% of total directional miles with speed ≤ 35 mph accounted for 30.9% of total travel time on urban interstate highways.

The integration of the private sector's emerging big travel-time data and the FHWA's Highway Performance Monitoring System has provided a powerful resource to monitor the travel time-related performance of the nation's highways. As the United States is transforming the Federal-Aid Highway Program into a performance-based program with enhanced accountability and transparency, this integrated resource will help states and metropolitan areas to monitor their performance and progress towards achieving targets, and enable the FHWA not only to draw high-level national highway performance profiles but also to pinpoint the exact where, when, and how much the challenges are.

Future analyses could be carried out to maximize the full potential of this integrated resource. Some examples could be an investigation of previous years (2013–2015) and future years' data to find any trends, conducting corridor level, metropolitan level, or multi-state regional level analysis, and looking at truck and passenger car data respectively. Other performance measures such as delay and vehicle hours of travel could also be explored.

Acknowledgments

The author would like to thank Dr. Tianjia Tang for his guidance on the technical analysis and his generous comments. The author also appreciates the constructive comments received from FHWA Office of Policy and Governmental Affairs, Office of Transportation Performance Management, Office of Operations, and anonymous reviewers.

References

1. Performance Measurement, *National Performance Management Research Data Set (NPMRDS) Technical Frequently Asked Questions*. Office of Freight Management and Operations, Federal Highway Administration, Washington, DC. https://ops.fhwa.dot.gov/freight/freight_analysis/perform_meas/vpds/npmrdsfaqs.htm. Accessed November 17, 2017.
2. *Performance Network*, Office of Highway Policy Information, Federal Highway Administration, Washington, DC. https://www.fhwa.dot.gov/policyinformation/tables/performancenetwork/. Accessed November 17, 2017.
3. *Highway Performance Monitoring System (HPMS) Field Manual (December 2016)*. Office of Management and Budget (OMB) Control No. 2125–0028. Office of Highway Policy Information, Federal Highway Administration, Washington, DC. https://www.fhwa.dot.gov/policyinformation/hpms/field-manual/. Accessed November 17, 2017.
4. Operations Performance Measurement Program. Urban Congestion Reports. Office of Operations, Federal Highway Administration, Washington, DC. https://ops.fhwa.dot.gov/perf_measurement/ucr/. Accessed November 17, 2017.
5. Urban Mobility Information. *Urban Mobility Scorecard*. Texas A&M Transportation Institute, Texas A&M University System, College Station, TX. https://mobility.tamu.edu/ums/. Accessed November 17, 2017.
6. *INRIX Global Traffic Scorecard*. INRIX, Inc., Kirkland, WA. http://inrix.com/scorecard/. Accessed November 17, 2017.
7. *TomTom Traffic Index*. TomTom International BV, Amsterdam, Netherlands. https://www.tomtom.com/en_gb/trafficindex/. Accessed November 17, 2017.
8. Travel Monitoring. *Traffic Volume Trends*. Office of Highway Policy Information, Federal Highway Administration, Washington, DC. https://www.fhwa.dot.gov/policyinformation/travel_monitoring/tvt.cfm. Accessed November 17, 2017.

The Standing Committee on Statewide Transportation Data and Information Systems (ABJ20) peer-reviewed this paper (18-00678).

All the contents of the paper constitute the opinion of the author, and the author alone; they do not represent the views and opinions of the author's employers, supervisors, nor do they represent the view of organizations, businesses, or institutions the author is, or has been a part of.

TRR
JOURNAL OF THE TRANSPORTATION RESEARCH BOARD

Article

Transportation Research Record
2018, Vol. 2672(42) 71–80

DOI: 10.1177/0361198118787362
journals.sagepub.com/home/trr
SAGE

A Latent Class Pattern Recognition and Data Quality Assessment of Non-Commute Long-Distance Travel in California

Adam W. Davis[1], Elizabeth C. McBride[1], and Konstadinos G. Goulias[1]

Abstract

This study analyzes 8-week long-distance travel records from the California Household Travel Survey for completeness and identifies general types of non-commute long-distance tours using Latent Class Analysis. Likely due to the difficulty of gathering data of this kind, there has been relatively limited study of non-commute long-distance travel, despite the substantial contribution to many households' greenhouse gas emissions and travel expenses. The California Household Travel Survey includes a valuable long-distance 8-week travel dataset, but this study identifies several possible shortcomings in the dataset. Of particular importance is a severe underreporting of shorter trips, which may result from a mix of respondent forgetfulness and survey fatigue. Despite the issues with the data, latent class cluster analysis was able to identify five distinct, informative patterns of long-distance travel. This analysis shows that long-distance tours for vacation, business travel, medical, and shopping are substantially distinct in terms of their travel characteristics and correspond to different combinations of other activities in the tour, and they are done by different types of households. The method used here to identify the typology of long-distance travel can be easily expanded to include a variety of other explanatory variables of this type of behavior in more focused data collection settings.

Long-distance travel is an important aspect of human travel as it provides people access to experiences and opportunities that they cannot find near home, but it also comes at considerable costs, both to individual travelers and to climate, local environments, and transportation infrastructure (*1–4*). Even though long-distance travel is relatively infrequent, a single cross-country trip by plane can represent a substantial share of a household's annual travel budget and greenhouse gas (GHG) emissions. Indeed, despite the fact that less than half of all Americans fly each year (and less than 20% make more than one round trip by plane) (*5*), commercial aviation accounts for 6.6% of U.S. total transportation greenhouse gas emissions and a much larger share of the GHGs produced by private transportation (*6*).

California considers long-distance travel important for policy analysis at the state level and includes a long-distance travel log to the California Household Travel Survey (CHTS) (*7*). This part of the survey was intended to derive prototypical travel profiles and understand the range and variation of long-distance travel statewide. An extensive analysis of CHTS data, with focus on long commute trips, revealed a strong relationship with residential location characteristics (*3*, *8*). Using CHTS these studies found that long-distance travel and cost of housing are strongly related, and higher income and employed persons travel more (*8*).

Though there has been substantial research into long-distance travel, many shortcomings remain. The literature contains a variety of definitions for "long distance" travel, including ones based on distance (e.g., longer than 50 miles, 100 miles, or longer than 100 kilometers) and travel time (e.g., 40 minutes). Long-distance travel researchers have also considered a variety of indicators including number of long-distance trips, activity before and/or after commute, mode used, time of day of trip, and destination (*3*, *9–11*). Few studies have directly addressed the significance of overnight stays and trip chaining directly, which are particularly important for long-distance travel, as high travel time requirements make it necessary to combine multiple destinations into a single trip. Long-distance trips are often chains of trips with different purposes such as business and leisure, visit friends and sightseeing, and education and vacation (*12*).

[1]Department of Geography and GeoTrans Lab, University of California, Santa Barbara, CA

Corresponding Author:
Address correspondence to Adam W. Davis: awdavis@geog.ucsb.edu

Although substantial research has been carried out on long-distance travel in general, and particularly into long-distance commuting, there has been less focus on non-commute trips, perhaps partly because of issues with data reliability. Reliable long-distance travel data can be difficult to collect because these trips are relatively infrequent and may be particularly clustered within certain sections of the population. In the CHTS 14% of respondents gave inconsistent answers about the completeness of their long-distance travel log, and there is substantial underreporting of long-distance trips, particularly those that are relatively short and repetitive. Bierce and Kurth found that people who made repetitive long-distance trips (like long-distance commuters) did not report all their trips in the 8-week long-distance data of CHTS (*13*), and there is evidence that longer trips may be underreported as well. In addition, individual people may disagree about what counts as a long-distance trip; in the CHTS, some respondents only report the individual trip segments that covered a long distance, whereas others provided information for every stop on tours that took them far away from home. It is not hard to think of potential explanations for these shortcomings in the data: frequent travelers may find the repetitive entries tedious; respondents may forget minor long-distance trips made over the 8-week period; and if a single member of the household is responsible for filling out the long-distance log, that person may be especially unlikely to remember all their housemates' travels.

This paper explores the data quality issues present in the CHTS 8-week long-distance travel log and proposes a statistical model-based typology for non-work long-distance trips based on a Latent Class Analysis of tours extracted from this dataset. This analysis is based on tour mode used, distance traveled, number of overnight stays, travel purpose, and destination types chosen by Californians. Also incorporated is an experiment with the use of social media data (Foursquare) to describe destination characteristics and their significance in explaining long-distance tour behavior.

Data Sources

This paper primarily relies on data from a simplified travel log in the CHTS of 2012–13, in which respondents were asked to report all the trips they had made of 50 miles or longer in the 8 weeks preceding the day they were assigned a full travel day diary. CHTS contains travel reports of approximately 42,000 households and members within these households, with a wide array of household demographics and detailed data from the single-day place-based diary. This study correlates some of these data with long-distance travel of 9,332 households with relatively clean non-commute records as explained below (for a detailed description of CHTS see NUSTATS [*7*]). The primary benefit of this dataset is that it spans a longer period than the 24 hours included in the more detailed CHTS daily diary, which means it captures a much larger range of infrequent long-distance trips, particularly those with overnight stays away from home. Also utilized are location data harvested from Foursquare (*14*).

Long-Distance Data Quality Assessment

The 8-week long-distance travel log in the CHTS provides valuable data, but because it was measured separately from the main travel diary, this dataset is less complete and consistent. A major cause of the uncertainty is the retrospective nature of the CHTS long-distance component (*15*). Respondents are likely to forget some details of trips they made in the more than two months before the survey, and may tire of repeatedly describing their long-distance commute trips. The limitations and biases in these data necessitate additional quality checking. The long-distance component of the CHTS contains two major sources of error and uncertainty: explicit uncertainty at the level of individual households, which arises from some respondents declining to fill out the log in full or providing inconsistent information (which affects roughly 14% of the respondents); and implicit uncertainty about the dataset as a whole, which is identified by comparing aggregate travel counts from the long-distance log against scaled aggregate results from the daily diary.

The CHTS household dataset contains two indicators that pertain to long-distance travel:

- "Do you have a completed Long Distance Log to refer to?"
- "We would like to gather a list of all long-distance trips made by anyone in your household in the eight weeks prior to your travel day. Remember that a long-distance trip is any one-way trip of more than 50 miles. Tell me about the first long distance trip."

Responses to these questions should align with each other and with the actual record of long-distance trips, but for many households they do not, as shown in Table 1. Of the 42,431 households for which usable data are available, 22,692 reported having a complete long-distance log, 16,965 did not have a complete log, and 2,448 reported never having received a copy of the log. Most long-distance trips were reported by households that gave consistent answers to questions about long-distance travel, and 86.6% of households are in this category (cells in bold). Of concern are the several thousand respondents who reported long-distance travel, but based their reports on an incomplete log, and those who reported never having received a long-distance travel log.

Implicit uncertainty in this dataset arises from a substantial undercounting of some types of long-distance trips. The 8-week travel log is designed to capture information about relatively infrequent travel that a daily diary would miss, and appears to miss shorter, more frequent trips, notably

Table 1. Relationship Between Reported Long-Distance (LD) Log Completeness and Reported LD Trips

Do you have complete LD log?	Tell me about your first LD trip	Households with any LD trip	Households without any LD trips	LD Trips	Trips/HH with Trip
1-Completed	1-Yes	**14,667**	1	59,174	4.03
1-Completed	2-No LD Trips	61	**7,952**	186	3.05
1-Completed	8/9-DK/RF	11	0	26	2.36
2-Not Completed	1-Yes	2,694	1	7,430	2.76
2-Not Completed	2-No LD Trips	110	**14,136**	329	2.99
2-Not Completed	8/9-DK/RF	23	1	28	1.22
3-Never Received	1-Yes	359	0	843	2.35
3-Never Received	2-No LD Trips	6	2,081	10	1.67
3-Never Received	8/9-DK/RF	2	0	3	1.50
8/9-DK/RF	1-Yes	62	0	140	2.26
8/9-DK/RF	2-No LD Trips	5	248	5	1.00
8/9-DK/RF	8/9-DK/RF	8	3	8	1.00

Table 2. Ratio of Long-Distance Trips per Day Predicted by Daily Diary to 8-week Log (Longest Trip per Household per Day, Only Including Trips Away from Home)

Distance	Ratio (from home)	Ratio (trip length)
50–75 miles	6.29	6.81
75–100 miles	5.01	5.05
100–200 miles	4.34	3.94
200–500 miles	2.97	2.28
500–1000 miles	2.38	2.35
1000–3000 miles	1.88	2.05
over 3000 miles	1.34	1.91

long-distance commuting (*13*). By comparing the data in the long-distance log with long trips recorded in daily diaries, the overall scale of this dataset's distance bias can be estimated.

To estimate the distance bias, the daily diary and the 8-week log are first placed on roughly equal footing in terms of data specificity and time span. Each person submitted an individual diary, whereas the 8-week log was filled out at a household level, so some trips appear multiple times in the daily diaries. As discussed above, a significant number of households excluded trip-by-trip detail from the 8-week logs, providing only the general arc of the tour instead of information about each leg and/or only including the trip away from home but not the return trip. To address these inconsistencies, this comparison includes only the longest trip with a destination over 50 miles from home made by each household on a given day in either dataset. Although route distances for trips in the daily diaries have been computed, this comparison is based on straight-line distance from home, to allow inclusion of trips overseas. Once the maximum daily distance for each household is extracted, the distances are grouped into bands and the number of household-days that occur in each band calculated. Totals from the 8-week log are divided by 56 (the number of days in 8 weeks) to convert them to per-day estimates. The results from this comparison are shown in Table 2.

Table 2 shows the ratio between the number of long-distance trips in the daily diary and the number of long-distance trips per day in the 8-week log. This ratio is calculated both for the maximum distance from home and for the longest single trip. As the table shows, the daily diary predicts between 6 and 7 times as many trips per household per day as the long-distance log in the 50–75 mile range, and this undercounting persists to a lesser extent in all distance bands regardless of whether maximum trip length or maximum distance from home is used. The inconsistent data quality among households must be considered when performing household-level analysis on these data (e.g., to estimate the total number of long-distance business trips made by California residents), but it presents a less significant limitation to the ability to explore the range of long-distance travel types at a tour level using probabilistic models of clustering.

Data Processing

The latent class cluster analysis is based on long-distance tours, rather than single trips, because each trip within a tour is dependent on the other trips for mode and scheduling, and this is likely even more important for long-distance tours, as it is harder to engage in all the activities at tour destinations on separate tours (much easier to take an additional trip to the grocery store instead of going after work than to visit a museum in a distant city if you weren't already there for a business trip). The aim was to investigate tour characteristics and destination choice at the level of single tours. To do this, each tour was assembled in a single record that contains a mix of household-level variables (income, household size, number of household cars, etc.), tour-level variables (total distance by each mode, total duration, and presence/absence of multiple purposes), and variables that pertain to the tour's primary leg (single trip purpose and measures of destination attractiveness for the primary destination). For this analysis, the primary destination was considered to be the destination at which members of the household stayed for the most time during the tour.

Because the aim was to consider as wide a range of tours and destinations as possible, a generous approach was taken to identifying tours and imputing trips for long tours with missing information. Long-distance tours are sequences of long-distance trips made by a household with end-to-end continuity (the destination location of each trip is the origin of the next trip), beginning with a trip from home and ending with a trip back to home. The pool of potential tours was expanded by running the tour identification process on households' 8-week long-distance logs sorted both by reported long-distance trip number and by date, keeping the results of whichever method produces a larger number of complete home-based tours for a given household (trip number ordering worked better for 91% of households). This analysis also includes tours with incomplete records, but steps are taken to account for uncertainty in duration and distance: in tours missing the from-home or to-home trip, the remaining distance was imputed by adding the mode-specific mileage from whichever of those trips is present is double-counted in the tour total. This tour identification process leaves 23,511 full or partial tours for analysis.

The following steps were taken to identify the primary leg of each tour to extract the primary purpose and destination characteristics:

- Disqualify trips ending at an airport, with their purpose coded as "Return Home", or ending within 30 miles of home. Some tours have all their destinations eliminated by this process, and 21,584 tours remain.
- 2,128 tours have multiple destinations; the longest-duration destination was selected for each of these.
- 522 tours have multiple destinations with equally long durations. To break these ties, the destination farthest from home was selected.
- 76 ties remain, which were resolved by selecting the first tour in order.
- Lastly, for this analysis, all commute tours were eliminated.

The aim is to investigate tour characteristics and destination choice at the level of single tours. To do this, each tour is collapsed to a single record that contains a mix of household-level variables (income, household size, number of household cars, etc.), tour-level variables (total distance by each mode, total duration, and presence/absence of multiple purposes), and variables that pertain to the tour's primary leg (single trip purpose and measures of destination attractiveness for the primary destination). For this analysis, the primary destination was considered to be the destination at which members of the household stayed for the most time during the tour. Table 1 shows a list of the variables analyzed here and their averages. Table 2 shows the frequencies of the categorical variables.

This study also experimented with the derivation and use of place attractiveness indicators from Foursquare. For each destination, the space covered by the fifty closest reporting locations (checkins) was estimated. This space is different depending on the place characteristics (e.g., low density environment) and the willingness of Foursquare users to check in. The second variable is the number of individuals checking in these locations. The text of the Foursquare report was also analyzed to identify unique topics characterizing each destination, but not reported here because it produced a wide variety of groups of words reported by Foursquare uses (these groups are called topics) and did not lend themselves to typical statistical modeling. The average ratings by the fifty Foursquare checkins were also used as an indicator of location attractiveness. Each household's residential area is also classified by its residence business density into center, suburb, exurb, and rural.

Latent Class Method Description

This analysis relies on *latent class cluster analysis* to identify distinct types of long-distance tours. This technique includes a *J*-category latent variable with each category representing a cluster; uses many "dependent" or clustering *criteria* variables; uses a mixture of multiple types of criteria variables (e.g., continuous, categorical, ordered, count); uses and tests the effect of covariates of many different types; is more flexible than many other clustering algorithms; is a model-based clustering approach providing probabilistic membership of observations in clusters; and provides convenient interpretable output. Model description of this method is provided in Vermunt and Magidson (*16*), and more details about this application can be found in Goulias et al. (*14*). This type of cluster analysis aims to find the categories of a variable representing different types of tours and assess each tour by its membership in a class (the category of the latent class variable).

Statistical goodness-of-fit measures for latent class cluster models are the typical chi-square statistics in the cross-categorical data analysis. The first measure is the likelihood ratio chi-square, G^2 or L^2. It has a chi-square distribution with degrees of freedom given by the number of "free" parameters (total number of different response patterns – the number of estimated model parameters − 1 if there are no covariates in the model). It represents the opposite of an R^2 in regression because it is the amount of unexplained associations among the criteria variables by the model. Therefore, higher values indicate models that do not fit the data well and lower values represent better fitting models. When two models are nested (i.e., they differ only in the number of estimated parameters), the difference between the G^2 of these two models was created. This difference is chi-square distributed and can be used for hypotheses testing. A test of this

Table 3. Latent Class Model Fit Diagnostics

	LL	BIC(LL)	PARAMETERS	CLASSIFICATION ERROR
1-CLUSTER	−227363	454941	22	0.0000
2-CLUSTER	−213929	428403	56	0.0469
3-CLUSTER	−209021	418918	90	0.0863
4-CLUSTER	−205625	412459	124	0.0737
5-CLUSTER	−204569	410677	158	0.0844
6-CLUSTER	−204104	410080	192	0.0936
7-CLUSTER	−202920	408043	226	0.1417
8-CLUSTER	−202046	406626	260	0.1379

type cannot be utilized between models that differ in the number of clusters because they are not nested. The L^2, the Bayes Information Criterion (BIC), Akaike Information Criterion (AIC), and the Consistent Akaike Information Criterion (CAIC) are computed to measure goodness of fit and to take into account model parsimony, penalizing models with many parameters. The lower the BIC, AIC, or CAIC values, the better the model estimation (*16*).

There are many advantages using this method for clustering. First, the latent class cluster method for identifying clusters is designed for combinations of continuous and discrete criteria variables, while other methods such as the k-means method are defined for continuous variables only. Second, the method used here allows for probabilistic membership of each observation in each cluster. This provides flexibility in observation classification that other methods do not. Third, post-processing of the cluster data using regression is not required because the method used allows the inclusion of covariates. There are other advantages of latent class methods in general and the specific implementation used here as illustrated in Vermunt and Magidson (*16*).

Results and Discussion

The purpose of this model is to produce a general typology of long-distance tours and identify what types of people make these tours and what types of destinations attract them. This model investigates the distribution of a set of long-distance tour characteristics (namely distance, purpose, duration, and destination region) and identifies five types of long-distance tours. Active covariates are characteristics of households that help predict what types of tours they make. For this analysis, of particular interest was how scheduling, income, and the characteristics of travelers' home neighborhoods affected the types of trips they made. After the model has converged, mean values for each class for other household characteristics, as well as some destination characteristics, were also extracted.

A set of criteria variables was chosen to be clustered, and the latent class clustering process was run over a range of class numbers. Each additional class substantially improves the model's likelihood of producing the distributions of and relationships between the observed indicator variables, but these improvements are counterbalanced and ultimately overwhelmed by increases to the number of parameters being estimated and a decrease in the model's ability to show clean breaks between the clusters. Because the aim was to produce a parsimonious model that is relatively easy to interpret, a model was selected with a relatively small number of classes and a low classification error that represents a substantial improvement in log likelihood and AIC/BIC over the base models. Once a model is developed that works well enough for the indicator variables, active covariates were added one at a time to improve the model's ability to place observed tours into classes. Fit statistics and classification errors for models with the final specification and 1–8 clusters are shown in Table 3.

A model was selected with five latent classes, four sets of indicator variables, and four sets of active covariates. Models with more classes had better likelihood and BIC scores, but also substantially more classification errors and were less clear to interpret. The model finds a clear distinction between long-distance tours made by car and those made by other modes, and makes other clear breaks based on primary purpose and duration. Shorter car tours for personal business and other purposes are separated from longer tours, which frequently involve air travel. Another key distinction is made between long business trips (e.g. flying to New York for a meeting) and vacation/multipurpose trips, which may involve more modes. Households with high incomes who live in urban areas are responsible for a much larger share of business trips, whereas rural residents are responsible for more of the single-day trips (likely because they must travel longer distances to access a wide range of opportunities). Income is also a key factor in distinguishing between general-purpose short trips and longer trips made for fun (vacations, entertainment trips, and outdoor activities). Day of the week is another primary distinguishing factor for predicting class membership. Household size, number of children, and number of trips in the daily diary were not strongly associated with any of the trip classes. Class means for these models are shown in Table 4.

Five tour classes identified using this method are as follows.

Table 4. Cluster Variable Means and Cluster Percentages for 5-cluster Latent Class Solution

	Cluster Number	Day Trips	Long Wknd.	Pass.	Business	Vacation
Indicator Variables	Travel Distance (miles)					
	Car Driver	159.0	446.4	3.7	21.1	928.2
	Car Passenger	3.3	8.4	221.2	6.3	258.0
	Ground Transit	0.6	1.9	33.7	5.8	164.8
	Air	2.7	12.8	24.0	2,995.7	2,454.7
	Tour Duration					
	Single-day	73.7%	9.5%	45.3%	3.6%	4.1%
	One overnight	13.3%	16.9%	13.3%	9.2%	2.0%
	Two–six overnights	7.2%	62.1%	33.4%	59.8%	37.5%
	Seven+ overnights	5.8%	11.5%	8.0%	27.5%	56.4%
	Destination in California	96.7%	87.2%	93.5%	17.3%	11.4%
	Primary Purpose					
	Business (meeting/convention/seminar)	13.9%	9.4%	9.2%	33.4%	8.8%
	Combined business and pleasure	2.1%	3.3%	1.7%	4.0%	2.8%
	School-related activity	2.4%	1.3%	5.2%	1.5%	1.8%
	Visit friends/family/relatives	30.6%	38.1%	26.7%	36.6%	35.9%
	Medical	8.2%	2.3%	5.9%	0.1%	0.7%
	Personal Business (e.g. Shopping)	11.1%	3.4%	7.1%	1.6%	2.7%
	Vacation/sightseeing	7.6%	25.0%	16.6%	17.4%	38.9%
	Outdoor recreation	7.3%	9.2%	13.0%	1.3%	3.2%
	Entertainment	10.6%	5.3%	12.4%	2.2%	2.7%
	Drive someone else / DK / RF	6.3%	2.8%	2.3%	1.9%	2.6%
Active Covariates	Day of the Week of Tour Start					
	Mon	9.7%	9.5%	9.1%	14.5%	12.3%
	Tue	11.3%	8.2%	8.5%	14.0%	11.9%
	Wed	10.7%	10.4%	10.7%	17.1%	14.1%
	Thu	11.0%	14.9%	11.6%	17.4%	16.4%
	Fri	13.3%	30.4%	22.8%	15.0%	15.5%
	Sat	27.5%	15.9%	23.0%	10.6%	16.9%
	Sun	16.6%	10.8%	14.3%	11.3%	12.9%
	Household Trips in Daily Diary	10.1	10.1	11.5	11.5	9.5
	Household Annual Income					
	Under $50,000	23.5%	17.4%	27.3%	8.5%	19.3%
	$50,000 to $74,999	20.1%	18.6%	17.8%	10.8%	16.9%
	$75,000 to $99,999	19.1%	19.9%	17.4%	12.6%	16.6%
	$100,000 to $149,000	20.8%	23.8%	20.7%	25.0%	23.0%
	$150,000 and above	16.6%	20.4%	16.9%	43.1%	24.3%
	Household Block Group Category					
	Urban Center	19.8%	22.3%	23.1%	44.3%	27.7%
	High Density Suburb / Exurb	27.5%	27.1%	27.4%	29.5%	26.8%
	Low Density Suburb / Exurb or High Density Rural	26.1%	25.0%	25.1%	20.7%	25.1%
	Low Density Rural	26.6%	25.7%	24.4%	5.4%	20.3%
Inactive Covariates	Household Size	2.6	2.5	2.7	2.5	2.5
	Children in Household	0.5	0.5	0.6	0.5	0.4
	Household is Hispanic	16.5%	13.8%	16.2%	9.3%	12.3%
	Employment Density Around Household (emp/km^2)	934	1,013	1,068	2,176	1,230
	Destination Characteristics (Foursquare)					
	Median Checkins	1016	966	1069	1477	969
	Median User Count	468	463	509	720	465
	Convex Hull Area of 50 nearest POIs (log km^2)	2.74	2.80	2.65	1.97	2.51

Class 1: Day Trips, 37.7% of Tours

The largest long-distance tour class is made up of mostly relatively short single-day tours made mostly by car. These tours are made by a wide range of households for a wide range of purposes, from necessary tasks like household replenishment / personal business, and medical care, to more recreational purposes like visiting friends and family, entertainment. Since relatively few of these tours are work related, most are made late in the week and on the weekend. These tours are short, and almost all visited destinations are within California. City-dwellers are somewhat less likely to make this sort of tour (note: all density categories have roughly the same population within the CHTS sample), but in all other ways, households who make this sort of tour are not notably distinct from the rest of the dataset. Destinations visited by these tours are like those visited by tours in the other general-purpose class (see 3 below) and slightly higher than those visited by vacation classes 2 and 5.

Class 2: Long Weekends, 31.8% of Tours

The next-largest class of tours is made up of short recreational trips made mostly by car. Unlike tours in class 1, these tend to feature a small number of overnight stays, but like class 1, most of these tours remain in California (though somewhat more of them visit neighboring states). A very large share of these tours start on Fridays, which strongly suggests that these are short vacations taken over a weekend. These tours represent a mix of purposes, but vacation and visiting friends/family are notably popular. Unsurprisingly, these optional trips are made by households that skew wealthier and suburban (like long vacation class 5). Outdoor recreation is a somewhat common purpose for these tours, and their destinations are generally slightly lower density.

Class 3: Passenger Trips, 13.0% of Tours

The classifying method separates passenger trips from all other categories. In general, these tours remain in California, and they are general like tours in class 1. These tours are somewhat more likely to feature likely group activities (e.g., outdoor recreation) and trips for medical purposes (which may require someone else to drive).

Class 4: Business Trips, 9.6% of Tours

Class 4 is in some ways the most clearly defined of the classes, in terms of mode mix, purpose, and household characteristics, as it overwhelmingly corresponds to trips by air outside of California for business purposes with a moderate number of overnight stays. These tours generally start during weekdays, but have a much more mixed selection of start dates than the other tours. These tours are primarily made by wealthy urban households and are made to destinations that are notably higher in density than those visited by other tour types.

Class 5: Vacations, 8.1% of Tours

The final class of long-distance tours contains most of the longer-duration tours in this dataset, and more than half include at least 7 overnights. These tours can generally be categorized as long vacations outside of California. Unlike the other tour classes, these tours are made by a mix of modes (plane, car, and transit). Vacation and visiting friends and family are the most common primary purposes of these tours. Since these trips are likely made for personal enjoyment, they tend to skew wealthier like the long weekend trips in class 2, but are much less uniformly made by wealthy households than class 4. The people who make these tours are more likely to live in cities, but much less so than class 4. These tours include visiting destinations that are slightly denser than in classes 1–3, but much less dense than those in class 4.

Tour classes 1–3 are predominantly made by car, and classes 4 and 5 contain nearly all the trips by air in the dataset. Unlike the other tour purposes, which are strongly tied to one or two of the classes, visiting friends and family is ubiquitous across classes. Whereas long-distance travel purposes like business trips, medical treatment, outdoor activities, and other purposes have specific sorts of destinations and tend to have a fairly distinct pattern of duration, travel distance, and mode, the locations of friends and family are much less constrained in space and the amount of time a person wants to spend.

Household characteristics display mixed results. On the one hand, income is emerging as important for many reasons (e.g., substantial expenditure for traveling with family, necessity of presence at a destination for some type of business). On the other hand, household size and number of children are generally uninformative in distinguishing among patterns of long-distance travel. A key limitation, however, results from noise in the data. Characteristics of the primary trip-maker would likely be useful in clustering, but this information is not consistently present in the log, and it is less clear how to incorporate the characteristics of other trip makers (who are not identified in the 8-week log). Number of people on the tour (also known as party size in long-distance travel) could also be highly useful, but recording quality of this variable is very inconsistent in the dataset. Land use around destination and around household show a surprisingly limited relationship between most kinds of long-distance travel and local land use around home and destination except for business flight trips (and to a lesser degree, single-day car trips).

Also investigated were the combinations of tours for households that made two tours in the 8-week travel log. Vacation tours are much more likely to be combined with other types of tours, which makes sense, as vacations require

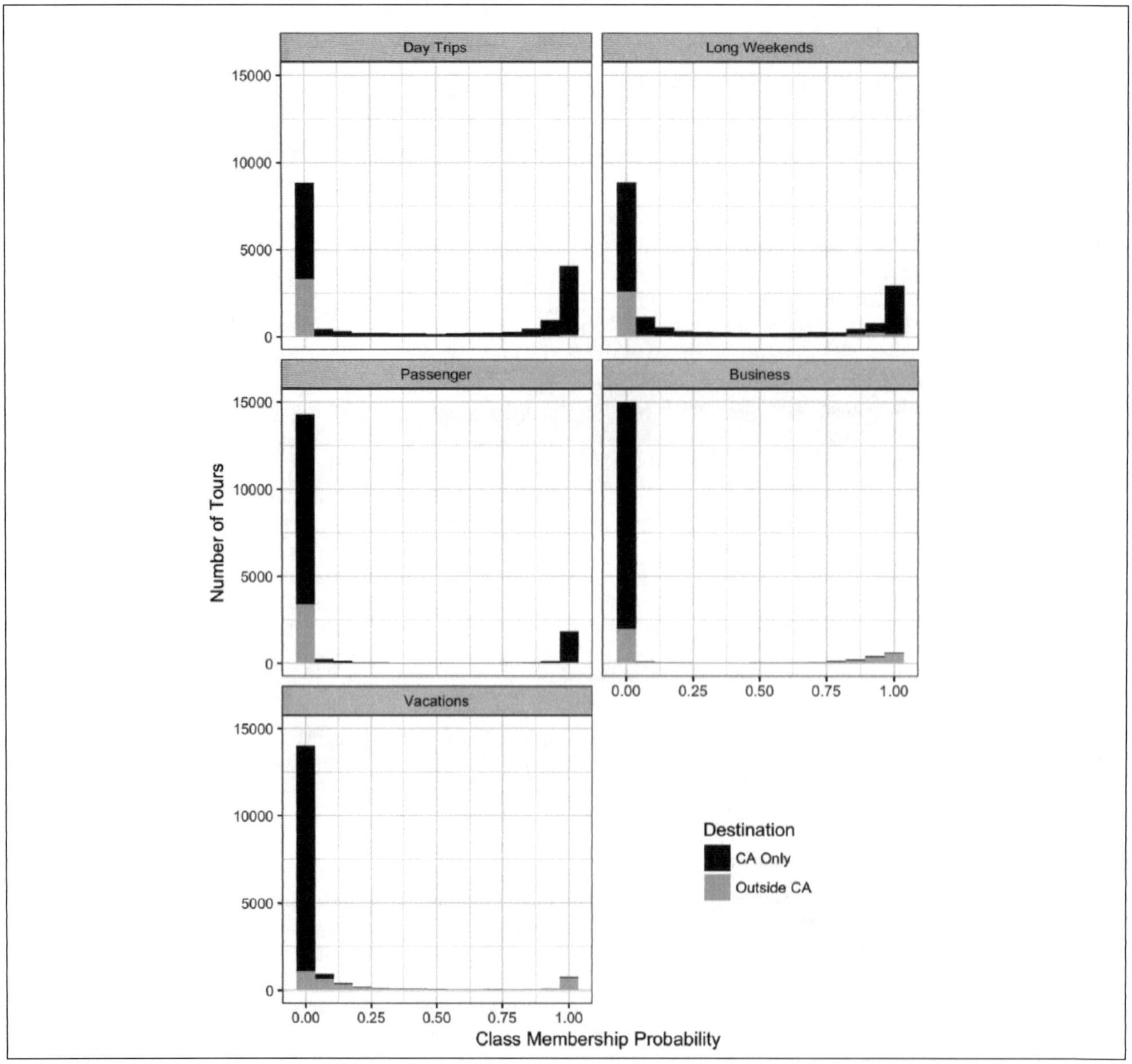

Figure 1. Estimated class membership probability histograms.

enough time and money that people are unlikely to take multiple within an 8-week period. In contrast, day trips and passenger trips were quite likely to be paired up with other tours in the same category. Several households made two business trips, but these did not co-occur to the degree that might be suggested by their uniqueness of their household land-use characteristics (Table 4). This indicates that it may be worthwhile to consider the correlations between tour categories within a household, perhaps with a longitudinal chain formation model that explores the sequencing of tour types in the reporting period (this remains a possible future task).

Latent class clustering does not produce absolute class assignments; rather, it provides an estimated probability of each observation belonging to each tour. These probabilities are particularly useful because they indicate how well the model can distinguish between classes. In this case, the model performs very well, as most observations belong to one of the clusters with a high degree of certainty (Figure 1). Although tour destination within California is the single most significant indicator in the latent class model, other variables still play a role in class assignment, as shown in Figure 1, where observations inside and outside of California

Table 5. Tour Class Probabilities for Households that Reported Multiple Long-distance Tours

Tours reported (8-wk log)	Number of households	Day trips (class 1)	Long weekends (class 2)	Passenger trips (class 3)	Business trips (class 4)	Vacations (class 5)
1	5193	30.2%	34.7%	12.9%	9.7%	12.6%
2	2222	35.3%	34.2%	12.0%	10.2%	8.2%
3	1069	39.6%	31.1%	14.1%	8.8%	6.4%
4	485	45.1%	28.4%	12.4%	10.2%	3.8%
5	161	42.8%	24.8%	19.1%	9.4%	3.8%
6	92	50.1%	24.9%	10.1%	10.4%	4.5%
7	41	54.2%	22.2%	8.3%	13.5%	1.8%
8	33	52.8%	28.3%	7.5%	8.6%	2.8%
9	13	62.2%	24.4%	11.6%	0.6%	1.1%
10 +	23	59.2%	19.9%	14.7%	2.7%	3.4%
	For all tours	37.7%	31.8%	13.0%	9.6%	8.1%

receive a range of probability estimates for all classes except the two that entail long-distance flights.

How tour categories vary as a function of the number of tours and the types of tours made by households is also of interest. For example, do households that make multiple tours in the 8-week period also show heterogeneity in the type of tours they form? Table 5 presents the mean class probabilities for each type of tour made by households that made a given number of tours. Households who make more long-distance tours tend to make more day trips. These people are likely residents of lower-density areas with limited access to important opportunities. Business trips are particularly common among households that reported a few tours in the 8-week diary, which makes sense, since business trips should concentrate in households with people who work in jobs requiring travel. Vacations stand out both for their solitariness and for the degree to which the cost and time involved excludes other long-distance travel. For many households that only reported a single long-distance trip, that trip was a major vacation, and households with many long-distance tours have a lower total vacation probability across all their tours, which indicates that almost no one makes more than one vacation tour in the 8-week reporting period.

Conclusion

This paper shows the results of a pattern recognition approach that is able to extract meaningful and informative clusters from a relatively noisy long-distance tour data source. The noise in the data is the result of imperfect measurement, but the latent class cluster analysis yields five distinct types of non-commute long-distance tours representing a wide spectrum of long-distance travel patterns of Californians. These tours are heavily influenced by the number of overnight stays, as shown in European studies. The main purpose of the tour is another major classification driver, but distance and main mode are key to the distinction among the patterns of long-distance travel. This study also points to the need of tailoring policies to long-distance tours types. For example, vacation tours are less likely to be influenced by California policies because they are multi-modal, often extend beyond California's borders, and involve diversity of decision makers. On the opposite side of the spectrum, day trip tours and long weekend tours make up a substantially larger chunk of the dataset and may be much more amenable to changes in policy. Pricing of services and the introduction of mobility as a service are more likely to affect these tours in a substantial way and yield a high impact in curbing emissions and congestion. This also shows that focused studies are needed into this type of travel, with possibly a multi-day diary and/or destination-specific intercept surveys.

There are many ways to expand the study here, including a more detailed analysis of the household structure and long-distance tour formation and sequencing of tour types by households. In addition, more information about destinations can be derived from secondary data sources, such as social media sites, using quantitative and qualitative analysis. This may provide additional information about the combination of different types of tours by households when they allocate time and resources in vacation versus business-related travel. The study here showed only limited correlation between daily travel and long-distance tour formation using frequency of daily trips. This can be expanded to include other travel behavior indicators from the daily diary and study their influence.

Acknowledgments

Funding for this research was provided by the California Department of Transportation (CALTRANS) through the University of California Center on Economic Competitiveness in Transportation (UCCONNECT).

Author Contributions

The authors confirm contribution to the paper as follows: study conception and design by Davis, McBride, Goulias; data collection and processing by Davis, McBride; analysis and interpretation of results by Davis, McBride, Goulias; draft manuscript preparation by Davis, McBride, Goulias. All authors reviewed the results and approved the final version of the manuscript.

References

1. Christensen, L. Environmental Impact of Long Distance Travel. *Transportation Research Procedia*, Vol. 14, 2016, pp. 850–859. https://doi.org/10.1016/j.trpro.2016.05.033
2. Gössling, S., C. B. Hansson, O. Hörstmeier, and S. Saggel. Ecological Footprint Analysis as a Tool to Assess Tourism Sustainability. *Ecological Economics*, Vol. 43, No. 2, 2002, pp. 199–211. https://doi.org/10.1016/S0921-8009(02)00211-2
3. Mitra, S. K. *Land Use, Land Value, and Transportation: Essays on Accessibility, Carless Households, and Long-distance Travel*. PhD dissertation. UC Irvine, 2016.
4. Pourabdollahi, Z., R. Tillery, M. Gawade, T. Hill, G. Mathews, and J. Worrell. Statewide Tourism Travel Demand Forecasting: A Behavior-Based Modeling Framework for the State of Florida. Presented at 96th Annual Meeting of the Transportation Research Board, Washington, D.C., 2017.
5. Gallup Inc. *Airlines*, 2016. Retrieved July 31, 2017. http://www.gallup.com/poll/1579/Airlines.aspx
6. United States Environmental Protection Agency. *Inventory of U.S. Greenhouse Gas Emissions and Sinks, 1990-2015*. (No. EPA 430-P-17-001). Washington, D.C., 2017. Retrieved from https://www.epa.gov/sites/production/files/2017-02/documents/2017_complete_report.pdf
7. NUSTATS. *2010-2012 California Household Travel Survey Final Report (p. 137)*, 2013. California Department of Transportation. Retrieved from http://www.dot.ca.gov/hq/tpp/offices/omsp/statewide_travel_analysis/Files/CHTS_Final_Report_June_2013.pdf
8. Mitra, S. K., and J. -D. M. Saphores. Determinants of Long-Distance Commuting: Evidence from the 2012 California Household Travel Survey. Presented at 96th Annual Meeting of the Transportation Research Board, Washington, D.C., 2017.
9. Axhausen, K. W. *Transportation Research Circular E-C026: Methodological Research for an European Survey of Long-distance Travel*. TRB, National Research Council, Washington, D.C., 2001. https://trid.trb.org/view.aspx?id=686708
10. Beckman, J. D., and K. G. Goulias. Immigration, Residential Location, Car Ownership, and Commuting Behavior: A Multivariate Latent Class Analysis from California. *Transportation*, Vol. 35, No. 5, 2008, pp. 655–671. https://doi.org/10.1007/s11116-008-9172-x
11. Holz-Rau, C., J. Scheiner, and K. Sicks. Travel Distances in Daily Travel and Long-Distance Travel: What Role is Played by Urban Form? *Environment and Planning A*, Vol. 46, No. 2, 2014, pp. 488–507. https://doi.org/10.1068/a4640
12. LaMondia, J. J., and C. R. Bhat. A Conceptual and Methodological Framework of Leisure Activity Loyalty Accommodating the Travel Context. *Transportation*, Vol. 39, No. 2, 2012, pp. 321–349. https://doi.org/10.1007/s11116-011-9342-0
13. Bierce, E., and D. Kurth. The Use of Three Surveys for Long Distance Travel Estimates in California. Presented at 93rd Annual Meeting of the Transportation Research Board, Washington, D.C., 2014.
14. Goulias, K. G., A. W. Davis, E. C. McBride, K. Janowicz, and R. Zhu. *Long Distance Travel in CHTS and Social Media Augmentation*. Department of Geography, UC Santa Barbara, 2017.
15. Federal Highway Administration. *Foundational Knowledge to Support a Long-Distance Passenger Travel Demand Modeling Framework*. (DTFH61-10-R-00036), 2015. Retrieved from https://www.fhwa.dot.gov/policyinformation/analysisframework/docs/national_model.pdf
16. Vermunt, J. K., and J. Magdison. Latent Class Cluster Analysis. In *Applied Latent Class Analysis* (J. A. Hagenaars and A. L. McCutcheon, eds.), Cambridge University Press, 2002.

The Standing Committee on National Transportation Data Requirements and Programs (ABJ10) peer-reviewed this paper (18-04945).

Article

TRR
JOURNAL OF THE TRANSPORTATION RESEARCH BOARD

Transportation Research Record
2018, Vol. 2672(42) 81–90

DOI: 10.1177/0361198118793495
journals.sagepub.com/home/trr
SAGE

Identifying Temporal and Spatial Characteristics of Residents' Trips from Cellular Signaling Data: Case Study of Beijing

Zufen Li[1], Lei Yu[2,3,4], Yong Gao[5], Yizheng Wu[6], Guohua Song[1], and Dapeng Gong[7]

Abstract

This study developed a method to extract valid temporal and spatial characteristics of residents' trips based on cellular signaling data to support urban transportation planning activity. The study first identified data triggered by active modes as redundant data by analyzing the characteristics of the trigger modes of cellular signaling data, which were then labeled and excluded from further analysis. Thus, only the data triggered by the passive mode were used in the study. Then, the temporal and spatial characteristics of residents' trips were extracted by mapping the cellular signaling data onto study regions, dividing them into traffic analysis zones (TAZs), and extracting the origin–destination (OD) matrix. Finally, real data from Beijing, China were used in a case study to verify the feasibility of the proposed method. The extracted temporal and spatial characteristics of residents' trips were compared with those from the 4th Comprehensive Transport Survey. It was observed that the results from the two data sources had a high correlation with correlation coefficients higher than 0.9.

High-quality travel data are important for urban transportation planning. Traditionally, travel data are obtained through a household survey. Not only is this expensive but its accuracy is highly dependent on the sample size of the survey and the quality of answers from respondents. In addition, the travel data obtained from household survey generally presents only a snapshot of the complete travel information at a certain time and, even worse, the travel survey usually takes several years to update. Compared with the traditional household survey, the method of using positioning data from mobile phones to collect travel data has some apparent advantages such as convenient and frequent data updating, low cost, and wide coverage.

With the increasing popularity of mobile phones, there have been studies on extracting travel data, such as the origin–destination (OD) matrix, travel trajectories, and trip mode, from mobile phone location data. Based on these extracted travel data, human mobility patterns have been identified (*1–3*). Existing studies on the extraction of travel data from mobile phone location data can be summarized as follows:

1) Before mobile phones were widely used, it was difficult to acquire mobile phone location data, so studies were mostly based on the data generated by simulations (*4*) and the accuracy could therefore not be assured. With the increasing popularity of mobile phones today, mobile phone location data, by which the OD flow based on cellular base stations can be generated, became easily obtainable (*5*). Combined with the traffic network, the OD flow based on cellular base stations could be converted into the OD flow based on traffic analysis zones (TAZs) or administrative districts (*6–8*). With the advancement of technology, the method of extracting the OD matrix has continuously been improved. Mobile phone

[1]MOE Key Laboratory for Urban Transportation Complex Systems Theory and Technology, Beijing Jiaotong University, Beijing, China
[2]Texas Southern University, Houston, TX
[3]Beijing Jiaotong University, Beijing, China
[4]Xuchang University, Xuchang, China
[5]Shenzhen Urban Transport Planning Center, Beijing, China
[6]Department of Civil & Environmental Engineering, University of California, Davis, CA
[7]Nanjing Institute of City & Transport Planning Co., Ltd., Nanjing, China

Corresponding Author:
Address correspondence to Lei Yu: yu_lx@tsu.edu

location data have been combined with other data sources, such as the data obtained from GPS or roadside detectors, to extract OD matrices (*9*, *10*).

2) Based on the studies on the extraction of OD matrices, the estimation of travel trajectories and trip chains of residents has been the subject of research. Alexander et al. estimated mobile phone users' home and work locations from call detail records and derived the trip chains for commuting trips. In addition to the studies on the trip chains of commuting trips, there have been more studies on the travel trajectories of trips for other purposes. Travel trajectories were generated by deriving location area sequences for single trips (*13*). The difference between the real human trajectory and the one obtained through the mobile phone data was evaluated by using different interpolation methods. It was concluded that linear interpolation could offer the best estimate for sedentary people whereas cubic was best for commuters (*14*).
3) Mobile phone location data include cellular signaling data and GPS data. There have been few studies on the identification of trip modes based on cellular signaling data (*15*), but use of GPS data to identify the trip mode has been extensively investigated. Many researchers used the decision tree method to identify the trip mode (*16*–*18*); the data used were mainly GPS data with low resolution, such as data at a 5-min time interval (*19*, *20*).

Based on the above discussions, it was found that most of the existing studies focused on extracting only the travel data using the mobile phone location data, but a few efforts attempted to further extract temporal and spatial characteristics of residents' trips from the travel data. In this context, this study strove to develop a method for identifying the ODs of residents' trips and for further extracting the temporal and spatial characteristics of residents' trips from the OD matrices by using cellular signaling data.

Methodology

The study first preprocessed the cellular signaling data to improve its quality and increase the efficiency of the data analysis. Then, the temporal and spatial characteristics of residents' trips were extracted by mapping the cellular signaling data onto study regions, dividing them into TAZs, and extracting the OD matrix.

Data Source and Preparation

The cellular signaling data were obtained from the global system for mobile communication (GSM), in which the mobile phone location was obtained through the base station. When the mobile phone was switched on, it would be connected to the base station and the base station location of the mobile phone would be recorded by the GSM. The GSM would update the base station location of the mobile phone when one of the following three events occurred: 1) a call was placed or received, 2) a short message was sent or received, and 3) the mobile phone switched the base station location. Further, the GSM could update the base station location of the phone at a constant interval.

About 15 million people use China Mobile Communications Corporation (CMCC) as their carrier in Beijing. In this study, the cellular signaling data relating to May 2nd, 2011 were provided by CMCC in Beijing. They included 24-h signaling data with about 11 million records of mobile phone location data from 90,000 mobile phone users who were randomly selected by CMCC.

The cellular signaling data include five fields, which are International Mobile Subscriber Identity (IMSI), TimeStamp, Longitude, Latitude, and Event ID. IMSI is the unique identifier for mobile phone users. TimeStamp indicates the updating time of the base station location. Longitude and Latitude describe the base station location. Event ID represents the event types of mobile phone users, as shown in Table 1. In this study, the trigger mode was classified into two modes based on whether the updating of the base station location was related to the behavior of mobile phone users: active or passive. The Calling, Called, Message, and Switching of the base station were related to the mobile phone user's behavior, so these four types of events were identified as the active modes. However, regular updating was conducted by the GSM at a constant interval to all the phones registered in CMCC, and this was unrelated to the phone user's behavior, so it was classified as the passive mode.

Preprocessing of Data

The purpose of preprocessing the data was to identify the redundant data and improve the quality and efficiency of the data analysis. In this study, the data preprocessing included two parts: 1) identifying the redundant

Table 1. Types of Trigger Modes and Events

Trigger mode	Event ID	Event
Active mode	0	Calling
	2	Called
	5	Message
	7	Switching of base station
Passive mode	4	Regular updating

data and 2) mapping the cellular signaling data onto the study regions.

Identifying the Redundant Data. This study identified redundant data by analyzing the characteristics of the trigger modes of the cellular signaling data, which were the time interval and the time distribution of the data.

First, the time interval of each event was calculated using

$$\Delta T = \frac{\sum_{i=1}^{N} (t_{i+1} - t_i)}{N} \quad (1)$$

where ΔT is the time interval of the event in minutes, t_i is the time of the ith record of the event, and N is the number of data for this event.

Then, the time distributions of the data triggered by active modes and the passive mode were represented by the percentages of the number of records for each hour, which were calculated by

$$\text{Active}_i = \frac{\text{AR}_i}{\sum_{i=0}^{23} \text{AR}_i} \times 100\% \quad (2)$$

$$\text{Passive}_i = \frac{\text{PR}_i}{\sum_{i=0}^{23} \text{PR}_i} \times 100\% \quad (3)$$

where Active_i(Passive_i) is the percentage of the number of active (passive) records in the ith hour and AR_i (PR_i) is the number of active (passive) records in the ith hour.

The calculated results of the time interval and the time distribution of the data are shown in Figure 1. Figure 1*a* shows that the time interval for regular updating, which is the passive mode, is longer than that for the message, but shorter than the average time interval of all active modes, which is represented by the dotted line in Figure 1*a*. Figure 1*b* shows that the number of records triggered by the active modes decreases at night, but the records triggered by the passive mode are relatively stable for the whole day.

Consequently, the data triggered by the active modes included redundant data generated by the frequent calls or short messages of mobile phone users, and the number of records triggered by the active modes was not as stable as the number triggered by the passive mode. Further, the average time interval of the passive mode was 55 min, a data resolution that could ensure accurate identification of the OD of residents' trips. Therefore, to improve the quality and efficiency of the data analysis, only the data triggered by the passive mode were investigated in the rest of this study.

Mapping Cellular Signaling Data onto Study Regions. The purpose of mapping cellular signaling data onto study regions was to add the location information of the regions to the cellular signaling data, so that the region location for each piece of data could be obtained. This study matched the cellular signaling data with the location information of study regions by using a GIS tool.

Division into TAZs

The division into TAZs is important, as TAZs are the basic spatial units in urban transportation planning, and they are used to define the locations of the OD of trips. The trip volume between each TAZ pair can be estimated by using the traffic demand forecasting model.

The city of Beijing is divided into 16 administrative districts. In this paper, the study region includes the entire city of Beijing. The study first calculated the number of mobile phone users and the area of each administrative district, in accordance with which the number of TAZs in each administrative district was defined.

Extraction of Temporal and Spatial Characteristics of Residents' Trips

Extraction of Temporal Characteristic of Residents' Trips. The temporal characteristic of residents' trips is represented by the fraction of the trip volume at different times of the day, which is defined as the trip volume in each hour divided by the trip volume for the whole day, shown by

$$F_i = \frac{Q_i}{\sum_{i=0}^{23} Q_i} \times 100\% \quad (4)$$

where F_i is the fraction of the trip volume in the ith hour and Q_i is the trip volume in the ith hour.

In this study, trip generation within an hour was identified. If the mobile phone user had crossed more than one TAZ within an hour, a mobile phone user trip was generated within that hour. On the other hand, if the mobile phone user stayed in a single TAZ during the hour, no trip was generated for that mobile phone user. In this way, both the trip volume and the trip generation in each hour were estimated.

Extraction of Spatial Characteristics of Residents' Trips. To extract spatial characteristics of residents' trips, this study first extracted the OD matrices and plotted the desire lines of residents' trips. Based on the extracted OD matrices and desire lines, other spatial characteristics were then extracted, including: 1) ratios of the trips that entered and left the 3rd ring road at the morning peak and evening peak; 2) trip productions and attractions of administrative districts; and 3) job–housing ratios of TAZs.

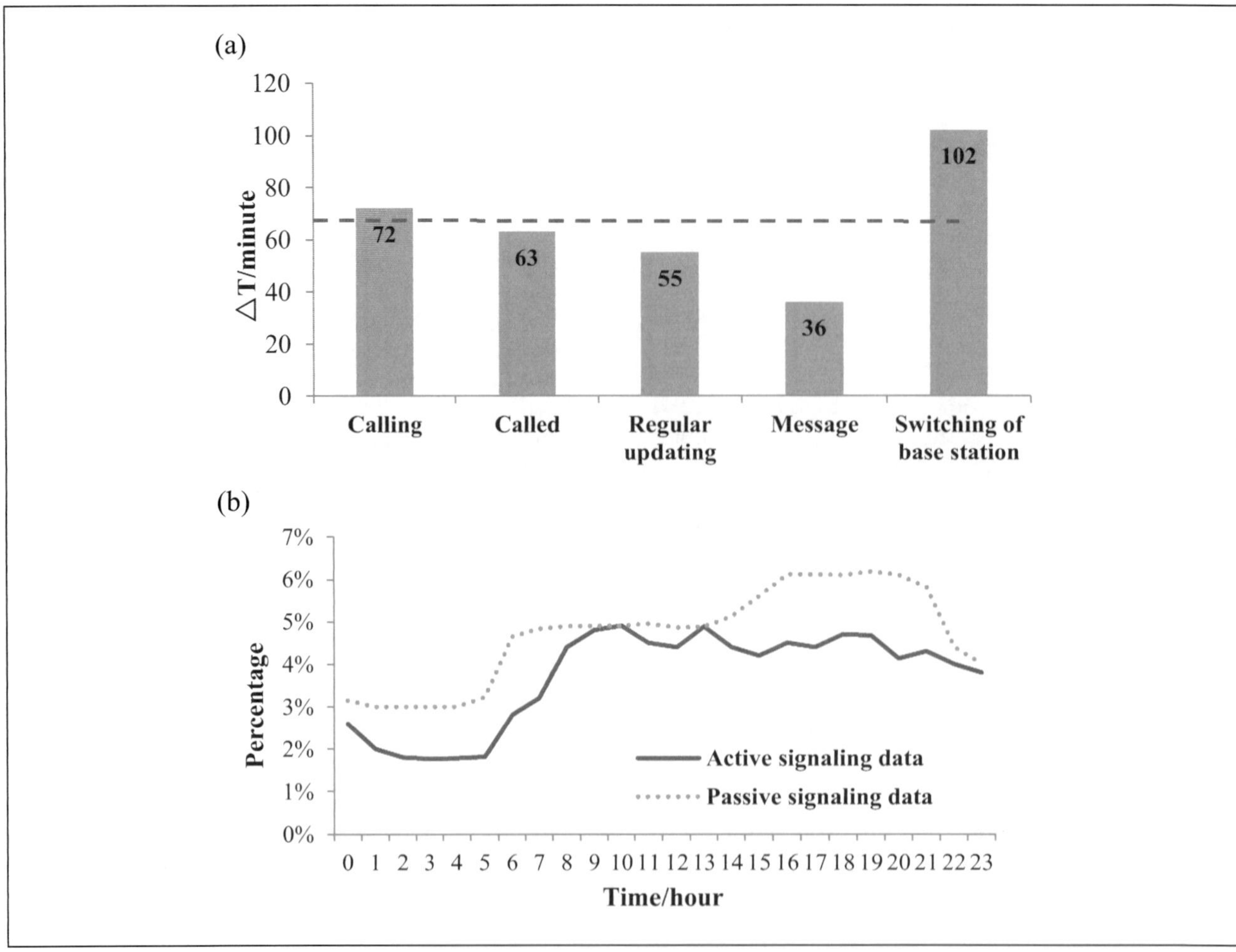

Figure 1. Characteristics of the trigger modes of the cellular signaling data, showing (*a*) time interval of each event and (*b*) time distribution of active modes and passive mode.

Extraction of OD Matrices and Desire Lines. The OD matrix includes the trip volume between each OD pair, by which the desire lines can be obtained. OD matrices and desire lines can describe the volume and the direction of trips.

The identification of ODs was in accordance with the temporal characteristic of residents' trips for the subject city. The temporal characteristic of residents' trips varies from city to city, and largely depends on the defined urban functions. As the real data for this study were collected in Beijing, China, the identification of ODs was virtually based on the temporal characteristic of residents' trips in Beijing. Commuting was the main trip purpose of the residents in Beijing, so the average travel times of commuting trips each day have to account for both directions, going to work and returning from work. Therefore, the OD of each of the two trips, going to work and returning from work, was identified. Figure 2 shows the temporal characteristic of residents' trips in Beijing, which was obtained from the 4th Comprehensive Transport Survey Report (*21*).

It is clear from Figure 2 that the peaks for trips are between 07:00 and 09:00 and between 16:00 and 18:00, and the trough for trips is between 22:00 and 05:00. The percentage of the total trip volume between 22:00 and 05:00 is lower than 5%. Based on the above results, the rule for identifying the OD of residents' trips was designed and applied. For the going to work trip, the TAZ where the mobile phone user stayed between 22:00 and 05:00 was identified as "O" (home), and the TAZ where the mobile phone user stayed between 09:00 and 11:00 was identified as "D" (work). For the same mobile phone user, if his or her "O" and "D" were different, a trip would be generated for this user. Similarly, for the returning from work trip, the TAZ where the mobile phone user stayed between 14:00 and 16:00 was identified as "O" (work) and the TAZ where the mobile phone user stayed between 22:00 and 05:00 was identified as

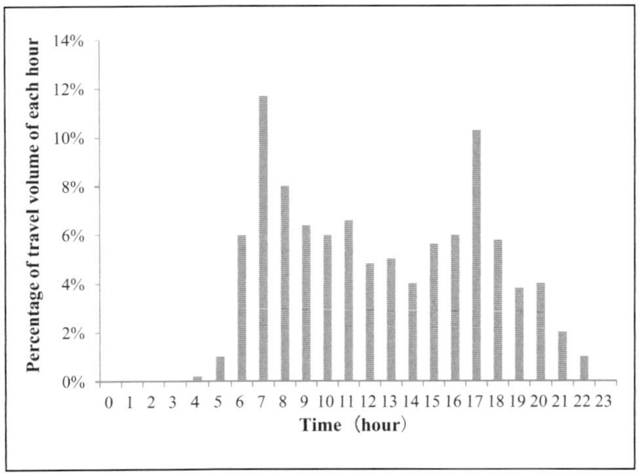

Figure 2. Temporal characteristic of residents' trips in Beijing.

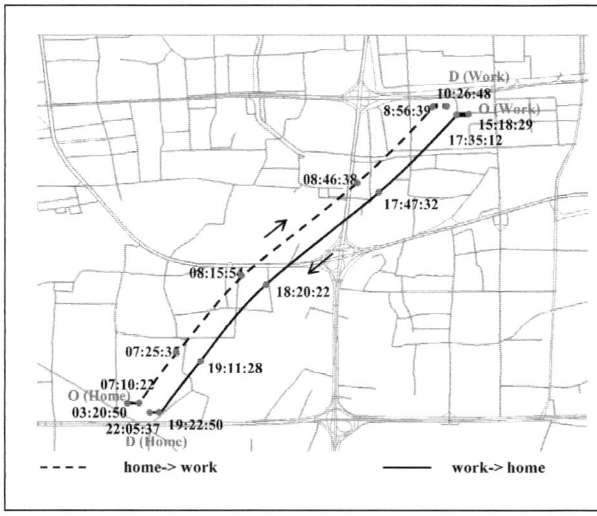

Figure 3. Identification of the OD of the selected resident's trips.

"D" (home). In this way, the OD matrices of the morning peak and evening peak could be established. Because the average commuting trips each day are two way, the commuting OD matrix for the whole day was approximately equal to the sum of the OD matrices for the morning peak and evening peak.

About 19% of users had different morning and afternoon/evening jobs in this study. The data from users who had different morning and afternoon/evening jobs were excluded, and the study only included analysis for users who had the same job in the morning and later in the day. To illustrate the process for identifying the OD of residents' trips, this study selected the records of one mobile phone user and mapped the user onto study regions with a GIS tool, as shown in Figure 3.

Trip Production and Attraction of Administrative Districts. The trip production and attraction of administrative districts represent the spatial distribution of the trip volume. Because the collected data in this study did not include all the residents in Beijing, the extracted OD matrices for the whole day were expanded based on each region's population to obtain the total trip productions and attractions for all residents. The trip productions and attractions for all residents were calculated using

$$P_i = \frac{\sum_{j=1}^{n} O_iD_j}{N} \times \text{pop}_i \tag{5}$$

$$A_i = \frac{\sum_{j=1}^{n} O_jD_i}{N} \times \text{pop}_i \tag{6}$$

where

P_i (A_i) is the trip production (trip attraction) of the ith administrative district,

O_iD_j (O_jD_i) is the OD flow between the ith (j^{th}) and j^{th} (i^{th}) administrative district,

n is the number of administrative districts,

N is the number of mobile phone users of the ith administrative district, and

pop_i is the population of the ith administrative district.

Ratio of Trips Entering and Leaving the 3rd Ring Road at Peak Hours. The definition of peak hours in this study is consistent with that of the 4th Comprehensive Transport Report. The period of 07:00 to 09:00 is defined as the morning peak and 17:00 to 19:00 is defined as the evening peak (*21*). Based on the extracted OD matrix of the morning peak, the trip volume and the ratio of trips of entering or leaving the 3rd ring road were both estimated. By using the same calculation method, the trip volume and the ratio of trips that entered and left the 3rd ring road at the evening peak were also estimated.

Job–Housing Ratios of TAZs. The job–housing ratio is an important indicator that describes the job−housing balance of a TAZ. Analysis of the job−housing balance is helpful for improving urban land use policy. The job−housing ratios of TAZs were calculated (*22*) by

$$\text{JR}_i = J_i/R_i \tag{7}$$

where JR_i is the job−housing ratio of the ith TAZ, J_i is the number of jobs of the ith TAZ, and R_i is the number of the ith TAZ's residents that have jobs.

In this study, the identification of home and work locations of mobile phone users was based on the identification of the OD of the morning trip, such as the going to work trip. The origin of the morning trip was the home location of the mobile phone user and the destination was the work location. So J_i and R_i in Equation 7 could be obtained from the OD matrix of the morning

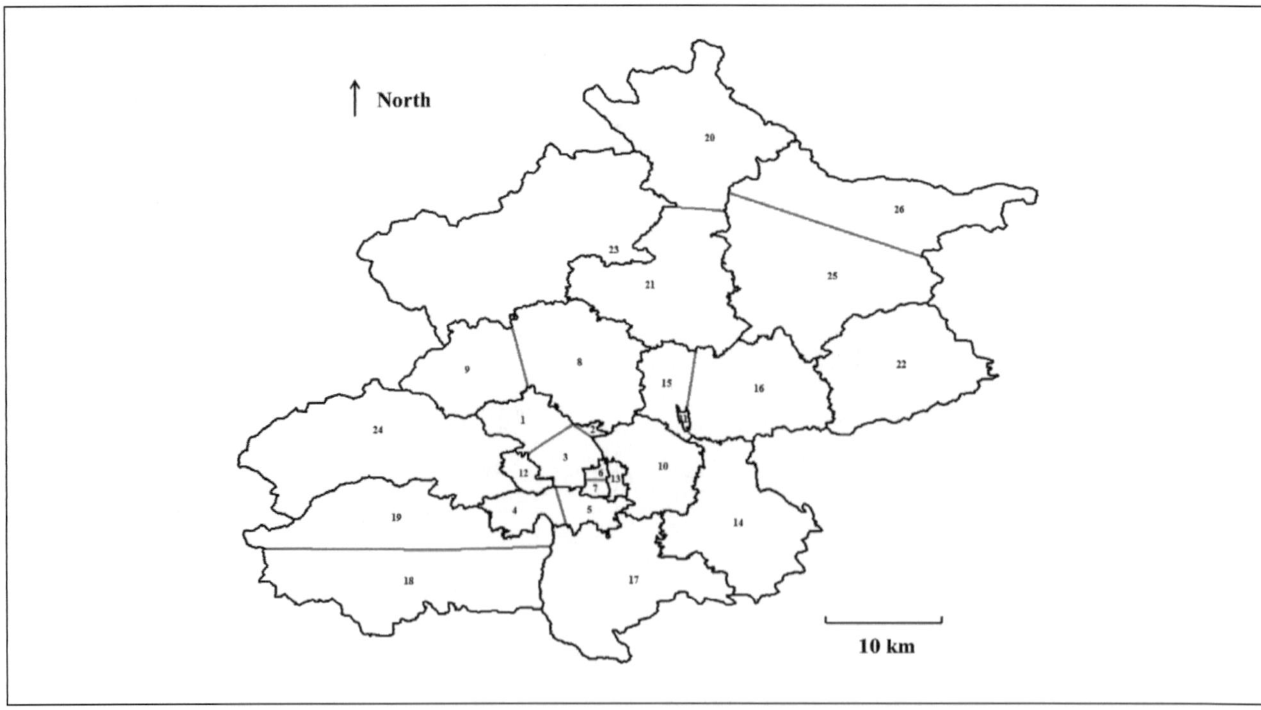

Figure 4. Division of TAZs in Beijing.

peak, J_i was the trip attraction of the ith TAZ, and R_i was the trip production of the ith TAZ.

Case Study

Introduction to Case Study

To examine the feasibility of the proposed method, the cellular signaling data collected in Beijing were used in the case study. After preprocessing of the data, the number of passive signaling data used in the case study was 2.3 million. In accordance with the number of mobile phone users and the area of each administrative district, the 16 administrative districts of Beijing were divided into 26 TAZs, as shown in Figure 4. Based on the TAZs, the temporal and spatial characteristics of residents' trips were extracted, and these were subsequently compared with those of the 4th Comprehensive Transport Survey Report (*21*). The 4th Comprehensive Transport Survey was conducted in 2010; it includes household survey, public transport survey, and screen line survey. It is an important data source for analyzing the characteristics of urban traffic development and for predicting traffic development in the future (*23*).

Temporal and Spatial Characteristics of Residents' Trips

Temporal Characteristic of Residents' Trips. By using the proposed method, the fraction of the trip volume for each

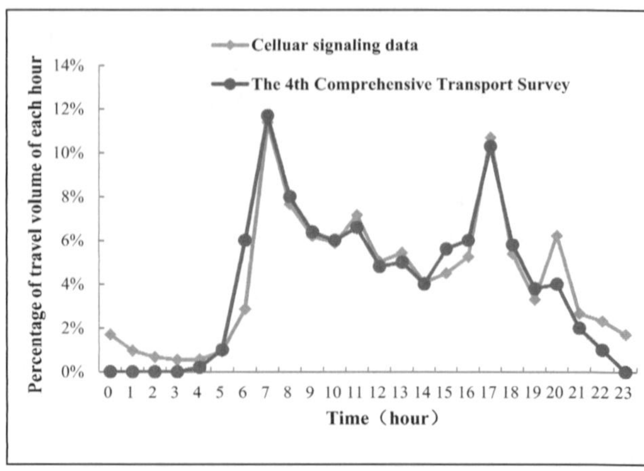

Figure 5. Comparison of the temporal characteristic of residents' trips from two different data sources.

hour was calculated. The result was compared with that of the 4th Comprehensive Transport Survey Report (*21*). Figure 5 shows that the two curves have a high consistency; the correlation coefficient is as high as 0.942. Between 20:00 and 23:00, the trip volume estimated was observed to be higher than that of the 4th Comprehensive Transport Survey Report. It was noted that some trips were delayed to after 20:00 because of the implementation of the no-driving policy based on the tail license number (*24*). As a result, the trip volume increased after

Table 2. Sample of OD Matrix for the Whole Day

O \ D	1	2	3	4	5	6	7	...	26
1	224	12	169	20	56	26	16	...	0
2	15	75	77	9	29	8	12	...	0
3	239	133	581	268	725	329	301	...	1
...	...	...	...					...	...
26	0	0	2	0	0	3	0	...	1

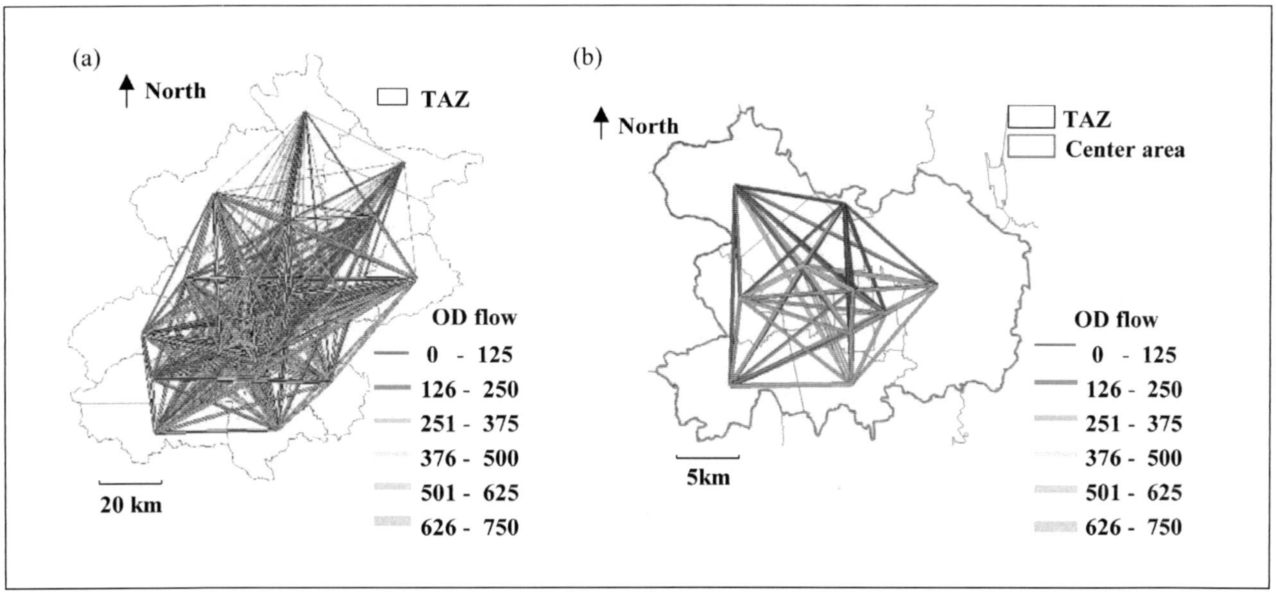

Figure 6. Desire lines for the whole-day OD flow, showing (*a*) all regions of Beijing and (*b*) the center area of Beijing.

20:00. The extracted temporal characteristic based on the cellular signaling data could identify this phenomenon which could not be captured by the 4th Comprehensive Transport Survey.

Spatial Characteristics of Residents' Trips

OD Matrix and Desire Lines. The OD matrices for the morning peak, the evening peak, and the whole day were obtained by using the proposed method. Table 2 shows part of the OD matrix for the whole day. Based on the OD matrices, the desire lines were plotted accordingly. Figure 6*a* shows the desire lines for the whole-day OD flows of 26 TAZs. Figure 6*b*, a zoomed-in map, shows the desire lines of the center area of Beijing, where the red line area is the center area. The center area was defined according to the population density and economic activities of the city (*21*), which includes the Hai Dian District, Xi Cheng District, Dong Cheng District, Chao Yang District, Feng Tai District, and Shi Jingshan District.

Ratios of Trips Entering and Leaving the 3rd Ring Road at Peak Hours. Using the proposed method, the ratios of trips that entered and left the 3rd ring road at peak hours were estimated. Figure 7*a* shows the result of the morning peak, at which point the difference between the result and that of the 4th Comprehensive Transport Survey (*21*) is only 0.1. Figure 7*b* shows the result of the evening peak, which is almost the same as that of the 4th Comprehensive Transport Survey (*21*).

Trip Production and Attraction of Administrative Districts. The trip productions and attractions were derived based on the cellular signaling data of each administrative district in Beijing, as shown in Figure 8. The results were compared with those of the 4th Comprehensive Transport Survey (*21*). The correlation coefficient of the trip productions of the two data sources is 0.932 and the correlation coefficient of the trip attractions is 0.953.

Job–Housing Ratio of TAZs. The job–housing ratios of TAZs were calculated, as shown in Figure 9. This shows

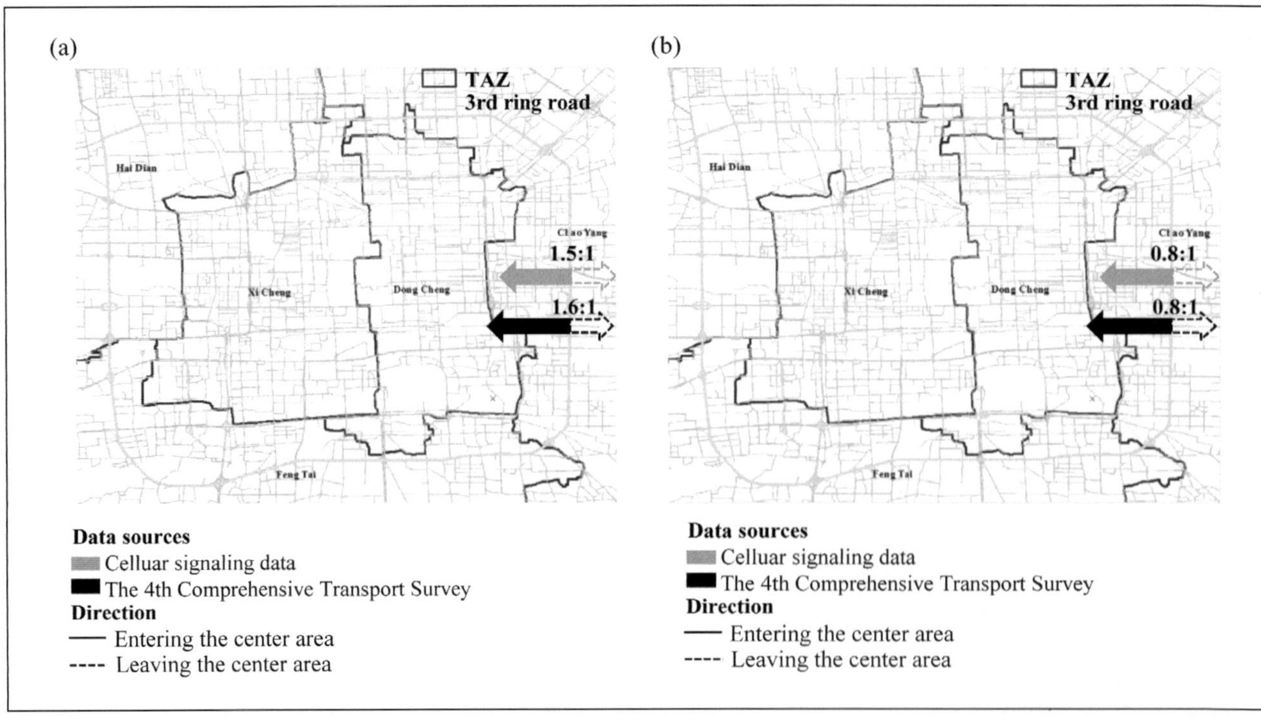

Figure 7. Comparison of the ratio of trips entering and leaving the 3rd ring road at peak hours from two different data sources: (*a*) morning peak and (*b*) evening peak.

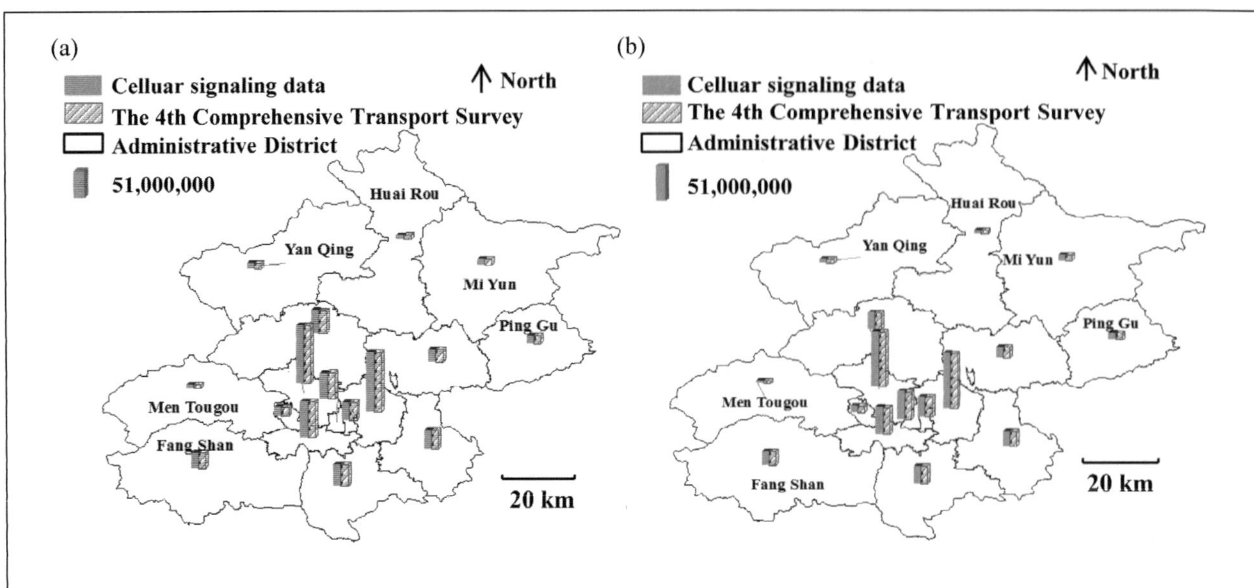

Figure 8. Comparison of trip production and attraction from two different data sources: (*a*) trip production and (*b*) trip attraction.

that the job–housing ratios of the center area are almost balanced, and the job–housing ratios of Tong Zhou, Da Xing, and Shun Yi are higher than those of other rural areas. Because the three administrative districts were planned as the satellite cities of Beijing with a relatively higher level of economic activities, employment in these regions was higher than that of other rural areas.

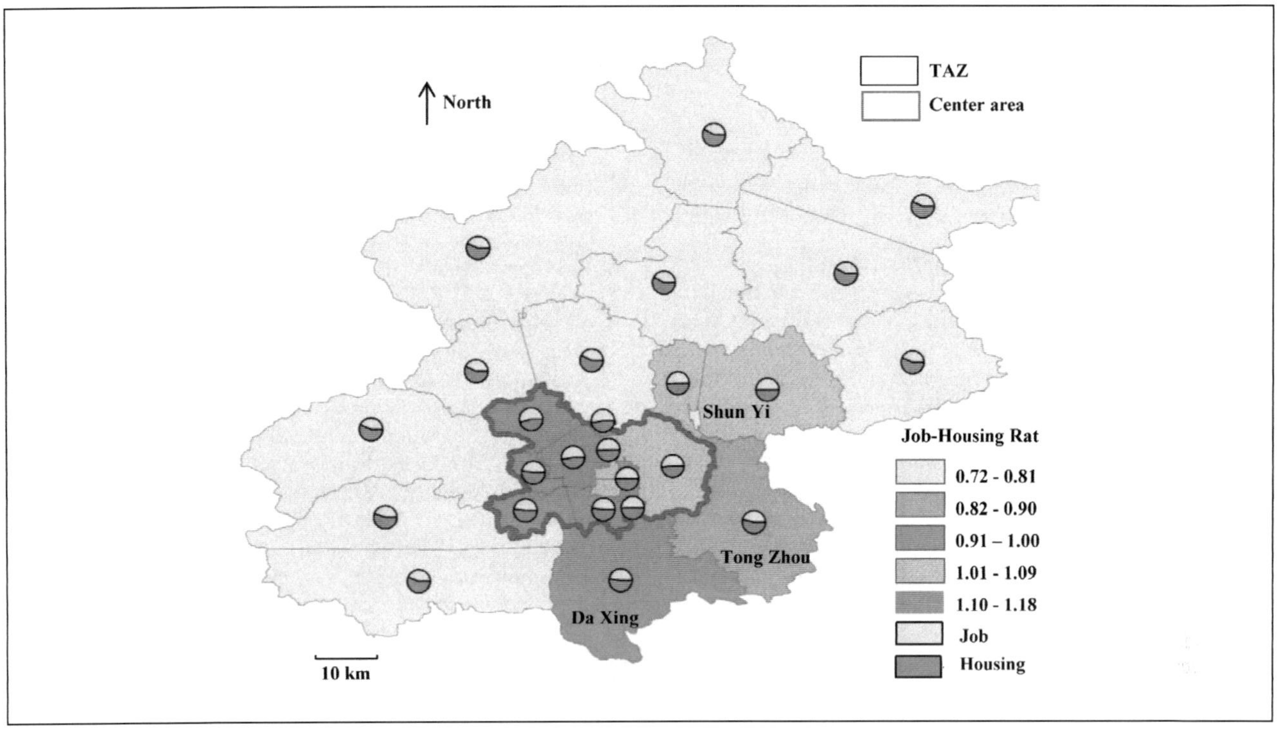

Figure 9. Job–housing ratios of TAZs.

Summary and Recommendations

This study developed a method for extracting temporal and spatial characteristics of residents' trips by using cellular signaling data. The paper used real data from Beijing in the case study to examine the feasibility of the method. The main findings of the research can be summarized as follows:

1) The data triggered by the passive mode can ensure accurate identification of the OD of residents' trips. This is because the time interval of the passive mode was shorter than the average time interval of all active modes and the amount of data triggered by the passive mode was more stable than that of active modes.
2) The OD of the trips was identified based on the temporal characteristic of residents' trips for the city.
3) A method was developed that was able to extract valid temporal and spatial characteristics of residents' trips and this was applied to the data from Beijing. The extracted results were highly correlated with those of the 4th Comprehensive Transport Survey as: *a*) the correlation coefficient of the temporal characteristic of residents' trips was 0.942 and *b*) the correlation coefficients of the trip productions and trip attractions were both higher than 0.9.
4) To develop some useful recommendations for land use planning for Beijing, the job–housing balance of the regions was analyzed based on the extracted OD matrices.

Limited by the data quantity, the method proposed in this paper has room for improvement. Further studies will focus on: 1) extracting more travel data, such as trip chains and trip modes, and 2) combining the cellular signaling data with other data sources to extract more accurate trip information, such as the float car data and the integrated circuit (IC) card data from transit systems.

Acknowledgments

This research was supported by the Natural Science Foundation of China (NSFC) # 51678045 and 51578052. This research was supported in part by the National Science Foundation (NSF) under grants #1137732.

Author Contributions

Study conception and design: Zufen Li, Guohua Song; data collection: Yong Gao, Yizheng Wu; analysis and interpretation of results: Zufen Li, Lei Yu, Dapeng Gong; draft manuscript preparation: Zufen Li, Lei Yu. All authors reviewed the results and approved the final version of the manuscript.

References

1. Phithakkitnukoon, S., T. Horanont, G. D. Lorenzo, R. Shibasaki, and C. Ratti. Activity-Aware Map: Identifying Human Daily Activity Pattern Using Mobile Phone Data. *Proc., 1st International Conference on Human Behavior Understanding*, Istanbul, Turkey, 2011, pp. 14–25.
2. Henpraserttae, A., S. Thiemjarus, and S. Marukatat. Accurate Activity Recognition Using a Mobile Phone Regardless of Device Orientation and Location. *Proc., 2011 International Conference on Body Sensor Networks*, Dallas, TX, IEEE, 2011, pp. 41–46.
3. Brezmes, T., J. Gorricho, and J. Cotrina. Activity Recognition from Accelerometer Data on a Mobile Phone. *Proc., International Work-Conference on Artificial Neural Networks*, ICANN, Salamanca, Spain, Springer, Berlin, Heidelberg, 2009, pp. 796–799.
4. Caceres, N., J. P. Wideberg, and F. G. Benitez. Deriving Origin Destination Data from a Mobile Phone Network. *IET Intelligent Transport Systems*, Vol. 1, No. 1, 2007, pp. 15–26.
5. White, J., and I. Wells. Extracting Origin Destination Information from Mobile Phone Data. *Proc., 11th International Conference on Road Transport Information and Control*, London, England, UK, 2002.
6. Caceres, N., L. M. Romero, and F. G. Benitez. Inferring Origin–Destination Trip Matrices from Aggregate Volumes on Groups of Links: A Case Study Using Volumes Inferred from Mobile Phone Data. *Journal of Advanced Transportation*, Vol. 47, No. 7, 2013, pp. 650–666.
7. Iqbal, M. S., F. Choudhury, P. Wang, and M. González. Development of Origin–Destination Matrices Using Mobile Phone Call Data. *Transportation Research Part C: Emerging Technologies*, Vol. 40, No, 1, 2014, pp. 63–74.
8. Calabrese, F., G. D. Lorenzo, L. Liu, and C. Ratti. Estimating Origin-Destination flows using opportunistically collected mobile phone location data from one million users in Boston Metropolitan Area. *IEEE Pervasive Computing*, Vol. 10, No, 4, 2011, pp. 36–44.
9. Yoo, B. S., S. P. Kang, K. Chon, and S. Kim. Origin-Destination Estimation Using Cellular Phone BS Information. *Journal of the Eastern Asia Society for Transportation Studies*, Vol. 6, 2005, pp. 2574–2588.
10. Sohn, K., and D. Kim. Dynamic Origin–Destination Flow Estimation Using Cellular Communication System. *IEEE Transactions on Vehicular Technology*, Vol. 57, No. 5, 2008, pp. 2703–2713.
11. Alexander, L., S. Jiang, M. Murga, and M. Gonzálezet. Origin–Destination Trips by Purpose and Time of Day Inferred from Mobile Phone Data. *Transportation Research Part C: Emerging Technologies*, Vol. 58, 2015, pp. 240–250.
12. Kung, K. S., K. Greco, S. Sobolevsky, and C. Ratti. Exploring Universal Patterns in Human Home-Work Commuting from Mobile Phone Data. *PLoS One*, Vol. 9, No. 6, 2014, pp. 1–15.
13. Schlaich, J., T. Otterstätter, and M. Friedrich. Generating Trajectories from Mobile Phone Data. Presented at 89th Annual Meeting of the Transportation Research Board, Washington, D.C., 2010, pp. 2–17.
14. Hoteit, S., S. Secci, S. Sobolevsky, C. Ratti, and G. Pujolle. Estimating Human Trajectories and Hotspots through Mobile Phone Data. *Computer Networks*, Vol. 64, No. 4, 2014, pp. 296–307.
15. Wang, H., F. Calabrese, G. D. Lorenzo, and C. Ratti. Transportation Mode Inference from Anonymized and Aggregated Mobile Phone Call Detail Records. *Proc., 13th International IEEE Conference on Intelligent Transportation Systems*, Funchal, Portugal, IEEE, 2010, pp. 3–8.
16. Reddy, S., M. Mun, J. Burke, D. Estrin, M. Hansen, and M. Srivastava. Using Mobile Phones to Determine Transportation Modes. *ACM Transactions on Sensor Networks (TOSN)*, Vol. 6, No. 2, 2010, pp. 1–27.
17. Widhalm, P., P. Nitsche, and N. Brändie. Transport Mode Detection with Realistic Smartphone Sensor Data. *Proc., 21st International Conference on Pattern Recognition (ICPR)*, Tsukuba, Japan, IEEE, 2012, pp. 573–576.
18. Manzoni, V., D. Manilo, K. Kloeckl, and C. Ratti. *Transportation Mode Identification and Real-Time CO2 Emission Estimation Using Smartphones.* Technical Report. SENSEable City Lab, Massachusetts Institute of Technology, Cambridge, MA, 2010.
19. Byon, Y., B. Abdulhai, and A. Shalaby. Real-Time Transportation Mode Detection via Tracking Global Positioning System Mobile Devices. *Journal of Intelligent Transportation Systems*, Vol. 13, No. 4, 2009, pp. 161–170.
20. Witayangkurn, A., T. Horanont, Y. Sekimoto, and R. Shibasaki. Trip Reconstruction and Transportation Mode Extraction on Low Data Rate GPS Data from Mobile Phone. *Proc., International Conference on Computers in Urban Planning & Urban Management, CUPUM*, Utrecht, Netherlands, 2013, pp. 1–19.
21. Beijing Transportation Research Center. *Beijing the 4th Comprehensive Transport Survey Summary Report.* Technical Report. Beijing Municipal Commission of Transport, Beijing, China, 2012.
22. Sun, B., X. Pan, and Y. Ning. Analysis on Influence of Job-Housing Balance on Commute Travel in Shanghai. *Urban Planning Forum*, Vol. 1, 2008, pp. 77–82.
23. Li, C., J. Guo, Z. An, and H. Zhao. Comments on Urban Comprehensive Transportation Survey: Learn Lessons from the 5th Beijing Comprehensive Transportation Survey. *Urban Traffic*, No. 2, 2016, pp. 29–34.
24. Beijing Municipal Government. *Circular on Rotation of Tail Number Plate of No-Driving Vehicles in Regional Rush Hours on Working Days.* http://www.ebeijing.gov.cn/Elementals/Bulletin/t1163981.htm. Accessed May 25, 2016.

The Standing Committee on Urban Transportation Data and Information Systems (ABJ30) peer-reviewed this paper (18-01076).

Article

TRR
JOURNAL OF THE TRANSPORTATION RESEARCH BOARD

Transportation Research Record
2018, Vol. 2672(42) 91–102

DOI: 10.1177/0361198118774170
journals.sagepub.com/home/trr
SAGE

Exploring the Relationships between the Non-Work Trip Frequency and Accessibility Based on Mobile Phone Data

Wenxiang Li[1], Ye Li[1], Xuegang (Jeff) Ban[2], Haopeng Deng[1], Hanyu Shu[1], and Dongcan Xie[3]

Abstract

Non-work trips have increased remarkably over recent years, even exceeding the amount of daily work-related trips. It is therefore important to understand the relationship between the non-work trip frequency and accessibility of the travelers. However, the conclusions of previous studies were inconsistent. The causes of this inconsistency have been speculated to be the limitations of sample size, aggregated accessibility, and home-based trips. To explore the actual relationships between the non-work trip frequency and accessibility, this study employs a more reliable method based on mobile phone data. First, positioning data of mobile phone users was used to identify activity points, home and workplace locations, and non-work trips of the users. Furthermore, the socioeconomic attributes and preferences of users were extracted from mobile phone user feature data. Second, accessibility was measured at the individual level based on a real-time travel-cost estimation method. Subsequently, the negative binomial regression model was employed to study the effects of accessibility, socioeconomic attributes, and preferences of users on their travel frequencies. A case study of the central business district of Lujiazui in Shanghai was conducted to prove the feasibility of the proposed method. The results show that the non-work trip frequency increases with accessibility. Moreover, users with higher consumption levels are more sensitive to accessibility when they decide the frequency of the non-work trips. However, the influence of accessibility on non-work trip frequency tends to weaken with the decrease of accessibility. These findings contribute to a more reliable transportation policy design and planning.

With the improvement of living standards, non-work trips (i.e. shopping, recreation, personal business, etc.) have increased remarkably during recent years. According to the Fifth Shanghai Comprehensive Transport Survey, the amount of daily non-work trips within Shanghai was 27 million, accounting for 52% of the total daily trips conducted in Shanghai during 2014. In contrast to work-related trips, non-work trips are elastic trips where the time of the activity is flexible, and the destination may be variable. Therefore, it is important to reveal the driving factors of these non-work trips and to investigate how these factors influence travel frequencies.

Many related studies (*1*, *2*) show that greater accessibility leads to more travel. In other words, people may travel more frequently if they live in more accessible locations since opportunities for activities can be taken at a comparatively lower cost. For example, trip frequencies might increase if distances to the relevant destinations are short (*3*). Otherwise, people may find it too time-consuming, cumbersome, and expensive to visit more distant locations regularly. Therefore, the "distance decay" phenomenon affects the attractiveness of large centers (*4*). However, several studies show inconsistent results, reporting instead that accessibility is independent of the trip or tour frequency. For example, Handy (*5*) showed that accessibility does not affect the frequency of shopping trips. Kitamura et al. (*6*) reported that

[1]Key Laboratory of Road and Traffic Engineering of the Ministry of Education, College of Transportation Engineering, Tongji University, Shanghai, P. R. China
[2]Department of Civil and Environmental Engineering, University of Washington, Seattle, WA
[3]Shanghai Urban Real Estate Appraisal Co., Ltd, Shanghai, P.R. China

Corresponding Author:
Address correspondence to Ye Li: jamesli@tongji.edu.cn

accessibility is neither significantly nor positively related to the total number of trips and tours.

The causes of this inconsistency have been speculated to be the data sources and the measures of accessibility. The information about travel and household/individual is based on household travel diary surveys, which are typically conducted on a weekday every 5 to 10 years and tend to under-report short trips. Furthermore, the sample can only cover 0.5 to 1% of all households in the city (*4*). With such limited time span and sample size, it is difficult to obtain an accurate relationship between accessibility and travel behaviors, particularly for non-work trips. In addition, the accessibility measures are mostly based on aggregate models (e.g. the gravity model or the logit destination choice models) developed from household survey data along with zone-based employment and transportation network data. As a result, people living in the same zones have the same accessibilities which neglect the heterogeneity between individuals. In addition, the travel time and distance between zones cannot always represent the actual travel time and distance of every person within this zone. Finally, all trips considered in the previous studies are home-based trips. However, a large portion of non-work trips in fact start from workplaces after work. Thus, the locations of workplaces should also be considered when calculating accessibility measures.

In this context, this study aimed to clarify the relationships between non-work trip frequency and accessibility using a more reliable method based on mobile phone data. This type of data contains spatial-temporal locations of millions of mobile phone users over an extended period and provides an alternative way of collecting multi-day travel data (*7*). To focus on a specific destination of the non-work trip, the central business district (CBD) of Lujiazui in Shanghai was selected as the study area since most people typically go to the city center for maintenance or discretionary activities. Non-work trip frequency in this study refers to the number of times a person visits the Lujiazui CBD for non-work purposes per month. Accessibility measures include accessibility by both private vehicle (auto) and public transit (transit) from both the home location and the workplace location. Additionally, given the actual travel time from home or workplace to the CBD for every person, the accessibility was calculated at the individual level rather than the aggregate level. In this way, shortcomings of previous studies could be overcome. Consequently, this study achieved an improved understanding and a clearer conclusion of the relationship between non-work trip frequency and accessibility.

The remainder of this paper is organized as follows. The second section reviews the literature related to this study. The data sources and methods of non-work trips identification, accessibility measures, and regression analyses are presented in the third section. The fourth section describes the results and presents further discussions of the relationship between non-work trip frequency and accessibility. The final section discusses the conclusions and limitations of this study and suggests future directions.

Literature Review

Trip Identification using Mobile Phone Data

A large amount of data is accessible from mobile phones, such as call detail records, locations, and trajectories of the mobile phone user. These data have motivated researchers from various fields to study both human activity and mobility. Gonzalez et al. (*8*) examined the six-month trajectories of 100,000 mobile phone users and found a high degree of regularity in their trajectories. Song et al. (*9*) reported that predictability in human mobility is stable at 93% for all regularly traveled distances above 10 km, which is consistent with the study by Gonzalez et al. Compared with travel survey data, mobile phone trace data provides new opportunities to examine individual mobility from an alternative perspective due to the lower collection cost, larger sample size, higher update frequency, and broader spatial and temporal coverage (*10*). As a result, many researchers have already recognized the importance of mobile phone data for transport analyses, such as the estimation of traffic parameters (i.e. OD matrices, traffic flow, travel time, and travel speed), modeling of individual activities and the exploration of interactions between urban environment and individual activities (*11*).

Previous research about the application of mobile phone data on trip identification can mainly be classified into three aspects: trip ends identification, trip mode detection, and trip purpose identification (*12*). The first aspect aims to distinguish the trip end from intermediate points of a trip. Many attributes, such as speed, dwelling duration, and distance between consecutive points are generally used as thresholds for the trip ending. For example, a trip end is detected if the individual spends more than 30 minutes within a range of 200 meters (*13*). The second aspect aims to detect the travel modes, and this is typically based on speed. Commonly used criteria include the average and maximum speed, the statistical mode of speed, and the average acceleration of each mode (*14*). Machine learning (*15*), probabilistic methods (*16*), and GIS tools (*17*) have also been applied to determine the utilized mode of transportation. Similar to mode detection, the identification of purpose has also been accomplished with three methods: criterion-based, probabilistic, and machine learning (*18*). It is generally the most difficult aspect since the trip purposes typically vary from person to person.

Accessibility Measurements

Accessibility can be defined and measured with different methods. These include well-known definitions (*19*) such as the ease of reaching any activity area using a specific transport system (*20*), potential opportunities for interaction (*21*), the freedom of individuals to decide whether to participate in different activities (*22*), and the overall benefits provided by a given transport system (*23*). According to Baradaran et al. (*24*) and Handy et al. (*25*), accessibility measures have often been classified into four categories: (a) travel-cost approach, (b) isochrones approach, (c) gravity-based approach, and (d) utility-based approach. A comparison of the four accessibility measures is presented in Table 1. Different measures have their own advantages and disadvantages. There is a need to select the appropriate approach according to the research interests, research objects, and data acquisition.

Relationships between Trip Frequency and Accessibility

Many studies have investigated the influence of accessibility on individual activities, such as vehicle ownership decisions, activity points, trip distances, activity participation, and trip frequencies (*26*). Næss (*27*) investigated the links between residential location and travel behavior based on data from a travel diary study of the metropolitan area of Copenhagen, Denmark. Multivariate regressions were adopted to study the relationship between trip frequencies for different activities and accessibility of residence relative to the downtown area. Seo (*4*) examined the effect of accessibility of urban and suburban locations on tour frequencies for both maintenance and discretionary activities of elderly inhabitants as well as of the non-elderly, using household travel survey data collected within the metropolitan area of Seoul, Korea. Ordered probit and binary probit models were used for maintenance and discretionary trips, respectively. The explanatory variables included: auto/transit accessibility by residential location and individual auto availability, household income level, age, gender, available time budget for non-work activities, and the number of drivers. Kockelman (*28*) estimated total non-work, home-based VKT (Vehicle Kilometers Traveled) per household to be linear and logit functions of accessibility, land-use mixing, and land-use balance, based on travel data from travel surveys conducted in the San Francisco Bay area in 1990. The results indicated that accessibility of opportunities is by far the best predictor of travel behaviors.

In summary, most of these studies have been based on zone-based travel survey data, which has limitations in terms of sample size, spatial and temporal scales, and update frequency. Furthermore, accessibility measures can be improved with more disaggregated data at a larger scale. The emerging techniques of trip identification using mobile phone data provide researchers with new opportunities to explore further the relationships between trip frequency and accessibility. However, the use of mobile phone data to investigate these questions has hardly been applied in previous research to date. Therefore, this study aimed to contribute a new method and arrive at a better explanation for these questions.

Data and Methodology

Data Preparation

Mobile Phone Positioning Data. Mobile phone positioning data is generated through either usage (calling, texting, or surfing the Internet) or physical movement of mobile phone handsets. Position information is obtained from a cellular network comprised of cell phone towers, which enables mobile phones to communicate with each other (*7*). The area served by a tower is called a cell. Initially, the position of a mobile phone is recorded as the Cell-ID to which it is currently connected, and each Cell-ID has its own latitude and longitude.

For the purpose of this study, China United Networks Communications Corporation of Shanghai (China Unicom) provided a dataset of mobile phone positioning data from a randomly drawn sample of 417,381 anonymous users. It was collected between March 1 and March 31, 2016. In total, 8,665 distinct positions were found within the dataset. They are actually cell phone towers in the 3G network. The operator reported that its positioning data has an average locational accuracy of 200 meters in urban areas. Therefore, the estimated positions of phone users ensure that the accuracy of related analysis will not be worse than that of traditional household travel surveys, which are based on the unit of "traffic analysis zone". The raw data included many redundant fields and exceptional data which were filtered out and deleted. The effective dataset, partitioned by the mobile identification number of each user (IMEI) and ordered by timestamp, was then obtained (Table 2).

Mobile Phone User Feature Data. To address the lack of socioeconomic and preference attributes within the mobile phone positioning data, the feature information of the corresponding 417,381 users was obtained from China Unicom. This information was collected when the user applied for a mobile phone number or was recorded by the background servers of mobile applications. It included: age, gender, consumption level (i.e. level of personal spending inferred based on phone bill), mobile phone model, mobile phone brand, social circle size (inferred based on contacts), and frequency of visiting different types of websites or apps (e.g. shopping, catering, tourism, finance, etc.). For privacy purposes, the real

Table 1. Comparison of the Four Accessibility Measures

Accessibility measures	Description	Formula	Advantages	Disadvantage
Travel-cost approach	The ease with which any land-use activity can be reached from a location using a particular transport system	$A_i = \sum_{j=1}^{n} \frac{1}{f(c_{ij})}$	• Quite easy to understand and to calculate; • Less demanding on data.	• Neglects variations across individuals; • Highly sensitive to the choice of demarcation area; • Neglects variations in the locations;
Isochrones approach	The number of opportunities that can be reached within a given travel time, distance, or generalized cost	$A_i = \sum_{j=1}^{n} w_j N_j$	• Relatively easy to understand and to calculate; • Easy to visualize.	• Neglects variations across individuals; • Highly sensitive to the size of range and the representation of opportunities; • Neglects variations in the locations;
Gravity-based approach	Potential of opportunities for interaction	$A_i = \sum_{j=1}^{n} \frac{N_j}{f(c_{ij})}$	• Able to differentiate between locations; • Represents the joint effect of transport systems and land-use patterns on accessibility; • Has been widely-used.	• Neglects variations across individuals; • Ambiguity regarding the magnitude of indicators.
Utility-based approach	Expected maximum utility from a random utility considering individual characteristics	$A_i^n = max(U_{ij}^n) = \frac{1}{\mu} ln \sum_{j=1}^{n} e^{\mu(v_j^n - c_{ij}^n)}$	• Captures individual differences • Captures the impact of all modes including auto, transit, and non-motorized options.	• Demands extensive data on locations and individuals' travel behavior.

Note: A_i = the measure of accessibility at location i; n = the number of included locations; $f(c_{ij})$ = the deterrence function; c_{ij} = a variable that represents travel cost between locations i and j; N_j = the opportunities in a zone j; w_j = 1 if $cj < c^*$ and 0 otherwise; c^* = the pre-determined range within which the activity opportunities are counted; U_{ij}^n = the utility of group n selecting the alternative; v_j = some measure reflecting the attraction of the alternative j; μ = a positive scale parameter.

Table 2. Sample Records of Effective Dataset of Positions of Mobile Phones

IMEI	TIMESTAMP	LON	LAT	N
48c3e280f7c301bebe284a368b331eb8	2016/3/01 00:22:12	121.47914	31.22088	1
48c3e280f7c301bebe284a368b331eb8	2016/3/01 02:22:33	121.48192	31.22311	2
...	...	...	...	...
48c3e280f7c301bebe284a368b331eb8	2016/3/01 22:43:32	121.47914	31.22088	18
24ae2f55a921460835ada61d12638013	2016/3/01 00:42:17	121.52317	31.27719	1
24ae2f55a921460835ada61d12638013	2016/3/01 01:05:57	121.52317	31.27719	2
...	...	...	...	...
24ae2f55a921460835ada61d12638013	2016/3/01 23:42:45	121.52261	31.27640	62

Note: IMEI = the mobile identification number of the user; TIMESTAMP = the time when the data record was collected; LON = longitude of the user's estimated location; LAT = latitude latitude of the user's estimated location; N = the serial number of each user's records ordered by TIMESTAMP.

Table 3. Sample Records of Effective Dataset of Mobile Phone User Features

IMEI	AGE	GEN	CL	SCS	VFS	VFC	VFT	VFE
0062f646c2a4f57e289a8e8e01da01c4	49	Male	Low	19	38	1	0	2
48c3e280f7c301bebe284a368b331eb8	30	Female	Middle	28	224	89	0	33
24ae2f55a921460835ada61d12638013	17	Male	High	76	553	30	123	110

Note: IMEI = the mobile identification number of the user; AGE = the age of the user; GEN = gender; CL = consumption level; SCS = social circle size; VFS = frequency of visiting shopping websites; VFC = frequency of visiting catering websites; VFT = frequency of visiting tourism websites; VFE = visiting frequency of visiting entertainment websites.

IMEI of each mobile phone user had been encrypted, but the IMEI available in our dataset is unique to each user, by which the user feature data could be related to the positioning data. After filtering irrelevant attributes, only the attributes possibly related to trip identification remained in the effective dataset (Table 3).

Non-Work Trip Identification

Inference of Individual Activity Points. In this study, the position a person stayed for more than 30 min was defined as an activity point (*29*), which represents the critical anchor point of individual trip. Given the spatio-temporal trajectories of the users from mobile phone positioning data, the activity points of users every day can be inferred. The data processing procedure is listed as follows:

Step 1: Re-ranking the data partitioned by IMEI and ordered by TIMESTAMP;
Step 2: For every user, filtering the repetitive and consecutive positioning data, only retaining the first (enter time) and last (departure time) records of every distinct position i;
Step 3: Calculating the stay time (ST_i) of each position via subtraction between the enter time and the departure time;
Step 4: Calculating the travel time and Euclidean distance between the consecutive positions i and $i+1$, then calculating the rough moving speed (V_i) for drift detection;
Step 5: Filtering the positioning data where $ST_i < 30$ min and $V_i > 50$ m/s, then the activity points of the user are finally obtained.

Identification of the Locations of Home and Workplace. Home and workplace are the two most representative types of activity points. Chen et al. (*30*) found that 97% of homes and 86% of workplaces identified via mobile phone data were within 1 km of the actual locations, respectively. Referring to the literature (*30–32*), a simple algorithm was developed to identify the locations of home and workplace based on the above activity points:

Step 1: Selecting the activity points with enter times between 22:00 p.m. and 6:00 a.m., 9:00 a.m. and 17:00 p.m., respectively, into new datasets $\{HomePoints1\}$ and $\{WorkPoints1\}$;
Step 2: Selecting the activity points from $\{HomePoints1\}$ and $\{WorkPoints1\}$ where the stay time is maximal for every day, respectively, into new datasets $\{HomePoints2\}$ and $\{WorkPoints2\}$; then,

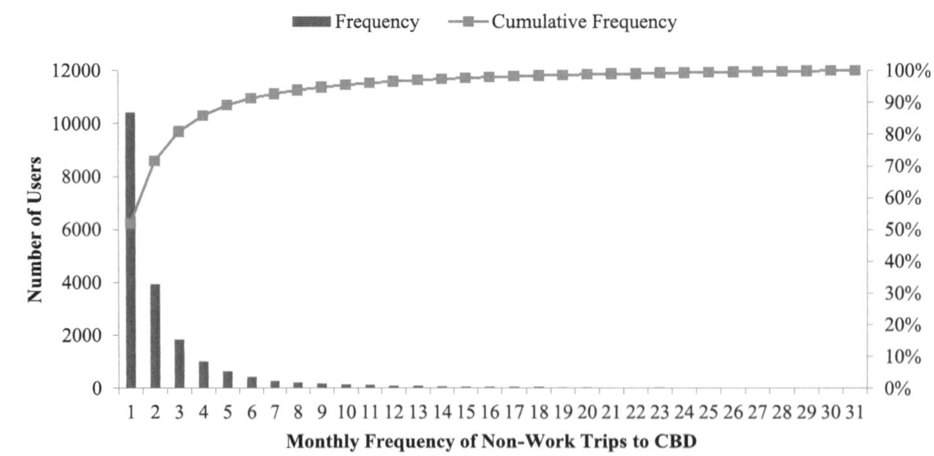

Figure 1. Distribution of the frequency of non-work trips to the CBD.

counting the frequencies of these points during a month;

Step 3: Selecting the activity points as home locations from {*HomePoints2*} where the frequency >20 days per month;

Step 4: Selecting the activity points as workplace locations from {*WorkPoints2*} where the frequency >15 days per month.

As a result, 144,975 users with stable homes and 110,440 users with stable workplaces could be identified from the sample. The number of users who have both stable homes and workplaces was 74,549.

Calculation of the Non-Work Trip Frequency. Generally, most people go to a city center for maintenance activities (shopping and personal business) or discretionary (leisure) activities because the city center offers a variety of facilities that can satisfy their diverse needs. Lujiazui CBD is such a city center in Shanghai near the Huangpu River. Therefore, the Lujiazui CBD of Shanghai was selected as the study area to focus on a specific destination of non-work trips. Consequently, non-work trip frequency in this study refers to the number of times of a person visits the Lujiazui CBD for non-work purposes per month, which could be calculated based on the activity points.

To identify the non-work trips to the Lujiazui CBD among all activities, the users with homes and workplaces located in the Lujiazui CBD were first filtered out, since their activity points are mostly within this area, which could lead to an overestimate of the frequency. As a result, the data of 69,784 users remained. The activity points of those users within the Lujiazui CBD were then selected. Furthermore, the activity points which are consecutive for one day were merged and their stay times were summed. To filter the temporary stay points, the activity points in Lujiazui CBD with stay time exceeding 60 min as non-work trip destinations were selected. Finally, only 20,339 users among the 69,784 users were identified as having been to the Lujiazui CBD for non-work purposes during this month. Since children and the elderly usually travel along with their families, their trip patterns may be unrepresentative. Thus, the data of users whose ages are less than 12 or above 80 were removed to make our results more reliable. As a result, the effective dataset contains a total of 20,259 users. The number of selected activity points during the month was calculated as the non-work trip frequency for each user. The distribution of the monthly frequency is shown in Figure 1.

Accessibility Measures

Definition of Accessibility. According to the concept of accessibility (*33*), there is a distinction between relative accessibility and integral accessibility. Relative accessibility is defined as the degree to which two places (or points) on the same surface are connected, while integral accessibility is defined as the degree of interconnection for a given point with all other points on the same surface (*20*). It is in this integral sense that the term "accessibility" is used in this paper. Since this study only focused on non-work trips to the Lujiazui CBD, accessibility was defined as the ease of reaching the Lujiazui CBD from either home or workplace using a specific transport system (i.e. auto or transit). This was influenced by the travel time from home or workplace to the CBD and the distance–decay coefficient which reflects the degree of unwillingness that people feel toward traveling long distances (*34*).

Estimation of Travel Time. Given the locations of both home and workplace of every user, the travel time to the CBD

Table 4. Descriptive Statistics of Explanatory Variables

Variable	Description	Min	Max	Mean	Std Dev.	Observations
Dependent Variable						
Y	Monthly frequency of non-work trips	1	148	3.40	6.70	20259
Independent Variable – Accessibility						
X_1	Accessibility measure by auto	3E-14	0.83	0.11	0.15	20259
X_2	Accessibility measure by transit	2E-22	0.06	2E-3	5E-3	20259
Independent Variable – Socioeconomic Attributes						
X_3	Age	12	80	38	13.99	20259
X_4	Gender: male	N/A	N/A	N/A	N/A	12721
X_5	Gender: female	N/A	N/A	N/A	N/A	7538
X_6	Consumption level: low	N/A	N/A	N/A	N/A	8307
X_7	Consumption level: middle	N/A	N/A	N/A	N/A	8507
X_8	Consumption level: high	N/A	N/A	N/A	N/A	3445
X_9	Social circle size	0	1260	46.32	53.03	20259
Independent Variable – Preference Attributes						
X_{10}	Frequency of visiting shopping websites	0	36907	367	1172	20259
X_{11}	Frequency of visiting catering websites	0	266	0.19	4.07	20259
X_{12}	Frequency of visiting tourism websites	0	8143	12.34	112.83	20259
X_{13}	Frequency of visiting entertainment websites	0	2308	4.54	44.92	20259

could be obtained using the Route Matrix API of Baidu Map, which computes the travel time and distance between origins and destinations in the real road network. The input parameters include the longitudes and latitudes of the origin (home or workplace) and the destination (the Lujiazui CBD), and the mode of transportation (by auto or transit). The output results are the shortest distance and travel time from origin to destination in real time.

The advantages of this type of method are reflected in three main aspects: (a) there is no need to obtain the road network data of the entire city to calculate the travel time; (b) both distance and travel time are more accurate than the zone-level travel impedance since it considers real-time traffic status of the road network; (c) this method can provide the travel time of any mode of transportation at any time.

Since most non-work trips happen during off-peak periods, the travel times at 10:00 a.m. from Monday to Sunday were calculated using the Route Matrix API. Considering the variation of traffic status, the estimation of travel time was an average value during a week.

Measurement of Accessibility. The accessibility measures used in this study had the following form, as can be frequently found in the accessibility-related literature (*35*).

$$A_i = \sum_{j=1}^{n} \frac{N_j}{e^{\beta T_{ij}}} \tag{1}$$

where A_i represents the accessibility for user i, N_j represents the total amount of attractions at each location j, T_{ij} represents the user i's travel time in minutes from location j to the CBD, and β represents the cost sensitivity parameter, which reflects the degree of unwillingness people feel toward traveling for an extended time. Since there are only two potential origins (home and workplace) and one attraction (the Lujiazui CBD) in this study, the equation can be transformed to:

$$A_i = \frac{1}{e^{\beta T_i^H}} + \frac{1}{e^{\beta T_i^W}} \tag{2}$$

where T_i^H is the travel time of user i from home to the CBD, and T_i^W is the travel time of user i from the workplace to the CBD. Based on Grengs et al. (*34*), the parameter β is substantially higher for non-work trips than it is for work trips (0.22 for non-work compared with 0.068 for work), reflecting a tendency for travelers to take shorter trips for non-work purposes.

Regression Analysis

Selection of the Explanatory Variables. To explore the relationships between non-work trip frequency and accessibility, socioeconomic attributes, and preferences of users, the related explanatory variables were selected based on the data prepared above. Table 4 shows the descriptive statistics of these variables for all 20,259 users in the dataset.

Regression Modeling. The count distribution of the monthly frequency of non-work trips in Figure 1 follows a Poisson shape. However, the data is over-dispersed,

Table 5. The Estimation Results of the Negative Binomial Regression Model

Variable	Description	Estimate	Std Error	Z value	Pr(>|z|)
a	Intercept	8.11E-01	3.10E-02	26.19	<2.00E-16***
X_1	Accessibility measure by auto	2.10E + 00	4.71E-02	44.597	<2.00E-16***
X_3	Age	−1.33E-03	5.41E-04	−2.46	0.013875*
X_4	Gender: male	1.05E-01	1.54E-02	6.82	9.08E-12***
X_6	Consumption level: low	1.60E-01	2.38E-02	6.737	1.61E-11***
X_7	Consumption level: middle	6.10E-02	2.24E-02	2.717	0.006592**
X_9	Social circle size	5.89E-05	1.53E-04	0.384	0.700881
X_{10}	Frequency of visiting shopping websites	3.00E-05	6.10E-06	4.919	8.68E-07***
X_{11}	Frequency of visiting catering websites	2.30E-03	1.74E-03	1.328	0.184317
X_{12}	Frequency of visiting tourism websites	5.81E-05	6.26E-05	0.928	0.353419
X_{13}	Frequency of visiting entertainment websites	5.83E-04	1.52E-04	3.834	0.000126***

Note: Signif. codes: 0 '***' 0.001 '**' 0.01 '*' 0.05 '.' 0.1 ' ' 1.
Null deviance: 21,270 on 20,258 degrees of freedom.
Residual deviance: 18,625 on 20,248 degrees of freedom.
AIC: 92,962, Theta: 1.2848, Std Error: 0.0156.
2 × log-likelihood: −92,938.3940.

with the variance higher than the mean. As a result, the Poisson-Gamma mixture form of the negative binomial (NB) distribution was used to fit the data. This form of the NB distribution allows for heterogeneity within the Poisson parameter, supporting a wider range of variance (i.e. dispersion) among individuals (*36*).

Assuming that the monthly non-work trip frequency obeys NB distribution with parameters r and p, that is, $Y \sim NB\ (r, p)$, then the resulting probability density function can be expressed as

$$\mathrm{P}(\mathrm{Y} = \mathrm{k}|X_1, X_1, \ldots X_{13}) = \binom{k + r - 1}{r - 1} \times p^r \times (1 - p)^k \quad (3)$$

The NB regression model is then derived as follows:

$$\log \mu = a + \sum_{i=1}^{13} b_i X_i \quad (4)$$

where μ represents the mean/expected frequency of non-work trips, a represents the intercept, and b_i represents the coefficient of the independent variable X_i. The maximum likelihood estimation process can generate estimates for all these parameters.

Results and Discussion

Analysis of Model Results

Since the Pearson's correlation between auto and transit accessibility measures is very high (0.934), this paper considers them separately and only presents the auto accessibility model in this section due to limitations of space. Estimation results of the NB regression model are presented in Table 5.

As shown in Table 5, except for the size of social circle and frequency of visiting catering websites and tourism websites, most other explanatory variables are significantly and positively related to the frequency of non-work trips. Age is negatively related to the non-work trips since elderly people tend to stay at home possibly due to physical problems. It can be concluded from the values of coefficients that accessibility is the most important factor of non-work trips among these variables. The better the accessibility, the more frequently people will travel. This result supports the common hypothesis "greater accessibility leads to more travel" for the non-work trips condition.

Further Discussion

It should be noted that the regression model in this study was only used to explain the relationships between the frequency of non-work trips and accessibility, socioeconomic attributes, and preferences of users. It probably cannot predict the trip frequency of a specific user, because individual behaviors are mostly influenced by subjective factors. To explore further such relationships between non-work trip frequency and accessibility of different groups of users, the users were divided into three groups based on their consumption levels: (a) low, (b) middle, and (c) high.

Accessibility by auto was transformed into grades 0 to 10 via normalization, and the average monthly frequencies at different accessibility grades for different groups were calculated. Since the frequency fit with the NB regression model, this paper presents the relationships between logarithmic average monthly frequencies versus the grade of accessibility for the three groups in a scatter diagram with linear regression models (as shown in

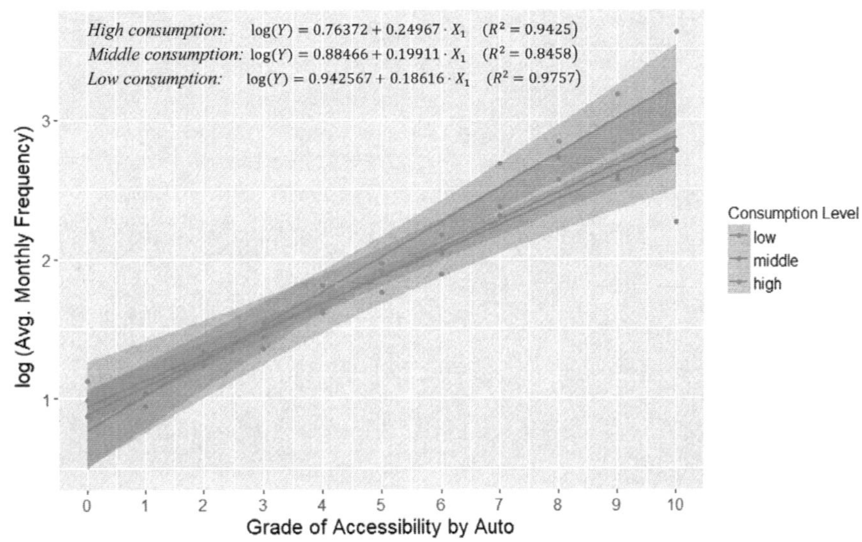

Figure 2. Relationships between non-work trip frequency and accessibility of users with different consumption levels.

Figure 2). It further verifies that the non-work trip frequency increases with accessibility. The slopes of those three models indicate that users with higher consumption levels are more sensitive to accessibility when they choose the frequency of their non-work trips. A possible explanation is that richer people have a smaller threshold value for the acceptable travel distance/time of non-work trips, while poorer people have a greater willingness to spend time traveling.

Furthermore, the users were also divided into three groups based on the monthly frequency of non-work trips: (a) low-frequency users ($Y=1$, i.e. once per month), (b) middle-frequency users ($1<Y\leq 4$, i.e. less than once per week), and (c) high-frequency users ($Y>4$, i.e. more than once per week). The spatial distributions of home and workplace of these three groups are presented in Figure 3. The isochrones (20 min, 40 min, and 60 min) of the Lujiazui CBD were also plotted in Figure 3. These are based on travel time from each center of a 100 m × 100 m grid to the CBD center computed by the Route Matrix API of Baidu Map. Figure 3 shows that most of the high-frequency users live (67%) or work (62%) at a location within 20 min from the CBD, while above half of the low-frequency users live or work outside of this area (i.e. above 20 min from the CBD). In other words, those users who live or work near the CBD are usually the potentially most frequent travelers to the CBD. This also reflects that the workplace locations have almost the same significant influence on the frequency of non-work trips as residence locations.

To study the effect of accessibility on the structure of different frequency groups, the percentages of the three groups were calculated in different accessibility conditions (as shown in Figure 4). The results for both home and workplace show that the percentage of high-frequency users tends to decrease with the decrease of accessibility, which is consistent with the conclusion of our regression model above. However, it can be found that the structure of three frequency groups tends to be stable when the accessibility level is worse than 30 min. In other words, the effect of accessibility on non-work trip frequency tends to weaken with the decrease of accessibility.

Conclusions and Future Directions

This study explored the relationships between the non-work trip frequency and accessibility from a new perspective. First, positioning data of mobile phone users was used to identify activity points, home and workplace locations, and non-work trips of the users. Furthermore, the socioeconomic attributes and preferences of users were extracted from mobile phone user feature data. Second, the accessibility was measured at the individual level based on a real-time travel-cost estimation method. The NB regression model was then employed to study the effects of accessibility, socioeconomic attributes, and preferences of users on their travel frequencies. The results show that this model fits the dependent variable and independent variables very well. The major findings of this study can be summarized as follows:

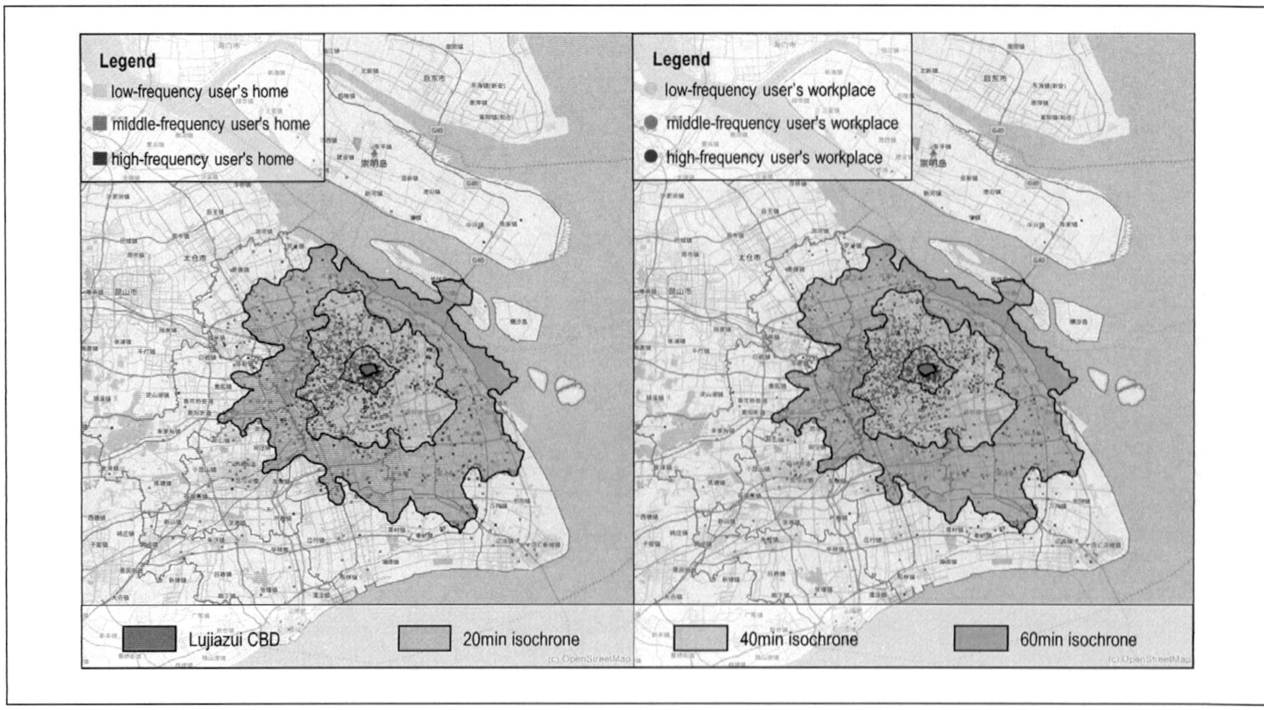

Figure 3. Spatial distributions of home and workplace for three frequency groups.

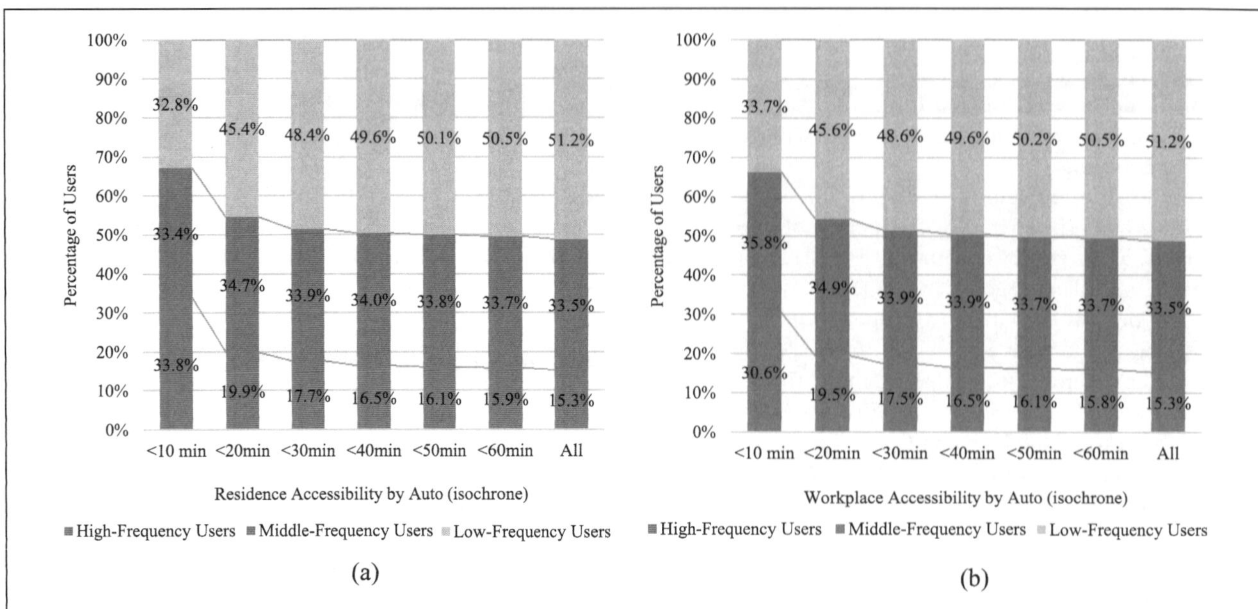

Figure 4. Structures of three frequency groups in different accessibility conditions: (a) accessibility by auto from residence; (b) accessibility by auto from workplace.

- Accessibility is strongly linked to the non-work trip frequency which increases with accessibility;
- Gender, consumption level, and frequency of visiting shopping websites and entertainment websites are all significantly and positively related to the non-work trip frequency;
- Different groups have different reactions to accessibility: users with higher consumption levels are

more sensitive to accessibility when they decide the frequency of the non-work trips;
- Not only users' residence locations but also their workplace locations can influence the frequency of non-work trips;
- The effect of accessibility on non-work trip frequency tends to weaken with the decrease of accessibility.

This study overcomes several shortcomings of previous studies, such as the limitations of sample size, aggregated accessibility, and home-based trips. Therefore, our findings contribute to a more reliable transport policy design and planning. For example, the relationships between the non-work trip frequency and accessibility could help with the forecasting of non-work travel demand for urban and land-use planning. Additionally, it provides a novel approach to measuring the individual accessibility, which may be used to evaluate the equity of accessibility. Generally, policies directed toward improving the accessibility of a CBD may result in more non-work travel to this area. Therefore, decision-makers should consider more about the traffic organization and parking spaces when implementing an accessibility-related transport policy.

However, some defects still exist in this paper, which need to be addressed in future studies. For example, the current analysis only focused on a CBD due to computational reasons. With the maturity of big data processing techniques, a future analysis may benefit by considering the study area at a greater scope. In addition, the non-work trips with different purposes should be separately considered, combining the built environment with the destinations given more data sources in future.

Acknowledgments

This study was supported by the National Natural Science Foundation of China (project nos. 71774118 and 91546115), the Science and Technology Commission of Shanghai Municipality (project no. 16511105204) and the China Scholarship Council.

Author Contributions

The authors confirm contribution to the paper as follows: study conception and design: Ye Li, Wenxiang Li; data collection: Haopeng Deng, Hanyu Shu; analysis and interpretation of results: Wenxiang Li, Xuegang Ban, Dongcan Xie; draft manuscript preparation: Wenxiang Li, Xuegang Ban. All authors reviewed the results and approved the final version of the manuscript.

References

1. Naess, P. Travelling Distances, Modal Split and Transportation Energy in Thirty Residential Areas in Oslo. *Journal of Environmental Planning and Management*, Vol. 38, No. 3, 1995, pp. 349–370.
2. Stead, D., and S. Marshall. The Relationships between Urban Form and Travel Patterns. An International Review and Evaluation. *European Journal of Transport and Infrastructure Research*, Vol. 1, No. 2, 2001, pp. 113–141.
3. Crane, R. On Form versus Function: Will the New Urbanism Reduce Traffic, or Increase It? *Journal of Planning Education and Research*, Vol. 15, No. 2, 1996, pp. 117–126.
4. Seo, S. E., N. Ohmori, and N. Harata. Effects of Household Structure and Accessibility on Travel. *Transportation*, Vol. 40, No. 4, 2013, pp. 847–865.
5. Handy, S. Regional versus Local Accessibility: Implications for Nonwork Travel. *Transportation Research Record: Journal of the Transportation Research Board*, 1993. 1400: 58–66.
6. Kitamura, R., T. Akiyama, T. Yamamoto, and T. Golob. Accessibility in a Metropolis: Toward a Better Understanding of Land Use and Travel. *Transportation Research Record: Journal of the Transportation Research Board*, 2001. 1780: 64–75.
7. Wang, T., C. Chen, and J. Ma. Mobile Phone Data as an Alternative Data Source for Travel Behavior Studies. Presented at 93rd Annual Meeting of Transportation Research Board, Washington, D.C., 2014.
8. González, M. C., C. A. Hidalgo, and A. Barabási. Understanding Individual Human Mobility Patterns. *Nature*, Vol. 453, No. 7196, 2008, pp. 779–782.
9. Song, C., Z. Qu, N. Blumm, and A. L. Barabási. Limits of Predictability in Human Mobility. *Science*, Vol. 327, No. 5968, 2010, pp. 1018–1021.
10. Calabrese, F., M. Diao, G. Di Lorenzo, J. Ferreira, Jr., and C. Ratti. Understanding Individual Mobility Patterns from Urban Sensing Data: A Mobile Phone Trace Example. *Transportation Research Part C: Emerging Technologies*, Vol. 26, 2013, pp. 301–313.
11. Caceres, N., J. P. Wideberg, and F. G. Benitez. Deriving Origin–Destination Data from a Mobile Phone Network. *IET Intelligent Transport Systems*, Vol. 1, No. 1, 2007, pp. 15–26.
12. Rojas IV, M. B., E. Sadeghvaziri, and X. Jin. A Comprehensive Review of Travel Behavior and Mobility Pattern Studies Using Mobile Phone Data. Presented at 95th Annual Meeting of the Transportation Research Board, Washington, D.C., 2016.
13. Ye, Y., Y. Zheng, Y. Chen, J. Feng, and X. Xie. Mining Individual Life Pattern Based on Location History. *Proc., 10th International Conference on Mobile Data Management: Systems, Services and Middleware*, Taipei, Taiwan, IEEE, 2009, pp. 1–10.
14. Pereira, F., C. Carrion, F. Zhao, C. D. Cottrill, C. Zegras, and M. Ben-Akiva. The Future Mobility Survey: Overview and Preliminary Evaluation. *Proceedings of the Eastern Asia Society for Transportation Studies*, Vol. 9, 2013, pp. 1–9
15. Wolf, J., W. Bachman, M. Oliveira, J. Auld, A. Mohammadian, P. Vovsha, and J. Zmud. *NCHRP Report 775: Applying GPS Data to Understand Travel Behavior, Volume II: Guidelines*. Transportation Research Board of the National Academies, Washington, D.C., 2014.

16. Nitsche, P., P. Widhalm, and S. Breuss. Supporting Large-Scale Travel Surveys with Smartphones maA Practical Approach. *Transportation Research Part C: Emerging Technologies*, Vol. 43, 2014, pp. 212–221.
17. Schuessler, N., and K. Axhausen. Processing Raw Data from Global Positioning Systems without Additional Information. *Transportation Research Record: Journal of the Transportation Research Board*, 2009. 2105: 28–36.
18. Lu, H., J. Yang, Z. Liu, N. D. Lane, T. Choudhury, and A. T. Campbell. The Jigsaw Continuous Sensing Engine for Mobile Phone Applications. *Proc., 8th ACM Conference on Embedded Networked Sensor Systems*, Zürich, Switzerland, ACM, New York, 2010, pp. 71–84.
19. Geurs, K. T., and B. van Wee. Accessibility Evaluation of Land-Use and Transport Strategies: Review and Research Directions. *Journal of Transport Geography*, Vol. 12, No. 2, 2004, pp. 127–140.
20. Dalvi, M. Q., and K. M. Martin. The Measurement of Accessibility: Some Preliminary Results. *Transportation*, Vol. 5, No. 1, 1976, pp. 17–42.
21. Hansen, W. G. How Accessibility Shapes Land Use. *Journal of the American Institute of Planners*, Vol. 25, No. 2, 1959, pp. 73–76.
22. Burns, L. D. *Transportation, Temporal, and Spatial Components of Accessibility*. Lexington Books, Lexington, Mass., 1979.
23. Ben-Akiva, M. Disaggregate Travel and Mobility Choice Models and Measures of Accessibility. In *Behavioural Travel Modelling* (Hensher, D. A., and P. R. Stopher), Croom Helm, London, 1979, pp. 639–653.
24. Baradaran, S., and F. Ramjerdi. Performance of Accessibility Measures in Europe. *Journal of Transportation and Statistics*, Vol. 4, No. 2/3, 2001, pp. 31–48.
25. Handy, S. L., and D. A. Niemeier. Measuring Accessibility: An Exploration of Issues and Alternatives. *Environment and Planning A*, Vol. 29, No. 7, 1997, pp. 1175–1194.
26. Astegiano, P., D. Akinc, C. M. J. Tampère, and M. Vandebroek. Quantifying the Influence of Accessibility to Personal Activities on Vehicle Ownership Decisions. Presented at 95th Annual Meeting of Transportation Research Board, Washington, D.C., 2016.
27. Næss, P. Accessibility, Activity Participation and Location of Activities: Exploring the Links between Residential Location and Travel Behaviour. *Urban Studies*, Vol. 43, No. 3, 2006, pp. 627–652.
28. Kockelman, K. Travel Behavior as Function of Accessibility, Land Use Mixing, and Land Use Balance: Evidence from San Francisco Bay Area. *Transportation Research Record: Journal of the Transportation Research Board*, 1997. 1607: 116–125.
29. Wallace, B., J. Barnes, and G. S. Rutherford. Evaluating the Effects of Traveler and Trip Characteristics on Trip Chaining, with Implications for Transportation Demand Management Strategies. *Transportation Research Record: Journal of the Transportation Research Board*, 2000. 1718: 97–106.
30. Chen, C., L. Bian, and J. Ma. From Traces to Trajectories: How Well Can We Guess Activity Locations from Mobile Phone Traces? *Transportation Research Part C: Emerging Technologies*, Vol. 46, 2014, pp. 326–337.
31. Li, W., X. Cheng, D. Yang, Z. Duan, and G. Guo. Mining Rail-transit Use Patterns from Mobile Phone Data: Case Study of Three Communities in Shanghai. Presented at 94th Annual Meeting of the Transportation Research Board, Washington, D.C., 2015.
32. Li, W., X. Cheng, J. Li, and D. Yang. Rail Transit User Classification Using Mobile Phone Data: A Case Study of Three Communities in Shanghai, China. Presented at 95th Annual Meeting of Transportation Research Board, Washington, D.C., 2016.
33. Ingram, D. R. The Concept of Accessibility: A Search for an Operational Form. *Regional Studies*, Vol. 5, No. 2, 1971, pp. 101–107.
34. Grengs, J., J. Levine, and Q. Shen. Intermetropolitan Comparison of Transportation Accessibility: Sorting Out Mobility and Proximity in San Francisco and Washington, D.C. *Journal of Planning Education and Research*, Vol. 29, No. 4, 2010, pp. 427–443.
35. Thomas, T., A. Mondschein, T. Osman, and B. D. Taylor. Nowhere to Run: Speed, Proximity, and their Relative Contributions to Accessibility. Presented at 95th Annual Meeting of the Transportation Research Board, Washington, D.C., 2016.
36. LaMondia, J. J., M. Moore, and L. Aultman-Hall. Modeling Inter-Trip Time Intervals Between Individuals' Overnight Long-Distance Trips. Presented at 94th Annual Meeting of the Transportation Research Board, Washington, D.C., 2015.

The Standing Committee on Urban Transportation Data and Information Systems (ABJ30) peer-reviewed this paper (18-01369).

Article

TRR
JOURNAL OF THE TRANSPORTATION RESEARCH BOARD

Transportation Research Record
2018, Vol. 2672(42) 103–114

DOI: 10.1177/0361198118793241
journals.sagepub.com/home/trr
SAGE

A Comparative Study between Private-Sector and Automated Vehicle Identification System Data through Various Travel Time Reliability Measures

Whoibin Chung[1], Mohamed Abdel-Aty[1], Juneyoung Park[2], and Raj Ponnaluri[3]

Abstract

Traffic data from private-sector sources is increasingly used to estimate the travel time reliability of major road infrastructure. However, there is as yet no study evaluating the difference in estimating travel time reliability between the private-sector data and automated vehicle identification (AVI) based on radio frequency identification. As ground truth data, the AVI data were collected from an AVI system using toll tags and aggregated into five-minute intervals. As one of the representative traffic information providers, data from HERE was obtained through the Regional Integrated Traffic Information System, calculated in five-minute intervals. For the comparison, four kinds of measures were selected and estimated on the basis of the day of the week, specific time periods, and time of day in five-minute, 15-minute, and one-hour intervals. The statistical difference in travel time reliability was assessed through paired t-tests. According to the results, AVI and HERE data are comparable based on day of the week, specific time periods, and time of day at one-hour intervals, whereas at five-minute and 15-minute intervals, HERE and AVI data are not generally comparable. Thus, when estimating travel time reliability in real time, travel time reliability derived from HERE data may be different from the true travel time reliability. Considering that private-sector traffic data are currently used to estimate travel time reliability measures, the measures should be harmonized on the basis of robust statistics to provide more consistent measures related to the true travel time reliability.

Many transportation researchers are interested in travel time reliability as it enables agencies to make preparations for uncertainty due to unexpected traffic demand, crashes, and adverse weather conditions. Travel time reliability can provide buffers to sustain reliable travel time for drivers, travelers, traffic operators, and even planners. In recent decades, the concept of travel time reliability has been defined and several metrics and models have been developed with regard to it from various perspectives (*1*). Based on the metrics and models, the diverse impacts of factors of nonrecurring congestion have been investigated. As a representative case, the Strategic Highway Research Program 2 (SHRP2) and the Federal Highway Administration (FHWA) have sponsored much research on travel time reliability (*2*).

Basically, these research studies have used three types of traffic data sources: infrastructure-based detectors such as loop and radar detectors, automated vehicle identification (AVI) systems such as Bluetooth readers, license plate readers, and radio frequency identification, and automated vehicle location (AVL) systems tracking a vehicle's location (*3*). Generally, the infrastructure-based detectors and the AVI systems are already used in the traffic management systems of many regions, whereas the AVL systems have not been deployed fully to provide sufficient data on a regional scale. As the data collection ability and coverage of the private sector AVL systems improve, some researchers are starting to use this private-sector traffic data to study travel time reliability and to analyze its performance measures.

In 2011, the United States Department of Transportation started to consider using private-sector data for national transportation performance management and several public agencies jointly developed and

[1]Department of Civil, Environmental and Construction Engineering, University of Central Florida, Orlando, FL
[2]Department of Transportation & Logistics Engineering, Hanyang University, Gyeonggi-do, South Korea
[3]Florida Department of Transportation, Tallahassee, FL

Corresponding Author:
Address correspondence to Whoibin Chung: jwhoibin@knights.ucf.edu

published guidelines for evaluating the accuracy of data on travel time and speed of commercial traveler information services (*4, 5*). Among traffic information providers, HERE, INRIX, and TomTom were selected as three highly qualified vendors in the I-95 Corridor Coalition's Vehicle Probe Project (VPP) (*6*). The VPP has validated the three vendors' data on freeways and arterials in four flow regimes (0–30 mph, 30–45 mph, 45–60 mph, more than 60 mph) by using Bluetooth technology. The data quality measures were average absolute speed error (AASE) and speed error bias (SEB) (*7*). The qualified traffic data require that the AASE should be less than 10 mph and the SEB should be within +/–5 mph in each of the four speed ranges. Nevertheless, not many studies have used and validated the private-sector data for travel time reliability.

One study investigated travel time reliability in work zones by using 15-minute traffic data of INRIX (*8*). As travel time reliability measures, 95th percentile travel time, a buffer index (BI) based on the average travel rate (minute/mile), and a planning time index (PTI) were used. Through the travel time reliability measures, it was statistically quantified that work zones have a negative impact on travel time reliability during non-peak periods as well as peak periods. Another study performed a detailed analysis of 13 travel time reliability measures based on 15-minute interval mean speed data from INRIX (*9*). It was recommended that ideal comparisons of reliability measures should use all 24 hours of the day, and time of day analysis should be conducted to find what time periods will be effective to improve travel time reliability through traffic management strategies. It was also shown that there is no single best performance measure for travel time reliability, and statistical range measures for travel time reliability have the lowest correlation with the average travel rate compared with other measures.

At the same time, users including public agencies may still question whether the travel time reliability performance measures can be estimated reliably under the condition that the processing algorithms and quality assessment methods of private data sources are unknown (*3*). Related to this question, one comparative study was conducted to analyze the effect of data source selection on travel time reliability assessment by using 15-minute aggregation data (*10*). The research analyzed travel time reliability derived from Bluetooth and INRIX data on Interstate 95 (I-95) and Interstate 207 (I-207) with high occupancy vehicle (HOV) lanes. According to the results, travel time reliability of I-95 is not statistically significantly different between the two data sources, but I-207 has significantly different travel time reliability because of HOV lanes. Thus, it was found that some reliability metrics are more sensitive to the data source than others.

TomTom's historical traffic data were evaluated in terms of travel time reliability through a comparative study in Calgary, Canada (*11*). Travel time reliability was measured by the 95th percentile travel time, BI, travel time index, and PTI. Although this study found that TomTom provides travel time reliability estimates with reasonable accuracy, the validation was not proven statistically since the sample size was not adequate.

Apart from travel time reliability, several comparison or evaluation studies have been conducted related to the data on travel time and speed of HERE. A comparison study of several data collection methods to estimate travel time on freeways and arterials in Florida was conducted. According to the research results, HERE provides more accurate travel times on freeways for oversaturated conditions than INRIX and the Bluetooth system, but INRIX and Bluetooth are better than HERE for uncongested periods (*12*). In the case of arterials, none of the methods were accurate. Furthermore, research compared speed data on arterials from Bluetooth, HERE, and INRIX in southeast Florida to find alternatives for transportation planning measures (*13*). The study showed that the data sets from Bluetooth and HERE were similar, but in the INRIX data speeds were 5 to 10 mph lower than Bluetooth and HERE.

Recently, Florida Department of Transportation (FDOT) has been trying to use multiple data sources for mobility performance measures, such as travel time reliability, travel time variability, vehicle hours of delay, and so on (*14*). In terms of data availability, cost effectiveness, and usability of the multiple data sources, the National Performance Measure Research Dataset and HERE, instead of TomTom and INRIX, were chosen for the mobility performance measures of Florida. The research plan was to evaluate and compare the estimated mobility performance measures, not with actual travel time collected by other truthful systems, but the existing model-based method. However, the evaluation and comparison results have so far not been confirmed.

The present study compares the travel time reliability of data from HERE with the actual truthful system, the AVI system, which differs from the previous research using Bluetooth. The AVI system uses toll tags, which provide much better, more stable and qualified, data than Bluetooth. For comparison, it explores travel time reliability performance measures based on several analysis scenarios including each weekday of the year, time period of an average weekday, day of the week, and time of day of an average weekday.

Study Locations

Six segments on Florida State Road 417 (SR 417) managed by Central Florida Expressway (CFX) Authority,

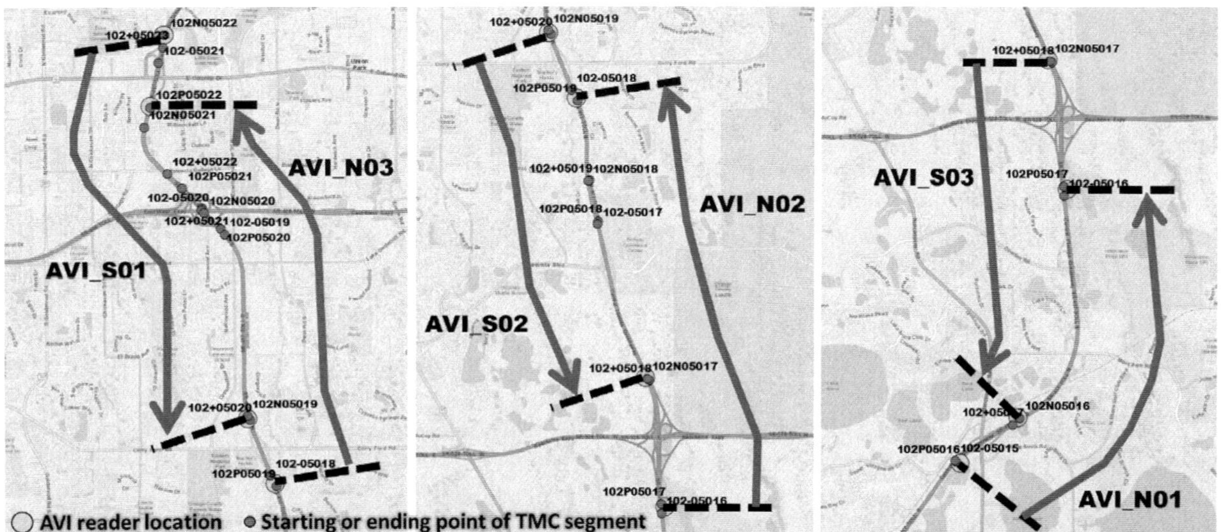

AVI Segment	Direction	Distance (Miles) AVI	Distance (Miles) HERE	Difference (Miles)	Here Segments
AVI_S01	SB	4.56	4.53	(0.038)	6
AVI_N03	NB	4.53	4.55	0.015	6
AVI_S02	SB	3.92	3.95	0.036	4
AVI_N02	NB	4.62	4.52	(0.109)	4
AVI_S03	SB	4.20	4.14	(0.054)	2
AVI_N01	NB	3.60	3.64	0.038	2

Figure 1. AVI and HERE segments on Florida State Road 417.

operating with the speed limit of 70 mph, were selected for the analysis because it was found that locations of AVI readers are practically identical with the starting or ending points of HERE segments (see Figure 1). Each AVI segment contains from two to six HERE segments and has an average length of about four miles. The node information of HERE segments was collected from the Regional Integrated Transportation Information System (RITIS) (*15*). Road-widening construction has been underway within AVI S01 and N03 segments since December 2015 (*16*).

Data Preparation

AVI data from 2016 were obtained from CFX's AVI system archiving the encrypted tag IDs and the passage timestamps of vehicles with toll tags since September 2012 (*17*). Uncapped raw AVI data, which is not adjusted by the speed limit, were archived for this research and used because more tangible travel time can be estimated as ground truth data. The uncapped raw AVI data were aggregated in five-minute intervals and their outliers were eliminated through the median absolute deviation (MAD) approach. The MAD approach provides high accuracy with low computational effort (*18*). The removal criterion of outliers becomes:

$$Median - b^{*}MAD < travel\ time(i) < Median + b^{*}MAD$$

where b is a threshold. The threshold value of 3 was applied very conservatively (*19*). In addition, it was necessary to confirm whether the count of the data used in each aggregation period satisfied the required sample size, which is estimated by the following equation (*20*):

$$n = \left(\frac{ts}{\varepsilon}\right)^2 \qquad (1)$$

where n = required sample size

s = standard deviation, which was estimated in each five-minute aggregation

ε = user-specified allowable error, in which 5 mph was applied

t = 1.96 at 95% confidence was used

If the number of data in five-minute increments is less than the required sample size, the corresponding time periods were removed. From one year of data, approximately 7.2% of five-minute traffic data were removed through the MAD approach.

The travel times of HERE, which were aggregated in five-minute intervals, were downloaded via the RITIS platform (*21*). The raw travel times are generated through data fusion processing of various data sources including state sensor data, probe vehicle data, GPS data, and historical data, but the traffic data processing algorithms are not published (*12*). The data is estimated at Traffic Message Channel (TMC) segments, which are divided at physical or logical geometric changes. Each

AVI segment is composed of several TMC segments (see Figure 1). On the basis of the AVI segments, each travel time of TMC segments was added at five-minute intervals in order to be compared with AVI data. Finally, all travel times of AVI and HERE were normalized by the distance of segments as follows (*22*, *23*):

$$Actual\ Travel\ Rate(minute/mile) = \frac{Actual\ Travel\ Time(minute)}{Distance\ of\ segments(mile)}$$

Travel Time Reliability Measures

Based on the previous research (*1*, *9*, *10*, *23*, *24*), travel time reliability metrics were selected within four classifications: statistical range measures, buffer time measures, tardy trip measures, and probabilistic measures. Currently, several agencies are using different travel time reliability measures that reflect their own mobility policies.

These measures can also be distinguished by robust statistics, which are insensitive to the effects of outliers or events, and non-robust statistics. The robust statistics are based on medians instead of means and use more information from the center than from the outlying data (*25*). A skew statistic, width statistic, BI based on median, and probabilistic measures use robust statistics.

Statistical Range Measures

Statistical range measures include standard deviation (SD), coefficient of variation (CV), skew statistic (λ^{skew}) and width statistic (λ^{var}), which are an attempt to quantify travel time reliability from a statistical perspective. The CV is one metric to measure data variability, which can be used to identify links or corridors that experience greater travel time variation over long periods of time than other links (*4*). The skew statistic and the width statistic follow the concept that asymmetric, wider, and larger distribution relative to median will be able to be unreliable (*26*). Thus, the two statistics should be considered together for travel time reliability:

$$CV = SD/mean(\mu) \times 100$$

$$\lambda^{skew} = \frac{TT_{90th} - TT_{50th}}{TT_{50th} - TT_{10th}};\ \lambda^{var} = \frac{TT_{90th} - TT_{10th}}{TT_{50th}} \quad (2)$$

where TT_{90th}, TT_{50th}, and TT_{10th} stand for the 90th, 50th, and 10th percentile travel time, respectively. Although FHWA does not recommend the use of statistical range measures, since they are not easy for the public to understand, this study used these measures to recognize specific difference of travel time reliability between AVI and HERE.

Buffer Time Measures

As buffer time measures, BI based on average, BI based on median, and PTI were selected. The BI implies that travelers should allow an extra percentage of travel time to arrive at their destinations on time, and the PTI provides an expected travel time budget, which could be used as a trip planning measure for journeys that require punctuality (*23*). FHWA, Georgia Regional Transportation Authority, Georgia Department of Transportation, and Maryland State Highway Administration introduced BI and PTI to represent travel time reliability (*27*). FDOT and the National Transportation Operations Coalition are using BI (*5*). Washington State DOT chose PTI to provide the best time for travelers to commence their journeys (*28*):

$$BI_{mean} = \frac{TT_{95th} - Average\ Travel\ Time}{Average\ Travel\ Time} \times 100(\%)$$

$$BI_{median} = \frac{TT_{95th} - Median\ Travel\ Time}{Median\ Travel\ Time} \times 100(\%)$$

$$PTI = \frac{TT_{95th}}{TT_{free\ flow\ or\ posted\ speed\ limit}} \quad (3)$$

Tardy Trip Measures

Tardy trip measures can explain the unreliability of travel time through late-arrival trips. Misery index (MI) and on-time arrival (OTA) were used in this study. MI focuses on the extra delay that occurred during the worst trip (*23*). The OTA measure can be estimated by the proportion of travel times less than a designated travel time, which can be defined as "speed limit minus 10 mph" (OTA(a)) or "1/3 × speed limit" (OTA(b)) speed (*29*):

$$MI = \frac{Average\ Travel\ Time\ for\ the\ longest\ 20\%\ of\ trips - Average\ Travel\ Time}{Average\ Travel\ Time}$$

Probabilistic Measures

This study adopts the probabilistic measure used by the Dutch Ministry of Transport, Public Works and Water Management (*24*). It calculates the probability that the observed travel times occur more than α times a predefined travel time threshold, which in this case is the median travel time on a given time of day or day of the week. For this study, the parameter α is chosen as 1.2, which means the probability that travel time is larger than the median travel time + 20% (*24*):

$$PR(a) = P(TT_i \geq \alpha TT_{50th})$$

Analysis Scenarios

Travel time reliability measures can be calculated according to various viewpoints. For example, the travel time reliability measures of each segment or corridor can be aggregated by day of week (DOW), time period (TP) such as morning peak, evening peak, mid-day, and late night of an average weekday, and time of day (TOD) of an average weekday. They can also be separated and analyzed depending on events including weather, incidents, and so on, but the events were not distinguished in this study. With reference to previous research (*23*), several analysis scenarios were established, as follows:

- Average travel time reliability by DOW of the whole year (the travel time reliability measures are aggregated for each DOW and analysis section)
- Average travel time reliability by TP (AM Peak, Mid-day, PM Peak, and Late PM) of an average weekday
- Average travel time reliability by TOD in one-hour intervals of an average weekday
- Average travel time reliability by TOD in 15-minute increments of an average weekday
- Average travel time reliability by TOD in five-minute increments of an average weekday

To analyze the difference of travel time reliability measures between the two data sources, it is necessary to confirm whether travel time reliability measures derived from two data sources are equal statistically. As a general statistical method, the paired t-test was applied.

Travel Time Data Distribution of AVI and HERE

This study concentrated on travel rates during weekdays of 2016, which can provide obvious travel patterns with incidents such as crashes, road work, and adverse weather. After abnormal AVI data were removed through the MAD approach and the statistically effective sample size was confirmed, scatter plots were used to confirm whether the overall tendency of travel times between AVI and HERE are similar. Figure 2 shows travel time patterns by the direction of SR 417. Segments S01 and N03 are the most congested in both southbound and northbound directions, respectively, due to the road-widening construction since December 2015, and they have obvious morning and evening peak time periods. Other segments also have a traffic pattern in which travel rates increase during commuting time periods, but the magnitude of the increment is much less than on S01 and N03. Comparing scatter plots of AVI with HERE, it seems that both AVI and HERE have similar traffic patterns although some travel rates of HERE, which could have occurred under nonrecurring congestion during the day, were estimated lower during the day and were spread more during the late night and early morning than AVI data.

To observe additional features between AVI and HERE, cumulative distributions of travel rates by direction were used. For the clear view, this study used average travel rates for five-minute times of day of an average weekday to make the cumulative distribution. According to the cumulative distribution of all average travel rates for all times of day (see Figure 3*a* and *c*), it is evident that the range of the average travel rate differs between the southbound and northbound directions, so it is necessary to distinguish between both directions in the comparative study.

More specifically, HERE data seem to fall behind AVI data until before a point of the first tangency because HERE travel times were capped, but AVI travel times were not capped by the speed limit of 70 mph, which is a rate of travel of 0.867 minutes/mile. After the point of tangency, at which the impact of the adjusted speed disappears, HERE and AVI data are moving in the same trend although HERE underestimated some travel times that were radically increased due to events. Additionally, the cumulative distributions of average travel rates during the morning and evening peaks (see Figure 3*b* and *d*) focus on the phenomenon after the first intersecting points. It shows again that HERE and AVI have similar travel rate distribution during peak-hour periods, although HERE has a capability to estimate lower travel times than the actual travel times. Therefore, it is necessary that different traffic data sources be evaluated in terms of travel time reliability as well as travel times because they can have different data distributions depending on their own processing algorithms.

Analysis Results

By using travel times and rates from 2016 on SR 417, based on the analysis scenarios, four types of travel time reliability measures between AVI and HERE were compared through the paired t-test. As with the review of travel rate distributions, the paired t-test was conducted by distinguishing southbound and northbound directions. The null hypothesis is that there is no significant mean difference in travel time reliability performance measures between AVI and HERE.

According to the results of the paired t-test of all data, regardless of driving direction and segments (see Table 1), SD, CV, MI, and OTA(a) represent that AVI and HERE are statistically significantly different in all test scenarios. However, the skew statistics, the width statistics, BI based on median, and PR (1.2) show that AVI and HERE are not different until TOD in one-hour

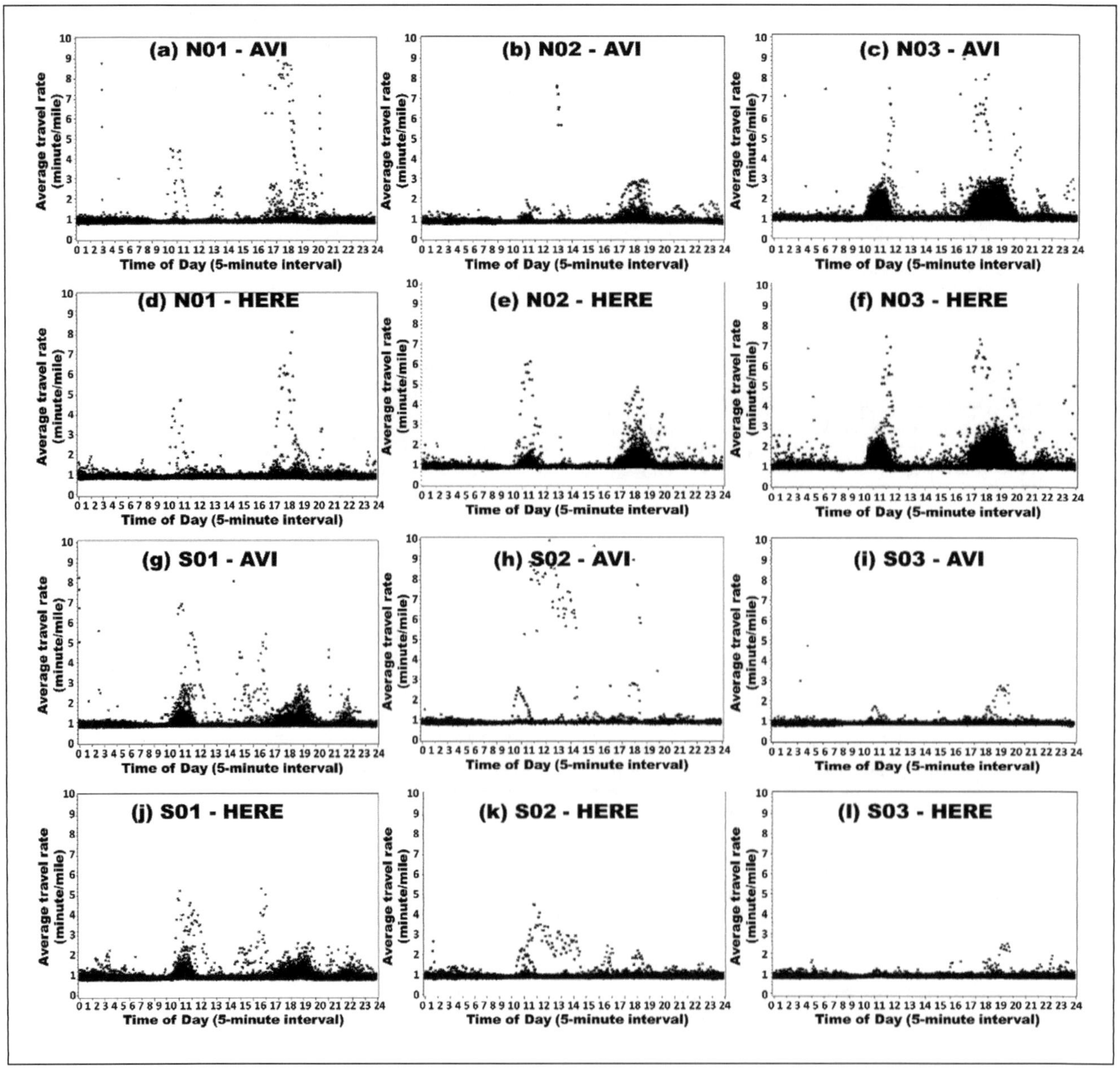

Figure 2. Scatter plots of five-minute travel rates for all segments.

increments. It seems that this kind of separation is caused by the characteristics of robust statistics. Considering various travel time distributions under different traffic conditions (*30*), this result shows that travel time reliability measures with robust statistics can explain the consistent relationship between different data sources having the same purpose.

In the next analysis, the travel time reliability measures were compared by driving direction. Table 2 shows the comparison results of the southbound direction and Table 3 the northbound direction. Among the statistical range measures of Tables 2 and 3, the SD and the coefficient of variance are statistically significantly different between AVI and HERE in all test scenarios, whereas the width statistic (λ_{var}) is not statistically different in most test scenarios except for TOD (5-min). However, the skew statistic (λ_{skew}) shows conflicting results in two different travel rate distributions. At least, the ratio of the range of travel times between 90th percentile travel time and 10th percentile travel time and the median is statistically consistent in AVI and HERE in all test scenarios except for TOD (5-min).

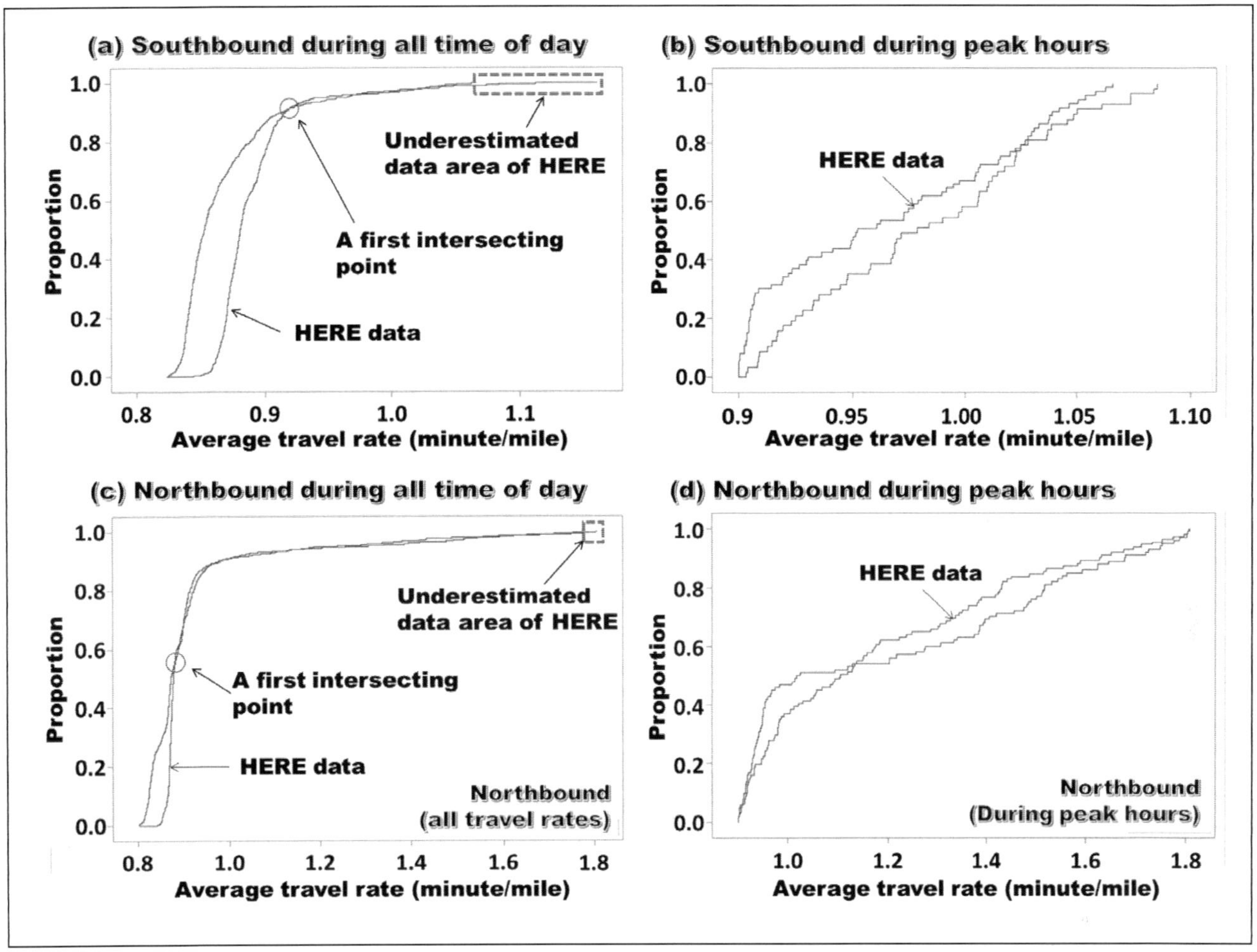

Figure 3. Empirical cumulative distributions of average travel rate for time of day in five-minute increments.

In the buffer time measures, BI based on mean and BI based on median show a consistent result in the two distributions. The BI based on mean has no difference between AVI and HERE in the only DOW and TP test scenarios, but the BI based on median has no difference in the DOW, TP, TOD (hour) and TOD (15-min). On the other hand, there is no statistical difference in PTI between AVI and HERE for the northbound direction's travel time distribution. The BI based on median using one of the robust estimators shows that there is no difference between AVI and HERE until the test scenarios from DOW to TOD (15-min). This is the same result as the width statistic. The only difference is that the width statistic uses 90th, 10th, and 50th percentile travel time, and BI is based on the median and uses the 95th and 50th percentile travel time.

Furthermore, no difference was found between AVI and HERE for all tardy travel measures in only the northbound direction's travel time distribution with DOW, TP, and TOD (hour) test scenarios. Finally, the probabilistic measure, PR(1.2), had the same results in three test scenarios, DOW, TP, and TOD (hour), for the travel time distribution of both directions. PR(1.2) also uses the 50th percentile travel time.

Finally, comparison of travel time reliability measures between AVI and HERE was conducted for each segment (see Table 4). For this test, DOW and TP test scenarios were not included because the sample size was too small. When the size of the interval of time of day is decreased, the number of measures, representing two distributions that are not different, is decreased. Based on Table 4, travel time reliability measures of HERE do not differ from AVI in most segments, except for SB02 in TOD (hour). The SB02 does not have any measure with p-value more than 0.05 at all TOD, which means that the AVI and HERE data in the SB02 segment are definitely different, or there may be some error between AVI and HERE.

Table 1. All Paired *t* Test Results of Travel Time Reliability Measures between AVI and HERE

	Statistical range measure				Buffer time measure			Tardy trip measure			
Test scenarios	SD	CV	λ_{skew}	λ_{var}	BI_{mean}	BI_{median}	PTI	MI	OTA(a)	OTA(b)	PR(1.2)
DOW											
p value	0.000	0.000	0.292*	0.060*	0.313*	0.502*	0.478*	0.002	0.623*	0.000	0.456*
t value	7.230	6.690	1.070	1.950	−1.030	0.680	−0.720	3.470	0.500	−4.440	0.760
Δ	0.812	0.224	0.171	0.038	−1.040	1.102	−0.015	0.041	0.003	−0.001	0.003
CI	0.582	0.155	−0.155	−0.002	−3.113	−2.216	−0.057	0.017	−0.009	−0.001	−0.006
	1.042	0.292	0.497	0.078	1.033	4.420	0.027	0.065	0.014	0.000	0.012
DF	29	29	29	29	29	29	29	29	29	29	29
TP											
p value	0.000	0.000	0.718*	0.699*	0.110*	0.586*	0.041	0.040	0.181*	0.021	0.442*
t value	6.120	5.880	−0.360	−0.390	−1.650	−0.550	−2.130	2.150	1.370	−2.440	−0.780
Δ	0.754	0.211	−0.054	−0.004	−1.516	−0.710	−0.037	0.021	0.014	0.000	−0.004
CI	0.502	0.138	−0.354	−0.024	−3.395	−3.348	−0.072	0.001	−0.007	−0.001	−0.015
	1.006	0.284	0.247	0.016	0.364	1.929	−0.002	0.041	0.035	0.000	0.007
DF	29	29	29	29	29	29	29	29	29	29	29
TOD (hour)											
p value	0.000	0.000	0.043	0.933*	0.003	0.421*	0.000	0.000	0.079*	0.000	0.180*
t value	7.290	7.360	−2.040	−0.080	−3.020	−0.810	−4.330	4.450	1.770	−3.800	−1.350
Δ	0.629	0.171	−0.178	0.000	−1.203	−0.444	−0.032	0.027	0.008	−0.001	−0.003
CI	0.458	0.125	−0.350	−0.009	−1.991	−1.533	−0.046	0.015	−0.001	−0.001	−0.006
	0.799	0.217	−0.006	0.009	−0.416	0.644	−0.017	0.039	0.018	0.000	0.001
DF	143	143	143	143	143	143	143	143	143	143	143
TOD (15-min)											
p value	0.000	0.000	0.000	0.565*	0.000	0.060*	0.000	0.000	0.001	0.000	0.017
t value	8.860	9.230	−4.430	−0.580	−5.590	−1.880	−8.530	6.620	3.300	−5.350	−2.400
Δ	0.454	0.120	−0.175	−0.001	−1.176	−0.474	−0.032	0.026	0.008	−0.001	−0.002
CI	0.353	0.095	−0.253	−0.005	−1.589	−0.969	−0.039	0.019	0.003	−0.001	−0.004
	0.555	0.146	−0.097	0.003	−0.762	0.021	−0.024	0.034	0.013	0.000	0.000
DF	575	575	575	575	575	575	575	575	575	575	575
TOD (15-min)											
p value	0.000	0.000	0.000	0.175*	0.000	0.000	0.000	0.000	0.000	0.000	0.000
t value	10.290	10.980	−7.390	−1.360	−9.450	−4.170	−15.440	8.070	5.580	−6.820	−3.690
Δ	0.318	0.081	−0.181	−0.002	−1.225	−0.614	−0.033	0.023	0.008	−0.001	−0.002
CI	0.258	0.066	−0.229	−0.004	−1.479	−0.903	−0.037	0.017	0.005	−0.001	−0.003
	0.379	0.095	−0.133	0.001	−0.970	−0.325	−0.029	0.029	0.011	0.000	−0.001
DF	1,727	1,727	1,727	1,727	1,727	1,727	1,727	1,727	1,727	1,727	1,727

Note: Δ = mean difference; CI = confidence interval; DF = degreed of freedom; OTA(a) = speed limit minus 10 mph; OTA(b)= 1/3 × speed limit.
*No rejection of the null hypothesis. There is no mean difference between paired measures at $\alpha = 0.05$.

Conclusion

This study compared HERE data with AVI data using toll tags for electronic toll collection and also providing sufficient sample size with high accuracy in terms of travel time reliability. Traffic data from INRIX and TomTom, as private-sector data, have been compared in previous studies in terms of travel time reliability, but HERE data has not been evaluated (*10, 11*). In addition, these studies evaluated the private data using data at 15-minute intervals because they use Bluetooth sensors that do not provide enough samples. In contrast, this study compared the reliability of travel time in more detail using AVI's five-minute interval data with statistically sufficient samples.

In order to understand the characteristics of the travel time data collected from AVI and HERE, data from three road sections by direction were combined. Scatter plots and cumulative empirical distributions were used to visualize travel rates of AVI with HERE. Through the scatter plots, the most congested sections and peak time periods were verified. The visualized cumulative distributions showed the difference between capped speeds of HERE data and uncapped speeds of AVI data. The difference disappeared at the first intersection points between the cumulative distributions of AVI and HERE.

The selected travel time reliability measures for this study were divided into four groups: statistical range

Table 2. Paired *t* Test Results of Travel Time Reliability Measures between AVI and HERE for the Southbound Direction

	Statistical range measure				Buffer time measure			Tardy trip measure			
Test scenarios	SD	CV	λ_{skew}	λ_{var}	BI_{mean}	BI_{median}	PTI	MI	OTA(a)	OTA(b)	PR(1.2)
DOW											
p value	0.000	0.000	0.045	0.262*	0.156*	0.828*	0.000	0.007	0.000	0.004	0.373*
t value	6.150	5.750	2.200	−1.170	−1.500	0.220	−6.000	3.160	7.870	−3.500	0.920
Δ	0.829	0.233	0.191	−0.006	−0.847	0.138	−0.036	0.027	0.007	−0.001	0.001
CI	0.540	0.146	0.005	−0.017	−2.057	−1.200	−0.049	0.009	0.005	−0.001	−0.002
	1.118	0.320	0.377	0.005	0.363	1.477	−0.023	0.045	0.009	0.000	0.005
DF	14	14	14	14	14	14	14	14	14	14	14
TP											
p value	0.002	0.002	0.115*	0.737*	0.658	0.589*	0.009	0.022	0.001	0.005	0.451*
t value	3.800	3.690	1.680	−0.340	−0.450	0.550	−3.010	2.570	4.470	−3.360	0.780
Δ	0.657	0.186	0.311	−0.003	−0.458	0.654	−0.033	0.024	0.008	−0.001	0.002
CI	0.286	0.078	−0.086	−0.021	−2.629	−1.880	−0.056	0.004	0.004	−0.001	−0.004
	1.028	0.295	0.709	0.015	1.712	3.187	−0.009	0.043	0.012	0.000	0.009
DF	14	14	14	14	14	14	14	14	14	14	14
TOD (hour)											
p value	0.000	0.000	0.947*	0.770*	0.021	0.486*	0.000	0.001	0.000	0.000	0.868*
t value	4.990	4.980	−0.070	−0.290	−2.360	−0.700	−6.430	3.530	3.900	−4.850	−0.170
Δ	0.529	0.147	−0.005	−0.001	−1.112	−0.366	−0.038	0.022	0.007	−0.001	0.000
CI	0.317	0.088	−0.168	−0.011	−2.051	−1.408	−0.049	0.010	0.003	−0.001	−0.003
	0.740	0.206	0.157	0.008	−0.173	0.676	−0.026	0.035	0.010	0.000	0.002
DF	71	71	71	71	71	71	71	71	71	71	71
TOD (15-min)											
p value	0.000	0.000	0.485*	0.399*	0.000	0.403*	0.000	0.000	0.000	0.000	0.828*
t value	6.570	6.670	−0.700	−0.840	−3.920	−0.840	−11.370	5.110	5.730	−6.740	0.220
Δ	0.400	0.108	−0.030	−0.002	−1.024	−0.244	−0.037	0.021	0.007	−0.001	0.000
CI	0.280	0.076	−0.114	−0.007	−1.539	−0.818	−0.043	0.013	0.004	−0.001	−0.002
	0.521	0.140	0.054	0.003	−0.509	0.330	−0.030	0.030	0.009	−0.001	0.002
DF	287	287	287	287	287	287	287	287	287	287	287
TOD (5-min)											
p value	0.000	0.000	0.282*	0.096*	0.000	0.031	0.000	0.000	0.000	0.000	0.784*
t value	7.630	8.030	−1.080	−1.660	−6.750	−2.160	−20.060	5.910	8.900	−8.540	0.270
Δ	0.285	0.074	−0.029	−0.003	−1.099	−0.368	−0.038	0.019	0.007	−0.001	0.000
CI	0.212	0.056	−0.082	−0.005	−1.419	−0.703	−0.042	0.013	0.005	−0.001	−0.001
	0.358	0.092	0.024	0.000	−0.779	−0.034	−0.034	0.026	0.008	−0.001	0.001
DF	863	863	863	863	863	863	863	863	863	863	863

Note: Δ = mean difference; CI = confidence interval; DF = degreed of freedom; OTA(a) = speed limit minus 10 mph; OTA(b) = 1/3 × speed limit.
*No rejection of the null hypothesis. There is no mean difference between paired measures at $\alpha = 0.05$.

measures, buffer time measures, tardy trip measures, and probabilistic measures. According to the predefined test scenarios, all travel time reliability measures were estimated and then tested through the paired t-test on whether the travel time reliability measures estimated from the two data sources are statistically different or not. The test was conducted in two groups: southbound travel times and northbound travel times.

According to the statistical test results of the two groups, it was confirmed that the results were different depending on elements of performance measures. It is shown that SD and CV, which are representative of non-robust estimators using an average, have statistically significant differences between AVI and HERE. In addition, most of the PTI, MI, and OTA using non-robust estimators, average travel time, did not provide consistent evaluation results in AVI and HERE, although BI based on mean travel times shows that the two data sources are not different in travel time reliability for DOW and time periods of an average weekday. Conversely, it was found that there is no statistical difference between AVI and HERE according to the test results of the width statistic, BI based on median, and PR (1.2), which are using the robust estimator, although the skew statistic did not yield a consistent conclusion in both distributions.

Considering the results of the previous research, robust statistics should be used for travel time having

Table 3. Paired *t* Test Results of Travel Time Reliability Measures between AVI and HERE for the Northbound Direction

	Statistical range measure				Buffer time measure			Tardy trip measure			
Test scenarios	SD	CV	λ_{skew}	λ_{var}	BI_{mean}	BI_{median}	PTI	MI	OTA(a)	OTA(b)	PR(1.2)
DOW											
p value	0.001	0.002	0.636*	0.053*	0.544*	0.532*	0.874*	0.024	0.912*	0.016	0.568*
t value	4.310	3.930	0.480	2.110	–0.620	0.640	0.160	2.530	–0.110	–2.730	0.580
Δ	0.796	0.214	0.151	0.081	–1.233	2.065	0.007	0.055	–0.001	–0.001	0.005
CI	0.400	0.097	–0.519	0.163	–5.487	–4.845	–0.081	0.008	–0.025	–0.001	–0.013
	1.192	0.331	0.821	–0.001	3.021	8.976	0.094	0.102	0.023	0.000	0.024
DF	14	14	14	14	14	14	14	14	14	14	14
TP											
p value	0.000	0.000	0.045	0.793*	0.113*	0.380*	0.240*	0.316*	0.354*	0.421*	0.304*
t value	4.780	4.540	–2.200	–0.270	–1.690	–0.910	–1.230	1.040	0.960	–0.830	–1.070
Δ	0.851	0.236	–0.419	–0.006	–2.573	–2.073	–0.041	0.018	0.020	0.000	–0.011
CI	0.469	0.124	–0.827	0.043	–5.836	–6.980	–0.113	–0.019	–0.024	–0.001	–0.032
	1.232	0.347	–0.010	–0.055	0.690	2.834	0.031	0.056	0.064	0.000	0.011
DF	14	14	14	14	14	14	14	14	14	14	14
TOD (Hour)											
p value	0.000	0.000	0.024	0.446*	0.049	0.593*	0.061*	0.003	0.289*	0.114*	0.173*
t value	5.370	5.430	–2.310	0.770	–2.000	–0.540	–1.900	3.040	1.070	–1.600	–1.380
Δ	0.729	0.195	–0.350	0.007	–1.295	–0.523	–0.026	0.031	0.010	0.000	–0.005
CI	0.458	0.123	–0.653	0.026	–2.583	–2.465	–0.052	0.011	–0.009	–0.001	–0.012
	1.000	0.266	–0.047	–0.012	–0.007	1.419	0.001	0.052	0.029	0.000	0.002
DF	71	71	71	71	71	71	71	71	71	71	71
TOD (15-min)											
p value	0.000	0.000	0.000	0.081*	0.000	0.088*	0.000	0.000	0.043	0.021	0.004
t value	6.160	6.480	–4.890	1.750	–4.020	–1.710	–3.980	4.640	2.030	–2.330	–2.910
Δ	0.507	0.132	–0.320	0.007	–1.328	–0.704	–0.026	0.032	0.010	0.000	–0.005
CI	0.345	0.092	–0.449	0.016	–1.978	–1.512	–0.040	0.018	0.000	–0.001	–0.008
	0.669	0.172	–0.191	–0.001	–0.678	0.105	–0.013	0.045	0.020	0.000	–0.002
DF	287	287	287	287	287	287	287	287	287	287	287
TOD (5-min)											
p value	0.000	0.000	0.000	0.011	0.000	0.000	0.000	0.000	0.001	0.004	0.000
t value	7.130	7.640	–8.290	2.550	–6.690	–3.580	–7.310	5.710	3.460	–2.880	–4.360
Δ	0.351	0.087	–0.332	0.006	–1.350	–0.859	–0.028	0.027	0.010	0.000	–0.005
CI	0.255	0.065	–0.411	0.011	–1.747	–1.330	–0.035	0.017	0.004	–0.001	–0.007
	0.448	0.110	–0.253	0.001	–0.954	–0.388	–0.020	0.036	0.016	0.000	–0.002
DF	863	863	863	863	863	863	863	863	863	863	863

Note: Δ = mean difference; CI = confidence interval; DF = degreed of freedom; OTA(a) = speed limit minus 10 mph; OTA(b) = 1/3 × speed limit.
*No rejection of the null hypothesis. There is no mean difference between paired measures at $\alpha = 0.05$.

compound distributions according to traffic conditions. There is no statistically significant difference between HERE and AVI in terms of travel time reliability using day of the week, time periods, and time of day in a one-hour unit. However, the travel time reliability measures calculated from the two different data sources at five-minute and 15-minute units can yield different results. It is obvious that HERE data as a real-time feed will differ from AVI data since the HERE data can be estimated for all time periods through unopened modeling methods including smoothing, filtering, and imputation using historical data when the collected data are insufficient. The differences will be revealed more obviously when the aggregation time span shortens. However, if raw traffic data without modeling is used, the differences are reduced although there might be irreducible errors. On the other hand, on the basis of the average-based BI, PTI, and OTA, which are currently used by public agencies in the U.S.A., AVI and HERE cannot estimate consistent travel time reliability measures.

Based on the results of this study, travel time reliability measures should be changed to use robust statistics such as median and percentiles. Thus, travel time reliability estimated through different data sources can be consistent from a macroscopic perspective such as transportation planning but not real-time systems, such as active traffic management strategies. This study compared two different data sources in terms of travel time

Table 4. Paired *t* Test Results (*p* value) of Travel Time Reliability Measures between AVI and HERE for Each Segment

	Statistical range measure				Buffer time measure			Tardy trip measure			
Test scenarios	SD	CV	λ_{skew}	λ_{var}	BI_{mean}	BI_{median}	PTI	MI	OTA(a)	OTA(b)	PR(1.2)
TOD (Hour)											
NB01	0.002	0.002	0.051*	0.664*	0.012	0.380*	0.600*	0.009	0.381*	0.002	0.406*
NB02	0.060*	0.043	0.091*	0.323*	0.042	0.093*	0.000	0.808*	0.050*	0.838*	0.053*
NB03	0.001	0.001	0.309*	0.079*	0.763*	0.462*	0.251*	0.031	0.086*	0.855*	0.972*
SB01	0.010	0.010	0.332*	0.061*	0.468*	0.185*	0.333*	0.005	0.112*	0.003	0.412*
SB02	0.000	0.000	0.001	0.003	0.000	0.000	0.000	0.028	0.000	0.004	0.000
SB03	0.136*	0.131*	0.452*	0.617*	0.145*	0.220*	0.000	0.730*	0.006	0.047	0.349*
TOD (15-min)											
NB01	0.000	0.000	0.000	0.360*	0.000	0.116*	0.328*	0.000	0.111*	0.000	0.159*
NB02	0.034	0.020	0.015	0.064*	0.000	0.003	0.000	0.495*	0.000	0.988*	0.000
NB03	0.001	0.001	0.009	0.000	0.846*	0.403*	0.043	0.003	0.003	0.795*	0.425*
SB01	0.000	0.000	0.164*	0.002	0.102*	0.005	0.144*	0.000	0.019	0.000	0.096*
SB02	0.000	0.000	0.000	0.000	0.000	0.000	0.000	0.001	0.000	0.000	0.000
SB03	0.038	0.035	0.163*	0.247*	0.009	0.030	0.000	0.605*	0.000	0.006	0.154*
TOD (5-min)											
NB01	0.000	0.000	0.000	0.128*	0.000	0.003	0.053*	0.000	0.019	0.000	0.080*
NB02	0.009	0.003	0.000	0.001	0.000	0.000	0.000	0.533*	0.000	0.913*	0.000
NB03	0.001	0.001	0.000	0.000	0.827*	0.371*	0.002	0.001	0.000	0.891*	0.292*
SB01	0.000	0.000	0.045	0.000	0.047	0.000	0.002	0.000	0.000	0.000	0.008
SB02	0.000	0.000	0.000	0.000	0.000	0.000	0.000	0.000	0.000	0.000	0.000
SB03	0.043	0.037	0.244*	0.050*	0.000	0.000	0.000	0.710*	0.000	0.004	0.093*

Note: OTA(a) = speed limit minus 10 mph; OTA(b) = 1/3 × speed limit.
*No rejection of the null hypothesis. There is no mean difference between paired measures at $\alpha = 0.05$.

reliability using six short segments on one corridor. This may not represent most of the freeways and expressways. For more general conclusions, more types of road segments should be added and analyzed. Also, changes in travel reliability due to events should be studied in the future.

Acknowledgments

The authors are grateful to the Florida Department of Transportation (FDOT) for supporting this study. The authors also appreciate the data provided by FDOT and the Central Florida Expressway Authority (CFX).

Author Contributions

The authors confirm contribution to the paper as follows: study conception and design: Whoibin Chung, Juneyoung Park; data collection: Whoibin Chung, Raj Ponnaluri; analysis and interpretation of results: Whoibin Chung, Mohamed Abdel-Aty, Juneyoung Park, Raj Ponnaluri; draft manuscript preparation: Whoibin Chung, Mohamed Abdel-Aty, Juneyoung Park, Raj Ponnaluri. All authors reviewed the results and approved the final version of the manuscript.

References

1. Taylor, M. A. Travel Through Time: The Story of Research on Travel Time Reliability. *Transportmetrica B: Transport Dynamics*, Vol. 1, No. 3, 2013, pp. 174–194.
2. TRB. *Strategic Highway Research Program 2 (SHRP 2) Reliability Research Reports.* http://www.trb.org/Publications/PubsSHRP2ResearchReportsReliability.aspx. Accessed July 15, 2017.
3. List, G., B. Williams, N. Rouphail, R. Hranac, T. Barkley, E. Mai, A. Ciccarelli, L. Rodegerdts, A. Karr, and X. Zhou. *SHRP 2 Report No. S2-LO2-RR2: Guide to Establishing Monitoring Programs for Travel Time Reliability*. Transportation Research Board of the National Academies, Washington, D.C., 2014.
4. Turner, S., J. Richardson, M. Fontaine, and B. Smith. *Guidelines for Evaluating the Accuracy of Travel Time and Speed Data.* Report No. TPF-5 (200). Virginia Department of Transportation, Richmond, VA, 2011.
5. Turner, S., K. Sadabadi, A. Haghani, M. Hamedi, R. Brydia, S. Santiago, and E. Kussy. *Private Sector Data for Performance Management.* Report No. FHWA-HOP-11-029. FHWA, U.S. Department of Transportation, Washington, D.C., 2011.
6. VPP. *Vehicle Probe Project.* http://i95coalition.org/projects/vehicle-probe-project/. Accessed July 25, 2017.
7. *Confidence Value and Smoothing.* I-95 Corridor Coalition, Vehicle Probe Project, 2010, p. 7. http://i95coalition.org/wp-content/uploads/2015/03/VP_Conf_value_White_Paper-4_14_2010_final-4.pdf?652af7. Accessed July 1, 2017
8. Edwards, M., and M. Fontaine. Investigation of Travel Time Reliability in Work Zones with Private-Sector Data.

Transportation Research Record: Journal of the Transportation Research Board, 2012. 2272: 9–18.
9. Chase, R. T. Jr., B. M. Williams, and N. M. Rouphail. Detailed Analysis of Travel Time Reliability Performance Measures from Empirical Data. Presented at 92nd Annual Meeting of the Transportation Research Board, Washington, D.C., 2013.
10. Haghani, A., Y. Zhang, and M. Hamedi. *Impact of Data Source on Travel Time Reliability Assessment*. Report No. UMD-2013-01. Mid-Atlantic Universities Transportation Center, College Park, MD, 2014.
11. Tahmasseby, S. Traffic Data: Bluetooth Sensors versus Crowdsourcing—A Comparative Study to Calculate Travel Time Reliability in Calgary, Alberta, Canada. *Journal of Traffic and Transportation Engineering*, Vol. 3, 2015, pp. 63–79.
12. Elefteriadou, L., A. Kondyli, and B. S. George. *Comparison of Methods for Measuring Travel Time at Florida Freeways and Arterials*. Transportation Research Center, University of Florida, Gainesville, FL, 2014.
13. Rapolu, S., and A. Kumar. *Comparing Arterial Speeds from "Big-Data" Sources in Southeast Florida (Bluetooth, HERE and INRIX)*. AECOM. https://www.trbappcon.org/2015conf/presentations/191_Travel%20Speed%20Data%20Comparison%20(HERE,INRIX%20and%20Bluetooth)%20-%20TRBApp%202015.pdf. Accessed July 25, 2017.
14. FDOT. *Use of Multiple Data Sources for Monitoring Mobility Performance*. Florida Department of Transportation, Tallahassee, Fla., 2015, p. 55.
15. CATT. *RITIS (Regional Integrated Transportation Information System)*. https://www.ritis.org. Accessed July 17, 2017.
16. CFX. *SR 417 Widening between Berry Dease Road and SR 408*. https://www.cfxway.com/for-travelers/projects/sr-417-widening-between-berry-dease-road-and-sr-408/. Accessed July 17, 2017.
17. Abdel-Aty, M., E. Radwan, Q. Shi, and L. Wang. *Efficient Utilization of the Existing ITS System and the Viability of a Proative Traffic Management System for the Central Florida Expressway Authority System*. U.S. Department of Transportation, Washington, D.C., 2014, p. 247.
18. Leys, C., C. Ley, O. Klein, P. Bernard, and L. Licata. Detecting Outliers: Do Not Use Standard Deviation Around the Mean, Use Absolute Deviation Around the Median. *Journal of Experimental Social Psychology*, Vol. 49, No. 4, 2013, pp. 764–766.
19. Miller, J. Reaction Time Analysis with Outlier Exclusion: Bias Varies with Sample Size. *The Quarterly Journal of Experimental Psychology*, Vol. 43, No. 4, 1991, pp. 907–912.
20. May, A. D. *Traffic Flow Fundamentals*. Prentice Hall, Englewood Cliffs, NJ, 1990.
21. Vandervalk, A. *Data for Mobility Performance Measures*. Florida Department of Transportation. http://www.fdot.gov/planning/statistics/symposium/2014/Fl-natlPerformData.pdf. Accessed July 17, 2017.
22. National Academies of Sciences, Engineering, and Medicine. *Cost Effective Performance Measures for Travel Time Delay, Variation and Reliability*. The National Academies Press, Washington, D.C., 2008.
23. Lomax, T., D. Schrank, S. Turner, and R. Margiotta. *Selecting Travel Reliability Measures*. 2003. https://static.tti.tamu.edu/tti.tamu.edu/documents/TTI-2003-3.pdf
24. Van Lint, J., H. J. Van Zuylen, and H. Tu. Travel Time Unreliability on Freeways: Why Measures Based on Variance Tell Only Half the Story. *Transportation Research Part A: Policy and Practice*, Vol. 42, No. 1, 2008, pp. 258–277.
25. *Robust Statistics*. Encyclopedia.com. http://www.encyclopedia.com/computing/dictionaries-thesauruses-pictures-and-press-releases/robust-statistics. Accessed July 27, 2017.
26. Van Lint, J., and H. Van Zuylen. Monitoring and Predicting Freeway Travel Time Reliability: Using Width and Skew of Day-To-Day Travel Time Distribution. *Transportation Research Record: Journal of the Transportation Research Board*, 2005. 1917: 54–62.
27. *Travel Time Reliability: Making It There on Time, All the Time*. Vol. FHWA-HOP-06-070. Office of Operations, Federal Highway Administration, Washington, D.C. https://ops.fhwa.dot.gov/publications/tt_reliability/brochure/. Accessed July 25, 2017.
28. WSDOT. *Best Time to Leave*. Washington State Department of Transportation, Olympia, WA. http://www.wsdot.com/traffic/seattle/traveltimes/reliability/. Accessed July 25, 2017.
29. Elefteriadou, L., and X. Cui. A Framework for Defining and Estimating Travel Time Reliability. Presented at 86th Annual Meeting of the Transportation Research Board, Washington, D.C., 2007.
30. Guessous, Y., M. Aron, N. Bhouri, and S. Cohen. Estimating Travel Time Distribution Under Different Traffic Conditions. *Transportation Research Procedia*, Vol. 3, 2014, pp. 339–348.

The Standing Committee on Urban Transportation Data and Information Systems (ABJ30) peer-reviewed this paper (18-01554).

All opinions expressed in this paper are those of the authors and do not necessarily reflect those of the Florida Department of Transportation.

Article

TRR
JOURNAL OF THE TRANSPORTATION RESEARCH BOARD

Transportation Research Record
2018, Vol. 2672(42) 115–126

DOI: 10.1177/0361198118774232
journals.sagepub.com/home/trr
SAGE

Carsharing Trip Characteristic Analysis: Do Users Choose Carsharing Rather Than Taxi to Economize?

Ying Hui[1], Cuiyi Liu[1], and Mengtao Ding[1]

Abstract

Carsharing and taxi are both shared or public car services in urban transportation systems, which means they have a lot in common. In particular, the narrow price gap between carsharing and taxi or public transit in China makes them similar. Trip characteristics analysis and market segmentation are needed, to investigate why and when travelers choose carsharing rather than taxi. In this study, vehicle GPS data and operation order data of a round-trip carsharing system in Hangzhou, China, is used to obtain information on 13,338 valid trips. The trips are divided into three groups based on the travel cost comparison with taxi. Then, an artificial neural network model is developed to analyze group characteristics. The trip characteristics concluded from the model, and analysis on typical service stations, reveal that carsharing has its price advantage on simple long-distance trips during off-peak hours, which is called the regular market. Carsharing's extended market, in which travel cost is higher than by taxi, covers two types of trips: One involves short driving distance (around 20 km) and long stopping time, and tends to occur in areas in which it is difficult to hail a taxi during peak hours; the other involves very short driving distance (around 10 km), more activity spots and very low travel cost (around 25 CNY). The results of this study can help carsharing operators to extend and adjust their business. Also, these results can contribute to help city administrators to reach better decisions.

In recent years, carsharing has become an important transportation trend in urban areas. ("Carsharing" in this study specifically refers to time-charging rental service and does not include other online ride-hailing services like Uber and Lyft.) With its rapid growth, carsharing in China tends to be a new mode of traveling and attracts users from other public transport modes. This results from the travelers' attitude to vehicles, shifting from owning to renting a vehicle service: On the one hand, besides the parking problems in urban areas, it has become increasingly difficult and expensive to own a car; on the other hand, using a vehicle service has become more efficient, according to the experience in North America (*1*). In addition, users can often get discounts when they use a carsharing service.

However, carsharing is not the only way to use transportation services. Public transport, taxi and car rental are among the alternative services available. Out of these modes, taxi service is the most similar to carsharing and both share many features:

- Individuals use them because of their occasional need for a vehicle (*2*).
- Individuals enjoy the use of a vehicle without owning one.
- All costs are distributed across use, and there are no fixed costs (*3*).
- Whether there is an available car is not guaranteed beforehand.
- The duration of use would not be too long.
- The travel is usually within the urban area.
- Individuals enjoy space privacy while using the service.

A study pointed out that carsharing members use it to substitute for modes that most resemble driving (*4*), but there are few studies comparing taxi and carsharing services. It is quite necessary to conduct this comparison, especially in China. The cost of taking a taxi in China is cheaper than in many other countries (*5*), but the difficulty of hailing a taxi is becoming a severe problem in big cities. Fortunately, competition exists. It is important for carsharing operators, transportation engineers, as

[1]Key Laboratory of Road and Traffic Engineering of the Ministry of Education, College of Transportation Engineering, Tongji University, Shanghai, China

Corresponding Author:
Address correspondence to Cuiyi Liu: soteo@tongji.edu.cn

well as city managers to understand this phenomenon. Thus, what the regular market of carsharing services is, why travelers would prefer to choose carsharing outside of its regular market, and what the extended market of carsharing services is, are questions worth studying. This paper fills this gap as it develops an artificial neural network model to conduct an analysis of the impact factors, and it evaluates the regular market and the extended market. The results of this research can help carsharing service operators to recognize their market competitiveness compared with taxi services, and to develop more specific strategies.

In this study, the GPS data and operation order data of "Fun Carsharing," the first operational electric carsharing system in China, will be used. "Fun Carsharing" is a round-trip carsharing system in Hangzhou. In this system, the rental vehicle must be returned to the same service station it was rented from. Therefore, one order's GPS data can perfectly record one complete trip, which makes it possible to identify and analyze trip chains.

The paper is structured as follows. First it provides a review of studies on carsharing user behavior and the application of GPS data in the transportation field. Then the dataset used in this study and the data processing work are introduced. Concepts and methodology are described, followed by presentation of the results and discussion. The final section comprises the conclusions and future directions.

Literature Review

Behavior Studies on Carsharing

Recent studies on carsharing user behavior can be sorted into two groups according to their methods.

Many studies adopted questionnaire survey. At first, the potential users are simply described as social activists, environmental protectors, and innovators (*6*). Later studies gave more details. Surveys in Japan (*7*) and North America (*8*, *9*) found that short travel distance is a positive factor, whereas the conclusion is totally opposite in Shanghai, China (*5*), which implies that the condition of carsharing in China may be different from other countries. Other common positive factors are lower income, younger age, private car ownership, and environment consciousness. In addition, gender, duration, travel period, public transport policy, and convenience are also factors that affect (*8–11*).

Questionnaire survey is limited by its sample size. With the development of data acquisition and mining technology, people have started to analyze the pattern of use of carsharing and the classification of users based on big data (*12–15*). Detailed carsharing trip characteristics, such as distance, duration, and frequency, are concluded, and they differ in different cities.

These impact factors inspire the variable selecting in this paper. However, the main limitation of the above studies is that there are few studies on the comparison of carsharing with other travel modes, especially with taxi services, which have a great deal in common with carsharing.

Application of Vehicle GPS Data in the Transportation Field

Vehicle GPS data contains information about the time, position and speed of a vehicle. In recent years, GPS data has been successfully applied in the transportation field. Due to the privacy concerns of individuals, taxi GPS data is most common, but a few carsharing GPS data are also used.

Carsharing GPS data is used to analyze the conditions of service station (*13*) and trip chain characteristics (*14*, *16*). Certainly, taxi GPS data is also used to extract travel information (*17*, *18*), and reveal the operation conditions of taxi services (*19*, *20*). However, due to the large number of taxi vehicles, also human daily life mobility and activity information can be reflected by the passenger flow of taxis (*21*). In addition, taxi GPS data can be used to analyze, and provide advice about more efficient routes to taxi drivers (*22*). Moreover, anomaly detection and prediction are important applications of taxi GPS data (*23–26*).

In summary, though the number of carsharing vehicles limits the use of carsharing GPS data, the existing methods of taxi GPS data analysis inspire the methodology of this study. This paper combines operation order data and GPS data to obtain more details about the trips that operation order data cannot support, and also to overcome the limited sample size issue of the questionnaire survey method.

Dataset

Overview

The data used in this research is provided by the Hangzhou EVnet Co., the operator of the "Fun Carsharing" carsharing system in Hangzhou City. The operation data, as well as the vehicle GPS data, cover the period from December 2013 to June 2015, containing 31,446 rental records. The return interval of the GPS data is approximately 30 s, even though the car is not booked or moving, which makes it possible to recognize user activity spots. The two kinds of data can be linked by their common column "ORDERID" (Table 1).

Table 1. Data Frame

GPS data		
Column	Data type	Explanation
ORDERID	CHAR	Identification of one rent process, unique
CARNUMBER	VARCHAR	License plate number of vehicle
ADD_TIME	DATETIME	Return time of data
RENTED	BOOLEANS	1 = 'been rented'; 0 = 'for rent'
SPEED	INT	Instantaneous speed of return time
LONGITUDE	FLOAT	Longitude of vehicle position
LATITUDE	FLOAT	Latitude of vehicle position
Operation Data		
Column	Data type	Explanation
ORDERID	CHAR	Identification of one rent process, unique
SERVICE STATION	VARCHAR	Position of service station that car was rented at
CARTYPE	VARCHAR	Different type of car has different charging standard
CARNUMBER	VARCHAR	License plate number of vehicle
PHONE NUMBER	CHAR	Phone number of member
START_TIME	DATETIME	Time member picks the car
END_TIME	DATETIME	Time member returns the car
DRIVING DISTANCE	FLOAT	Distance of car movement Rename as "D" in the followings
PAYMENT	FLOAT	The amount of money that member has paid Rename as 'P_{CS}' in the followings
RADIUS	FLOAT	The maximal distance between every single data point and service station Rename as "R" in the followings
DURATION	DATETIME	The duration time of rental
ACTIVE MEMBER	INT	1 = 'active member'; 0 = 'inactive member' Rename as 'ACTIVE' in the followings

Data Preprocessing

Errors often exist in the GPS data. To increase the accuracy and reliability of the analysis, data preprocessing is necessary. Ding (*14*) achieved some data cleaning jobs and cleared data records using the following principles:

- The position of the first record of the rented state did not correspond to the service station
- The average velocity between adjacent records is higher than 200 km/h (*27*)
- The change of average velocity does not satisfy the inequality (*28*)

$$\frac{\left|\bar{v}_{(i,i+2)} - \bar{v}_{(i,i+1)}\right|}{\frac{\left(T_{(i+1,i+2)} - T_{(i,i+2)}\right)}{2}} \leq 10 \qquad (1)$$

After that, the total number of valid orders is 28,553.

To reduce the impact of the GPS random error on travel distance, the GPS data is used to calculate the whole travel distance, and this is compared with the "DRIVING DISTANCE" in the order data. This can diminish the effects and potential problem caused by GPS drifting. Considering that different GPS devices still have 20% deviation in a round (*29*), then 20% is taken as the allowable error to select records with acceptable precision.

Moreover, to emphasize the service as an urban traffic mode for daily trips and to distinguish it from the conventional car rental mode, the selected orders are limited to a "RADIUS" within 50 km, which is just a bit further than the diameter of Hangzhou city, and with a "DURATION" of less than 24 hours.

Finally, a dataset with 13,338 orders is selected for the research.

Concepts and Methodology

Concepts and Coding

Before introducing the data processing and analysis process, it is necessary to define several new indices to label and describe the trip and to complete the following analysis. It is necessary to highlight that here the whole process of a rental is considered as a complete trip, which is different from the FHWA's definition in the 2001 National Household Travel Survey (*30*). In addition to the terms D, P_{CS}, R and ACTIVE mentioned above (Table 1), others are listed as follows, and a diagram helps to understand the indices better (Figure 1).

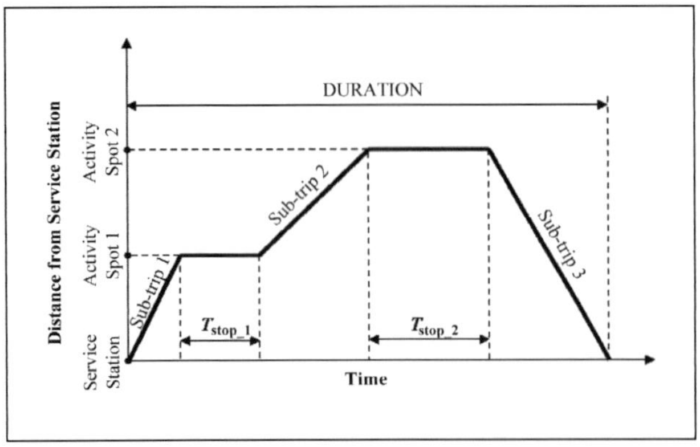

Figure 1. Trip with two activity spots.

Activity: Intervening stops lasting more than 720 s (*14*), such as work, business meeting, long-time shopping, entertainment, and so forth.

Stop time (T_{stop}): The duration time of the user's stop due to activities.

Sub-trip: Trip between two activity spots or the service station.

Number of sub-trips: Equals the number of activity spots plus 1. Number of sub-trips is renamed as T in the following. And, before modeling, T has been discretized to become "T1_1_3", here

$$T1\_1\_3 = \begin{cases} 1, T = 1 \\ 2, 1 < T \leq 3 \\ 3, T > 3 \end{cases} \tag{2}$$

Dwelling: Short-time dwelling lasting no more than 720 s during the sub-trip, such as dropping children at school, transfer, loading or unloading, refueling, and traffic factors like congestion and traffic control.

Dwelling time: Total duration time of the user's dwelling.

P_T: Payment if the user takes a taxi to complete the same trip. Taxis in Hangzhou adopt a block meter rate. The per kilometer charging standard changes with travel distance. Also, every 4 min of dwelling time will be regarded as 1 km. This study regards every trip between two activity spots as one single taxi trip. The distance calculated from the GPS data between these two activity spots is used as the taxi travel distance, and the dwelling time during the trip is used as the dwelling time of a taxi ride. Thus, the payment, if the user takes a taxi to complete the same entire trip, can be calculated as

$$P_{Ti} = \begin{cases} 11, D_i \leq 3\text{km} \\ 11 + 2.5*(D_i - 3 + \text{Dwellingtime}/4), 3\text{km} < D_i \leq 10\text{km} \\ 28.5 + 3.75*(D_i - 10 + \text{Dwellingtime}/4), D_i > 10\text{km} \end{cases} \tag{3}$$

$$P_T = \sum_i^T P_{Ti} \tag{4}$$

Proportion of the longest stop time: The proportion of the longest stop time (on each activity spot) of "DURATION". The longest stop time proportion is abbreviated below as PLST.

$$PLST = \frac{max\{T_{stop\_i}, \text{i} = 1, 2, \ldots, Number\ of\ Activity\ Spots\}}{DURATION} \tag{5}$$

CENTER: If the service station is in the central area of Hangzhou city, code 1; otherwise, code 0.

PEAK: If there is record of movement between 7:00 a.m. and 9:00 a.m. or 4:30 p.m. and 6:30 p.m., code 1; otherwise, code 0.

NIGHT: If there is record of movement between 9:00 p.m. and 5:00 a.m., code 1; otherwise, code 0.

Methodology

This paper (except for the discussion section) will include three main steps. First, group segmentation; data will be divided into three groups. Second, variable selection; large sample test will be used in this step. Third, modeling.

Step 1: Group Segmentation. Financial factors, for example travel cost, are mentioned frequently in existing research findings (*31–33*). Assuming that the users have considered the cost of both carsharing and taxi services beforehand, the trips are grouped by comparing the P_{CS} and P_T.

Considering that the pre-estimation of travel cost by people always slightly deviates from reality, a double-constraint tolerance range is set to judge whether taking a taxi is more economic. In this step, trips whose P_T is 20% higher and costing 10 CNY more than P_{CS} are assorted into Group I. Likewise, trips whose P_T is 20% lower and costing 10 CNY less than P_{CS} are assorted into Group III. The remaining trips are assorted into Group II. The result of group segmentation is demonstrated in the overview part of the next section.

Analysis of Group I and Group II will focus on the trip chain and member attribute characteristics. Analysis of Group III will place emphasis on the competitive advantage of the regular market of carsharing, based on the assumption of the rational-economic individual in economic theory.

Step 2: Large Sample Test. Before modeling, it is necessary to select variables suitable to incorporate into the model. A large sample test is used to test whether the sample has significant difference with the population. If so, it can be stated that this index affects the group's choice behavior, to a certain degree.

In this step, eight indices will be tested, including D, R, T, PLST, ACTIVE, CENTER, PEAK and NIGHT. Taking the D of Group I as an example, the process of a large sample test is as follows:

- The Null Hypothesis(H_0) of all the tests is "there is no significant difference between sample and population." Accordingly, the Alternative Hypothesis(H_1) is "there is significant difference between sample and population."
- Test Statistic z of both type of variables is calculated as

$$z = \sqrt{n}\frac{\overline{x} - \mu_{\overline{x}}}{\sigma} \tag{6}$$

where, n is the size of Group I, $\overline{x}$ is the mean of D of Group I, $\mu_{\overline{x}}$ is the mean of D of the population, σ is the standard deviation of D of the population.

- The value of 0.01 is chosen as the Significance Level α.

If the absolute value of z is larger than the critical value $u_{1-\alpha}(2.326)$, refuse H_0, take D as an influenced factor to Group I, otherwise, accept H_0, D is not an influenced factor to Group I.

All the test results and discussions of these eight indices of the three groups are shown in the next section. Additionally, variables that have significant difference in at least two groups are capable of being predictors of the model.

Step 3: Artificial Neural Network Modeling. As there are high correlations among all the factors (Table 2), the common discrete choice model would not work well here. Therefore, the artificial neural network (ANN) is introduced.

The ANN model finds a hyperplane in multidimensional data space to split the data. It helps to determine differences among data groups. There are advantages in adopting ANN. First, ANN can deal with relatively large numbers of predictors and predictors with high correlation among them. Next, for carsharing service operators, with their operation, the more data they obtain, the better model they gain. Finally, the ANN model can work well in this study, benefiting from the large size of the data; particularly since about 30% of the data will be used for testing, not training, and if there is not enough data, the model would not be effective.

The main output of the ANN is two matrices of weight. One is from the input layer to the hidden layer; the other is from the hidden layer to the output layer. There is no direct weight from the input layer to the output layer. The hidden and output layers are calculated as

$$H_j = \frac{1}{1 + e^{-(\sum_{i=1}^{n} w_{ij}x_i + a_j)}} \tag{7}$$

$$O_k = \sum_{j}^{m} H_j w_{jk} + b_k \tag{8}$$

where, w_{ij} is the weight from input layer to the hidden layer, w_{jk} is the weight from hidden layer to output layer, a and b are bias. Accordingly, $H_j \propto w_{ij}$ and $O_k \propto w_{jk}$. Thus, it can be surmised whether it is a positive effect or not from the two matrices.

The modeling work is conducted using the SPSS 22.0 software (IBM Corp., Armonk, NY, USA).

Results and Discussion

Overview of Group Segmentation and Large Sample Test

It is evident that the proportion of Group III (38%), whose P_{CS} is clearly lower than P_T, is the largest. However, an interesting finding is that the proportion of

Table 2. Correlations Table

Correlation coefficient	R	ACTIVE	CENTER	PEAK	NIGHT	T	D	PLST
R	1.000	.011	−.004	−.033**	.034**	.300**	.810**	−.173**
ACTIVE	.011	1.000	.048**	.157**	−.070**	.001	−.045**	.068**
CENTER	−.004	.048**	1.000	−.002	.023**	−.024**	−.012	−.073**
PEAK	−.033**	.157**	−.002	1.000	.131**	−.030**	−.057**	.065**
NIGHT	.034**	−.070**	.023**	.131**	1.000	.020*	.037**	−.006
T	.300**	.001	−.024**	−.030**	.020*	1.000	.662**	.132**
D	.810**	−.045**	−.012	−.057**	.037**	.662**	1.000	−.297**
PLST	−.173**	.068**	−.073**	.065**	−.006	.132**	−.297**	1.000

**Correlation is significant at the 0.01 level (2-tailed).

*Correlation is significant at the 0.05 level (2-tailed).

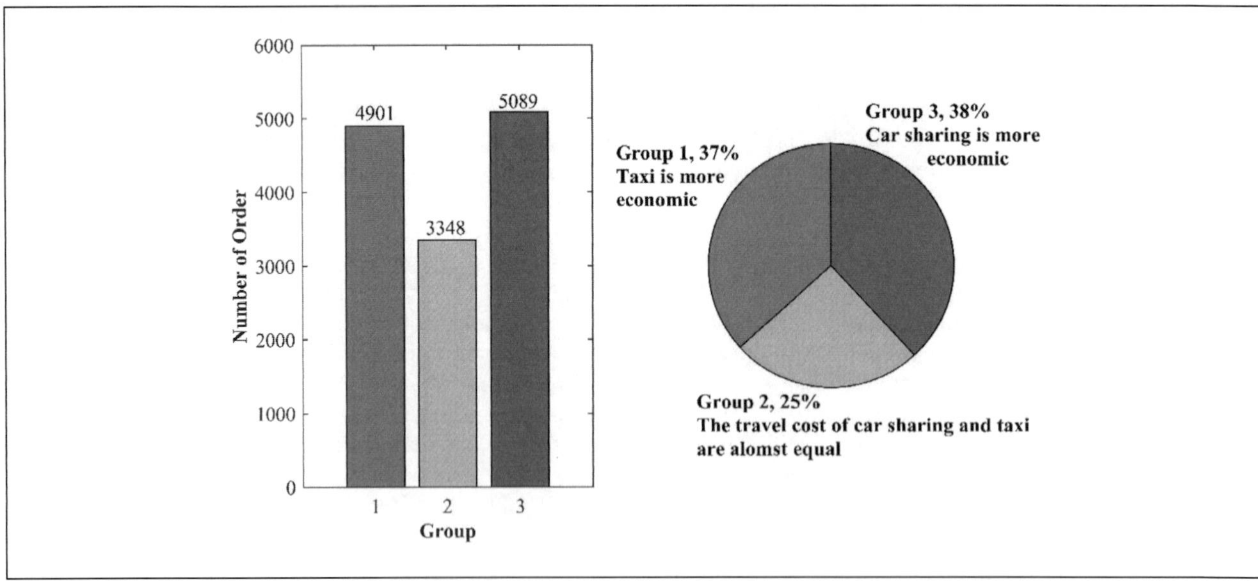

Figure 2. Group statistics and description.

Table 3. Classification and Percent Correct

		Predicted			
Sample		Group I	Group II	Group III	Percent correct
Training	Group I	2883	277	269	84.1
	Group II	685	1020	641	43.5
	Group III	222	294	3004	85.3
	Overall percent	40.8%	17.1%	42.1%	74.3
Testing	Group I	1229	112	131	83.5
	Group II	284	434	284	43.3
	Group III	88	143	1338	85.3
	Overall percent	39.6%	17.0%	43.4%	74.2

Group I (37%), whose P_{CS} is obviously higher than P_T, is just a little bit lower than that of Group III. This is probably due to the users' strong preference for carsharing services, based on reasons associated with being conscious of convenience and the environment or their lifestyle and habits (*11*). Meanwhile, there is a moderate proportion of Group II (25%), whose P_{CS} and P_T are almost equal (Figure 2).

According to the result of the large sample test, D, R, T, PLST ACTIVE and PEAK are incorporated into the model. D, R and PLST of the three groups are significantly different from the population, respectively; whereas T, ACTIVE and PEAK of two of the groups are significantly different from the population; and among the three groups, it does not matter whether it is night time or not. Also, the result reveals that trips in Group I occur during peak hours more often than in the general population, whereas trips in Group III are opposite to those in Group I. In addition, there are more active members in Group II, in contrast to Group III.

Result of ANN Modeling

In the final model, P_{CS} is also selected because it still plays an important role. The correct percentage of the model falls sharply in the absence of P_{CS}.Table 4 and Figure 3*a* carry the same information. The width and color of the lines in Figure 3*a* represent the values in Table 4. The correct percentage of Group I and Group III is 83% and 85% (Table 3), respectively, which are adequate. The correct percentage of the Group II is merely 43%, but the Group II-Group II pair is still the highest (Figure 3*b* and Table 3). The reason why the correct percentage of the Group II is around half of that of

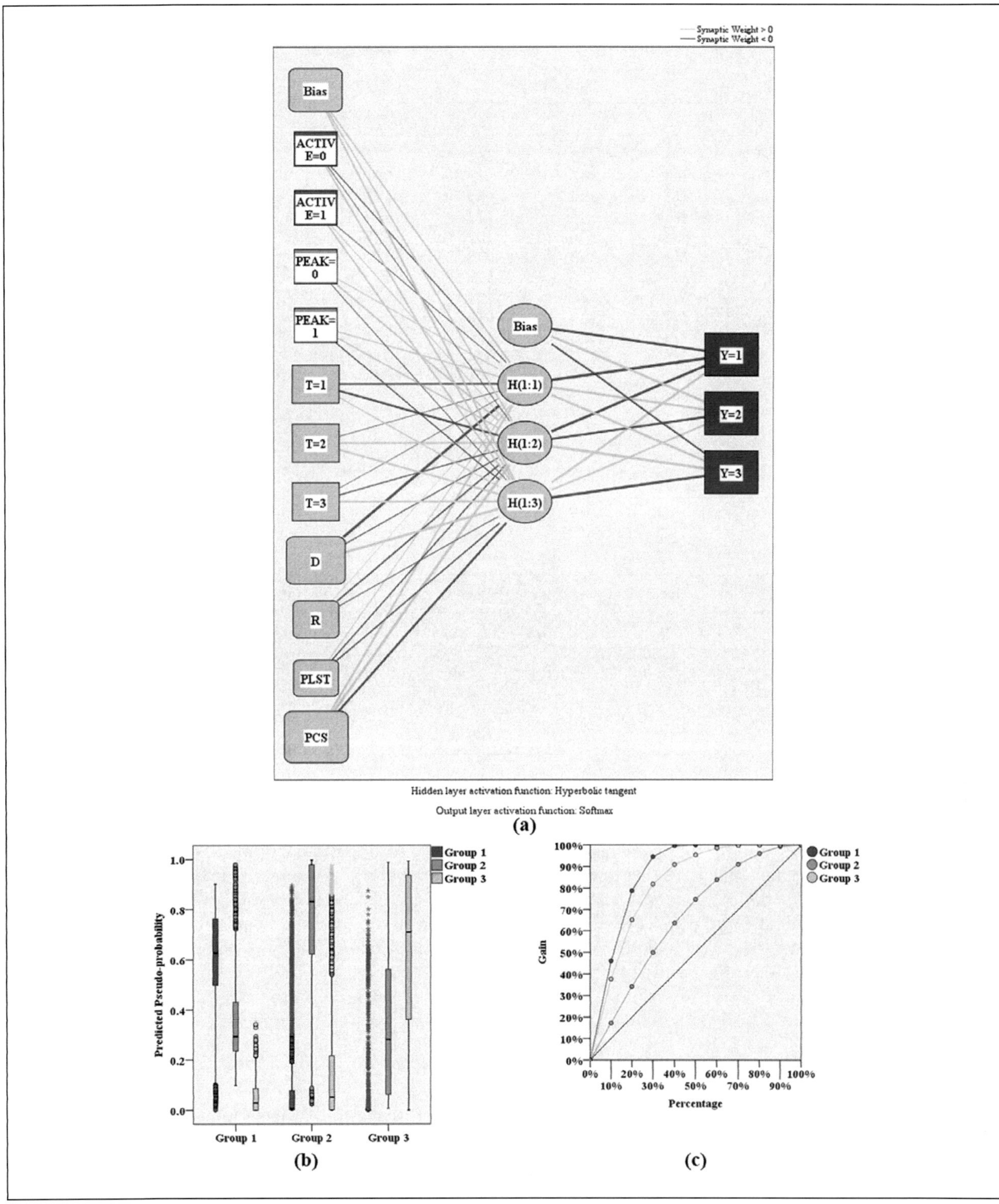

Figure 3. Showing (*a*) network diagram, (*b*) predicted-observed chart, and (*c*) cumulative gains chart.

Table 4. Parameter (Weight) Estimates

Predictor		Predicted						
		Hidden layer 1				Output layer		
		H(1:1)	H(1:2)	H(1:3)	H(1:4)	[Y21 = 1]	[Y21 = 2]	[Y21 = 3]
Input layer	[Bias]	−.319	−.542	.263	.271			
	[ACTIVE = 0]	−.563	−.443	.611	.047			
	[ACTIVE = 1]	−.575	−.334	.569	−.301			
	[PEAK = 0]	−.471	.189	.193	−.208			
	[PEAK = 1]	.636	.361	.102	−.479			
	[T1_1_3 = 1]	.214	−.470	−.528	.912			
	[T1_1_3 = 2]	−.249	.617	1.556	.132			
	[T1_1_3 = 3]	−.091	−.821	−.180	−1.029			
	R	.277	−.354	−.018	.126			
	D	−2.975	.596	−1.166	.937			
	PLST	.268	−.470	−.424	.014			
	P_{CS}	1.741	−.319	3.221	−.891			
Hidden layer 1	[Bias]					−1.142	.414	.013
	H(1:1)					2.442	.304	−2.876
	H(1:2)					.172	−.638	.374
	H(1:3)					2.748	−.445	−2.088
	H(1:4)					−1.119	−.442	.902

Table 5. Regular and Extended Market of Carsharing

Market		Driving distance	Activity spots	Cost	Peak hour	Others	Possible purpose
Regular		long	0–1	low	no	short time stop	picking-up; delivering
Extended	1	short	1–3	high	yes	long time stop; might occur at areas where hailing a taxi is difficult	commuting
	2	short	>3	low	not salient		not sure

the other groups might be that Group II has both maximum and minimum thresholds in the group segmentation step, whereas the other groups only have one threshold. Moreover, the predictive ability of the model is appreciable, according to the cumulative gains chart (Figure 3*c*), in which a larger area under the curves indicates a better performance by the model. Thus, the result of the modeling is still acceptable.

Discussion

Regular Market: Analysis of Group III. The characteristics of this group of trips could reveal the competitive advantages of the regular market of carsharing services compared with those of the taxi services.

The data shown in Table 4 reveal that the most salient positive effect is from D and T1_1_3 = 1, whereas the most salient negative affect is from P_{CS}, T1_1_3 = 3 and PEAK = 1. The fact that T1_1_3 = 1 has a positive effect whereas T1_1_3 = 3 has a negative effect implies that these kind of trips are simple, having less or even no activity spots. PEAK = 1 as a negative effect means that these trips tend to occur during off-peak hours.

The probability distribution plot (Figure 4*a*) reveal that most driving distances of Group III lie in the interval from 20 km to 60 km, and the total cost is less than 50 CNY. Considering that the radius of Hangzhou city is around 23 km, 20 to 60 km is long-distance travel for urban activity. Besides, the PLST of Group III is obviously lower, indicating that there is a relatively long driving time and short stop time, which correspond to "simple trip".

In conclusion, the regular market of carsharing services compared with taxi services is those simple long-distance trips during off-peak hours (Figure 4*b*), and the long-distance characteristic matches the survey conducted by Wang et al. (*5*) in Shanghai. The regular market may involve trips with the purpose of delivering and picking somebody up.

Extended Market: Analysis of Group I and Group II. The characteristics of these groups could reveal how carsharing

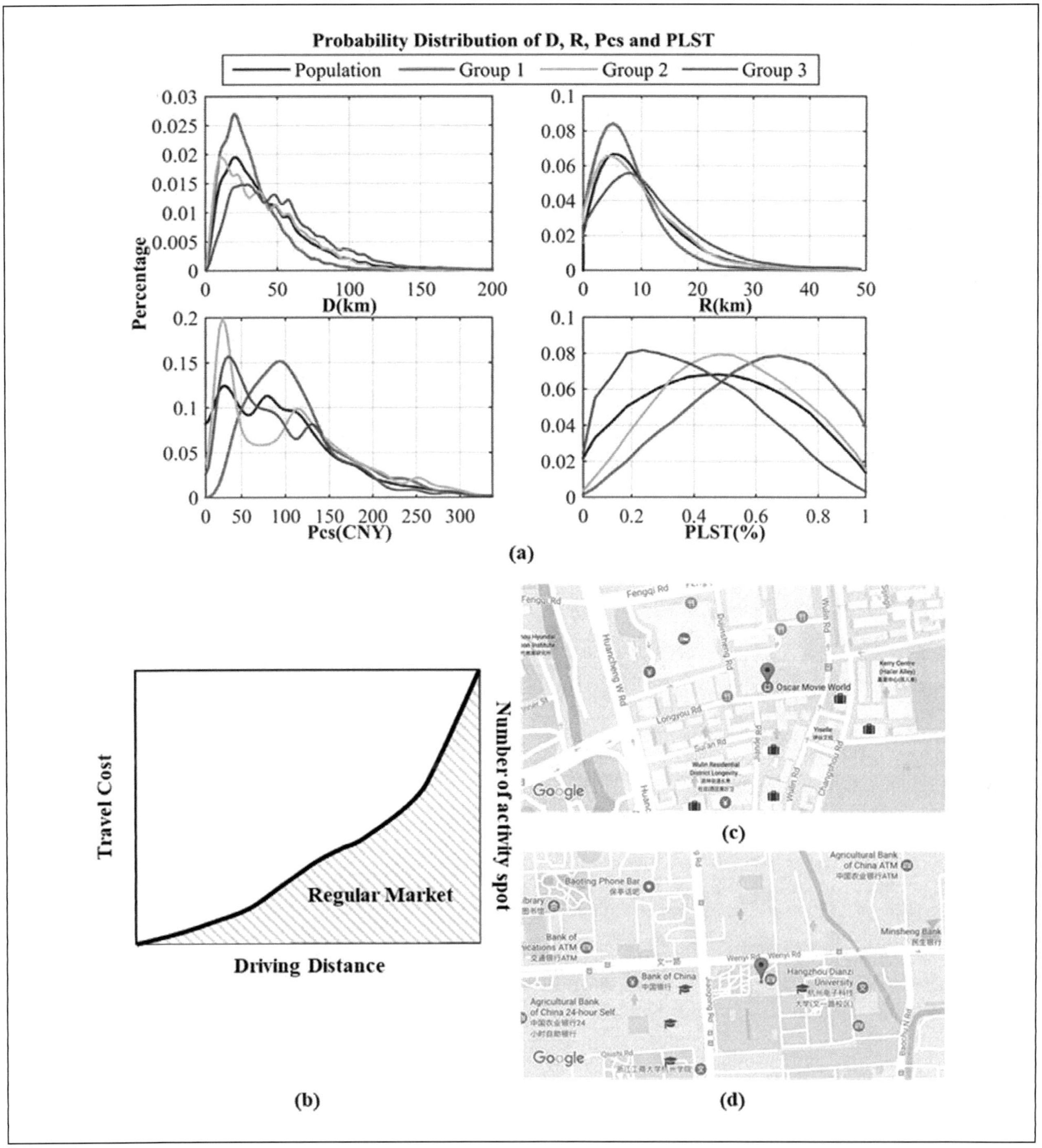

Figure 4. Showing (*a*) probability distribution plot of continuous variables, (*b*) regular market of carsharing (sketch), (*c*) Oscar movie world service station surroundings land use classification, and (*d*) Sunshine pier service station surroundings land use classification.

services get their extended market, and specifically, what these trips focus on.

The data for Group I shown in Table 4 reveal that the most salient positive effect is from P_{CS}, PEAK = 1 and T1_1_3 = 2, whereas the most salient negative effect is from D and T1_1_3 = 1. As for Group II, the most salient positive effect is from P_{CS} and T1_1_3 = 3, whereas the most salient negative effect is from D. These results imply that trips in Group I and Group II have more activity spots, but shorter travel distance. Compared with Group III, these trips are complex. Moreover, PEAK = 1 is a positive effect for Group I, implying that these trips tend to occur during off-peak hours, whereas Group II does not show this characteristic.

The combination of the probability distribution plots (Figure 4*a*), indicates that most of the driving distances of these two groups of trips lie at about 20 km; most of the values of Group II are even smaller. However, the travel cost of these trips shows different characteristics. Most of the costs of Group I are around 100 CNY. Group II has two peak values, namely about 25 CNY and over 100 CNY, and the former is significantly predominant. It can be stated that trips in the extended market tend to be of short driving distances and the travel costs are both low and high. The reason why short driving distances would result in high travel cost is that the PLST of Group I is larger. The charging standard considers both distance and time. Thus, although they are not driving, they still charged for stopping.

To examine the services from a deeper perspective, the service stations whose proportion of the trips predominantly belongs to Group I is analyzed. The proportion of trips belonging to Group I in the following service stations are over 60%. One is Oscar Movie World in the Xiacheng District, the other is Sunshine Pier in the Xihu District. A search on the online map reveal that they are both located in the central area. Oscar Movie World lies in the most economically prosperous area of Hangzhou, where land use is mainly commerce, including companies, restaurants, hotels and malls (Figure 4*c*). Land use classifications around Sunshine Pier are mainly education, finance and residential (Figure 4*d*). The land use around the service station implies that there might be high demand of travel due to entertainment and commuting. However, using carsharing services is more convenient and reliable, considering that it is extremely difficult to hail a taxi in Hangzhou, especially during peak hours (*34*). Thus, these can be the evidence to support the analysis of Group I and Group II.

Summing up all the above information, the extended market consists of two types of trips:

- Type 1: short driving distance with long time stops at activity spots; travel during peak hours; might occur at areas where it is difficult to hail a taxi.
- Type 2: short driving distance with more activity spots; low cost.

What needs to be emphasized is that the characteristics of Type 1 are highly correspondent with commuting.

Summary

The analysis performed on the real operation order data shows that travel cost is not the key factor of mode choice, at least not between carsharing services and taxi services. In contrast with taxi services, around two-thirds of carsharing trips are not covered by its regular market, implying that competition exists among these trips. Table 5 gives a brief review and summary of the analysis of the carsharing markets.

Conclusions and Future Directions

This paper analyzed three groups of trips and identified the regular and extended markets of carsharing services compared with taxi services, based on data mining and modeling results. The conclusions of this research can be summarized as follows:

- The location of carsharing services is indeed different from that of taxi services in urban transportation systems because there is a specific regular market for carsharing services according to comparison with taxi services. This conclusion is consistent with Prettenthaler (*3*).
- Since about two-thirds of the trips are outside the carsharing regular market, the cost difference is not the key, or at least not the only, impact factor. In addition, the non-price advantages of carsharing services do not exist in suburban trips as expected, implying that there is competition between carsharing and taxi services in some trips in the central area.
- The carsharing market might expand due to its convenience and reliability, which coincides with the results reported by Prieto et al. (*10*) and Catalano et al. (*31*). For carsharing system operators, it is a good opportunity to expand and adjust their business. They should allocate more vehicles at service stations in areas where most trips fit the extended market. Additionally, they should notice that price battle is no longer the most effective strategy.
- City administrators can recognize the weakness of the existing taxi system and realize that the development of carsharing is necessary. As the extended market grows, if they want to control the development of carsharing for the extended market, then there must be improvement in the taxi and public transit systems for commuters, since public transport policy is an important impact factor (*10*). Alternatively, they can implement beneficial policies for carsharing to encourage travelers to use it, and improve the transportation system service.

In this study, there are still issues to be solved. First, the exact purpose and detailed information of Type 2 of the extended market is not clear. Questionnaire survey directed at members who undertake this kind of travel is needed. Second, there are trips that probably involve using carsharing as a mode of commuting (Extended market, Type 1). Whether this use pattern should be encouraged or not is still worth discussing. Third, similar comparison should be extended to other carsharing systems to verify the result of this work.

Author Contributions

The authors confirm contribution to the paper as follows: study conception and design: Ying Hui, Cuiyi Liu; data collection: Ying Hui, Cuiyi Liu, Mengtao Ding; analysis and interpretation of results: Cuiyi Liu, Ying Hui; draft manuscript preparation: Cuiyi Liu. All authors reviewed the results and approved the final version of the manuscript.

References

1. Martin, E. W., and S. A. Shaheen. Greenhouse Gas Emission Impacts of Carsharing in North America. *IEEE Transactions on Intelligent Transportation Systems*, Vol. 12, No. 4, 2011, pp. 1074–1086.
2. Katzev, R. Car sharing: A New Approach to Urban Transportation Problems. *Analyses of Social Issues and Public Policy*, Vol. 3, No. 1, 2003, pp. 65–86.
3. Prettenthaler, F. E., and K. W. Steininger. From Ownership to Service Use Lifestyle: The Potential of Car Sharing. *Ecological Economics*, Vol. 28, No. 3, 1999, pp. 443–453.
4. Lane, C. PhillyCarShare: First-year Social and Mobility Impacts of Carsharing in Philadelphia, Pennsylvania. *Transportation Research Record: Journal of the Transportation Research Board*, 2005. 1927: 158–166.
5. Wang, M., E. Martin, and S. Shaheen. Carsharing in Shanghai, China: Analysis of Behavioral Response to Local Survey and Potential Competition. *Transportation Research Record: Journal of the Transportation Research Board*, 2012. 2319: 86–95.
6. Burkhardt, J., and A. Millard-Ball. Who is attracted to Carsharing? *Transportation Research Record: Journal of the Transportation Research Board*, 2006. 1986: 98–105.
7. Kato, H., A. Inagi, and T. Igo. Awareness and Potential Choices of Carsharing: Comparative Analysis of Data from Four Japanese Cities. Presented at 91st Annual Meeting of the Transportation Research Board, Washington, D.C., 2012.
8. Costain, C., C. Ardron, and K. N. Habib. Synopsis of Users' Behaviour of a Carsharing Program: A Case Study in Toronto. *Transportation Research Part A: Policy and Practice*, Vol. 46, No. 3, 2012, pp. 421–434.
9. Wielinski, G., M. Trépanier, and C. Morency. What About Free-Floating Carsharing? A Look at the Montreal, Canada, Case. *Transportation Research Record: Journal of the Transportation Research Board*, 2015. 2536: 28–36.
10. Prieto, M., G. Baltas, and V. Stan. Car Sharing Adoption Intention in Urban Areas: What are the Key Sociodemographic Drivers? *Transportation Research Part A: Policy and Practice*, Vol. 101, 2017, pp. 218–227.
11. Schaefers, T. Exploring Carsharing Usage Motives: A Hierarchical Means-end Chain Analysis. *Transportation Research Part A: Policy and Practice*, Vol. 47, 2013, pp. 69–77.
12. Schmöller, S., and K. Bogenberger. Analyzing External Factors on the Spatial and Temporal Demand of Car Sharing Systems. *Procedia-Social and Behavioral Sciences*, Vol. 111, 2014, pp. 8–17.
13. Hui, Y., M. Dinga, C. Qiana, W. Wanga, and Q. Xu. Research on the Operational Characteristics of Car Sharing Service Stations: A Case Study of a Car Sharing Program in Hangzhou. *Transportation Research Procedia*, Vol. 25, 2017, pp. 4144–4160.
14. Hui, Y., M. Dinga, K. Zheng, and D. Lou. Observing Trip Chain Characteristics of Round-Trip Carsharing Users in China: A Case Study Based on GPS Data in Hangzhou City. *Sustainability*, Vol. 9, No. 6, 2017, pp. 949.
15. Morency, C., M. Trepanier, and B. Agard. Typology of Carsharing Members. Presented at 90th Annual Meeting of the Transportation Research Board, Washington, D.C., 2011.
16. Leclerc, B., M. Trépanier, and C. Morency, Unraveling the Travel Behavior of Carsharing Members from Global Positioning System Traces. *Transportation Research Record: Journal of the Transportation Research Board*, 2013. 2359: 59–67.
17. Tulic, M., D. Bauer, and W. Scherrer. Link and Route Travel Time Prediction Including the Corresponding Reliability in an Urban Network Based on Taxi Floating Car Data. *Transportation Research Record: Journal of the Transportation Research Board*, 2014. 2442: 140–149.
18. Hochmair, H. H. Spatiotemporal Pattern Analysis of Taxi Trips in New York City. *Transportation Research Record: Journal of the Transportation Research Board*, 2016. 2542: 45–56.
19. Qian, X., and S. V. Ukkusuri. Spatial Variation of the Urban Taxi Ridership Using GPS Data. *Applied Geography*, Vol. 59, 2015, pp. 31–42.
20. Yang, C., and E. Gonzales. Modeling Taxi Trip Demand by Time of Day in New York City. *Transportation Research Record: Journal of the Transportation Research Board*, 2014. 2429: 110–120.
21. Nam, D., K. K. Hyun, H. Kim, K. Ahn, and R. Jayakrishnan. Analysis of Grid Cell–Based Taxi Ridership with Large-Scale GPS Data. *Transportation Research Record: Journal of the Transportation Research Board*, 2016. 2544: 131–140.
22. Ge, Y., H. Xiong, A. Tuzhilin, K. Xiao, M. Gruteser, and M. Pazzani. An Energy-efficient Mobile Recommender System. *Proc., 16th ACM SIGKDD International Conference on Knowledge Discovery and Data Mining*. ACM, Washington, D.C., US, NY, US, 2010.
23. Chawla, S., Y. Zheng, and J. Hu. Inferring the Root Cause in Road Traffic Anomalies. *12th International Conference on Data Mining (ICDM), 2012 IEEE*. IEEE, Brussels, Belgium, Washington, D.C., US, 2012.

24. Pang, L. X., S. Chawla, W. Liu, and Y. Zheng. On Detection of Emerging Anomalous Traffic Patterns Using GPS Data. *Data & Knowledge Engineering*, Vol. 87, 2013, pp. 357–373.
25. Zhang, J. Smarter Outlier Detection and Deeper Understanding of Large-scale Taxi Trip Records: A Case Study of NYC. *Proc., ACM SIGKDD International Workshop on Urban Computing*, ACM, Beijing, China, NY, US, 2012.
26. Donovan, B., and D. B. Work. Using Coarse GPS Data to Quantify City-scale Transportation System Resilience to Extreme Events. *Computer Science*, arXiv preprint arXiv:1507.06011, 2015.
27. Bohte, W., and K. Maat. Deriving and Validating Trip Destinations and Modes for Multiday GPS-Based Travel Surveys: Application in the Netherlands. Presented at 87th Annual Meeting of the Transportation Research Board, Washington, D.C., 2008.
28. Vidalakis, L. *Observing Travel Behaviour from GPS Data-A Tool Comparison Survey in the Torino Metropolitan Area.* Doctoral thesis. Interuniversity Department of. Regional and Urban Studies and Planning (DIST), Politecnico di Torino, 2013.
29. Jennings, D., S. J. Cormack, A. Coutts, and R. Aughey. Variability of GPS Units for Measuring Distance in Team Sport Movements. *International Journal of Sports Physiology and Performance*, Vol. 5, No. 4, 2010, pp. 565–569.
30. McGuckin, N., and Y. Nakamoto. *Trips, Chains and Tours-Using an Operational Definition.* Presented at National Household Travel Survey Conference, Washington, D.C., 2004.
31. Catalano, M., B. Lo Casto, and M. Migliore, Car Sharing Demand Estimation and Urban Transport Demand Modelling Using Stated Preference Techniques. *European Transport*, Vol. 40, No. 44, 2008, pp. 33–50.
32. Correia, G., and J. M. Viegas. Carpooling and Carpool Clubs: Clarifying Concepts and Assessing Value Enhancement Possibilities Through a Stated Preference Web Survey in Lisbon, Portugal. *Transportation Research Part A: Policy and Practice*, Vol. 45, No. 2, 2011, pp. 81–90.
33. Efthymiou, D., C. Antoniou, and P. Waddell, Factors Affecting the Adoption of Vehicle Sharing Systems by Young Drivers. *Transport Policy*, Vol. 29, 2013, pp. 64–73.
34. E, Ping. Beijing, Shanghai, and Hangzhou: Shanghai and Hangzhou. What is the Solution to These Blocking Cities for Taxi Problem? *Xiao Kang Index of Chinese Public Service*, No. 3, 2013, pp. 60–61.

The Standing Committee on Urban Transportation Data and Information Systems (ABJ30) peer-reviewed this paper (18-02383).

Article

TRR
JOURNAL OF THE TRANSPORTATION RESEARCH BOARD

Transportation Research Record
2018, Vol. 2672(42) 127–136

DOI: 10.1177/0361198118788462
journals.sagepub.com/home/trr
SAGE

Is There a Limit to Adoption of Dynamic Ridesharing Systems? Evidence from Analysis of Uber Demand Data from New York City

Raymond Gerte[1], Karthik C. Konduri[1], and Naveen Eluru[2]

Abstract

Recent technological advances have paved the way for new mobility alternatives within established transportation networks, including on-demand ride hailing/sharing (e.g., Uber, Lyft) and citywide bike sharing. Common across these innovative modes is a lack of direct ownership by the user; in each of these mobility offerings, a resource not owned by the end users' is shared for fulfilling travel needs. This concept has flourished and is being hailed as a potential option for autonomous vehicle operation moving forward. However, substantial investigation into how new shared modes affect travel behaviors and integrate into existing transportation networks is lacking. This paper explores whether the growth in the adoption and usage of these modes is unbounded, or if there is a limit to their uptake. Recent trends and shifts in Uber demand usage from New York City were investigated to explore the hypothesis. Using publicly available data about Uber trips, temporal trends in the weekly demand for Uber were explored in the borough of Manhattan. A panel-based random effects model accounting for both heteroscedasticity and autocorrelation effects was estimated wherein weekly demand was expressed as a function of a variety of demographic, land use, and environmental factors. It was observed that demand appeared to initially increase after the introduction of Uber, but seemed to have stagnated and waned over time in heavily residential portions of the island, contradicting the observed macroscopic unbounded growth. The implications extend beyond already existing fully shared systems and also affect the planning of future mobility offerings.

Technology has long been the driver for advances in our everyday means of transport. The development of data-enabled and location-aware mobile phones, combined with adoption of sharing economies (also referred to as collaborative economies or peer-to-peer networks), has opened the door for novel modal offerings like Uber, Lyft, and citywide bike sharing systems. These specific modal offerings are referred to generally as dynamic ridesharing or ridesourcing systems. They capitalize on the growth of mobile phone adoption and integrated global positioning (GPS) capabilities to actively link riders to drivers in an on-demand setting. Historically, ridesharing has filled unique niches in the transportation landscape (carpooling, preplanned trips, etc.) and has consistently been encouraged for its ability to minimize user costs and reduce congestion. However, the widespread adoption of conventional ridesharing has been linked with an inability to capture impromptu trips (*1*). Due to the fact that almost all trips need to be prearranged, conventional ridesharing has substantial limitations, and users that recognize its benefits are unwilling to make the switch because of inconvenience. With the growth of dynamic ridesharing, the ability to travel on demand in a manner similar to taxi trips opens a wealth of mobility options. In this paper, the temporal trends in one ridesharing offering, namely Uber, are explored.

As the current industry leader, Uber's growth is readily apparent. At the close of 2014, Uber operated in 230 cities worldwide; by June 2015 Uber had grown to operate in 300 cities (*2*, *3*). In only 6 years of existence (2009–2015), Uber reported facilitating over one billion trips. As of June 2016, Uber reported surpassing two billion total trips across 450 cities (*4*). The popularity of Uber can be associated with its inherent flexibility and user-perceived efficiency of use. When using the Uber

[1]Department of Civil and Environmental Engineering, University of Connecticut, CT
[2]Department of Civil, Environmental and Construction Engineering, University of Central Florida, FL

Corresponding Author:
Address correspondence to Raymond Gerte: raymond.gerte@uconn.edu

app, individuals are able to hail a ride remotely from anywhere within the service city, assuming a driver is available. This is an improvement over the call ahead or on-street hailing methods of its most similar competitor, the taxi, while also offering service in locations that may not have many viable alternatives. In a major metropolitan hub like New York City (NYC), the growth of Uber has been substantial. From June 2014 to June 2015, Uber's Manhattan monthly June pickups increased by over 250% with over one million more trips made using the dynamic ridesharing app (*5*).

Although Uber's growth is noteworthy, there has also been substantial controversy in regards to the ridesharing giant. Established taxi groups and regulators have filed numerous lawsuits against Uber and other dynamic ridesharing corporations for infringing on market share without the burden of for hire vehicle (FHV) licensure and regulation (*6*). There has also been considerable debate on the ethics of the dynamic pricing incorporated into Uber's app (*7*). Furthermore, there is ongoing discussion regarding Uber's safety, its policies concerning drivers, and how these policies compare with similar regulations governing other FHV companies (*8*). The growing impact of dynamic ridesharing is becoming an important piece in the planning of cities and regions around the United States and elsewhere. These systems are not only expanding and growing in the short term, but also have substantial implications for the future, namely in regards to the deployment of autonomous vehicles (AVs). AV offerings are projected to perform in a shared, on-demand manner, similar to ridesharing offerings. Furthermore, shared AVs have been identified as one of the key pieces in realizing the complete suite of AV benefits (*9*).

As such, understanding ridesharing can provide insight into how these dynamic systems are currently performing and being received by the public, while also offering a possible window into the impact of future modal offerings employing a similar concept. Substantial investigation into how dynamic ridesharing affects travel behaviors and integrates into existing transportation networks is lacking. To address these limitations, this study focuses on quantifying and understanding temporal trends in Uber demand within a major metropolitan center. In particular, the study explores whether growth in the adoption and usage of these modes is unbounded or there is a limit to the adoption of dynamic ridesharing. In order to pursue this research, trip pickup data made available by Uber through a Freedom of Information Law (FOIL) request by the analytics group FiveThirtyEight, was utilized. Weekly demand, aggregated spatially to a taxi zone level, was analyzed using a panel-based random effects model while also accounting for potential heteroscedasticity and autocorrelation issues. Demand is expressed as a function of various demographic, land use, and environmental factors to further explore the trends.

The remainder of this paper is organized as follows: the Existing Research section summarizes the literature and positions the current study; Data describes the data and presents the approach to data preparation; the Descriptive Analysis section includes a summary of the data; Panel Model outlines the model used; Results discusses the findings, and finally, the Conclusion section provides concluding remarks and offers some directions for future research.

Existing Research

The existing research on dynamic ridesharing demand, and analysis of demand trends in particular, is quite limited. These limitations can be attributed to the lack of readily available data from the providers of these modal offerings, and limitations in data that is publicly available.

The first vein of research aims to offer a more complete understanding of dynamic ridesharing system performance in relation to available supply and demand. Chen et al. and Guo et al. explore the characteristics of Uber's surge pricing and assess its impact on supply and demand (*7*, *10*, *11*). The authors collected four weeks of comprehensive data on all Uber vehicle types using the Uber App API and Client App API. The data gathered included information on the supply of vehicles, the demand for vehicles, and the associated surge price multiplier in both downtown San Francisco and Midtown Manhattan. They found that the majority of vehicles in their study areas were concentrated in commercial and tourist areas within each study zone. Further, they observed that surge pricing did in fact negatively affect demand while encouraging supply at the study scale. Correa et al. used the same data explored in this paper to quantify changes in Uber and taxi demand over space and time and found substantial significant growth in demand over time with a high correlation between taxi and Uber modes in the core of the city (*12*). Increases in roadway length, income, and job opportunities, as well as decreases in transit access time and vehicle ownership, were observed with increased demand for Uber.

Another major line of research is the comparison between dynamic ridesharing and traditional mode offerings such as taxi and transit. Rayle et al. found that dynamic ridesharing users are typically younger and more educated than those that use taxis (*13*). Furthermore, although ridesharing does remove potential trips from taxis, there is substantial evidence that users opted to switch from available transit modes citing travel time as the reason. Survey respondents using dynamic ridesharing were skewed toward no vehicle ownership with multiple users traveling together. Researchers Cramer and Krueger compiled a direct comparison of Uber's utilization rate (percentage of time drivers are serving passengers) against that of taxis in five major U.S. cities (*14*). Their results showed that, as a whole, Uber is the more efficient mode of transportation compared with taxis. Poulsen et al. explored the relationship between Uber and green cabs in NYC and found that although demand for both modes is growing, Uber's is growing at a much faster rate, and green cabs perform better in poor neighborhoods (*15*). These studies highlight the fact that Uber is similar to existing taxi infrastructure, however, they highlight key differences that warrant further investigation into how people are using dynamic ridesharing on a citywide scale.

The final set of literature pertaining to dynamic ridesharing and demand is informal in the form of blog posts and news articles. These are mostly based on Uber demand information for all of NYC made openly available by researchers at FiveThiryEight; the same data was also used in this paper. FiveThirtyEight researchers have published numerous articles on the topic. Fischer-Baum and Bialik address how Uber's growth should be classified (*16*). They found that in the urban core of Manhattan, Uber trips seem to come almost entirely from taxis, with the sum of total trips across both modes remaining the same across study years. Schneider explored the trends in Uber demand and found that Uber pickups in the urban core are growing, and notes that these pickups increase more substantially in times of inclement weather compared with increases in taxi pickups (*5*).

The research highlighted above provides interesting insights into the demand for dynamic ridesharing, but there are substantial gaps. A key question that this research attempts to answer is whether growth in demand for these systems is subject to saturation, or is unbounded. To this end, the study explored demand over time, while also capturing the impact of demographic, built environment, and environmental variables. In order to explore this idea quantitatively, this study developed a panel-based random effects model that is similar in nature to the works by Faghih-Imani et al.; the authors studied the relationship between NYC Citi Bike ridership, land use, and urban form amongst other studies focused on Citi Bike and taxi travel time (*17*). Their research focused on three main types of variables: those related to the environment, temporal variables, and land use variables. A similar structure is also observed in an analysis of bus ridership with the incorporation of real-time location information by Brakewood et al. and work by Campbell and Brakewood (*18*, *19*). These studies, along with some of the variables described in the work above, served to motivate the modeling aspect of this research and helped guide selection of potential demographic, land use, and environmental factors that may influence Uber demand.

Data

As noted in the limitations regarding dynamic ridesharing research, restrictions in data availability and quality have had a substantial impact on the volume of research. Recent adoption of open-data initiatives (wherein data is made publicly available) and the practice of making data available in response to public requests (through FOIL and other similar mechanisms) have made much of this work possible. The Uber trip information was made available by Uber after a FOIL request and was gathered from FiveThirtyEight's GitHub repository (*20*). Although these data are the only source of information on Uber demand, they also have several limitations that dictated the data preparation and approach to analysis. First, Uber data was only available for 6 months in 2014 (April through September) and 6 months in 2015 (January through June). Within this, trip pickup location was recorded at a latitude/longitude level in 2014 and aggregated up to a predefined taxi zone level in 2015, including pickup date and time information. Due to the spatially aggregated nature of 2015 data, taxi zone was chosen as the spatial unit of analysis. Subsequently, all remaining data (demographic, land use, and environmental factors) were spatially aggregated to the taxi zone level. Taxi zone delineations are defined by the NYC Taxi and Limousine Commission and align with neighborhood boundaries in NYC. This analysis focused only on the taxi zones in the borough of Manhattan. There are 69 unique taxi zones in Manhattan, which includes three zones for ferries to Ellis Island, Liberty Island, and Governors Island, and four zones for major parks, namely, Central Park, Battery Park, Inwood Hill Park, and Highbridge Park. To add context, there are 288 census tracts located in the same study area. The boundaries of census tracts and taxi zones are coincident. However, taxi zones are relatively larger in spatial extent compared with tracts and on average each taxi zone is comprised of about four census tracts. Built environment variables were collected at the tax lot level and gathered from the NYC Planning Department's website (*21*). Information on transit stations/stops and bike share infrastructure was accessed through NYC's Open Data hub (*22*). Environmental variables like temperature and precipitation were gathered at the day level from 2014 through 2015 in Central Park from the National Climate Data Center hosted by the National Oceanic and Atmospheric Administration (*23*). All demographic data were collected at the census tract level from the American

Community Survey's 5-year estimates for 2014 and 2015 and aggregated to the taxi zone level.

Substantial data processing and data preparation was performed prior to the analysis. First, the taxi zone aggregated Uber pickups were summed to the week level because of the high level of variability at the day level. From the 12 months of initial data, there were 54 total weeks, however, weeks that fell at the beginning and end of the dataset were incomplete. As a result, these incomplete weeks were dropped, leaving 49 total complete weeks of demand data. For the remainder of the paper, the weekly pickup counts are used to constitute demand and will be referred to as such. Daily weather information is treated similarly. Built environment variables were assumed to not vary across the investigation period and the demographics of each taxi zone were assumed to be represented by the 2014 and 2015 ACS estimates, respectively.

Descriptive Analysis

The growth in overall demand for Uber is evident both from the literature presented above and its ubiquity in the societal lexicon. When analyzing the data at the borough level, substantial growth in Uber demand over time is easily observed. The total number of trips made in Manhattan from April 2014 through September 2014 was just under 140,000 per week. In 2015, that number rose to just over 430,000 trips per week from January through June. However, it is important to note that this growth was not evenly distributed. At the taxi zone level, the top five taxi zones in both 2014 and 2015 were as follows: Midtown Center, Union Square, Tribeca/Civic Center, and both East and West Village. All of these zones are located either in Midtown or lower Manhattan and together account for about 20% of the total trips in 2014 and 2015, respectively. In reviewing the shifts in demand by each taxi zone individually, differences were noted in zones that were more residential. These differences typically manifested as lower overall demand each week, but the growth of that demand appeared lower than that in commercial zones. This observation is reasonable because residential areas typically have less tourism or visitor traffic.

To better understand overall demand and the temporal trends, the taxi zones were classified into five groups based on a residential built environment measure derived from the land use dataset. The first group, residential taxi zones, are those zones that have a higher ratio of residential land use than commercial land use. The second group, commercial zones, are those that have a higher ratio of commercial land use than residential. These two groups were then further divided into those found in northern Manhattan (north of 86th Street on the east side of the island and north of Central Park) and those not. Taxi zones that were exclusively parks were treated as the fifth group. Figure 1 highlights these trends with respect to four of the land use areas classified above, excluding parks. From the map inset, it is shown that residential zones surround Central Park and the northern portion of the island, as well as in the south-east. The line shading for the demand plots ranges from dark (less residential/commercial) to light (heavily residential/commercial). Furthermore, residential- and commercial-dominant zones are shaded in blue and orange hues respectively. Therefore, more mixed-use areas, that is, places with an even split of residential and commercial land, are represented by the darker hues. Weekly trends show that there is a general increase in demand for all zones from 2014 to 2015. Looking only at the commercial-dominant taxi zones, a few notable trends are apparent. First, a banding effect can be observed wherein adjacent zones experience similar levels of demand. The upper bands with the most demand correspond to Midtown and surrounding taxi zones, whereas the remaining zones are distributed further downtown. Almost all commercial zones located in central and southern Manhattan have an upward trend even after noting seasonal decreases. Only one commercial-dominant taxi zone is found in the northern part of the island and that is Morningside Heights. Home to Columbia University, this zone exhibits atypical behavior when compared with the other commercial and residential zones. Interestingly, the substantial increases and decreases observed in 2015 line up with typical semester beginning and ending points.

On the other hand, focusing in on the residential areas in central and lower Manhattan, where in Figure 1 the darker lines indicate the more commercial (or mixed-use) of the residential areas. These zones have the largest increase in demand of the residential zones and have higher volatility and clear seasonal shifts. The lighter colored lines indicate heavily residential zones and have fewer weekly trips and less overall variability at the week level. Perhaps the most interesting observed trend within these zones is that demand appears to stagnate over time in the heavily residential areas. The results are similar in the residential zones in northern Manhattan. The more mixed zones have the highest demand and growth, whereas the most residential of the group fall within the middle of the graph. Here too, the lines below the 1,000 trip mark appear to decrease. The observations of flattening demand lend credence to the hypothesis that growth in Uber demand may not be unbounded and that it may be subject to a limit. We further explore this hypothesis using a random effects panel model.

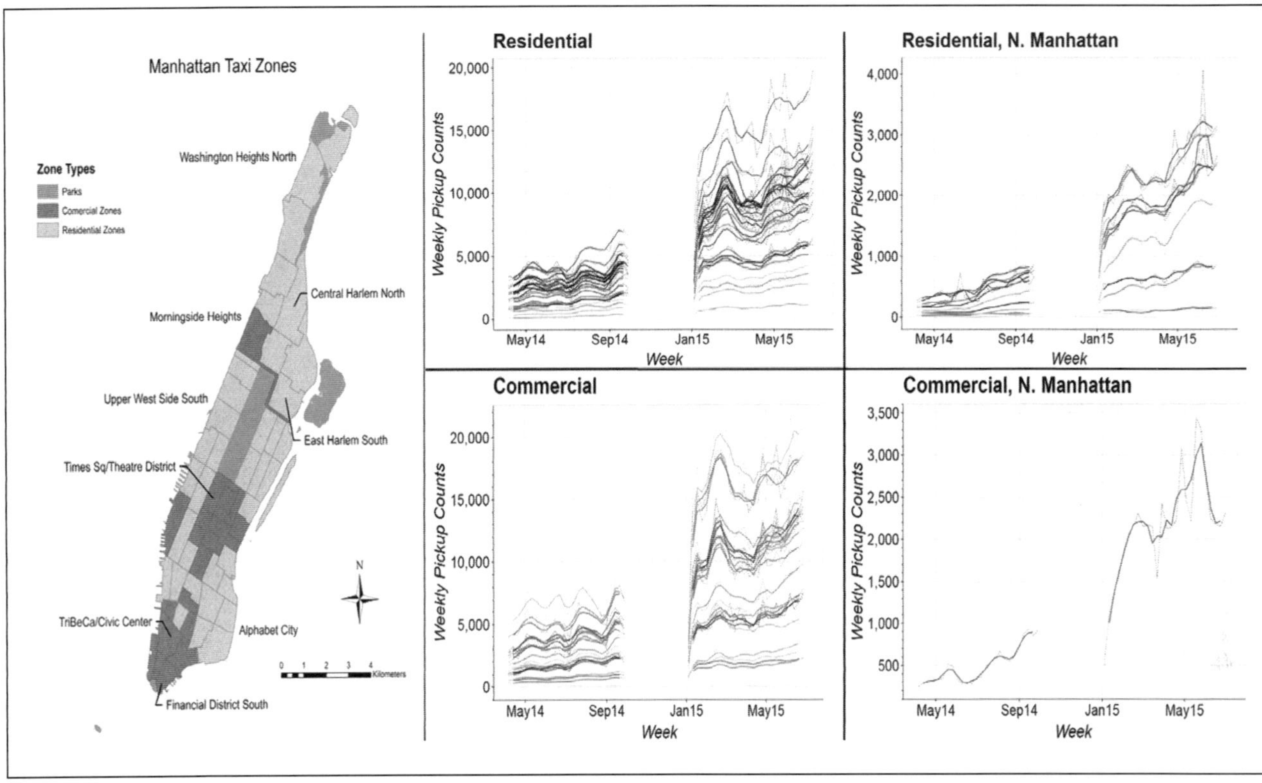

Figure 1. Distribution of residential and commercial taxi zones in Manhattan, along with trends in Uber's weekly pickup demand by zone.
Note: The demand lines are shaded by the ratio of residential floor area to total floor area for the residential-dominant zones, and by ratio of commercial floor area to total floor area for the commercial-dominant zones. Darker lines indicate a less intense relationship, whereas lighter lines indicate that a zone is almost entirely residential or commercial, respectively.

Panel Model

The nature of the weekly Uber demand data (i.e., panel) lends itself to a linear panel-based analysis (*24*). The demand data are available temporally by week for 25 complete weeks in 2014, and 24 complete weeks in 2015. However, not all elements of the demographic, land use, and environmental data used to explain the demand are time variant. As a result, a random effects structure was employed that allows consistent estimation of coefficients associated with time invariant variables.

Following the exposition of linear panel models in Cameron and Trivedi, an outline of the model structure and assumptions are presented below (*24*). Let $z = 1, 2, \ldots, Z$ represent an index for the unique taxi zones and $z = 1, 2, \ldots, Z$ represent an index for the week. The dependent variable, weekly pickup demand, is then modeled using a linear panel structure similar to the one below.

$$y_{zt} = \alpha_z + \beta' X_{zt} + \varepsilon_{zt}$$

where

y_{zt} = weekly pickup demand in taxi zone z in week t;
α_z = zone specific unobserved heterogeneity;
X_{zt} = vector of independent variables and time variables;
β = regression coefficients associated with the independent variables;
ε_{zt} = errors within each taxi zone z for time period t.

ε_{zt} is assumed to be strictly exogenous and normally distributed as $N(0, \sigma_\varepsilon^2)$. Further, specific assumptions about α_z result in the fixed effect and random effect variants. In a random effects model, used in the analysis, α_z is assumed to be random and distributed normally as $N(\alpha, \sigma_\alpha^2)$. Also, ε_{zt} and α_z are assumed to be independent. Further, heteroscedasticity and autocorrelation are allowed in specification of ε_{zt}.

The model was estimated using Stata software by means of the feasible generalized least squares routine (i.e., xtgls). In order to test the autocorrelation and heteroscedasticity specification of the error term, separate hypothesis tests were run. First, to test heteroscedasticity, a likelihood ratio test proposed by Poi and Wiggins was used that compares a heteroscedastic-treated panel model against the same model assumed to be homoscedastic (*25*). Second, to test autocorrelation, the procedure outlined in Wooldridge was adopted (*26*).

Table 1. GLS Random Effects Panel Model Results

Parameter	Random effects model	
	Coefficient	Standard error
Constant	−19056.180	−6274.937***
Time variables		
Summer dummy variable	−278.133	46.857***
Winter dummy variable	89.647	50.472*
Difference in weeks from initial week	93.323	6.446***
Environmental variables		
Weekly precipitation (in.)	25.892	6.262***
Weekly average max temperature (°F)	−10.143	1.174***
Built environment variables (taxi zone level)		
Count of Citi Bike stations	242.850	29.211***
Total built area (10,000 ft^2)	0.936	0.071***
Percent residential by floor area	64.247	7.003***
Percent retail by floor area	151.426	33.043***
Demographic variables (taxi zone level)		
Percent African American	−58.925	7.517***
Percent male	60.847	17.042***
Percent aged under 19	220.657	23.752***
Percent aged over 65	−0.213	0.047***
Percent of households with no vehicle	126.400	60.253**
Percent of households with one vehicle	131.103	62.971**
Average household size	−2991.058	363.236***
Percent of adults with at least some college experience	17.922	9.481*
Interaction variables		
Percent residential by floor area × Difference in weeks	−0.909	0.089***
Percent retail by floor area × Difference in weeks	9.581	0.653***

$^{*}p < 0.10$, $^{**}p < 0.05$, $^{***}p < 0.01$.

Discussion of Results

As noted earlier, a linear panel model was estimated using the generalized least squares approach in Stata software. The model allowed heteroscedasticity and autocorrelation in the specification of the disturbance term. Due to the presence of autocorrelation and gaps in the data (i.e., missing weekly data between 2014 and 2015), the standard procedure of maximum likelihood estimation was not applicable. The final panel contained 61 taxi zones over 49 weeks for a total of 2,989 observations. Parks and zones with missing independent variables were dropped from the model, including Ellis Island, Liberty Island, Governors Island, Randall's Island, Central Park, Battery Park, Inwood Hill Park, and Highbridge Park. Before running the panel model, the tests for heteroscedasticity and autocorrelation identified above were performed. The results indicated that both autocorrelation (with an F-test statistic of 331 with degrees of freedom of 1 and 60 and a p-value of 0.00) and heteroscedasticity (with a chi-squared statistic of 2242.47 with degrees of freedom of 60 and a p-value of 0.00) were present and, as such, the error term specification needed to accommodate these issues. In the interest of brevity, the complete listing of results from the statistical testing of autocorrelation and heteroscedasticity is not presented here.

A variety of explanatory factors, including demographic, land use, and environmental, were explored. Additionally, temporal variables were explored to capture the impact of unobserved time-varying factors. The final model is presented in Table 1. All the variables included are significant at least at the 10% level of significance.

Influence of Temporal Variables

A variety of temporal variables were explored to capture the influence of unobserved time-varying and seasonal factors. As the scope of the data was limited to only two 6-month periods, the extent of seasonal coverage was also limited to only include winter, spring, and summer. In the model, seasonal effects are reported with reference to spring. Spring was chosen because it was the only season present in both years of the data used in the analysis. It appears that Uber demand at the taxi zone level decreases during the summer as noted by the negative coefficient. This observation is also confirmed by the descriptive analysis wherein a clear drop was observed during the summer months of 2014. This may be attributed to individuals' preference to use non-motorized modes, such as walking or biking, when the weather is warmer compared with other seasons. While reviewing research on seasonal influences on motorized and non-motorized travel, Guo et al. found that during the summer, people make more discretionary trips via non-motorized modes (*27*). On the other hand, an opposite reasoning can be offered to explain the positive effect of the winter dummy variable on Uber demand: given the extreme weather that NYC experiences during the winter months, people may seek the comfort of a climate-controlled vehicle in order to escape the cold conditions.

Furthermore, the difference in week from the initial week variable is defined as the number of weeks past the first complete week (i.e., week ending on April 13th, 2014) of demand data. Here, a positive influence is apparent and can be interpreted to mean that as time progresses, the pickup demand within taxi zones increases. This finding is in line with both the observed trends at the descriptive level, as well as research and reports presented earlier on the growth of Uber overall.

One of the primary hypotheses was to explore the potential tapering off of Uber demand with time. To this end, the impact of higher order versions of the difference in week variable were explored after controlling for the other explanatory variables. For example, a negative coefficient associated with a second order influence will lend evidence in support of the hypothesis. It was found that higher order versions of the difference in week variable were not significant. However, this contradicts the observations from the trend analysis by taxi zones that showed evidence of a tapering effect for residential zones. To further explore the tapering-off effect, interaction variables between difference in week and other explanatory factors, namely demographic and land use variables were explored. The results for the interaction variables are discussed at the end of this section.

Influence of Environmental Variables

The environmental variables were included to extend interpretation beyond just seasonality and to quantify how weekly weather trends influence zonal level Uber demand. The positive influence related to total weekly precipitation was expected and is intuitive. As the overall total of weekly precipitation increases, demand for Uber also increases. This is plausible because rain/sleet/snow move people away from other non-automobile modes and toward trip making in the comfort of an on-demand vehicle. A substantial amount of users in Manhattan use non-automobile modes like Citi Bike, walk, and transit among others (*17*, *19*). These riders may potentially opt for an Uber trip if it means avoiding exposure to the elements. As the average high temperature by week increases, the demand for Uber decreases. Increased temperatures, especially those outside of the normal weather condition, may be prompting individuals to trend toward non-motorized travel in a similar manner to the summer influence.

Influence of Built Environment Variables

Unfortunately, in the final model formulation, the influence of number of bus stops and subway stops at the taxi zone level was found to be statistically insignificant. However, the large positive influence of Citi Bike stations at the taxi zone level is an interesting result. At a high level, the positive influence makes sense as Citi Bike stations are primarily located in lower Manhattan and these taxi zones correspond to the areas with the highest trip attraction. However, upon closer inspection, it can be seen that even with the inclusion of other built factors like total floor area, which are more direct determinants of trip attraction, Citi Bike stations have a significant and sizeable impact. One plausible reason for this is that the users of the on-demand shared Citi Bike system may also be inclined to use shared on-demand services provided by Uber. It is also plausible that Uber is used by such individuals to complete trips that Citi Bike could not. This is consistent with research by Faghih-Imani et al.; in their analysis of Citi Bike and taxi modes, the authors found that for trips of a substantial distance, taxi modes proved to be more suitable option – a similar scenario could be at work here (*17*).

The total built area is assumed to capture the overall intensity of each taxi zone, and the positive relation to Uber demand follows from our expected results. With more built area, there is more to see, and places to live. This is also consistent with other findings that Uber demand is highest in the core of Manhattan and in tourism-driven taxi zones. The percent of residential floor area has a positive influence on Uber pickup demand. Residents, just like visitors and tourists, make trips and there are substantial amounts of demand in these residential zones – referring to Figure 1, it can be seen that residential zones like East and West Village have some of the highest demand of all observed taxi zones.

Variables representing other land use, namely, percentages of overall commercial, retail, garage (parking), and office areas were explored. These remaining land use percentages were also included in the model, however, the only other type that proved significant was the percentage of retail by floor area. This strong positive interaction shines light on previously noted trends regarding Uber's propensity to serve tourism and central business districts. Within Manhattan, these locations coincide with higher proportions of retail area and, as such, higher Uber demand.

Influence of Demographic Variables

Understanding how demographic makeup of taxi zones influence Uber demand can provide valuable insights into how different groups are adopting dynamic ride-sharing services. When exploring the link between Uber demand and the percentage of the population that is African American in a taxi zone, a negative relationship was observed. This observation raises questions on equity. A better understanding of the factors influencing (or lack thereof) African Americans to use Uber is needed.

The second demographic factor explored is that of gender on Uber demand. Here the percentage of males within the population was found to have a positive relationship with Uber pickup demand. A possible explanation for this could be safety. As mentioned previously, Uber has been dealing with issues regarding safety (*8*). As a result, male riders may feel more comfortable using the dynamic ridesharing system than their female counterparts.

Age has been an important topic in regard to Uber adoption, and previous research indicates that Uber users tend to skew younger (*13*). When holding the typical adult age range as the constant categorical factor and exploring the effects on individuals under 19 and over 65, our findings fall in line with existing research. The percentage of the population under 19 years of age is more positively associated with demand for Uber than their older counterparts. On the other hand, senior individuals exhibit a lower demand for Uber. A propensity for technology and existing mode bias are potential explanations for this relationship.

Vehicle ownership was investigated over conventional income variables due to findings in the Uber literature. In San Francisco, survey respondents that utilized Uber identified as owning few vehicles (*13*). As such, we investigated the influence of household vehicle ownership on Uber demand. It was observed that both no-vehicle and one-vehicle households exhibit heightened Uber demand compared with households with two or more vehicles. Overall vehicle deficiency could explain this, as households still need to make trips to fulfill their activity needs.

The modeled effects of household size in regard to demand indicate that larger households have a substantially lower demand for Uber than smaller ones. As household size increases, traveling typically becomes more difficult and potentially also expensive especially for a mode like Uber that has limited capacity per vehicle.

The final significant demographic variable explored was aimed at quantifying the relationship of education on Uber demand. Although only marginally significant, the findings indicate that individuals with higher education exhibit higher demand for Uber. Once again in line with preliminary survey findings, this can be can be attributed to a higher propensity for adopting and using technology by those associated with higher levels of education.

Influence of Interaction Variables

As noted earlier, the main hypothesis was to explore whether there is a potential tailing-off effect in the growth of Uber demand with time. As noted before, higher order versions of the weekly difference variable were found to be insignificant despite the evidence from the descriptive analysis. In an effort to further explore the tailing-off effect, interaction variables between the weekly difference variable and different demographic and land use variables were investigated to see if there were any potential negative shifts in the demand for specific tax zones or population groups, or tax zones and population groups, over time. This exploration provided substantial evidence in support of our proposed hypothesis. We found that the interaction between percent retail and weekly difference had a substantial positive impact on Uber demand, suggesting that Uber demand in high retail areas is growing with time. As Uber becomes more readily adopted, a greater proportion of users are making trips from and to areas with a higher retail concentration using the service. However, when looking at the interaction of the percentage of residential area by weekly difference there is a clear negative impact. Although the magnitude of the relationship is small, the negative coefficient indicates that residential areas over time have a waning or decline in overall Uber demand. There are a couple of possible reasons for this observation. First, some residents may have a preference toward existing modes (and bias against Uber) and as such are not going to adopt new systems. Second, if Uber only replaces single-occupancy vehicle (SOV) trips or borrows from other non-automobile modes, or does both, it will not improve congestion on roadways. As a result, those who are conscious of travel time performance and are price conscious may prefer the cheaper alternative with similar travel time performance (e.g., subway, bus).

Conclusion and Future Research

With the substantial growth in dynamic ridesharing offerings over the last few years, a comprehensive investigation into understanding the underlying factors affecting demand is of interest. Using multiyear data from an existing ridesharing service, namely Uber from NYC, the temporal trends in demand were explored. Existing research supports the growing adoption of Uber, but here it was hypothesized that although macroscopic trends may suggest unbounded increases in the demand, meso/microscopic levels may offer a different picture. It is posited that the demand for ridesharing services has a limit. To explore these underlying factors and test our hypothesis, a random effects panel model was developed.

The findings indicate that there has been a significant growth over time, but that seasonality in demand exists with an increase in demand in the winter months and a decrease in demand in the summer months. Weather was also noted to have a statistically significant impact, with increased precipitation positively influencing demand and warmer temperatures decreasing the weekly demand. At a built environment level, the presence of transit was not determined to have a significant impact, however bike share infrastructure had a positive influence. Similarly, the more built area a taxi zone had, as well as a higher percentage of residential- and retail-based floor area, the higher the demand for Uber service. The influence of zonal level demographics was also investigated. The percentage of population that is male, under 19 years of age, has some college education, and has one or no household vehicles showed positive impacts on weekly Uber demand. The percentage of the population that is African American, over 65, and from larger households all had negative impacts on Uber demand. Within the model, weekly difference had a positive effect on Uber demand; it was included at a higher order to capture the observed tailing-off (or leveling) over time, however the results were insignificant. Interaction variables were then included to test our hypothesis for observed declines in demand over time with respect to various built environment variables. A positive influence of retail area across time was determined. Perhaps the most substantial finding, a negative relationship was observed when exploring the influence of percent residential area across time. This negative relationship aligns with the decreasing trends observed in the descriptive analysis. Together, these results offer substantial evidence in support of a leveling in Uber demand over time in residential-dominant areas.

These last findings have a substantial impact beyond the scope of just Uber demand analysis and have potentially important policy implications for the overall adoption of dynamic shared modes. The significant tapering off within residential zones indicates that even as the demand grows at a citywide scale, residents' inertial modal biases and familiarity with existing modes may limit the dynamic shared ride adoption by all individuals within a subarea (*28*). A second policy implication that can be drawn here is that users are typically motivated to change modes if the alternative provides an improvement over existing offerings (e.g., lower costs or shorter travel times). In the current operational framework, Uber and other dynamic shared ride providers most often replace SOV auto-modes, further contributing to the effects of congestion and hindering overall adoption (*13*, *16*). Further investigation into these effects is warranted.

This research presents a substantial first step in understanding dynamic ridesharing in a real-world context, but it is important to recognize some of the limitations of this work. First, holding Uber pickups as a stand-in for demand could be problematic as it inherently omits any demand by users who attempted to request rides but were not able to have those requests filled. It also inherently omits any user without a smartphone or knowledge of the Uber interface, which may be disproportionately represented by elderly and low-income groups. Second, with data only on the pickup event of each trip, a rich understanding of how these trips integrate into the existing network is not possible. Third, the spatially aggregated nature of the data limited this analysis to the neighborhood level offered by taxi zones. The data quality and structure issues noted with the Uber data motivates an important discussion. As open data becomes more readily available, it is vital for open-data providers to adopt a standard of practice for quality and structure that could in turn promote additional research endeavors.

Furthermore, the panel model used in this research assumes strong exogeneity, which may not be supported by the data and alternate model forms must be explored. A comprehensive spatial analysis will help to further explore and characterize the spatial relationships with Uber demand. Moreover, to supplement and confirm the findings presented here, we anticipate further exploration through a primary survey that will collect disaggregate-level information on users and their preferences. Lastly, a comprehensive study of the temporal trends in all modes including taxis, subways, buses, and other dynamic rideshare providers from the region could help understand the tradeoffs between on-demand modes like Uber, and existing shared ride services will be explored in future work.

Author Contributions

The authors confirm contribution to the paper as follows: study conception and design: RG, KK; data collection: RG; analysis and interpretation of results: RG, KK, NE; draft manuscript preparation: RG. All authors reviewed the results and approved the final version of the manuscript.

References

1. Furuhata, M., M. Dessouky, F. Ordóñez, M. Brunet, X. Wang, and S. Koenig. Ridesharing: The State-of-the-art and Future Directions. *Transportation Research Part B: Methodological*, Vol. 57, 2013, pp. 28–46.
2. Harding, S., M. Kandlikar, and S. Gulati. Taxi Apps, Regulation, and the Market for Taxi Journeys. *Transportation Research Part A: Policy and Practice*, Vol. 88, 2016, pp. 15–25. ISSN 0965-8564, http://dx.doi.org/10.1016/j.tra.2016.03.009.

3. Lazo, L. *Uber Turns 5, Reaches 1 Million Drivers and 300 Cities Worldwide. Now What?* 2015. https://www.washingtonpost.com/news/dr-gridlock/wp/2015/06/04/uber-turns-5-reaches-1-million-drivers-and-300-cities-worldwide-now-what/?utm_term=.b30a6f5dc40d. Accessed July, 2017.
4. Tepper, F. *Uber has Completed 2 Billion Rides.* 2016. https://techcrunch.com/2016/07/18/uber-has-completed-2-billion-rides/. Accessed July, 2017.
5. Schneider, T. *Analyzing 1.1 Billion NYC Taxi and Uber Trips, with a Vengeance.* 2015. http://toddwschneider.com/posts/analyzing-1-1-billion-nyc-taxi-and-uber-trips-with-a-vengeance/. Accessed July, 2017.
6. Sun, C., and P. Edara. Is Getting an Uber-Lyft from a Sidecar Different from Hailing a Taxi? *Transportation Research Record: Journal of the Transportation Research Board*, 2015. 2536: 60–66.
7. Chen, L., A. Mislove, and C. Wilson. Peeking Beneath the Hood of Uber. *Proc., 2015 ACM Conference on Internet Measurement Conference - IMC 15*, Tokyo, Japan, ACM, 2015, pp. 495–508.
8. Picchi, A. *The Rising Safety Issues That Could Throttle Uber.* 2014. http://www.cbsnews.com/news/the-rising-safety-issues-that-could-throttle-uber/. Accessed July, 2017.
9. Fagnant, D. J., and K. Kockelman. Preparing a Nation for Autonomous Vehicles: Opportunities, Barriers and Policy Recommendations. *Transportation Research Part A: Policy and Practice*, Vol. 77, 2015, pp. 167–181.
10. Chen, M. K., and M. Sheldon. *Dynamic Pricing in a Labor Market: Surge Pricing and the Supply of Uber Driver-Partners.* University of California, Los Angeles, 2015.
11. Guo, S., Y. Liu, K. Xu, and D. M. Chiu. Understanding Ride-on-Demand Service: Demand and Dynamic Pricing. *Proc., 2017 IEEE International Conference on Pervasive Computing and Communications Workshops (PerCom Workshops)*, Kona, Hawaii, IEEE, New York, 2017, pp. 509–514.
12. Correa, D., K. Xie, and K. Ozbay. Exploring the Taxi and Uber Demand in New York City: An Empirical Analysis and Spatial Modeling. Presented at 96th Annual Meeting of the Transportation Research Board, Washington, D.C., 2017.
13. Rayle, L., D. Dai, N. Chan, R. Cervero, and S. Shaheen. Just a Better Taxi? A Survey-Based Comparison of Taxis, Transit, and Ridesourcing Services in San Francisco. *Transport Policy*, Vol. 45, 2016, pp. 168–178.
14. Cramer, J., and A. Krueger. Disruptive Change in the Taxi Business: The Case of Uber. *American Economic Review*, Vol. 106, No. 5, 2016, pp. 177–182.
15. Poulsen, L. K., D. Dekkers, N. Wagenaar, W. Snijders, B. Lewinsky, R. R. Mukkamala, and R. Vatrapu. Green Cabs vs. Uber in New York City. *Proc., 2016 IEEE International Congress on Big Data (BigData Congress)*, San Francisco, Calif., IEEE, New York, 2016, pp. 222–229.
16. Fischer-Baum, R., and C. Bialik. *Uber is Taking Millions of Manhattan Rides Away From Taxis.* 2013. https://fivethirtyeight.com/features/uber-is-taking-millions-of-manhattan-rides-away-from-taxis/. Accessed July, 2017.
17. Faghih-Imani, A., S. Anowar, E. J. Miller, and N. Eluru. Hail a Cab or Ride a Bike? A Travel Time Comparison of Taxi and Bicycle-Sharing Systems in New York City. *Transportation Research Part A: Policy and Practice*, Vol. 101, 2017, pp. 11–21.
18. Brakewood, C., G. S. Macfarlane, and K. Watkins. The Impact of Real-Time Information on Bus Ridership in New York City. *Transportation Research Part C: Emerging Technologies*, Vol. 53, 2015, pp. 59–75.
19. Campbell, K. B., and C. Brakewood. Sharing Riders: How Bikesharing Impacts Bus Ridership in New York City. *Transportation Research Part A: Policy and Practice*, Vol. 100, 2017, pp. 264–282.
20. FiveThirtyEight. *Uber Trip Data* [Data Files]. 2015. https://github.com/fivethirtyeight/uber-tlc-foil-response/tree/master/uber-trip-data. Accessed July, 2017.
21. NYC Planning. *PLUTO* [Data File]. 2016. https://www1.nyc.gov/site/planning/data-maps/open-data/dwn-pluto-mappluto.page. Accessed July, 2017.
22. NYC Open Data [Data File]. 2016. https://opendata.cityofnewyork.us/. Accessed July, 2017.
23. National Centers for Environmental Information. *NOAA* [Data File]. https://www.ncdc.noaa.gov/data-access. Accessed July, 2017.
24. Cameron, A. C., and P. K. Trivedi. *Microeconometrics: Methods and Applications.* Cambridge University Press, 2005.
25. Wiggins, V., and B. Poi. *Testing for Panel-Level Heteroskedasticity and Autocorrelation StataCorp FAQs*, StataCorp, 2003. https://www.stata.com/support/faqs/statistics/panel-level-heteroskedasticity-and-autocorrelation/. Accessed July, 2018.
26. Wooldridge, J. M. *Econometric Analysis of Cross Section and Panel Data.* MIT Press, Cambridge, Mass., 2010.
27. Guo, J., C. Bhat, and R. Copperman. Effect of the Built Environment on Motorized and Nonmotorized Trip Making: Substitutive, Complementary, or Synergistic? *Transportation Research Record: Journal of the Transportation Research Board*, 2007. 2010: 1–11.
28. Lee, J. H., and K. G. Goulias. A Decade of Dynamics of Residential Location, Car Ownership, Activity, Travel and Land Use in the Seattle Metropolitan Region. *Transportation Research Procedia*, Vol. 23, 2017, pp. 420–439.

The Standing Committee on Urban Transportation Data and Information Systems (ABJ30) peer-reviewed this paper (18-03265).

Article

TRR
JOURNAL OF THE TRANSPORTATION RESEARCH BOARD

Transportation Research Record
2018, Vol. 2672(42) 137–147

DOI: 10.1177/0361198118792124
journals.sagepub.com/home/trr
SAGE

Resampling Methods for Estimating Travel Time Uncertainty: Application of the Gap Bootstrap

Bhaven Naik[1], Laurence R. Rilett[2], Justice Appiah[3], and Lubinda F. Walubita[4]

Abstract

To a large extent, methods of forecasting travel time have placed emphasis on the *quality* of the forecasted value—how close is the forecast point estimate of the mean travel time to its respective field value? However, understanding the reliability or uncertainty margin that exists around the forecasted point estimate is also important. Uncertainty about travel time is a fundamental factor as it leads end-users to change their routes and schedules even when the average travel time is low. Statistical resampling methods have been used previously for uncertainty modeling within the travel time prediction environment. This paper applies a recently developed nonparametric resampling method, the gap bootstrap, to the travel time uncertainty estimation problem, especially as it pertains to large (probe) data sets for which common resampling methods may not be practical because of the possible computational burden and complex patterns of inhomogeneity. The gap bootstrap partitions the original data into smaller groups of approximately uniform data sets and recombines individual group uncertainty estimates into a single estimate of uncertainty. Results of the gap bootstrap uncertainty estimates are compared with those of two popular resampling methods—the traditional bootstrap and the block bootstrap. The results suggest that, for the datasets used in this research, the gap bootstrap adequately captures the dependent structure when compared with the traditional and block bootstrap methods and may thus yield more credible estimates of uncertainty than either the block bootstrap method or the traditional bootstrap method.

In recent years, technological advances have enabled the collection and diffusion of real-time traffic information and this, combined with growing traffic volume and congestion, has triggered an increasing interest in traffic modeling (*1, 2*). The focus for researchers and users of predicted traveler information has been twofold: (i) traffic condition level analysis, and (ii) traffic condition uncertainty analysis. Research in the area of traffic condition level analysis (i.e., first moment of the traffic condition series) has been plentiful with primary focus on using the available traffic data to compute a prediction of the mean (speed, flow rate, or travel time) for a given time period and on ensuring that the difference between the predicted value and its real-time value for the same time period is as small as possible. In particular, this entails the ability to provide a forecasted point estimate of a traffic variable in close comparison with values encountered in real-time (*3*).

On the other hand, traffic condition uncertainty analysis refers to the second moment of the traffic condition series. It deals with any information on the quality (distributional properties) of the condition level value. This quality of confidence measure (or conditional variance) is often also referred to as the reliability of the forecast. A measure of the reliability is important because "drivers claim this influences their decisions that are based on the available travel time information" (*4, 5*) and it has been proven to have a notable impact on motorists' travel decisions, leading to a more efficient use of the traffic network (*6*). In effect, drivers are unlikely to follow the information if they do not believe in its validity (*7*). Therefore, the provision of a measure of reliability could not only increase user comfort by reducing the error risk associated with the traffic information, it could also (for the traffic

[1]Ohio University, Athens, OH
[2]University of Nebraska-Lincoln, Lincoln, NE
[3]Virginia Transportation Research Council, Charlottesville, VA
[4]Texas A&M Transportation Institute (TTI), The Texas A&M University System, College Station, TX

Corresponding Author:
Address correspondence to Bhaven Naik: naik@ohio.edu

engineer) be used to assess the predicted values for model selection. Intuitively, having a measure of the reliability of a forecast will make the job of the transportation professionals who operate the systems considerably easier. In statistical terms, this would mean quantifying the uncertainty (error bars or statistical intervals) associated with a given forecast.

Several researchers have shared insights into the uncertainty margin that exists around a predicted point (or condition level) estimate including Eisele et al. (*8*), Mazloumi et al. (*3*), and Van Lint (*9*). These research efforts assume that datasets are independent and identically distributed. However, this is generally not the case for Intelligent Transportation Systems' (ITS) data that are typically used in the prediction process. The ITS data, as will be described in later sections of this paper, have been observed to have nonstationary distribution and exhibit a general periodicity. Studies by Guo, Huang, and Williams have addressed this periodicity by adopting seasonal autoregressive integrated moving average (SARIMA) models at the condition level modeling stage to remove the autocorrelation structure from the vehicular traffic flow series (*10–12*). Next, the generalized autoregressive conditional heteroscedasticity (GARCH) approach is applied to model uncertainty in the univariate vehicular traffic flow series. Unlike previous studies on uncertainty that assume conditional variance is constant, the GARCH approach accounts for the heteroscedasticity observed in the traffic condition series. Other applications of the GARCH approach include modeling relative velocity (*13*), and also link travel time variability (*14*). These GARCH based traffic condition uncertainty models inherently assume that the second-order property of the traffic conditional series is time varying (or heteroscedastic in nature). It is worthwhile to note that research on traffic condition uncertainty is increasingly becoming popular because many prediction applications have issues with traditional methods (*9*) and is quickly gaining momentum because of its importance in many transportation applications (*5, 6*).

Based on the above background, a methodology that (i) will provide an accurate prediction of the mean travel time for a link, (ii) will give a measure of the reliability in the predicted estimate, and (iii) is based on nonparametric statistical theory is the focus of this paper. This approach will allow transportation agencies to better predict traffic parameters, and their associated reliability metrics, when the assumptions behind standard techniques do not apply.

Background and Problem Definition

The model prediction task for a given statistic $\hat{t}$ (such as the mean of a sample dataset) typically consists of two steps:

- *Step 1—Developing an initial model using a set of data.* This model (essentially a mean regression function) defines the relationship between inputs and the output. The model could be parametric (linear regression, autoregressive) or nonparametric (neural network) in nature.
- *Step 2—Estimating future values (predictions) of interest relative to current input.*

The predicted parameter, $\hat{t}$, is reflective of the "true" population parameter, t. However, it is unlikely to be exact. The value of the "true" population parameter is usually unknown and it is important that empirical measurements or estimates of the sample parameter are accompanied by a statement about their corresponding uncertainty (*15*). For travel time, this is traditionally expressed mathematically as

$$t = \hat{t} \pm k\sigma_t^2 \quad (1)$$

where

t = "true" travel time mean (s);
$\hat{t}$ = sample travel time mean (s);
σ_t^2 = variance (s^2); and
k = multiplier parameter (s^{-1}).

Uncertainty in Travel Time Prediction

The sources of travel time related uncertainty can be categorized into two aspects—one related to input data quality and one related to model structure. Uncertainty related to the input data quality focuses largely on errors that emanate from failures in traffic monitoring equipment, communication between the field and traffic management center, and the traffic management archiving system (*16, 17*). Typically, uncertainties caused by input data quality are addressed by "cleaning" the data.

The uncertainty in the model structure relates to the noise in the parameter (or weight) selection process. For the reason that a model (with specified parameter) is based on a selected random sample of data, if the current input is outside of the domain, then the prediction would be uncertain.

Bootstrapping

Methods such as maximum likelihood estimation, approximate Bayesian, and the Delta method have previously been used for confidence estimation with nonlinear transportation models. However, these methods have been found to have potential sources of failure that may limit their applicability for transportation applications (*18–20*). Bootstrapping is a more commonly used data-driven simulation technique for computing statistical measures of accuracy for estimates. The idea behind bootstrapping is that the available dataset is nothing but a particular realization of some unknown probability distribution (*21, 22*). The method creates many replicates ("phantom samples") of the original data and then re-estimates the distribution of an estimate or test-statistic on each replicate sample to obtain an estimate of the standard error of the predicted values. This standard error is the basis for the reliability metric in many applications—including those related to transportation parameter forecasting. This paper demonstrates the application of a recently developed gap bootstrap method. A comparison against two existing methods (ordinary and block bootstrap) with specific application to link travel time prediction is then provided.

Ordinary Bootstrapping. The ordinary bootstrap is the simplest and most general bootstrapping technique (*23*). It is commonly adopted in most applications in which the data are (or are assumed to be) independent and identically distributed (i.i.d.). Given an original sample $x_n = \{(x_1, y_1), (x_2, y_2), \ldots, (x_n, y_n)\}$, the bootstrap procedure implies that an unknown distribution F is estimated by constructing an empirical distribution $\hat{F}_n$ based on the n original observations. As illustrated in Figure 1*a*, the bootstrap samples $x_n^{*b} = \{(x_1^{*b}, y_1^{*b}), (x_2^{*b}, y_2^{*b}), \ldots, (x_n^{*b}, y_n^{*b})\}$ are repeatedly drawn with replacement from the estimated empirical distribution until a total of B bootstrap samples are obtained (*23, 24*). Then, for each bootstrap sample, the bootstrap version $\hat{\theta}_B^*$ of an estimator $\hat{\theta}$ is calculated. The ordinary bootstrap uncertainty estimate (standard error) is then obtained (*25*) as

$$\text{standard error(se)} = \left\{\frac{1}{B-1}\sum_{b=1}^{B}\left[\hat{\theta}^{*b} - \bar{\theta}^*\right]\right\}^{1/2} \tag{2}$$

where,

$$\text{bootstrap mean } \bar{\theta}^* = \frac{1}{B}\sum_{b=1}^{B}\hat{\theta}^{*b} \tag{3}$$

Block Bootstrapping. When implementing the ordinary bootstrapping method, each bootstrap sample is drawn from data that are "scrambled" in such a way that aspects of any dependence structure in the original dataset are lost. The ordinary bootstrap is therefore not appropriate for dealing with dependent data (such as link travel times) because the resampling does not capture the dependence structure. Block bootstrapping is currently "the best known method for implementing bootstrapping with dependent data" (*26*).

The basic concepts of block bootstrapping are similar to those of ordinary bootstrapping discussed earlier.

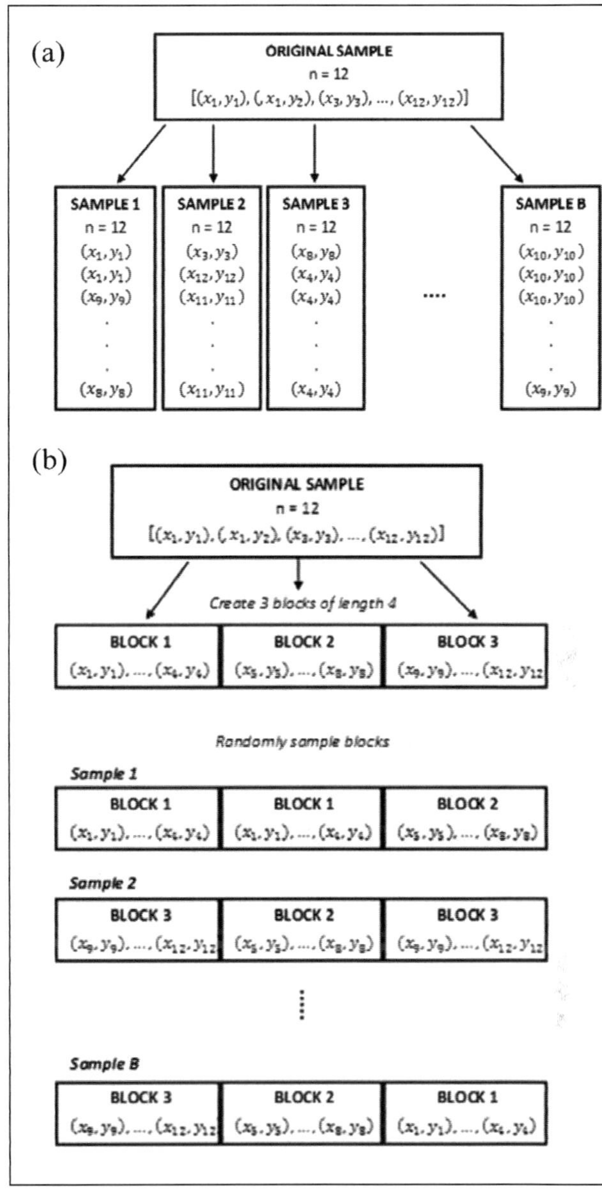

Figure 1. Illustration of bootstrap methods: (*a*) ordinary bootstrapping; and (*b*) block bootstrapping.

However, with block bootstrapping, a bootstrap sample is drawn by dividing the data into contiguous blocks that are randomly sampled (as opposed to sampling individual observations in the ordinary bootstrap method), and laying these blocks end-to-end in the order sampled. This method is illustrated in Figure 1*b*. The blocks could be created as nonoverlapping or overlapping with fixed or varying lengths. The four common block bootstrapping methods available are the moving block bootstrap (*27*), the nonoverlapping block bootstrap (*28*), the circular block bootstrap (*29*), and the stationary block bootstrap (30). Synopses of these techniques can be found in the publication by Lahiri (31).

Gap Bootstrapping. Gap bootstrapping is a procedure that is appropriate for datasets that can be partitioned into approximately exchangeable partitions (32). The only previous applications of the gap bootstrap evaluate uncertainties in point estimates of origin–destination split proportions (32, 33).

To illustrate the gap bootstrap technique for travel time estimation, denote the time series of mean travel times (for a given link) by the $p \times q$ matrix X and denote the estimator based on the entire data by $\hat{\theta}$. The rows of X are then partitioned into m groups $X_i, i = 1, 2, 3, \ldots, m$ such that there are $k = T/m$ observations in each column of the partition matrices. Let $\hat{\theta}_1, \hat{\theta}_2, \hat{\theta}_3, \ldots, \hat{\theta}_m$ be the mean travel times predicted from the submatrices X_1, X_2, X_3,..., X_m respectively. The main result of the gap bootstrap is that the usual covariance matrix estimator obtained from $\hat{\theta}_1, \hat{\theta}_2, \hat{\theta}_3, \ldots, \hat{\theta}_m$ provides a consistent estimate of the covariance matrix of $\hat{\theta}$ and the rate of convergence is $\sqrt{T}$. Further, the estimator

$$\text{bootstrap mean } \bar{\theta} = \sum_{i=1}^{m} \hat{\theta}_i / m \tag{4}$$

is an asymptotically efficient estimator of θ. The component-wise variance of the m groups is calculated by performing ordinary bootstrapping within each subset. Assuming a constant covariance among the subsets, the final gap bootstrap uncertainty estimate of $\bar{\theta}$ is given by

$$\text{Var}(\bar{\theta}) = \left(\sum_{i=1}^{m} \text{Var}(\hat{\theta}_i) + \sum_{i \neq j} \text{Cov}(\hat{\theta}_i, \hat{\theta}_j) \right) / m^2 \tag{5}$$

where

$\text{Var}(\hat{\theta}_i)$ = the ordinary bootstrap estimator for the variance of $\hat{\theta}_i$,

$\text{Cov}(\hat{\theta}_i, \hat{\theta}_j)$ = the variance estimate of $(\hat{\theta}_i - \hat{\theta}_j)$ given by,

$$\sum_{i \neq j} \left(\boldsymbol{\hat{\theta}}_i - \boldsymbol{\bar{\theta}} \right) \left(\boldsymbol{\hat{\theta}}_j - \boldsymbol{\bar{\theta}} \right)^t / \boldsymbol{m}(\boldsymbol{m} - 1) \tag{6}$$

Study Methodology

A conceptualization of the methodology to calculate the predicted travel time values and additional computation of an estimate of the standard error (or confidence interval) discussed above is presented in Figure 2. It may be seen that the methodology consists of five steps.

Step 1: Initial sample selection—A sample dataset $x_n = \{(x_1, y_1), (x_2, y_2), \ldots, (x_n, y_n)\}$ of size n is selected from the population $x_N = \{(x_1, y_1), (x_2, y_2), \ldots, (x_N, y_N)\}$ of size N.

Step 2: Bootstrapping—"Phantom" samples of the sample travel time dataset (now assumed as the "population") are created. A "bootstrap pairs" approach that considers each training case as a sampling unit and samples with replacement from the training set cases to create a bootstrap sample is adopted (25).

Step 3: Fitting ensemble of models—An ensemble of models is fitted on each bootstrap dataset. This study used neural network models. However, it should be noted that the approach is generic and other model types could be used. Essentially, for a given bootstrap sample $\{(x_1^{*b}, y_1^{*b}), (x_2^{*b}, y_2^{*b}), \ldots, (x_n^{*b}, y_n^{*b})\}$, a model is fitted by

$$\text{minimizing} \sum_{i=1}^{n} \left[y_i^{*b} - y(x_i^{*b}; w) \right]^2 \tag{7}$$

subject to $0 < w \leq 1$

where

w = set of model parameters (or weights),

$y(x_i; w)$ = predicted value for input x_i and model parameter w; and

b = bootstrap sample ($b = 1, \ldots, B$).

A multi-layer feed forward neural network with five input nodes and a hidden layer with 15 neurons and 15 output nodes was adopted. The Levenberg-Marquardt algorithm was used to train the neural networks. The neuron activation functions used are the log-sigmoid and the linear function for the hidden and output layers, respectively. Specific details on the design of the neural network model are presented elsewhere (34).

Step 4: Calculate predicted values—Models trained in Step 3 are each used here to predict the travel times for a given test dataset. Therefore, if B ensemble models were fitted in Step 3, then for a given set of values there would be B sets of predictions available (i.e., B predictions for period Δt). Note that the neural network in this study was trained on 127 days of data records and tested on 3 days of data (test data).

Step 5: Calculate the mean and standard error - A mean for a given time period Δt, and for all the B predictions from Step 4 is computed in this step. The standard deviation of the sampling distribution of means (i.e., standard error) is also computed.

Demonstration Case Study

The data used in this study were from a corridor located northeast of downtown San Antonio, Texas, along Interstate 35 (IH-35) as indicated in the dashed box shown in Figure 3*a*. This corridor section is part of one of

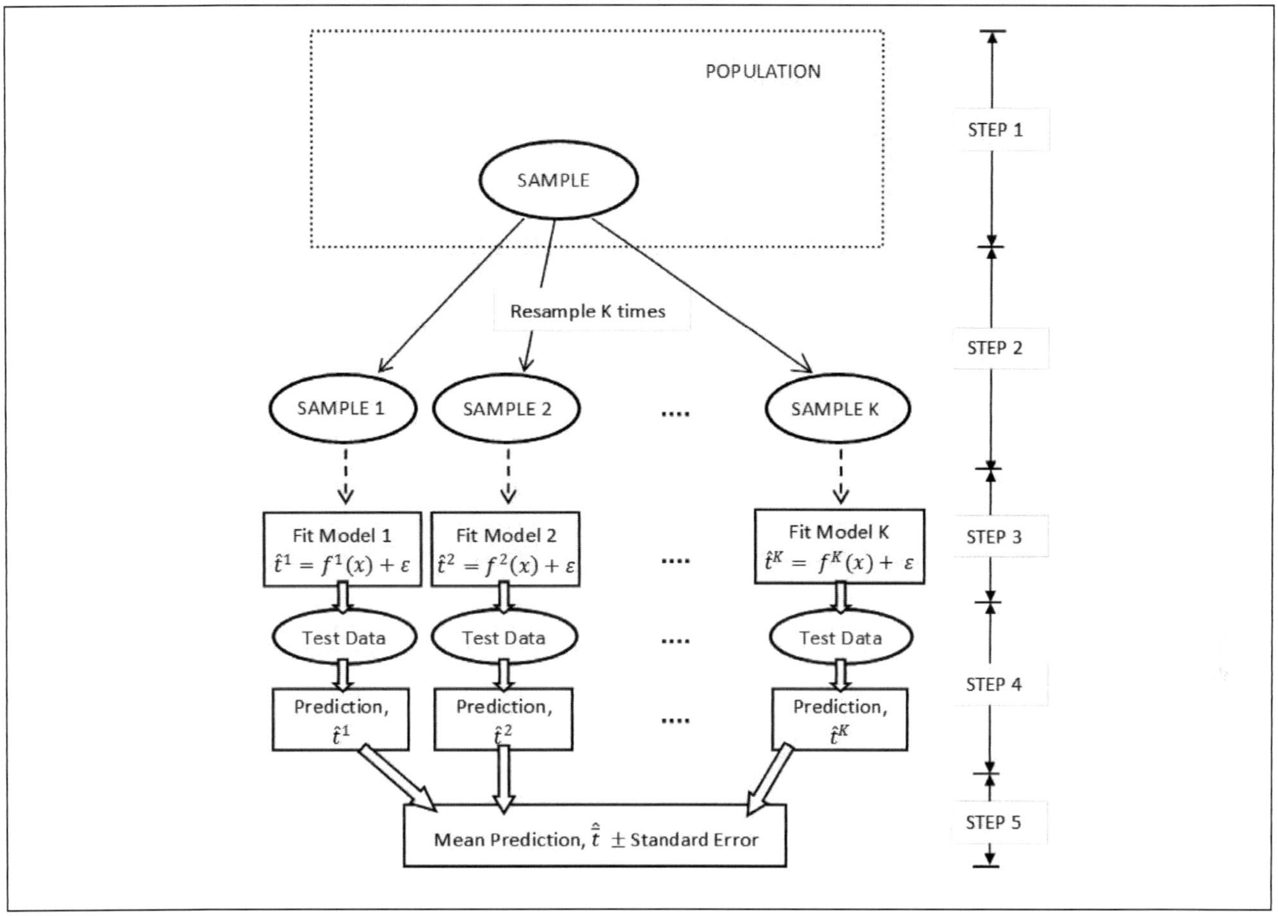

Figure 2. Schematic of the proposed gap bootstrap methodology.

the busiest interstates in Texas. The IH-35 merges into the IH-10 (north leg of the San Antonio downtown loop) and connects the cities of San Antonio and Austin, Texas.

Figure 3*b* shows the specific corridor, which is approximately 1.39 miles and has three lanes with three links. There were four main lane dual loop detectors (stations 159.500 to 160.892) including two on-ramps and one off-ramp on the test bed. Speed, volume, and occupancy data from the loop detectors for the morning peak period (7:00–9:00) for all weekdays between April 1, 2007 and September 30, 2007 were obtained. The data were aggregated at 2-min intervals and, therefore, 7,930 observations were available per link. The link travel times were estimated as a function of the speed, volume, and occupancy using a methodology developed by Vanajakshi et al. (*35*).

Preliminary Data Analyses

Point data such as that from loop detectors have been shown to have a nonstationary distribution and exhibit a general periodicity (*36*). For example, a given link may exhibit a repeating pattern such as a large peak at 8:00 a.m., a smaller peak at around 4:00 p.m., and an absolute minimum around 2:00 p.m. Additionally, the data are also weakly dependent. That is, the dependence diminishes as the aggregation interval becomes large (i.e., traffic characteristics on Monday at 7:30 a.m. influence the traffic characteristics on Monday at 7:35 a.m., but do not influence traffic characteristics on the next Monday at 7:35 a.m.). It was necessary to explore the characteristics of the data—temporally and spatially—to select the most appropriate estimation and prediction models for both the mean and the confidence interval measures.

The preliminary analyses were performed to explore the characteristics of the observed speed, volume, occupancy, and travel time data. Note that, although each data type was explored as an example, only results for the volume data and travel times are presented here.

Periodicity

A periodic behavior (repeating pattern) can be observed when each day's data are plotted side-by-side as shown

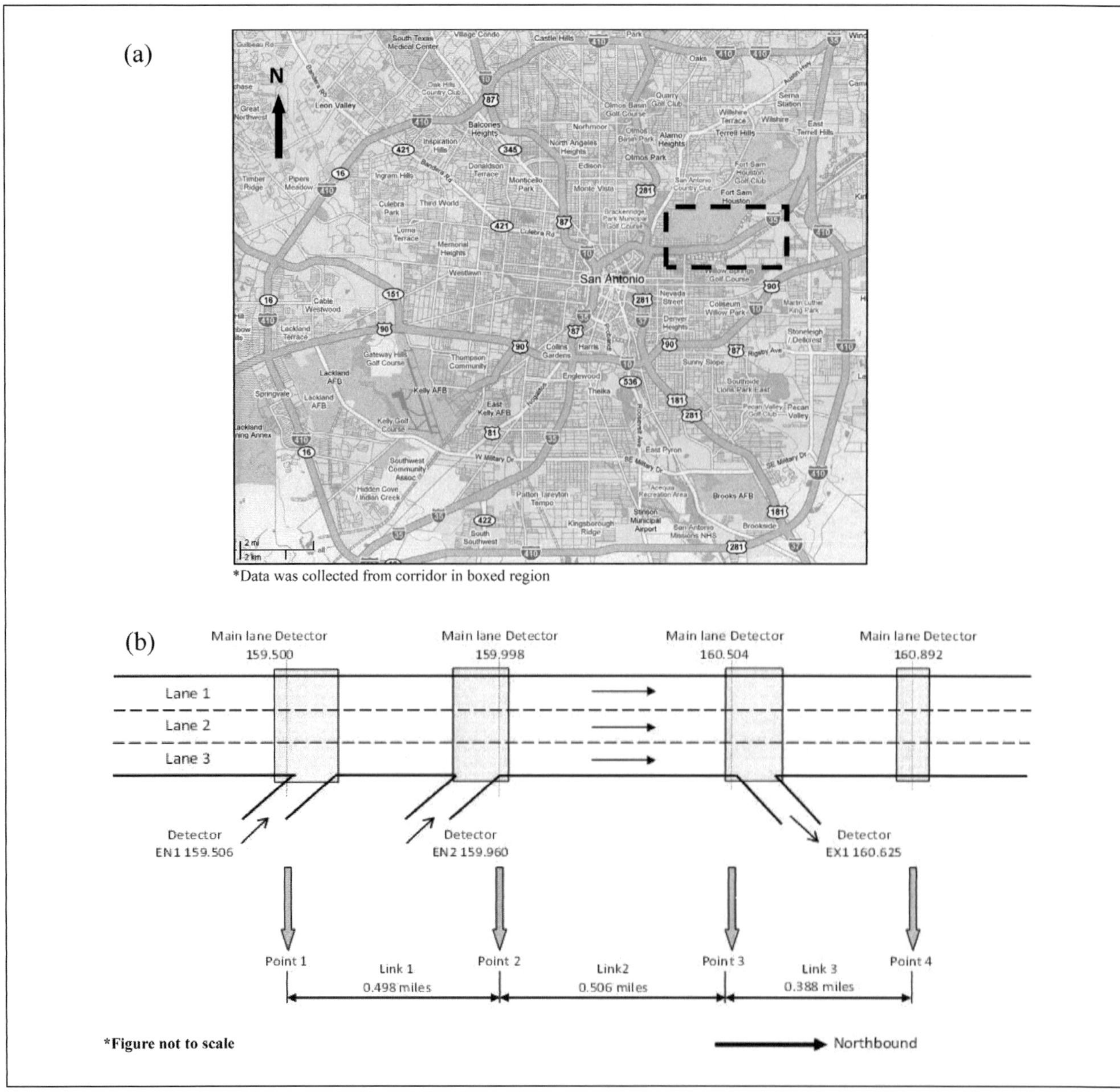

Figure 3. Data collection test bed, showing (*a*) the location on a regional map of San Antonio, Texas, and (*b*) a schematic diagram of the test bed at San Antonio, Texas.

in Figure 4. Figure 4, *a* and *b*, show the 2-min aggregated data for the morning peak (7:00–9:00 a.m.) as a function of the time of the day. A noticeable pattern can be observed. It should be noted that it was difficult to view the entire data (130 days) when plotted at once, so only data for 5 days (305 data points) were plotted as a demonstration example.

Dependence and Nonstationarity

An autocorrelation function, or a correlogram, is a plot of the autocorrelation coefficients against the time lags and these were created for the volume and travel time data as shown in Figure 4, *c* and *d*, respectively. The autocorrelation coefficients measure the correlation between observations at different lags or distances apart (*37*). The autocorrelation coefficient r_k at time lag k is computed (*37*) as

$$r_k = \frac{\sum_{t=1}^{N-k}(x_t - \bar{x})(x_{t+k} - \bar{x})}{\sum_{t=1}^{N}(x_t - \bar{x})^2} \tag{8}$$

where

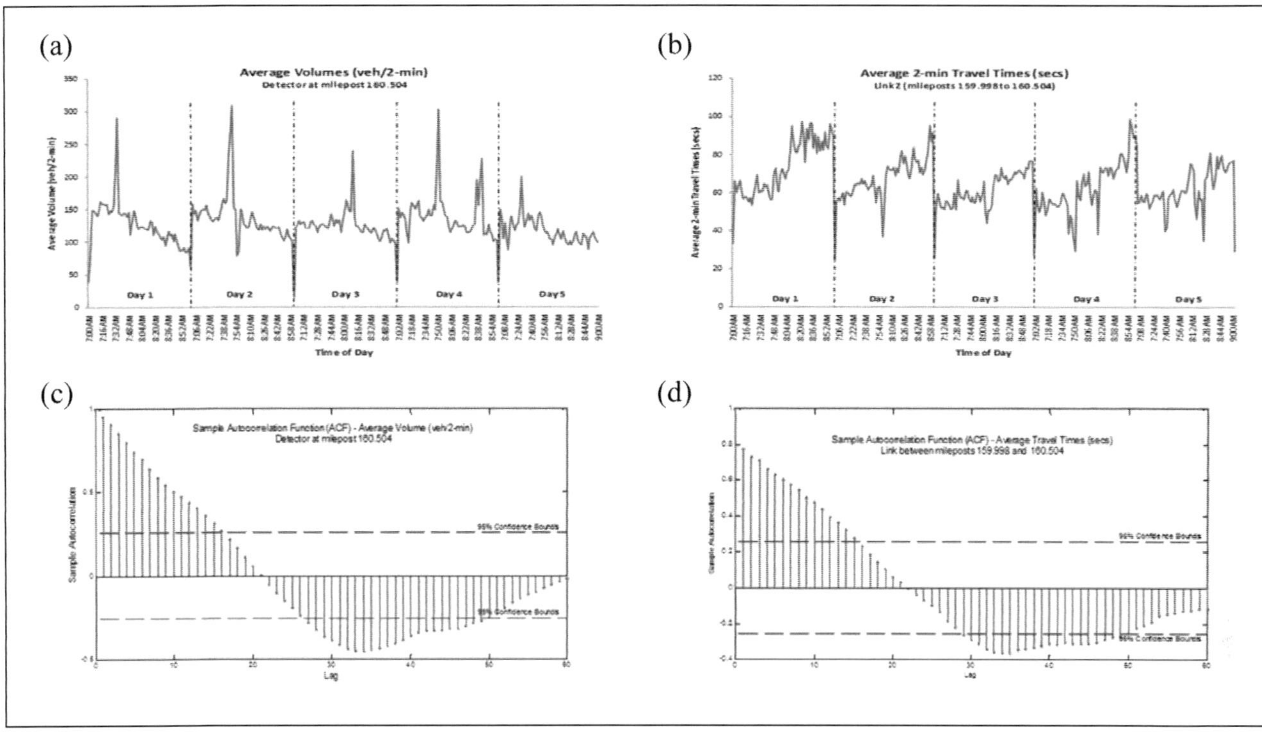

Figure 4. Results of exploratory data analyses, showing (*a*) peak hour traffic volume data, (*b*) peak hour travel time data, (*c*) correlogram of traffic volume data, and (*d*) correlogram of travel time data.

$$\bar{x} = \frac{\sum_{t=1}^{N} x_t}{N} = \text{mean};$$

r_k = correlation coefficient at lag k;
x_t = observation at current time;
x_{t+k} = observation at one lag ahead; and
N = size of series.

In Figure 4, *c* and *d*, the dashed horizontal lines are the confidence bounds: $\pm 1.96\sqrt{n}$ (1.96 is the .975 quantile of the standard normal distribution). The 60-lags represent the 07:00–09:00 a.m. time period with each lag denoting a 2-min interval. The correlogram illustrates that more than 95% of the autocorrelation coefficients are outside the confidence bounds. This indicates that there is time dependence in the data (*38*). That is, data from the current time step are correlated with the data from the previous time step. As an example, travel time at 7:12 a.m. is correlated with travel time at 7:10 a.m. Additionally, the correlation coefficients tend to decay slowly as the time lag increases and are indicative of the data being nonstationary. That is, the statistical characteristics (mean and variance) of the data change over time.

In conclusion, the exploratory analyses suggest that the data (point and interval) exhibit a periodic behavior, are time dependent, and have a nonstationary distribution. Standard historical statistical methods do not apply and any modeling with this data must account for this characteristic behavior. For such data sets, Lahiri et al. found that the gap bootstrap method outperformed other commonly used resampling methods in closeness to the true standard errors based on Monte-Carlo simulation (*32*).

Implementation

Mean travel times were forecasted for fifteen 2-min time periods (30 min ahead) using the methodology shown in Figure 2. An overall mean travel time (at each 2-min interval) as well as its corresponding bootstrap estimate of uncertainty (standard error) was then computed. The overall means and standard error values were computed for the three test days.

Ordinary bootstrap estimates were computed using 250 realizations of the original travel time data that were randomly sampled. The block bootstrap estimates were calculated by dividing the original data into 25 blocks, each consisting of 305 rows (i.e., travel time values for 5 days). Each bootstrap sample was then formed by randomly drawing the blocks with replacement and laying them end-to-end. The overall mean and standard error of the predicted travel times (block bootstrap estimates) were computed as calculated from 250 such samples.

The gap bootstrap estimates were obtained by dividing the original data into 42 subsets of independent data, in which the elements of the *i*th subset are the *i*th 2-min travel times for each of the 125 days in the training dataset. That is, the first subset consists of 7:00 a.m. travel times, the second subset consists of 7:02 a.m. travel times, the third subset consists of 7:04 a.m. travel times, and so on. The predicted overall mean travel times for each 2-min period were obtained by averaging the 2-min predictions from the 42 independent subsets. Component-wise variance estimators of the 42 subsets were calculated by performing ordinary bootstrapping within each subset. These were calculated using 250 realizations of each subset by resampling across the rows of the subsets. Based on the preliminary analysis, it was assumed there is constant covariance among the subsets and the final gap bootstrap uncertainty estimates were computed by combining the component-wise variance estimates with the usual covariance matrix estimator obtained from the subsets.

Results and Analysis

The predicted overall mean travel times and corresponding estimates of the standard error are presented in Table 1. It can be observed that the mean travel time values computed using the three bootstrapping techniques were within 7% of each other. This suggests that the choice of the resampling technique will not affect the predicted point estimate.

However, the estimated standard errors are clearly different. In general, the uncertainty estimates are lower for the ordinary bootstrap than those of the block and gap bootstrap. In addition, the gap bootstrap estimates are the largest. To illustrate these results, the standard error values computed on test day 1 are shown in Figure 5. The block bootstrap uncertainty estimates were on average only 1.2 times larger than the ordinary bootstrap estimates. As expected, the gap bootstrap provided substantially larger estimates of the standard error for the predicted travel times than the ordinary bootstrap (on average 2.7 times) and the block bootstrap (on average 2.3 times). These results are not surprising given the heteroscedastic nature of the data. In this situation, standard techniques would underestimate the confidence interval of the estimate.

It is hypothesized that the more conservative confidence interval would be beneficial to both system operator and system users because it more accurately captures the nature of the data and associated prediction values.

Summary and Conclusions

To a large extent, methods of forecasting travel time have placed emphasis on the *quality* of the forecasted value.

Table 1. Predicted Means and Standard Errors

	Prediction interval (min)														
Method	2	4	6	8	10	12	14	16	18	20	22	24	26	28	30
Test day 1															
Ord	26 (0.616)	26 (0.617)	26 (0.618)	26 (0.615)	26 (0.616)	26 (0.616)	26 (0.615)	26 (0.617)	26 (0.617)	26 (0.616)	26 (0.615)	26 (0.616)	26 (0.615)	26 (0.614)	26 (0.617)
Blk	26 (0.725)	26 (0.754)	26 (0.748)	26 (0.756)	26 (0.727)	26 (0.706)	26 (0.702)	26 (0.701)	26 (0.721)	26 (0.730)	26 (0.714)	26 (0.716)	26 (0.710)	26 (0.728)	26 (0.738)
Gap	28 (2.120)	28 (2.745)	28 (1.614)	27 (2.425)	28 (1.564)	28 (1.314)	28 (1.548)	26 (1.998)	27 (1.461)	27 (1.540)	28 (1.484)	28 (1.690)	27 (1.353)	28 (1.098)	27 (1.258)
Test day 2															
Ord	27 (0.823)	27 (0.824)	27 (0.819)	27 (0.824)	27 (0.824)	27 (0.821)	27 (0.824)	27 (0.823)	27 (0.822)	27 (0.823)	27 (0.822)	27 (0.824)	27 (0.824)	27 (0.825)	27 (0.824)
Blk	27 (0.850)	27 (0.891)	27 (0.870)	27 (0.879)	27 (0.858)	27 (0.846)	27 (0.843)	27 (0.843)	27 (0.848)	27 (0.846)	27 (0.831)	27 (0.833)	27 (0.834)	27 (0.838)	27 (0.843)
Gap	29 (1.753)	28 (1.852)	28 (2.304)	27 (2.245)	27 (1.547)	28 (1.882)	27 (1.783)	26 (1.298)	27 (1.161)	27 (1.704)	27 (1.911)	27 (1.427)	27 (1.489)	27 (1.817)	27 (1.491)
Test day 3															
Ord	25 (1.586)	25 (1.590)	25 (1.590)	25 (1.587)	25 (1.584)	25 (1.590)	25 (1.585)	25 (1.587)	25 (1.584)	25 (1.586)	25 (1.585)	25 (1.588)	25 (1.587)	25 (1.591)	25 (1.587)
Blk	25 (1.550)	25 (1.366)	25 (1.335)	25 (1.338)	25 (1.332)	25 (1.310)	25 (1.305)	25 (1.287)	25 (1.310)	25 (1.307)	25 (1.281)	25 (1.275)	25 (1.264)	25 (1.275)	25 (1.255)
Gap	27 (3.232)	26 (4.723)	28 (4.217)	27 (2.409)	27 (2.726)	26 (2.850)	27 (2.566)	27 (2.829)	27 (2.201)	26 (2.860)	27 (2.720)	27 (3.065)	27 (3.649)	27 (2.812)	27 (4.375)

Note: Bootstrap method: Ord = Ordinary; Blk = Block; and Gap = Gap. Predicted travel times are in seconds. Value in parenthesis is the estimated uncertainty (standard error).

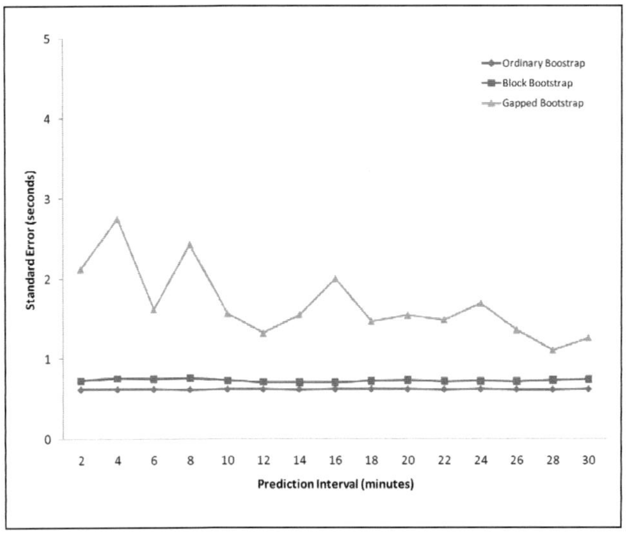

Figure 5. Estimated standard errors for predicted travel time.

However, understanding the reliability or uncertainty margin that exists around the forecasted point estimate is also important, especially understanding that it has been shown to lead end-users to change their routes and schedules even when the average travel time is low (*6*). This paper demonstrates the application of the recently developed gap bootstrap method and compares it against two existing methods (ordinary and block bootstrap) with specific application to link travel time prediction. The method is unique in that, unlike the techniques used in past research, it accounts for datasets that are not independent and identically distributed. These characteristics are common within ITS data which are typically used in the travel time prediction process.

From the results and analysis presented in this paper, it was concluded that uncertainty estimates from the ordinary and block bootstrap methods are similar. In comparison with the ordinary bootstrap, the block bootstrap does not appear to adequately capture the dependence structure (identified in the exploratory analyses) in a dataset. Therefore, although the ordinary bootstrap is not suitable for dependent data, the block bootstrap does not seem to provide much advantage when applied to the dataset used in this study. The estimates from the gap bootstrap are larger than the ordinary and block bootstrap estimates. It is hypothesized that the difference between the gap bootstrap estimate and those from the other two resampling methods are a result of the gap bootstrap explicitly accounting for the dependent structure evident in the dataset. Moreover, unlike the ordinary bootstrap, which is suitable only for data that are independent, it appears the gap bootstrap can adequately address uncertainties for both independent and dependent structured datasets.

Because the true uncertainty estimate of travel times is unknown for the test network, further work using simulation data is required to validate the findings. Such research will allow for comparing uncertainty estimates obtained with the various resampling methods as well as other time series based methods with the true values of the uncertainty based, for example, on Monte-Carlo simulation.

This paper makes a strong case for the gap bootstrap uncertainty estimator given that the existing methods, including the ordinary and block bootstrap, might not be appropriate. The empirical example in this paper, in which the exchangeability of partitions might have been only approximate, led to fairly conservative estimates of uncertainty. In many applications, having a slightly conservative estimate of uncertainties is not a major disadvantage and, in transportation engineering in particular, it is more desirable than the converse.

As indicated earlier, the GARCH approach has also been applied to model uncertainty in the univariate vehicular traffic flow series (*10–12*). The proposed gap bootstrap approach accounts for datasets that are not independent and identically distributed (which is common for ITS data) whereas the GARCH approach accounts for the heteroscedasticity observed in the traffic condition series. The authors suggest further research that would involve a direct comparison of the performance of the proposed gap bootstrap method and the existing GARCH based methods. The comparison should be made under different conditions such as locations, weather, and traffic patterns. An investigation of this kind will provide insights on which is a better method under what set of prevailing conditions.

One potential application of the proposed methodology is in advanced traffic management systems (ATMS) in which, given the number of travel time prediction models available, engineers are faced with the challenge of selecting and implementing the most effective modeling scheme. As a general guide to selecting one of the numerous models for implementation, it would be instructive to have a means of assessing the reliability of these models by being able to accurately quantify the uncertainties in the predictions in terms of variances or confidence intervals but also accounting for the structure of the ITS data which has a dependent structure.

Acknowledgments

The authors would like to thank Dr. Clifford Spiegelman and Dr. Soumen Lahiri for providing background information and technical expertise on the technique used in this research. The authors are also grateful to the Texas Department of Transportation for the use of the Transguide data.

Author Contributions

The authors confirm contribution to the paper as follows: study conception and design: Naik 50%, Rilett 50%; data collection: Naik 50%, Appiah 50%; analysis and interpretation of results: Naik 25%, Rilett 25%, Appiah 25%, Walubita 25%; draft manuscript preparation: Naik 25%, Rilett 25%, Appiah 25%, Walubita 25%. All authors reviewed the results and approved the final version of the manuscript.

References

1. Mori, U., A. Mendiburu, M. Álvarez, and J. A. Lozano. A Review of Travel Time Estimation and Forecasting for Advanced Traveller Information Systems. *Transportmetrica A: Transport Science*, Vol. 11, No. 2, 2015, pp. 119–157.
2. Du, L., S. Peeta, and Y. H Kim. An Adaptive Information Fusion Model to Predict the Short-Term Link Travel Time Distribution in Dynamic Traffic Networks. *Transportation Research Part B*, Vol. 46, No. 1, 2012, pp. 235–252.
3. Mazloumi, E., G. Currie, G. Rose, and M. Sarvi. Statistical Confidence Estimation Measures for Artificial Neural Networks: Application in Bus Travel Time Prediction. Presented at 89th Annual Meeting of the Transportation Research Board, Washington, D.C., 2010.
4. Mahmassani, H. S., and Y. H. Liu. Dynamics of Commuting Decision Behavior Under Advanced Traveler Information Systems. *Transportation Research Part C*, Vol. 7C, 1999, pp. 91–107.
5. Van Berkum, E. C., and P. H. J. Van Der Mede. *The Impact of Traffic Information: Dynamics in Route and Departure Time Choice.* Delft University Press, Delft, the Netherlands, 1993.
6. Wang, X., and A. Khattak. Role of Travel Information in Supporting Travel Decision Adaption: Exploring Spatial Patterns. *Transportmetrica A: Transport Science*, Vol. 9, No. 4, 2013, pp. 316–334.
7. Yang, H. *An Analysis of Route Choice Behavior under Advanced Traveler Information Systems: A Neural Network Approach.* ITS, UC Davis, 1992.
8. Eisele, W. L., B. Naik, and L. R. Rilett. Estimating Route Travel Time Reliability from Simultaneously Collected Link and Route Vehicle Probe Data and Roadway Sensor Data. *International Journal of Urban Sciences*, Vol. 19, No. 3, 2015, pp. 286–304.
9. Van Lint, J. W. C. *Reliable Travel Time Prediction for Freeways.* PhD dissertation. The Netherlands TRAIL Research School, 2004.
10. Guo, J., B. M. Williams, and B. L. Smith. Data Collection Time Intervals for Stochastic Short-Term Traffic Flow Forecasting. *Transportation Research Record: Journal of the Transportation Research Board*, 2008. 2024: 18–26.
11. Guo, J., and B. M. Williams. Real Time Short Term Traffic Speed Level Forecasting and Uncertainty Quantification Using Layered Kalman Filters. *Transportation Research Record: Journal of the Transportation Research Board*, 2010. 2175: 28–37.
12. Guo, J., W. Huang, and B. M. Williams. Real Time Traffic Flow Outlier Detection Using Short-Term Traffic Conditional Variance Prediction. *Transportation Research Part C*, Vol. 50, 2015, pp. 160–172.
13. Kamarianakis, Y., A. Kanas, and P. Prastacos. Modeling Traffic Volatility Dynamics in an Urban Network. *Transportation Research Record: Journal of the Transportation Research Board*, 2005. 1923: 18–27.
14. Sohn, K., and D. Kim. 2009. Statistical Model for Forecasting Link Travel Time Variability. *Journal of Transportation Engineering*, Vol. 135, No. 7, pp. 440–453.
15. Ellison, S. L. R., M. Rosslein, and B. Williams. *Quantifying Uncertainty in Analytical Measurement*, 2nd ed. EURACHEM/CITAC Guide, 2000.
16. Robinson, S., and J. W. Polak. ILD Data Cleaning Treatments and their Effect on the Performance of Urban Link Travel Time Models. Presented at 85th Annual Meeting of the Transportation Research Board, Washington, D.C., 2006.
17. Vanajakshi, L. D., and L. R. Rilett. Loop Detector Data Diagnostics Based on Conservation-of-Vehicles Principle. *Transportation Research Record: Journal of the Transportation Research Board*, 2004. 1870: 162–169.
18. Zapranis, A., and E. Livanis. Prediction Intervals for Neural Network Models. *Proc., WSEAS International Conference on Engineering Education*, Greece, 2005.
19. Bishop, C. M. *Neural Networks for Pattern Recognition.* Clarendon Press, Oxford, 1995.
20. Nix, D. A., and A. S. Weigend. Learning Local Error Bars for Nonlinear Regression. *Proc., Advances in Neural Information Processing Systems (NIPS 7)*, MIT Press, Cambridge, 1995.
21. Singh, K., and M. Xie. *Bootstrap: A Statistical Method.* Unpublished manuscript, Rutgers University, USA, 2008. http://www.stat.rutgers.edu/home/mxie/RCPapers/bootstrap.pdf.
22. Heskes, T. Practical Confidence and Prediction Intervals. In *Advances in Neural Information Processing Systems* (Mozer, M., M Jordan, and T. Petsche, eds.), MIT Press, Cambridge, Vol. 9, 1997, pp. 176–182.
23. Chernick, M. R. *Bootstrap Methods: A Practitioner's Guide.* John Wiley & Sons Inc., New York, NY, 1999.
24. Davison, A. C., and D. V. Hinkley. *Bootstrap Methods and their Applications.* Cambridge University Press, New York, NY, 1997.
25. Tibshirani, R. A Comparison of Some Error Estimates for Neural Network Models. *Neural Computation*, Vol. 8, 1996, pp. 152–163.
26. Hardle, W., J. Horowitz, and J. Kriess. Bootstrap Methods for Time Series. *International Statistics Review*, Vol. 71, No. 2, 2003, pp. 435–459.
27. Kunsch, H. R. The Jackknife and the Bootstrap for General Stationary Observations. *Annals of Statistics*, Vol. 17, 1989, pp. 1217–1261.
28. Carlstein, E. The Use of Subseries Methods for Estimating the Variance of a General Statistic From a Stationary Time Series. *Annals of Statistics*, Vol. 14, 1986, pp. 1171–1179.
29. Politis, D., and J. P. Romano. A Circular Block Resampling Procedure for Stationary Data. In *Exploring the*

Limits of Bootstrap (Lepage, R., and L. Billard, eds.), Wiley, New York, 1992, pp. 263–270.
30. Politis, D., and J. P. Romano. The Stationary Bootstrap. *Journal of the American Statistical Association*, Vol. 89, 1994, pp. 1303–1313.
31. Lahiri, S. N. Theoretical Comparison of Block Bootstrap Methods. *Annals of Statistics*, Vol. 27, 1999, pp. 386–404.
32. Lahiri, S. N., C. Spiegelman, J. Appiah, and L. Rilett. Gap Bootstrap Methods for Massive Data Sets with an Application to Transportation Engineering. *Annals of Statistics*, Vol. 6, No. 4, 2012, pp. 1552–1587.
33. Appiah, J. *Quantifying Uncertainties in Synthetic Origin-Destination Trip Matrix Estimates*. PhD dissertation. University of Nebraska-Lincoln, 2009.
34. Naik, B. *A Generalized Nonparametric Approach for Uncertainty Evaluation in Travel Time Prediction Models*. PhD dissertation. University of Nebraska-Lincoln, 2010.
35. Vanajakshi, L. D., B. M. Williams, and L. R. Rilett. Improved Flow-Based Travel Time Estimation Method from Point Detector Data for Freeways. *Journal of Transportation Engineering*, Vol. 135, No. 1, 2009, pp. 26–36.
36. May, A. D. *Traffic Flow Fundamentals*. Prentice Hall, Upper Saddle River, NJ, 1990.
37. Chatfield, C. *The Analysis of Time Series: An Introduction*. Chapman and Hall/CRC, Boca Raton, FL, 2003.
38. Brockwell, P. J., and R. A. Davis. *Introduction to Time Series and Forecasting*, 2nd ed. Springer-Verlag, NY Inc., 2002.

The Standing Committee on Urban Transportation Data and Information Systems (ABJ30) peer-reviewed this paper (18-03453).

The contents of this paper reflect the views of the authors, who are responsible for the facts and the accuracy of the information presented here.

Article

TRR
JOURNAL OF THE TRANSPORTATION RESEARCH BOARD

Transportation Research Record
2018, Vol. 2672(42) 148–158

DOI: 10.1177/0361198118772956
journals.sagepub.com/home/trr
SAGE

Reconstructing Vehicle Trajectories to Support Travel Time Estimation

Zheng Li[1], Robert Kluger[1], Xianbiao Hu[2], Yao-Jan Wu[1], and Xiaoyu Zhu[3]

Abstract

The primary objective of this study was to increase the sample size of public probe vehicle-based arterial travel time estimation. The complete methodology of increasing sample size using incomplete trajectory was built based on a *k*-Nearest Neighbors (*k*-NN) regression algorithm. The virtual travel time of an incomplete trajectory was represented by similar complete trajectories. As incomplete trajectories were not used to calculate travel time in previous studies, the sample size of travel time estimation can be increased without collecting extra data. A case study was conducted on a major arterial in the city of Tucson, Arizona, including 13 links. In the case study, probe vehicle data were collected from a smartphone application used for navigation and guidance. The case study showed that the method could significantly increase link travel time samples, but there were still limitations. In addition, sensitivity analysis was conducted using leave-one-out cross-validation to verify the performance of the *k*-NN model under different parameters and input data. The data analysis showed that the algorithm performed differently under different parameters and input data. Our study suggested optimal parameters should be selected using a historical dataset before real-world application.

Travel time plays a significant role in traffic planning, traffic management, and advanced traveler information systems (ATIS). In the past decades, there has been an increasing trend of using large public probe vehicle datasets for arterial travel time estimation (*1, 2*). A public probe vehicle dataset, or sometimes a "passive" dataset (*3*), comprises probe vehicle data that are collected from the public (navigation app, etc.) or public transport (taxis, transits, etc.). Because probe vehicle data are generally collected via crowdsourcing, the data can support travel time estimation on a large temporal and spatial scale but at a relatively low cost.

Although probe vehicle travel time estimation has plenty of advantages over traditional methods, it has several limitations. First, accurate travel time estimation requires a relatively high penetration rate and sampling rate. The penetration rate is defined as "the flow fraction of vehicles (unique devices) reporting to the probe dataset as compared with the total flow of vehicles along a road" and sampling rate is "the average rate at which any device reports its position and velocity" (*4*). Because probe vehicles are samples from all vehicles on the road, the travel time estimation result may not be statistically significant if the penetration rate is low, leading to a low confidence level in the estimates. Also, a low sampling rate is likely to result in poor accuracy of travel time estimation. Most current probe vehicle datasets have a relatively low sampling rate and penetration rate, which limit their applications. Public probe vehicle datasets also suffer from uneven temporal-spatial sample distribution. For example, more data are collected on major arterials but less on low-grade sections; and more data are collected during peak hours but less collected during off-peak times, or even none may be collected in the late night hours.

As a result of the limitation of current probe vehicle datasets, many current studies can be classified into two research areas. The first research area aims to estimate travel time when the sampling rate or the penetration rate is low (*1, 2, 5–8*). Another research area focuses on improving the accuracy of travel time estimation (*9–12*). Although there are already several research reports on travel time estimation using probe vehicle data, most of the previous studies were built on the scenario in which

[1]Department of Civil Engineering & Engineering Mechanics, The University of Arizona, Tucson, AZ
[2]Department of Civil, Architectural and Environmental Engineering, Missouri University of Science and Technology, Rolla, MO
[3]Metropia Inc., Tucson, AZ

Corresponding Author:
Address correspondence to Yao-Jan Wu: yaojan@email.arizona.edu

the probe vehicle data have a relatively high penetration rate but low sampling rate. Travel time estimation when probe vehicle data have a good sampling rate but poor penetration rate is a more relevant problem, given the way probe vehicle data are collected today.

Accurate urban link travel time estimation can provide more useful information to agencies, researchers, and travelers. This study aims to increase the sample size of probe vehicle data by including partial vehicle trajectories. Partial trajectories already exist in the probe vehicle datasets but are currently treated as outliers or discarded in travel time estimation. If the partial trajectories can be utilized, the expected benefits are: (1) effectively increasing the sample size available for travel time estimation, (2) estimating more accurate travel time with low penetration rates of probe vehicles, and (3) reducing the costs of data collection.

Literature Review

Travel Time Data Sources

Traditionally, travel time estimation for an urban area relies on fixed sensors, including loop detectors (*13–15*), automated vehicle identification (*16–18*), Bluetooth/Wi-Fi devices (*19–21*), microwave sensors (*22*), and so on. All the above-mentioned data collection methods require corresponding sensors installed to retrieve data. Once the sensor is installed, it can continuously record data on the monitored road section. However, the cost of installing and maintaining fixed sensors is relatively high because many sensors are needed to achieve the appropriate accuracy level or cover a large research area.

An alternative approach is to measure travel time by mobile traffic sensors, for example, floating cars (*23*), probe vehicles (*24*), cellular data, and so on. Vehicles equipped with tracking devices (GPS or mobile phone) can be used for collecting travel time at any location without roadside equipment. However, mobile sensors are still costly because stabilized data collection needs operational vehicles running on the study area all the time. Hence, they can only cover a limited number of routes for a limited duration of time (*6*). As a result of cost considerations, there are only a few traffic studies using mobile sensors.

Many public vehicles (e.g., taxis, transit, etc.) are equipped with GPS devices. These public vehicles, to some extent, are probe vehicles, and they can collect travel time data on most of the network links during their service time with a low cost. In addition, with the popularity of mobile phones, trajectory data that are collected from mobile phones can also be used for travel time estimation. The appearance of these new data sources provides the possibility for large-scale and long-term travel time estimation. Along with the growth and availability of probe vehicle datasets, numerous studies have been conducted on travel time estimation using public datasets. Zhan et al. successfully estimated hourly travel time using New York City taxicab origin and destination trip data (*2*). Jenelius and Koutsopoulos discussed a statistical model for urban road network travel time estimation using vehicle trajectories obtained from low-frequency GPS probes. A case study was conducted on an arterial network in Stockholm, Sweden using taxi fleet data (*6*).

According to the location where data are collected, travel time can be classified into travel time on freeways or travel time on arterials. Whereas vehicular flow on freeways is often treated as uninterrupted flow, flow on arterials is much more complicated because it can be affected by signal delay, queue delay, pedestrians, and entry vehicles. On highway or urban environments, as travel time depends on the origin and destination, ATIS normally use methods that calculate travel time at a link or section level, rather than a trip level (*25*). Feng et al. proposed that the distribution of link travel time in an urban area can be approximated using mixtures of normal distributions. Although historical travel time data are available, probe vehicle data can be used to identify current traffic statement based on Bayes Theorem (*26*).

Probe Vehicle Travel Time Estimation Methods

Probe vehicles equipped with GPS systems can collect position, speed, and time stamp data every few seconds (*27*). Theoretically, probe vehicles can provide all the information needed to calculate travel time on any area at any time. However, because of the shortcomings of current probe vehicle datasets, this approach still has many limitations with respect to applications of probe vehicle travel time estimation. The limitations primarily come from two aspects: low sampling rate and/or low penetration rate.

Low sampling rate has made it difficult to measure travel time directly because little information is known between every two continuous data points. As most current datasets have a low sampling rate, there are many papers that seek to calculate accurate travel time using a sparse probe vehicle dataset. Wan et al. proposed a method to reconstruct maximum likelihood trajectory of probe vehicles between sparse updates based on the expectation maximization algorithm (*1*). Another method is to use models to estimate travel time (neural networks, etc.). Zheng and Van Zuylen built a three-layer neural network model to estimate complete link

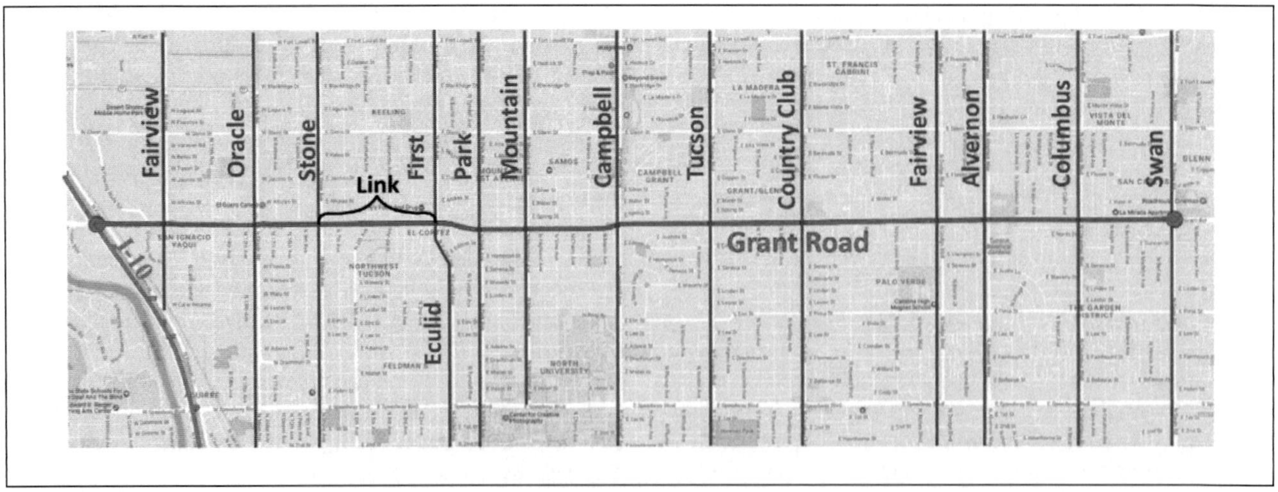

Figure 1. Study corridor.

travel time for individual probe vehicles traversing the link, and both simulation data and real-world data were used to verify the result of the model (*5*).

When penetration rate is low, probe vehicle samples cannot represent the entire population and the estimation may not be accurate. There is much research on the relationship between sample size and estimation error. Patire et al. analyzed the estimation error when sampling rate and penetration rate are different by a data fusion approach (*4*). Bucknell and Herrera analyzed estimation error of different combinations of penetration rate and sampling rate on highways using a NGSIM dataset (*7*). However, few research studies focus on how to increase the sample size of probe vehicle datasets. The appearance of public probe vehicle datasets increases the penetration rate of probe vehicles, which is important for the application of probe vehicle data. However, the problem of low penetration rate is still very common, and this means a way to increase probe vehicle sample size based on existing datasets is beneficial.

In summary, there are several studies on the penetration rate requirement for probe vehicle travel time estimation (*4, 7, 8, 28*), but only one study was identified that focused on increasing penetration rate (*29*). In addition, there are few valid methods to increase probe vehicle samples without adding new data sources. The primary objective of this paper is to outline and evaluate a process for increasing the number of usable samples when calculating travel time using probe vehicle trajectories. Currently vehicles must traverse an entire link for inclusion in travel time calculations; however, many vehicles do not traverse the entire link and thus are left out of the calculation. Including those vehicles' trajectories increases the sample size without requiring an increased penetration rate.

Data

Study Corridor

The study corridor is focused on Grant Road between I-10 and Swan Road in Tucson, Arizona. Grant Road is a major east–west direction arterial with annual average daily traffic of 36,000 vehicles per day (*30*). The study corridor is shown in Figure 1 with primary cross-streets labeled. Most of the roads are five lanes in total, with two lanes in each direction and a two-way left-turn lane. At the time of data collection, the only six-lane sections extended from Fairview Ave. to Stone Ave. and starting at Swan Ave. heading eastward. All study links have a speed limit of 40 mph (64 km/h). The links in the study refer to the one-direction segment between each of the contiguous primary cross-streets on Grant Road. For example, the eastbound Oracle–Stone link refers to the road on Grant Road between Oracle Road to Stone Road in the east direction as shown in Figure 1. Data collection was conducted in both directions.

Probe Vehicle Trajectories

Vehicle trajectory data are used to extract travel time information and to further build a historical database. The data are collected by a smartphone app, "Metropia"; when a user starts a trip using the app, the internal GPS module built into the smartphone is activated and starts to record the second-by-second data. These data, including detailed position such as latitude, longitude, heading, timestamp, velocity, and corresponding link in the roadway network, are collected at a fine time interval and sent back to the cloud server, where they are stored and will be used for further analysis.

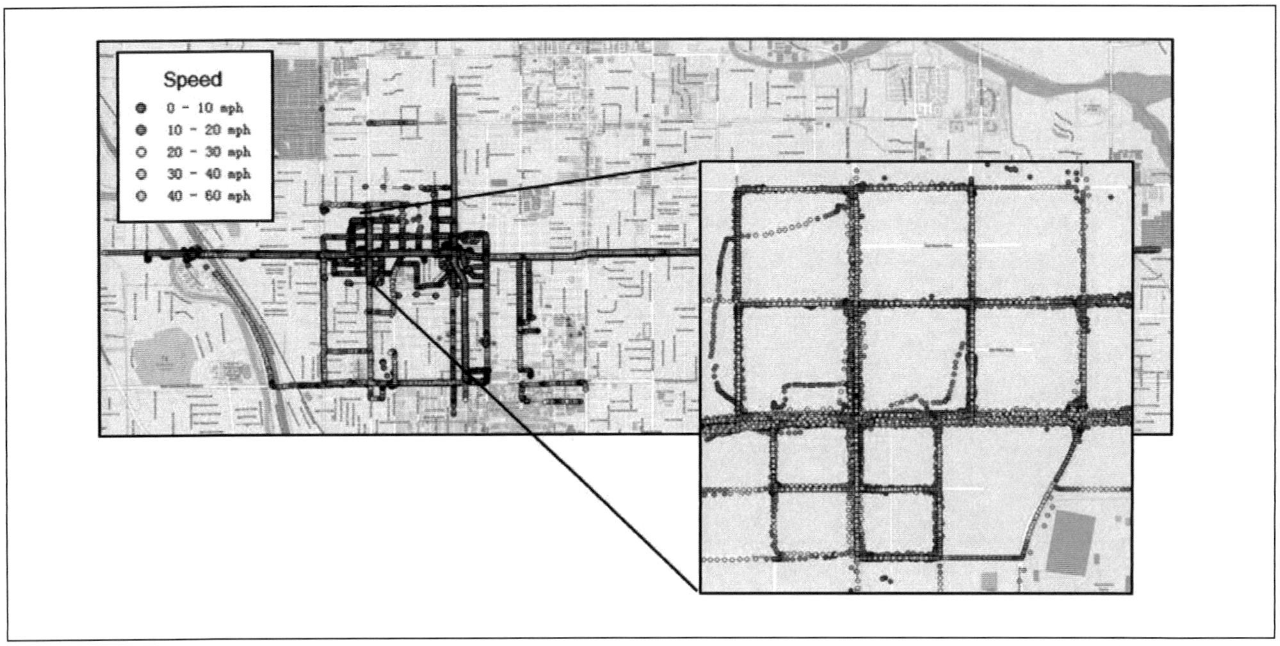

Figure 2. Visualization of example probe vehicle data (7:30–9:30 AM, Nov 17th, 2015).

The original data were collected from January 1 to December 31, 2015. There were 57,645,478 GPS points collected from 1,837 users and 43,315 trips in Tucson. The sampling rate of probe vehicle data is 1 second. In addition, roadway geometry data were acquired for the entire city of Tucson. An example of probe vehicle raw data on Grant Road is shown in Figure 2.

The data underwent an extensive selection and cleaning process including the following steps:

- Selection of trajectories at the study corridor
- Data cleaning through heuristic checks
- Map matching, to pair probe vehicle to roadway links
- Movement determination to identify travel movements
- Dimension reduction of trajectories to distance-time attributes

Methodology

Definition of Trajectories

There are two types of trajectories: complete trajectories and incomplete trajectories. Complete trajectories are defined as those from probe vehicles that passed through both an upstream and downstream intersection surrounding a link, whereas incomplete trajectories are from the probe vehicles that only traverse part of the link.

The concept of this research is to utilize incomplete trajectories in travel time estimation. For this paper, three types of incomplete trajectories are defined based on their relationship to the two intersections surrounding a link, as shown in Figure 3. The first type of trajectory is the one that passes the upstream intersection but exits the link before reaching the downstream intersection. The second type passes through neither intersection surrounding the link. The third type enters the link after the upstream intersection, and passes through the downstream intersection.

k-Nearest Neighbors Regression

The *k*-Nearest Neighbors regression algorithm (*k*-NN) is a non-parametric technique and it has been widely used in travel time estimation. Handley et al. used flow, occupancy, and other variables as inputs of the *k*-NN algorithm to estimate travel time on freeways (*31*). Robinson and Polak successfully used single loop detector data as inputs of *k*-NN to estimate travel time within an urban area. They compared different parameters of the *k*-NN algorithm and the results of the *k*-NN algorithm with other algorithms such as neural network. They also inferred that there is a high potential to use the probe vehicle GPS data as the input of the *k*-NN algorithm (*14*). Zhou et al. applied sparse probe vehicle data as the input of the *k*-NN algorithm to estimate link travel time in an urban area. The study suggested that the *k*-NN

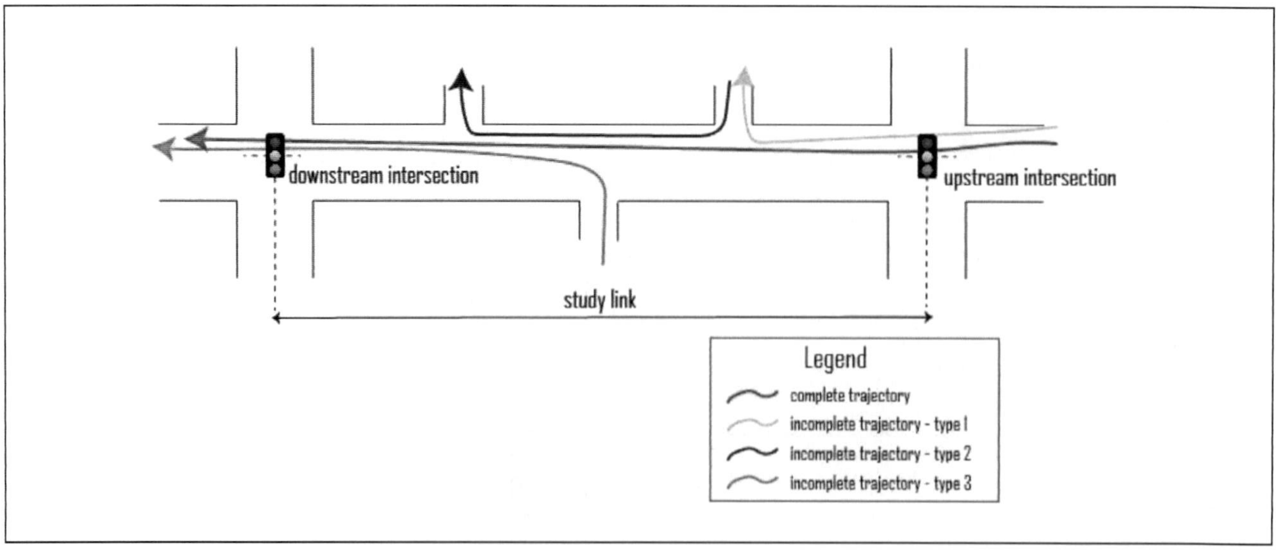

Figure 3. Definitions of complete and incomplete trajectories.

algorithm performed better than the neural network model (*32*).

k-NN has many advantages over other regression algorithms in terms of probe vehicle data analysis. The assumption under *k*-NN regression is that target value is represented by *k* closest samples. Compared with parametric techniques such as linear regression, *k*-NN has no target function. The lack of a target function makes it a more suitable approach to modeling to probe vehicle data in urban areas, because probe vehicles are greatly affected by surrounding environment (road geometry, signal timing, time of day, and other vehicles, etc.). Because of that variation in road environment, a fixed target function may not be able to fit the data well.

The problem in regression is to predict labels $y' \in \mathbb{R}^d$ for new patterns $x' \in \mathbb{R}^q$ based on a set of N observations, that is, labeled patterns $\{(x_1, y_1), \ldots, (x_n, y_n)\}$. For an unknown pattern x', *k*-NN regression computes the mean of the function values of its *k*-nearest neighbors, shown in Equation (1) (*33*).

$$f_{k-NN}(x') = \frac{1}{K} \sum_{i \in \mathcal{N}_k(x')} y_i \quad (1)$$

where $\mathcal{N}$ is neighborhood set, set $\mathcal{N}_k(x')$ containing the indices of the k-nearest neighbors of x'.

The aim is to use *k*-NN to estimate link travel time using one or more incomplete trajectories of a probe vehicle traveling on the link. Because the incomplete trajectory did not cross the whole link, the link travel time cannot be directly calculated. However, there are likely historical complete trajectories under similar traffic conditions that can be used to represent the incomplete trajectory. *k*-NN regression uses the *k* most similar complete trajectories to represent the link travel time, $f_{k-NN}(x')$. For example, if the link travel time of the *k* most similar complete trajectories are $y_{i1}, y_{i2}, y_{i3}, y_{i4}, y_{i5}$ separately, the predicted link travel time of the incomplete trajectory is, using Equation 1, $\frac{(y_{i1} + y_{i2} + y_{i3} + y_{i4} + y_{i5})}{5}$.

The distance between an incomplete trajectory and each complete trajectory is used to find the *k*-nearest neighbors. The procedure to calculate the distance between an incomplete trajectory and a complete trajectory is shown using Figure 4. The length of the study link is L. After the dimension reduction, vehicles trajectories have a position on the link at each timestamp. Dimension reduction makes the *k*-NN algorithm simpler and allows for direct comparison of travel time-related attributes rather comparing the exact path taken. There are two trajectories; the longer one is a complete trajectory and the shorter one is an incomplete trajectory. The study link is divided into n segments (n is 8 in Figure 4) on average and each segment has a length of $\frac{L}{n}$. In Figure 4, the complete trajectory passed all the segments and the incomplete trajectory passed five segments, three of which were fully passed. T_i is the segment travel time for segment i that the complete trajectory has fully traversed, and t_i is the segment travel time for segment i that the incomplete trajectory has fully traversed. The i segments used are those the incomplete trajectory fully passed. Segment travel time is only calculated when a probe vehicle's trajectory has fully traversed the segment. For example, the red, dashed line in Figure 4 is the trajectory that did not fully pass any segments and those red portions are not used to calculate segment travel time. The red part of the trajectory represents the vehicle entering

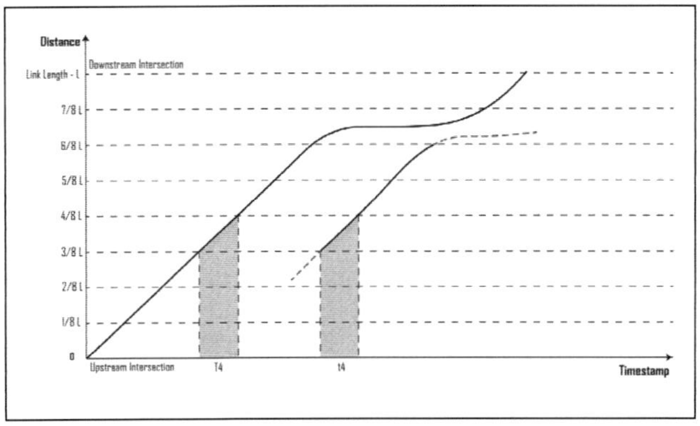

Figure 4. Trajectory similarity calculation.

and/or exiting the link, so it is unlikely to reflect actual traffic conditions.

The distance between two trajectories S is defined by Equation (2):

$$S = \sum_{i \in M} (T_i - t_i)^2 \tag{2}$$

where S is the distance between two trajectories, T_i is the time interval for the complete trajectory to pass the i_{th} segment; $t_i$is the time interval for the incomplete trajectory to pass the i_{th} segment; i is the sequence of segments; and M is the set of segments that the incomplete trajectory has fully passed.

Validation of Method

Theoretically, there is no ground truth for an incomplete trajectory because the link travel time for the incomplete trajectory cannot be calculated. However, a method was developed to evaluate the algorithm by using an incomplete trajectory generated from a complete trajectory as the input. A complete trajectory was cut-off to simulate an incomplete trajectory with a known ground truth. Using this method, the performance of the algorithm can be evaluated based on the ability to accurately recreate trajectories.

Leave-one-out cross-validation (LOOCV) (*34*) was utilized to evaluate the algorithm with the historical complete trajectory dataset. In each round of LOOCV, one complete trajectory was converted into an incomplete trajectory and this incomplete trajectory was input into the k-NN algorithm. Two measures of accuracy were used to verify the algorithm's performance: mean absolute error (MAE) and mean absolute percentage error (MAPE). MAE shows the average error of each round in LOOCV. As link travel time is related to link length, MAE shows an average time difference between algorithm output and ground truth but it cannot reflect the performance difference between links. MAPE shows the error as a percentage and the performance can be compared between links. The definitions of the two measures are shown in Equations (3) and (4), respectively.

$$MAE = \frac{1}{N}\sum_{i=1}^{N} |g_i - e_i| \tag{3}$$

$$MAPE = \frac{1}{N}\sum_{i=1}^{N} \frac{|g_i - e_i|}{g_i} \tag{4}$$

where N is the number of LOOCV times, and g_i is the ground truth of link travel time in the i_{th} LOOCV, and e_i is the estimated link travel time in i_{th} LOOCV.

Sensitivity analysis was conducted to determine the preferred inputs for the k-NN algorithm, using the LOOCV approach. Numerous inputs can be tested, but four were selected for preliminary testing. The selected parameters for this research were the length of the incomplete trajectories, incomplete trajectory type generated in LOOCV, and the values of k and n in the k-NN regression algorithm.

Results

Effectiveness of Sample Size Increasing

Incomplete trajectories in November 2015 were used as the input and complete trajectories from January to October 2015 were used as the historical dataset. Although a specific month was selected, the sensitivity analysis can indicate how well this approach may hold during other months. The sample size comparison is shown in Figure 5. After the implementation of the k-NN algorithm, around half of the links have a sample size that has increased more than 30% and some links even have a sample size increased more than 100%. There are two links that have no improvements, westbound through movement from First Ave. to Stone Ave. and eastbound through movement from Dodge Blvd. to Alvernon Way. A visual inspection of these two links shows that the performance of the algorithm is related to link geometry characteristics and land use around the link. Land use on the north side from First Ave. to Stone Ave. is primarily residential, which may explain the shortage of incomplete trajectories. The link of Dodge Blvd. to Alvernon Way was very short and there were very few access points, and this could be the reason why incomplete trajectories were not captured. These results show that sample size can be substantially increased, depending on roadway geometry features such as access points.

Sensitivity Analysis of the k-NN Model

Sensitivity analysis was conducted to determine the preferred inputs for the k-NN algorithm, using the LOOCV

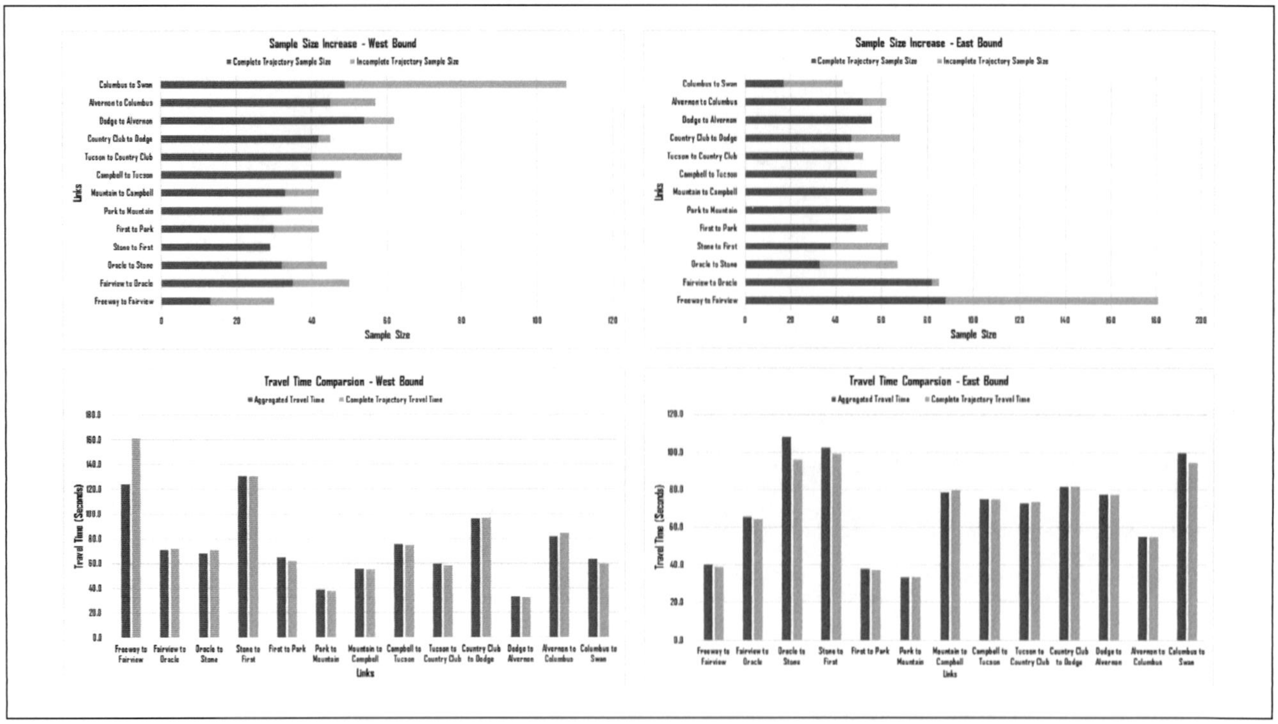

Figure 5. Sample size and travel times by links.

approach. Four parameters were selected for preliminary testing. The selected parameters for this research were the length of the incomplete trajectories, incomplete trajectory type generated in LOOCV, and the values of k and n in the k-NN regression algorithm. These parameters were selected because of their importance in laying the groundwork for future application of this method.

To identify the algorithm's sensitivity to incomplete trajectory length, the size of the incomplete trajectories was varied between 5% and 90%. Complete trajectories were converted as type 2, which is the type of incomplete trajectories that enter after the upstream intersection and leave before the downstream intersection. Only complete trajectories that were in peak hours were used. The performance of the algorithm continuously drops with the shortening of incomplete trajectory length. Both MAPE and MAE reach to their minimum values when incomplete trajectory length is 90% of the link length. The minimum and maximum value of MAPE is 6.8% and 33.3%, respectively, and the minimum and maximum value of MAE is 5.7s and 22.1s, respectively. MAPE and MAE were negatively associated with incomplete trajectory length. As longer incomplete trajectories contain more information about the complete trajectories the error is smaller, on average, for those longer trajectories. These results are shown in Figure 6.

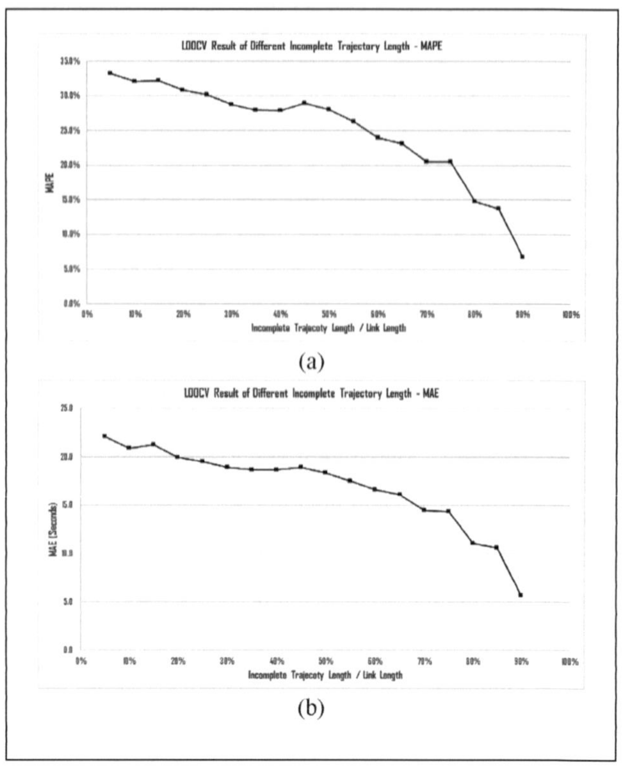

Figure 6. Sensitivity analysis on varying incomplete trajectory lengths; (*a*) MAPE (*b*) MAE.

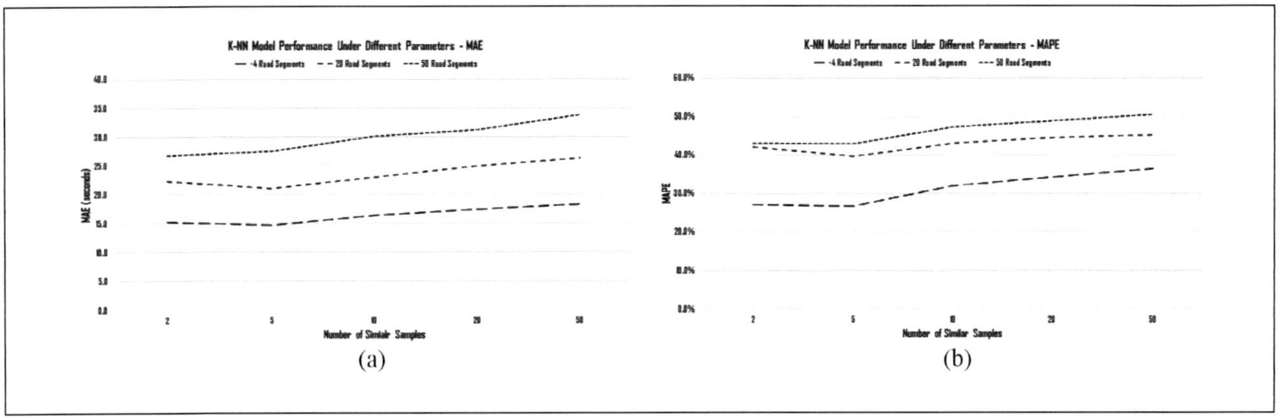

Figure 7. *k* and *n* sensitivity analysis: (*a*) MAE sensitivity and (*b*) MAPE sensitivity.

To understand how the algorithm performs based on the trajectory type, the MAE and MAPE were calculated for each of the three types independently, shown in Table 1. Complete trajectories were converted into 50% of the link length. Only complete trajectories that were in peak hours were used. Incomplete trajectory type 1 performed worst with an MAE of 18.3s and an MAPE of 39.1%. Incomplete trajectory type 2 and type 3 performed almost the same with an MAE around 7s and an MAPE around 20.7%. Incomplete trajectory type 2 and type 3 performed better than type 1 on average. It can be inferred from the result that incomplete trajectory that contains queue information performs better.

Finally, two input parameters to k-NN regression were simultaneously varied in the number of similar segments (k) and the segment length (n). The value of k was varied between 2 and 50. This was done for three different scenarios, breaking the links into 4, 20, and 50 segments. The MAE and MAPE are shown in Figure 7. With the increase in the number of similar samples, the performance of the algorithm improved at first. When the number of similar samples reached a threshold, the algorithm performed worse with the increase of the number of similar samples. The result was similar to results found in many other k-NN applications. We suggest that an appropriate number of similar samples needs to be selected before real-world application of the algorithm.

The results of the sensitivity analysis show that one can successfully recreate the complete trajectories using incomplete trajectories in certain cases using k-NN regression. Although there is no ground truth to evaluate the impact the algorithm has on the accuracy of the travel time calculation, the fact that the complete trajectories were able to be recreated likely means the travel time calculation would not change significantly for segments that already have a sufficient number of samples. However,

Table 1. Algorithm Performance by Incomplete Trajectory Type

	Type 1	Type 2	Type 3
MAE/Second	18.3	6.5	7.5
MAPE	39.10%	20.50%	20.90%

for segments where the sample size is sparse, including incomplete trajectories should be able to help provide more accurate estimates by virtue of the increased sample size. In addition, it can be used in more specific scenarios where sample size may be an issue, such as movement-based travel time calculation.

As the model could be affected by different traffic conditions, such as free-flow or congested conditions, the model result was analyzed under different times of day. As the time of day cannot be simulated, input data were classified into several categories by time of day. Morning peak was defined as 7:30 AM to 9:30 AM on weekdays and evening peak was defined as 4:00 PM to 6:00 PM on weekdays. Note that, even though the input data are classified as being at peak, the training set still covers all time periods because the assumption is that a trajectory can reflect all traffic conditions.

Table 2 shows that the algorithm performed better during peak hours. The average of MAPE during peak hour is around 14% and the average of MAE during peak hour is about 10s. The algorithm performed worse during non-peak hour with a MAPE of 26.4% and a MAE of 12.4s. The reason the algorithm performed better during peak hour may be because vehicles have similar trajectories during peak hour, as traffic is more congested. Vehicle trajectories during non-peak hour may depend more on drivers' behavior. Similar trajectories during non-peak hour may be caused by similar driving behavior rather than traffic condition. These

Table 2. LOOCV Result of Different Time of Day

Time of day	Morning peak	Evening peak	Peak	Non-peak	All
Sample size	372	113	485	470	955
MAPE	13.6%	15.6%	14.0%	26.4%	20.1%
MAE/seconds	10.3	9.49	10.1	12.4	11.2

results may also indicate how well one could expect the method to perform at different times of year. As the method performs better in congested conditions, one could expect the months with more frequent congestion to see better performance. For Tucson, those months are when the University of Arizona is in session and during the winter when the population grows because of the pleasant climate.

Conclusion

This study proposed a k-NN based travel time calculation method using incomplete trajectories to generate additional travel time samples. Incomplete trajectories were compared with historical complete trajectories and link travel times of incomplete trajectories were represented by these similar complete trajectories. Both the feasibility of the model and the sensitivity of the model were considered in the study. The case study showed that sample size increased 42% on average after implementing k-NN method. The results showed that the effect of sample size increasing is related with road geometry and driving behavior of the study area. As the corridors all had pre-timed signals, the conclusion only apply to pre-timed control, and corridors with other control methods require further analysis and validation.

The sensitivity analysis of the k-NN algorithm showed that the algorithm performed differently under different parameters and input data. The research into the key parameters and input data concluded the following:

(1) Both the number of similar samples and the number of road segments influence the accuracy of the algorithm. The study suggests that although a small number of road segments can improve the performance of the algorithm in the study, a small number of road segments would also reduce the accuracy of similarity calculation between trajectories.

(2) The length of incomplete trajectory has a positive correlation with performance of the algorithm. The study suggests that longer trajectory contains more traffic information so complete trajectories found can better estimate the travel time of incomplete trajectories.

(3) Incomplete trajectories that contain queue information performed better. Link travel time is the summary of free-flow travel time and delay, in which delay is the key element that decides link travel time. Queue information is highly correlated with delay so the algorithm performed better when the input data reflected queue information.

The k-NN algorithm developed in this study can help increase public probe vehicle-based travel time collection without collecting extra data. The study also gives some suggestions on the performance of the algorithm under different parameters and input data. This study suggests that, before real-world application, optimal parameters need to be selected using a historical dataset for an accurate result. In future studies, quantitative analysis on the sample size increasing potential under different road geometry and driving behavior can be conducted. The result can help decision makers choose a better strategy to collect more data. In addition, the performance of the model can be further verified in a full connected vehicle environment. The authors recommend that future studies could focus on those issues.

Acknowledgments

This research was supported by The University of Arizona, Metropia, Inc., Pima Association of Governments, and the City of Tucson. The authors would like to thank to the reviewers for their comments and valuable suggestions for improving the quality of the article.

Author Contributions

The authors confirm contribution to the paper as follows: study conception and design: Yao-Jan Wu, Zheng Li; data collection: Xianbiao Hu, Xiaoyu Zhu; analysis and interpretation of results: Zheng Li, Robert Kluger, Yao-Jan Wu, Xianbiao Hu; draft manuscript preparation: Zheng Li, Robert Kluger. All authors reviewed the results and approved the final version of the manuscript.

References

1. Wan, N., A. Vahidi, and A. Luckow. Reconstructing Maximum Likelihood Trajectory of Probe Vehicles between

Sparse Updates. *Transportation Research Part C: Emerging Technologies*, Vol. 65, 2016, pp. 16–30. https://doi.org/10.1016/j.trc.2016.01.010.
2. Zhan, X., S. Hasan, S. V. Ukkusuri, and C. Kamga. Urban Link Travel Time Estimation Using Large-Scale Taxi Data with Partial Information. *Transportation Research Part C: Emerging Technologies*, Vol. 33, 2013, pp. 37–49. https://doi.org/10.1016/j.trc.2013.04.001.
3. Turner, S. M., W. L. Eisele, R. J. Benz, and D. J. Holdener. *Travel Time Data Collection Handbook*. Report FHWA-PL-98-035. Texas Transportation Institute, 1998.
4. Patire, A. D., M. Wright, B. Prodhomme, and A. M. Bayen. How Much GPS Data Do We Need? *Transportation Research Part C: Emerging Technologies*, Vol. 58, 2015, pp. 325–342.
5. Zheng, F., and H. Van Zuylen. Urban Link Travel Time Estimation Based on Sparse Probe Vehicle Data. *Transportation Research Part C: Emerging Technologies*, Vol. 31, No. 111, 2013, pp. 145–157. https://doi.org/10.1016/j.trc.2012.04.007.
6. Jenelius, E., and H. N. Koutsopoulos. Travel Time Estimation for Urban Road Networks Using Low Frequency Probe Vehicle Data. *Transportation Research Part B: Methodological*, Vol. 53, 2013, pp. 64–81. https://doi.org/10.1016/j.trb.2013.03.008.
7. Bucknell, C., and J. C. Herrera. A Trade-Off Analysis between Penetration Rate and sampling Frequency of Mobile Sensors in Traffic State Estimation. *Transportation Research Part C: Emerging Technologies*, Vol. 46, 2014, pp. 132–150. https://doi.org/10.1016/j.trc.2014.05.007.
8. Argote-Cabañero, J., E. Christofa, and A. Skabardonis. Connected Vehicle Penetration Rate for Estimation of Arterial Measures of Effectiveness. *Transportation Research Part C: Emerging Technologies*, Vol. 60, 2015, pp. 298–312. https://doi.org/10.1016/j.trc.2015.08.013.
9. Zhang, Y., and A. Haghani. A Gradient Boosting Method to Improve Travel Time Prediction. *Transportation Research Part C: Emerging Technologies*, Vol. 58, 2015, pp. 308–324. https://doi.org/10.1016/j.trc.2015.02.019.
10. Cao, P., T. Miwa, & T. Morikawa. Modeling Distribution of Travel Time in Signalized Road Section Using Truncated Distribution. *Procedia-Social and Behavioral Sciences*, Vol. 138, 2014, pp. 137–147. https://doi.org/10.1016/j.sbspro.2014.07.189.
11. Seo, T., and T. Kusakabe. Probe Vehicle-Based Traffic Flow Estimation Method without Fundamental Diagram. *Transportation Research Procedia*, Vol. 9, 2015, pp. 149–163. https://doi.org/10.1016/j.trpro.2015.07.009.
12. Hellinga, B. R., & L. Fu. Reducing bias in probe-based arterial link travel time estimates. *Transportation Research Part C: Emerging Technologies*, Vol. 10, No. 4, 2002, pp. 257–273. https://doi.org/10.1016/S0968-090X(02)00003-7.
13. Coifman, B., and M. Cassidy. Vehicle Reidentification and Travel Time Measurement on Congested Freeways. *Transportation Research Part A: Policy and Practice*, Vol. 36, No. 10, 2002, 899–917. https://doi.org/10.1016/S0965-8564(01)00046-5.
14. Robinson, S., and J. Polak. Modeling Urban Link Travel Time with Inductive Loop Detector Data by Using the k-NN Method. *Transportation Research Record: Journal of the Transportation Research Board*, 2005. 1935: 47–56.
15. Wu, C., C. Wei, D. Su, M. Chang, & J. Ho. Travel Time Prediction with Support Vector Regression. *Proceedings of the 2003 IEEE International Conference on Intelligent Transportation Systems*, Vol. 2, 2004, pp. 1438–1442. https://doi.org/10.1109/ITSC.2003.1252721.
16. Park, D., and L. R. Rilett. Forecasting Multiple-Period Freeway Link Travel Times Using Modular Neural Networks. *Transportation Research Record: Journal of the Transportation Research Board*, 1998. 98: 63–170.
17. Li, R., & G. Rose. Incorporating Uncertainty into Short-Term Travel Time Predictions. *Transportation Research Part C: Emerging Technologies*, Vol. 19, No. 6, 2011, pp. 1006–1018. https://doi.org/10.1016/j.trc.2011.05.014.
18. Sherali, H. D., J. Desai, & H. Rakha. A Discrete Optimization Approach for Locating Automatic Vehicle Identification Readers for the Provision of Roadway Travel Times. *Transportation Research Part B: Methodological*, Vol. 40, No. 10, 2006, pp. 857–871. https://doi.org/10.1016/j.trb.2005.11.003.
19. Wang, Y., Y. Malinovskiy, Y. Wu, and U. K. Lee. *Error Modeling and Analysis for Travel Time Data Obtained from Bluetooth MAC Address Matching*. Research Project Agreement No. 61-8390, Final Research Report. Department of Civil and Environmental Engineering University of Washington, 2011.
20. Haghani, A., M. Hamedi, K. Sadabadi, S. Young, and P. Tarnoff. Data Collection of Freeway Travel Time Ground Truth with Bluetooth Sensors. *Transportation Research Record: Journal of the Transportation Research Board*, 2010. 2160: 60–68.
21. Park, S., A. Saeedi, D. S. Kim, & J. D. Porter. Measuring Intersection Performance from Bluetooth-Based Data Utilized for Travel Time Data Collection. *Journal of Transportation Engineering*, Vol. 142, No. 5, 2016, 1–9. https://doi.org/10.1061/(ASCE)TE.1943-5436.0000836.
22. Yeon, J., L. Elefteriadou, and S. Lawphongpanich. Travel Time Estimation on a Freeway Using Discrete Time Markov Chains. *Transportation Research Part B: Methodological*, Vol. 42, No. 4, 2008, pp. 325–338. https://doi.org/10.1016/j.trb.2007.08.005.
23. Byon, Y. J., A. Shalaby, and B. Abdulhai. GISTT: GPSGIS Integrated System for Travel Time Surveys. Presented at 85th Annual Meeting of the Transportation Research Board, Washington, D.C., 2006.
24. Boyce, D. E., A. M. Kirson, and J. L. Schofer. Advance: The Illinois Dynamic Navigation and Route Guidance Demonstration Program, Chapter 11. In *Advanced Technology for Road Transport: IVHS and ATT* (I. Catling, ed.), Artech House, London, 1994, pp. 247–270.
25. Cheu, R. L., C. Xie, and D. H. Lee. Probe Vehicle Population and Sample Size for Arterial Speed Estimation. *Computer-Aided Civil and Infrastructure Engineering*, Vol. 17, No. 1, 2002, pp. 53–60. https://doi.org/10.1111/1467-8667.00252.
26. Feng, Y., J. Hourdos, & G. A. Davis. Probe Vehicle Based Real-Time Traffic Monitoring on Urban Roadways. *Transportation Research Part C: Emerging Technologies*,

Vol. 40, 2014, pp. 160–178. https://doi.org/10.1016/j.trc.2014.01.010.
27. Li, Y., and M. McDonald. Link Travel Time Estimation Using Single GPS Equipped Probe Vehicle. *Proc., IEEE 5th International Conference on Intelligent Transportation Systems*, IEEE, Singapore, 2002, pp. 932–937.
28. Srinivasan, K., and P. Jovanis. Determination of Number of Probe Vehicles Required for Reliable Travel Time Measurement in Urban Network. *Transportation Research Record: Journal of the Transportation Research Board*, 1996. 1537: 15–22.
29. Liu, H. X., and W. A. Ma. Virtual Vehicle Probe Model for Time-Dependent Travel Time Estimation on Signalized Arterials. *Transportation Research Part C: Emerging Technologies*, Vol. 17, 2009, pp. 11–26.
30. Pima Association of Governments. *PAG Traffic Counts*. 2014. http://www.pagregion.com/Default.aspx?tabid=909. Accessed May 11, 2016.
31. Handley, S., P. Langley, and F. A. Rauscher. Learning to Predict the Duration of an Automobile Trip. *Proc., 4th International Conference on Knowledge Discovery and Data Mining*, Conference Proceedings. New York, 1998, pp. 219–223.
32. Zhou, X., Z. Yang, W. Zhang, X. Tian, and Q. Bing. Urban Link Travel Time Estimation Based on Low Frequency Probe Vehicle Data. *Discrete Dynamics in Nature and Society*, Vol. 2016, 2016, pp. 1–13.
33. Kramer, O. *Dimensionality Reduction with Unsupervised Nearest Neighbors*. Intelligent Systems Reference Library 51. Springer, New York; Berlin, 2013.
34. Rizzo, M. L. *Statistical Computing with R*. CRC Press, Boca Raton, Fla., 2007.

The Standing Committee on Urban Transportation Data and Information Systems (ABJ30) peer-reviewed this paper (18-03631).

TRR
JOURNAL OF THE TRANSPORTATION RESEARCH BOARD

Article

Transportation Research Record
2018, Vol. 2672(42) 159–169

DOI: 10.1177/0361198118792323
journals.sagepub.com/home/trr
SAGE

Exploring the Spatial Dependence and Selection Bias of Double Parking Citations Data

Jingqin Gao[1,2], Kun Xie[3], and Kaan Ozbay[1,2,4]

Abstract
Parking violation citations, often used to identify factors contributing to parking violation behavior, offer one of the most valuable datasets for traffic operation research. However, little has been done to examine their spatial dependence caused by location-specific differences in features such as traffic, land use, etc., and potential selection biases resulting from different effects of traffic enforcement. This study leveraged extensive data on double parking citations in Manhattan, New York City, in 2015, along with other relevant datasets including land use, transportation, and sociodemographic features. Moran's I statistics confirmed that double parking tickets were spatially correlated so that spatial lag and spatial error models were proposed to account for the spatial dependence of parking tickets to avoid biased estimates. To investigate whether selection bias exists in issuing tickets, we estimated the effects of parking ticket density and police precinct distance, when controlling for variables such as commercial area, truck activity, taxi demand, population, hotels, and restaurants. Parking ticket density and police precinct distance were used as indicators of the enforcement levels and coverage and were found to be statistically significant. This indicated the existence of selection bias due to heterogeneity in enforcement levels or coverage across different regions. Moreover, patrol patterns of traffic enforcement officers revealed that the majority had less than three daily patterns. These findings can assist with proper usage of the citation data by recommending that researchers and agencies consider spatial dependence as well as selection bias, and provide insights for parking violation management strategies.

For urban areas, parking can be extremely costly, imposing a substantial burden on drivers. In 2016, Americans spent an average of 17 hours searching for parking at an estimated cost of 72 billion dollars in wasted time, fuel, and emissions (*1*). Other indirect parking challenges for drivers include the risk of parking fines, which add up to 2.6 billion dollars annually (*1*). Parking violation management is increasingly a concern for many megacities, such as New York City (NYC), due to the conflict between limited on-street parking supply and rigid parking demand. This high demand and competition for parking space often leads to illegal parking behaviors, such as double parking, which contribute significantly to traffic congestion and accidents. Several recent studies have also identified the effects of parking in the context of commercial vehicles and proposed innovative yet practical approaches to alleviate these negative effects (*2–7*).

Every year, over 10 million parking violation tickets are issued in NYC, and these citations offer a valuable dataset for researchers to develop parking models that can help to explore hotspots and to investigate the impacts of contributing factors. Conventional statistical analysis relies on the assumption that each parking citation is independent (*8*). However, this assumption could be violated if spatial autocorrelation exists in the data, and thus spatial models are desired. Additionally, the increase in volume and variety of emerging data sources, such as GPS-equipped vehicles, enables more precise estimation of the effects of potential contributing factors by providing richer data for modeling (*9*).

This also provides great opportunities for a deeper investigation of parking citation data, particularly on the

[1]C2SMART Tier 1 University Transportation Center, Tandon School of Engineering, New York University, Brooklyn, NY
[2]UrbanMITS Lab, Department of Civil and Urban Engineering, Tandon School of Engineering, New York University, Brooklyn, NY
[3]Department of Civil and Natural Resources Engineering, University of Canterbury, Christchurch, New Zealand
[4]Center for Urban Science + Progress, Tandon School of Engineering, New York University, Brooklyn, NY

Corresponding Author:
Address correspondence to Jingqin Gao: jingqin.gao@nyu.edu

potential underlying bias. Many studies in other disciplines have shown that record-based data, such as crime data, is not a complete population of all occurrences, nor does it yield a representative random sample (*10*). Although this point has not been fully proven in parking citations, parking citations may underestimate the true occurrences of parking violations as most of such activities may not be ticketed. Correlation between total numbers of tickets may be proportional to different vehicle types (i.e., commercial versus passenger vehicle) or the number of enforcement officers that monitor a certain area. In many cases, the data on the incidence of parking violations may be skewed to areas with high-intensity enforcement, when in fact there may be a higher incidence in other areas. The fact that parking ticket data might not represent the actual population of parking violations accurately may be due to the selection bias resulting from heterogeneous enforcement intensity. Selection bias in this case means certain areas are more likely to be ticketed by the enforcement officers than others. Because selection bias can be a problem when using ticket data, its existence and impact need to be explored.

As an extension of two previous studies (*11*, *12*), this study leverages one year of geocoded parking tickets issued in 2015 in Manhattan, NYC, with a focus on double parking, which is a phenomenon typical of urban areas. The objectives of this study are two-fold. The first goal is to find out whether spatial dependence exists in the double parking citation data. If so, models that do not address spatial dependence would lead to biased estimation. Second, this study investigates whether selection bias exists in issuing parking tickets. We estimated the effects of two indicators of police enforcement intensity, parking ticket density and police precinct distance, when controlling for other variables such as land use, transportation, or sociodemographic features.

Literature Review

Various studies have been conducted for different cities using parking citation data. Wang and Gogineni (*13*) performed an empirical investigation of commercial vehicle parking violations in NYC during May 2014 to identify the factors that may influence parking violation behaviors. Three models: Poisson regression, negative binominal, and zero-inflated negative binominal model, were used to analyze violation frequency. Their results indicated that land use and value, road type, on-street parking price, population, and employment densities are related to commercial parking tickets. However, although the study considered spatial distribution patterns of the violations, it did not account for potential spatial dependence structure and possible bias while conducting regression analysis.

Wenneman et al. (*14*) established an ordinary least squares (OLS) regression model to examine the relationship between density of commercial vehicle parking citations issued in Toronto, freight trip generation, and built environment factors. Though the model claimed to achieve an adjusted R-squared value of 0.68, it did not consider the possible spatial dependence of the parking ticket data.

Kawamura et al. (*15*) conducted hot spots analysis using parking citations issued over 12 months in Chicago. The results from the regression model with trial-and-error and the backward elimination process showed that concentrations of food businesses and wealthy neighborhoods with a high percentage of residents with tenure over 5 years can present problems for truck parking. One important contribution of this model is that it assumed the log of the density of tickets issued to passenger vehicles reflects the number of parking enforcement and police officers in the area. The authors believed that, in the areas that were patrolled heavily, truck parking violations had a higher probability of being ticketed compared with areas with a low level of enforcement.

Gao and Ozbay (*11*) studied the spatial distribution of NYC double parking violation records in 2014 for all vehicle types including commercial vehicles. The study showed there were more violation records for double parking in commercial districts and mixed commercial/residential districts. Furthermore, field data collected in another study by Gao and Ozbay (*12*) confirmed that violation ticket data highly underestimated the real occurence of double parking violations. For example, a street block with a record of 208 double parking tickets per year had about 40 unticketed violations on a single day during the morning peak hour (*12*).

Smith and Steif (*16*) used more than 1.4 million parking violation tickets in Philadelphia in 2013 to calculate the probability of at least one ticket being issued by hour and by street. They claimed that a more general probability cannot be estimated due to unobserved factors, including the patrol patterns of traffic enforcement officers.

Although all of the above studies claimed that citation data is very valuable, most studies have mainly focused on its relationship with land use, parking supply, and sociodemographic features (*13–15*). There has been little discussion of its spatial dependence caused by location-specific differences and potential selection biases resulting from different traffic enforcement levels or coverage. Only a few studies (*12*, *15*, *16*) have discussed the potential bias in citation data, and only one study (*15*) has attempted to analyze and quantify the bias due to parking enforcement. The results of the latter model found a strong association between "enforcement level" and parking violations tickets by trucks. Thus, there is a need to investigate spatial dependence and bias in the parking data, especially selection bias due to police precinct locations or patrolling patterns of traffic enforcement officers.

Data Preparation

Parking Violation Tickets

Parking violation ticket data is released by NYC Department of Finance monthly and shared via the NYC open data portal (*17*). A batch geocoding program was developed based on U.S. Census Bureau Geocoding Services Web Application Programming Interface (API) (*18*) and Google Geocoding API (*19*) to convert the original address information in citation data into geographic coordinates. The geocoding rate can reach 95.3% after reformatting the address and correcting special cases with direction abbreviation or distance information (i.e. "200 feet W/O an intersection").

In total, 10,905,102 violation tickets were issued in NYC in the year 2015. The lowest daily parking ticket totals occurred on January 27th (North American Blizzard), July 5th (the day after Independence Day), and December 27th (weekend after Christmas day). Among the total of nearly 11 million tickets issued, 633,050 (6.3%) were for double parking. Table 1 lists descriptive statistics for all parking violations and double parking violations. The result lends support to the claim that, unlike the general trend for all parking violations, commercial vehicles account for a significant portion of the double parking violation tickets (45.2%). This may be because of inadequate parking spaces for commercial vehicles in a high-density urban network and the desire of the commercial vehicles to park as close as possible to their delivery spot. The result also highlights that more than half of the double parking tickets were issued in Manhattan (58.8%). About 85% to 90% of the tickets were issued during weekdays, and the hourly distribution pattern is consistent with work hours (Figure 1).

Table 1. Descriptive Statistics of All Parking Tickets/Double Parking Tickets

	All parking violations	Double parking violations
Total number of tickets in 2015	10,905,102	695,369
%Commercial vehicle tickets	19.4%	45.2%
%Passenger cars tickets	72.5%	47.3%
%Tickets issued in Manhattan	34.0%	58.8%
%Tickets issued in Brooklyn	20.7%	17.5%
%Tickets issued in Queens	18.3%	7.8%
%Tickets issued in Bronx	10.1%	14.2%
%Tickets issued in Staten Island	0.9%	0.2%
%Tickets issued during weekday	85.7%	89.9%
%Tickets issued during weekend	14.3%	11.1%

Geographical Analysis Units

The map of Manhattan is uniformly split into 6,204 equal-sized grid cells (300 feet × 300 feet). The approximate width of a street block in Manhattan is close to 300 feet and the length of it (800˜900 feet) can be integer multiples of 300 feet. Advantages of using equal-sized grid cells as the basic geographical units include: (a) providing street-to-street resolution, (b) easing the bias from different analysis unit sizes, and (c) incorporating easily with data such as land use features (*9*). Recent studies (*9, 20*) show the implementation of a cell-structured modeling framework in transportation research. Parking tickets, along with sociodemographic, land use, transportation, and enforcement data, were aggregated for each cell using spatial analysis tools in ArcGIS (*21*). Figure 2 demonstrates the density of double parking tickets at the grid-cell level in Manhattan in 2015.

Land Use, Sociodemographic, Transportation, and Enforcement Data

Various factors contributing to parking violations have already been identified by previous studies. The most commonly used factors are land use and sociodemographic features like land use type and population (*13–15*). This study obtained detailed and categorized information on land use zoning for commercial, residential, mixed, and park usage from NYC Department of Planning (NYCDCP) (*22*). A Visual Basic for Applications (VBA) program was developed to compute the ratio for each zoning category in every grid cell (*9*). Sociodemographic information based on the 2011 census survey was retrieved from U.S. Census Bureau (*23*) and disaggregated into each grid cell.

Additionally, Gao and Ozbay (*12*) and Kawamura et al. (*15*) pointed out that places of interest such as hotels or food sale locations could be potential contributing factors for double parking. Therefore, hotel information from New York State Department of Transportation (NYSDOT) open data portal (*24*) and geo-processed restaurant information (*25*) was also utilized.

Road network features and traffic information (*13, 15*) are commonly used as well. In this study, taxi pick-ups and drop-offs obtained from NYC Taxi and Limousine Commission (*26*) were used as estimates of traffic demand. Road network information like sidewalks, bicycle paths, and vehicle miles traveled (VMT) were obtained from NYCDCP (*22*), NYC Department

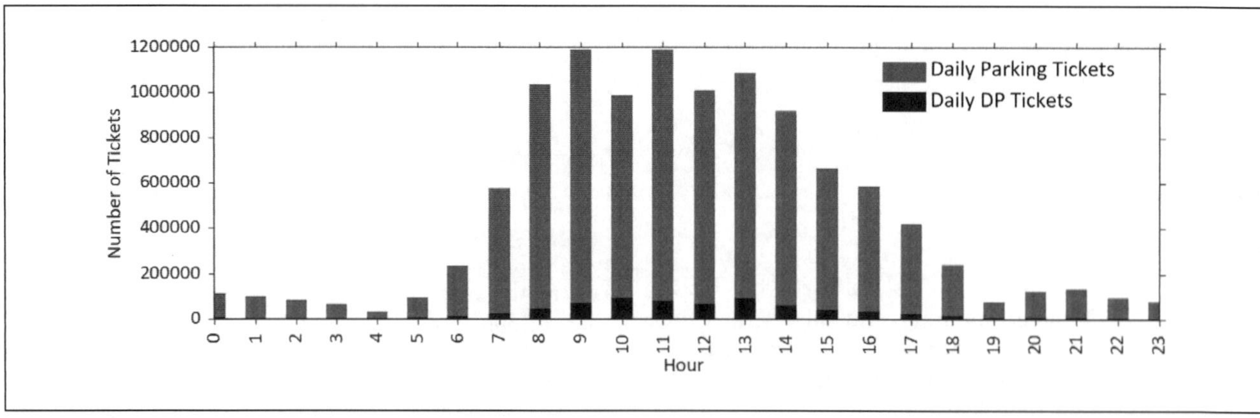

Figure 1. Hourly distribution of all parking/double parking tickets.

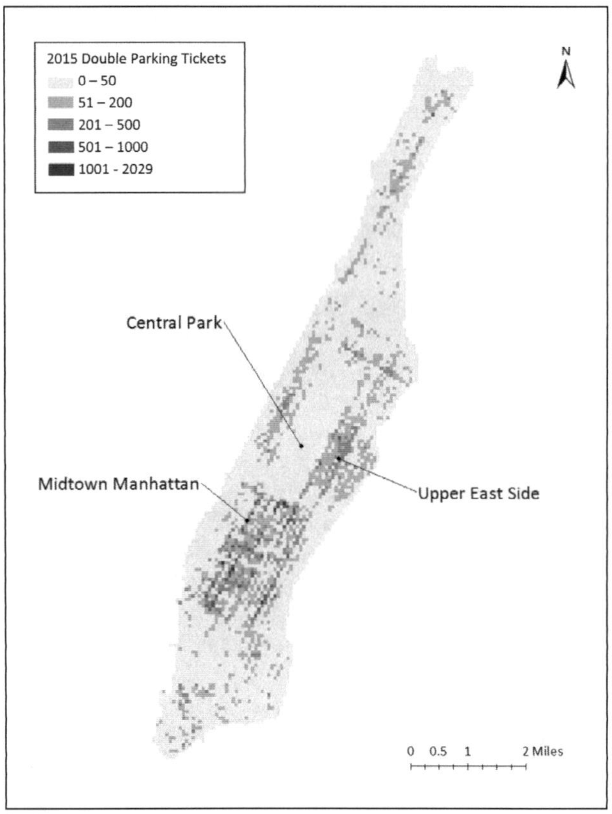

Figure 2. Double parking ticket density at grid-cell level in Manhattan in 2015.

of Transportation (*27*), and NYSDOT (*28*), respectively. It should also be noted that the VMT for each grid cell was estimated from the length and the average daily traffic of the road segments in the cell (*9*). A MapReduce program for expressing distributed and parallel computations was developed to process the massive taxi trip records in 2015 (>20 Gigabytes) (*9*). Public transit information, such as bus and subway station geographic information system (GIS) data and the ridership for each subway station, was generated using open data from the Metropolitan Transportation Authority (*29*).

There has been an inconclusive debate about whether traffic enforcement and patrol patterns have an impact on parking tickets. The number of tickets issued to a street block or an area may be biased by the level of enforcement and coverage. To investigate these factors, this study also collected the following data for each grid cell: (a) the density of tickets for all parking violations, and (b) the distance to the nearest police precinct station. The former is assumed to reflect the level of enforcement in the area. This variable was computed by counting the total number of parking tickets (all violation types) in each police precinct and then dividing this number by the precinct area, and then these values were assigned to each grid cell in the study region. The distance to the nearest police precinct station was computed for each grid cell. Police precinct station addresses retrieved from the New York Police Department website (*30*) were geocoded onto the map of our study area. Table 2 summarizes the data descriptions and descriptive statistics.

Methodology

Spatial Dependence Test

Spatial dependence tests, such as global Moran's I and local Moran's I (*31*, *32*) statistic tests, are widely used for measuring spatial autocorrelation. If a spatial dependence exists in the data, then commonly used assumptions for statistical models such as independent observation of the data will no longer hold, and the estimates from these models will be biased and inefficient. This type of bias may be corrected by using spatial models.

Global Moran's I statistic, proposed by Moran (*31*), illustrates an indication of "the degree of linear association

Table 2 Data Descriptions and Descriptive Statistics (6,204 Grid Cells)

Variable	Description	Mean	Standard deviation
Dependent variable			
Double parking tickets	Annual double parking tickets	47.20	104.08
Land use			
Commercial ratio	The ratio of commercial zone area to the whole area	0.29	0.40
Residential ratio	The ratio of residential zone area to the whole area	0.50	0.44
Mixed ratio	The ratio of mixed zone area to the whole area	0.06	0.22
Park ratio	The ratio of park area to the whole area	0.14	0.31
Hotel density	Number of hotels after spatial processing	0.04	0.23
Hotel distance	Distance to the nearest hotel (in miles)	0.95	1.45
Restaurant density	Number of restaurants after spatial processing	1.25	2.53
Enforcement			
Parking ticket density	Number of parking tickets for all violation types per 1000 square feet	3.22	2.22
Police precinct distance	Distance to the nearest police precinct station (in miles)	0.41	0.27
Sociodemographic			
Population	Total population	241.83	151.22
Population under 14	The population under 14 years	30.13	24.44
Population over 65	The population 65 years and over	32.10	25.70
Male	The population of males	113.86	70.73
Female	The population of females	127.95	82.21
Median age	Median age of population	1.58	0.99
Median income	Median income per household (in thousands of dollars)	68.64	48.01
Employed	Number of population employed	129.51	87.47
Unemployed	Number of population unemployed	11.77	10.37
Transportation			
VMT	Annual vehicle miles traveled (in thousands of vehicle miles)	309.66	501.49
Truck ratio	The average ratio of truck flow to total flow	0.04	0.05
Subway ridership	Annual subway ridership after spatial processing (in thousands of trips)	245.76	392.20
Bus stop density	Number of bus stops after spatial processing	0.36	0.23
Sidewalk	Total length of sidewalks (in miles)	0.07	0.07
Bicycle path	Total length of bicycle paths (in miles)	0.02	0.03
Taxi pick-ups	Annual taxi pick-ups (in thousands of trips)	21.25	38.57
Taxi drop-offs	Annual taxi drop-offs (in thousands of trips)	20.30	29.99

between a vector of observed values and a weighted average of the neighboring values that underlie the specification of spatial autoregressive processes (*33*)." Local Moran's I is developed based on the assumption that global Moran's I is a summation of individual cross-products and computes a measure of spatial association for every observation (*32*). The purpose of local Moran's I statistics is to capture local patterns that the global Moran's I may not observe.

Global Moran's I was used first to measure the spatial dependence of double parking tickets. The z-score of Moran's I (Z_I) and pseudo *p*-value (*34*) obtained from permutation test is used to assess the significance of Moran's I. Z_I can be computed as:

$$Z_I = (I - E[I])/SD[I]. \quad (1)$$

where $E[I]$is the expectation of I and $SD[I]$is the standard deviation of I. A positive Z_I indicates the observation distribution is spatially clustered (*8*) and a pseudo *p*-value less than 0.05 confirms that I is statistically significant at the confidence level of 95% (*35*). More details about global Moran's I practice can be found in previous studies by Xie et al. (*8*, *36*).

GeoDa (*34*), an open-source spatial analysis software, was used to test whether double parking tickets were spatially correlated. The global Moran's I test was conducted using several weight matrices (threshold distances, *k*-nearest neighbor) and 9,999 permutations were performed to compute the pseudo *p*-value. The results of global Moran's I test are presented in Table 3. All the *z*-scores are positive and the pseudo *p*-values are found to be less than 0.05 using different weight matrix. Since the result indicates a strong spatial autocorrelation on double parking tickets, it will lead to biased estimations and unreliable statistical inferences if such spatial dependence is ignored.

This study also conducted a local spatial autocorrelation analysis based on the local Moran local indicators of spatial association (LISA) (*32*) statistics. While the global Moran's I measures spatial autocorrelation globally, LISA investigates individual locations and identifies hot spots and cold spots (*34*). Local Moran's I for the observation z_i, z_j in cell i,j with weight matrix w_{ij}, and N observations (*32*) can be computed as:

Table 3. Global Moran's I Test Results

Weight matrix	I	E[I]	SD[I]	Z_I	Pseudo *p*-value
Threshold distance-300 feet	0.3321	–0.0002	0.0090	36.9521	0.0001
Threshold distance-800 feet	0.2581	–0.0002	0.0041	62.7201	0.0001
4-nearest neighbor	0.3318	–0.0002	0.0089	37.2234	0.0001
8-nearest neighbor	0.3262	–0.0002	0.0063	51.8185	0.0001

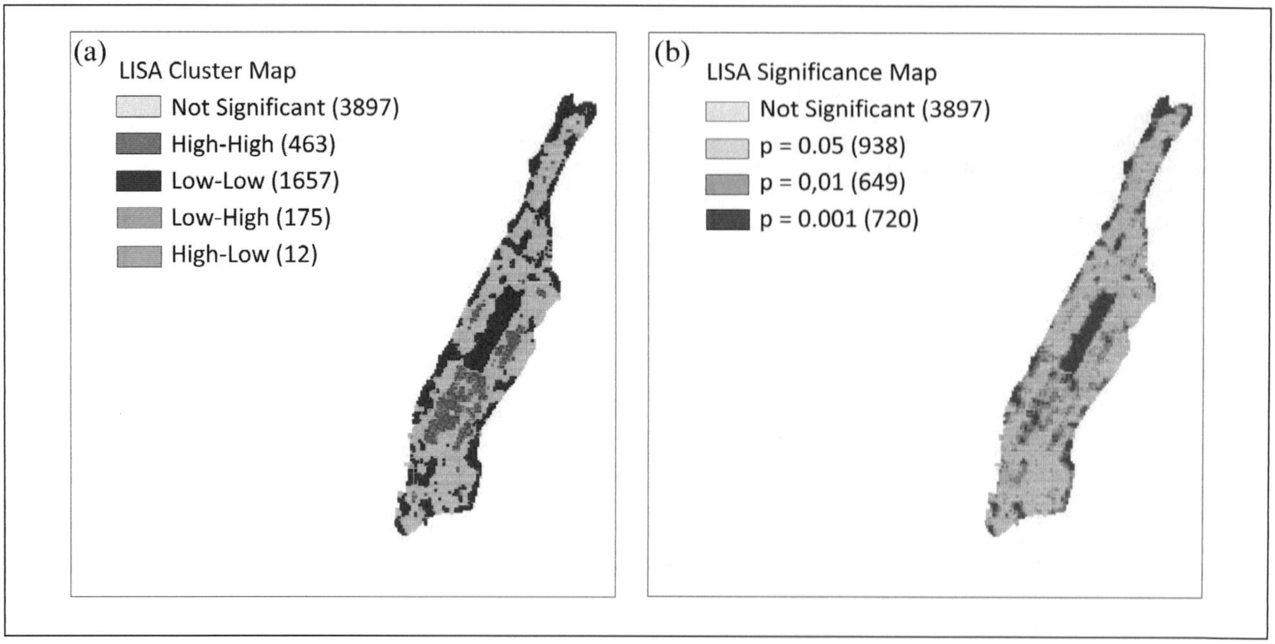

Figure 3. Local indicators of spatial association (LISA) cluster and significance map.

$$I_i = \frac{z_i}{\left(\frac{\sum_i z_i^2}{N}\right)} \sum_j w_{ij} z_j \quad (2)$$

The high–high and low–low locations in the LISA cluster map (Figure 3*a*) suggest clustering of similar high or low values, whereas high–low and low–high clusters indicate spatial outliers (*37*). In this study, low–low spatial clusters, such as Central Park, are cold spots, while high–high clusters, such as the Upper East Side, can be regarded as hotspots for double parking. On the other hand, the LISA significance map is also a good indicator that shows the significance levels of the clusters (*37*). Figure 3*b* shows the significance map using *k*-nearest neighbor (8 neighbors).

Global and local Moran's I statistics confirmed that double parking tickets were spatially correlated so that spatial lag and spatial error models in the next section were proposed to account for the spatial dependence of parking tickets to avoid biased estimates.

Model Specification

We assume the spatial dependence identified from the previous section results from two aspects: (i) location-specific differences in features such as traffic, land use type, or population density, and (ii) selection bias due to enforcement activities. The second aspect—selection bias due to enforcement activities—is tested by introducing two new variables that were described in the data preparation section, namely, parking ticket density and police precinct distance. Parking ticket density is assumed to be an indicator of level of enforcement, and police precinct distance is assumed to be an indicator of enforcement coverage. To observe the effect of the selection bias, we keep the variables in the first aspect as control variables, and variables in the second aspect as experimental variables. The number of double parking tickets in each grid cell is used as the dependent variable. Besides a standard linear regression model using OLS estimation by minimizing the sum of squared prediction errors, a spatial error and a spatial lag model are examined since they can account for spatial dependence and unobserved spatial factors. The idea is that if traffic, land use, socio-demographic features, and unobserved spatial factors are controlled, but the two experimental variables still show a strong association with the number of tickets issued, then a selection bias exists in the ticket data.

After diagnosing multicollinearity for the variables using variance inflation factors (VIF) (*38*), the same set of control variables were selected for all three models so that valid model comparison can be made. The selected control variables include commercial ratio, hotel distance, restaurant density, total populations, truck ratio, and taxi drop-offs.

In spatial analysis, there are two primary types of spatial dependence: spatial error and spatial lag (*39*). The former states that the spatial error terms across different spatial units are correlated. In other words, omitted variables at one location can affect the dependent variable of itself and its neighboring locations (*8*, *36*). The latter, spatial lag specification, allows spatial dependence through both spatial error correlation effects and spatial spillover effects (*8*, *36*). The spatial error model can be expressed as follows:

$$y = x\beta + u \tag{3}$$

$$u = \lambda W u + \varepsilon \tag{4}$$

where ε is a vector of independent normally distributed errors, while u is a vector of spatially autocorrelated errors, controlled by the term λWu, W is the spatial weight matrix, and the constant λ is the spatial autoregressive parameter that represents the strength of the spatial autocorrelation.

The spatial lag model accounts for the spatial autocorrelation first, so that classic assumptions in Equation 5 can still be kept for each error in error matrix ε considering them independent:

$$y = \rho W y + X\beta + \varepsilon \tag{5}$$

where ρWy is a spatially lagged dependent variable, ρ is a spatial autoregressive parameter. The maximum likelihood estimation (*40*) was used to calibrate the spatial error model and the spatial lag model.

It should be noted that R^2 is no longer adequate to measure the goodness-of-fit of spatial models because the residuals of spatial models are not independent of one another (*8*). Therefore, alternative performance measurements based on likelihood estimation, such as AIC (*41*) or BIC (*42*), were used. The former introduces parameter number as a penalty term while the latter combines parameter number and sample size into the penalty term.

Modeling Results, Discussions, and Limitations

Coefficient estimates and statistical indicators, as well as model assessment measures, are reported in Table 4. Considering street resolution and convergence time, the weight matrix with 300-feet distances was used in spatial lag and spatial error models.

Both spatial models have lower AIC and BIC values, which means they are substantially improved by considering spatial dependence. The autoregressive parameters ρ in the spatial lag model and λ in the spatial error model

Table 4. Model Results and Assessment

	Standard model		Spatial lag model		Spatial error model	
Variables	Coefficient	*p*-value	Coefficient	*p*-value	Coefficient	*p*-value
Intercept	−12.712	<0.001	−14.472	<0.001	−11.501	0.006
Experimental variables						
Parking ticket density	5.313	<0.001	3.044	<0.001	5.199	<0.001
Police precinct distance	−18.433	<0.001	−14.665	0.002	−18.112	0.001
Control variables						
Commercial ratio	8.829	0.014	6.630	0.060	10.927	0.004
Hotel distance	6.606	<0.001	6.053	<0.001	6.592	<0.001
Restaurant density	7.316	<0.001	7.204	<0.001	7.311	<0.001
Total population	0.040	<0.001	0.037	<0.001	0.034	<0.001
Truck ratio	1.210	<0.001	1.225	<0.001	1.283	<0.001
Taxi drop-offs	0.886	<0.001	0.825	<0.001	0.862	<0.001
Autoregressive parameter						
ρ			0.224	<0.001	-	-
λ			-	-	0.245	<0.001
Model assessment						
R^2	0.225		0.259		0.263	
AIC	73683.300		73486.300		73469.900	
BIC	73743.800		73553.600		73530.500	

are highly significant, which provides confirmatory proofs that the double parking citations are spatially correlated. This finding suggests that spatial dependence should always be considered when conducting traffic operation research using parking ticket data. Ignorance of spatial dependence may lead to biased estimates.

Statistic indicator *p*-value was used to test the significance of experimental variables. Both experimental variables—parking ticket density and police precinct distance—were found to be statistically significant at 95% level (*p*-values <0.05) in all three models. Parking ticket density, an indicator of level of enforcement, was found to have a positive impact on the number of double parking citations. The distance to the nearest police precinct was confirmed to be negatively associated with number of double parking tickets. In other words, the further away from the police station, the less likely a parking ticket would be issued. It is worth pointing out that the spatial heterogeneity resulting from unobserved factors can be handled by the error term in our spatial models. As land use, transportation, and sociodemographic features are controlled, and unobserved spatial factors are also taken into account in spatial models, both the experimental variables still show a strong correlation with the dependent variable. This indicates that parking tickets are closely related to enforcement intensity, and further confirms the existence of selection bias in parking data.

To further investigate potential characteristics of traffic enforcement, one month of geocoded parking tickets for all types of parking violations (not only limited to double parking tickets) in September 2015 were analyzed. "Issuer code", which represents a unique code identifying the issuing officer, was used. For the same issuer code, the time sequence and location of issued parking tickets are used as a proxy for the individual officer's patrol route. In September 2015, 4,589 issuers issued parking tickets, and 68% of issuers issued more than one ticket. Their daily patrol routes covered one (43% of issuers), two (24%), three (18%), and four (10%) police precincts, respectively. Just 5% of the daily patrol routes covered five or more precincts. In this study, 50 issuers were randomly selected and their daily patrol routes were analyzed using GIS tools. The result shows that 80% of the issuers had less than three patrol patterns during the whole month, and 25% of the issuers preferred patrolling on avenues (major streets) to minor streets. Figure 4 demonstrates the patrol pattern of a particular issuer by day in the study month. Three different patterns were found according to the number of police precincts covered. In this study, we did not attempt to correct such bias by quantifying individual behaviors—a quite difficult endeavor. However, on the basis of the currently available data, considerable care must be taken when utilizing citation data due to the selection bias identified in this study.

While the goal of the study is achieved, the current approach has limitations. Firstly, the variable "total ticket density" may not be completely true as a proxy of level of enforcement, especially if the study area is extended to other boroughs that have low parking citations in general. Secondly, the spatial models are not capable of accounting for unobserved, non-spatial factors. Unfortunately, unlike the long-studied crime or medical data model, parking models are limited to the available quantitative data. Future research efforts are needed in both data collection and methodology. Once more "unreported" parking data (i.e. identified from traffic cameras) becomes available, methods such as synthetic population-based approach (*10*) can be used to acquire a "ground truth" that contains representative samples of parking violations.

Conclusions and Future Work

This paper leveraged extensive data to investigate the spatial dependence of double parking citations via a grid-cell-structured geographic framework and analyzed the existence of selection bias resulting from different levels or coverage of traffic enforcement. Large quantities of real-world data from multiple sources, including parking tickets, land use, place of interest, sociodemographic, enforcement, and transportation were collected and geoprocessed. The descriptive analyses of the parking citations show that, although the majority of parking tickets are issued to passenger vehicles, double parking tickets do not follow this general trend. In fact, half of the double parking citations were issued to commercial vehicles. The possible reasons for this finding can be: (a) the lack of parking spaces in a highly congested urban network, (b) the desire of commercial vehicle drivers to park close to the delivery spot, even if no parking space is available, to satisfy their delivery schedule, and (c) a combination of these two and other operational factors.

The global and local Moran's I statistics were in complete agreement that double parking citations were spatially dependent and such spatial dependence should not be neglected when using citation data. As a result, spatial lag and spatial error models were proposed to account for the spatial dependence of parking tickets to avoid biased estimates.

To investigate whether selection bias exists in issuing parking tickets, the effects of parking ticket density and police precinct distance were estimated while controlling for variables such as commercial area, truck activity, taxi demand, population, hotels, and restaurants. When parking ticket density was used as an indicator of the level of enforcement it was found to have a positive

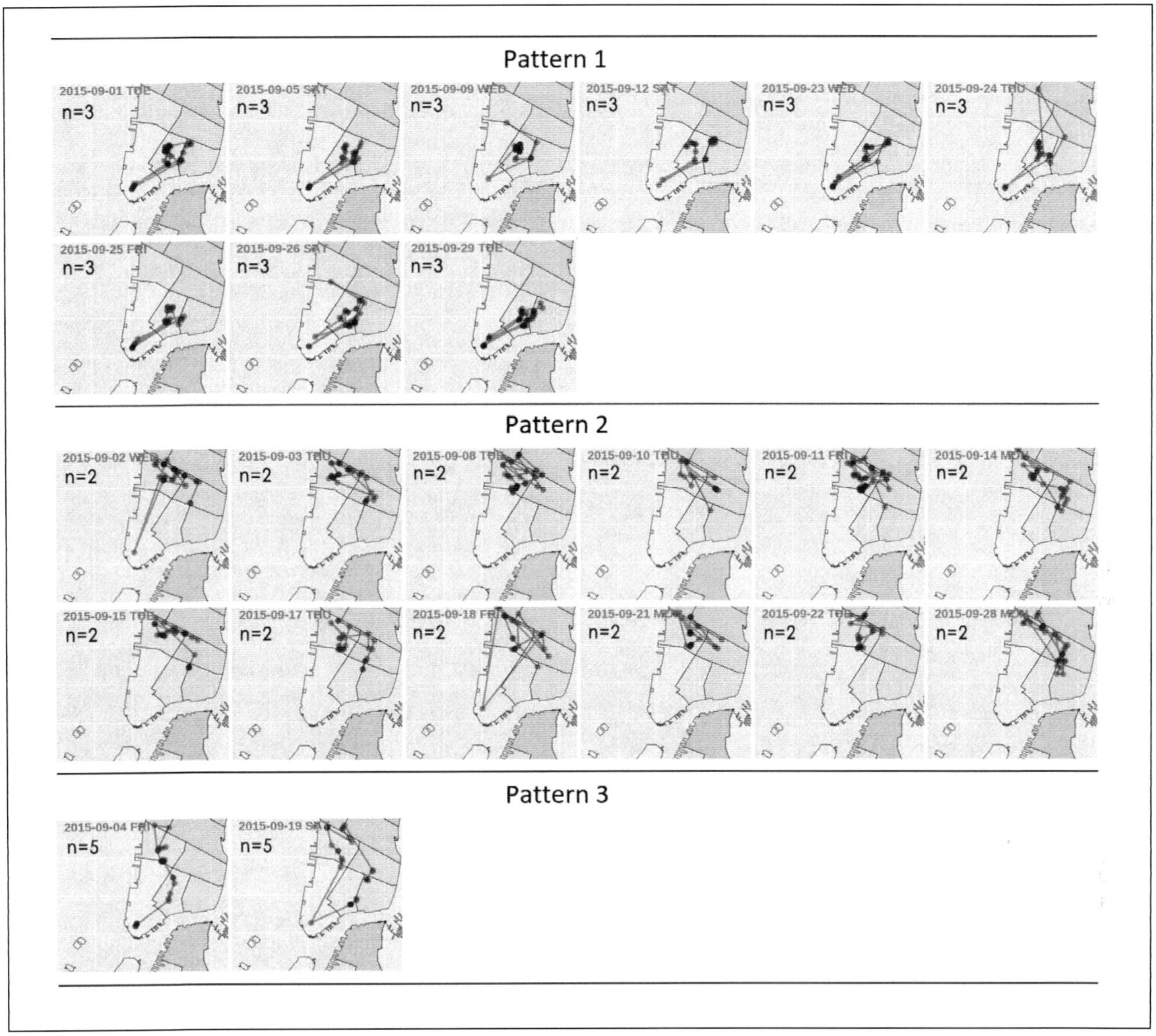

Figure 4. Demonstration of patrol patterns of traffic enforcement officers.
Note: n = number of police precincts covered by an issuer. A covered precinct is a precinct having more than two parking tickets issued per day by the same issuer.

impact on the number of double parking tickets. When police precinct distance is used as an indicator of the enforcement coverage, it was found to be negatively correlated with the number of double parking tickets. Both of the experimental variables were found to be statistically significant, confirming the assumption that a certain selection bias caused by enforcement intensity exists in the parking ticket data. Thus, it is recommended that researchers and policy makers be aware of this bias when using double parking ticket data to develop operational and tactical strategies to address the problem of double parking.

In addition, this study further contributed to the literature by investigating the spatial bias potentially caused by patrol patterns of traffic enforcement. This result highlighted that the majority of issuers have less than three daily patrol patterns in the studied month, and some of them have personal preferences, such as patrolling more on major streets than minor streets. This further underlines the fact that considerable care must be taken when utilizing the citation data. Unfortunately, this kind of bias is challenging to quantify and correct. Machine learning techniques such as unsupervised path clustering and more accurate GPS data collection may be applied as part of future research efforts to quantify the effectiveness of selection bias due to patrolling activities. Investigating passenger vehicle and commercial vehicle citations separately, and examining all types of parking violations, can also be part of the future work.

Acknowledgments

The research work presented in this paper was partially supported by C2SMART Tier 1 University Transportation Center (c2smart.engineering.nyu.edu), and UrbanMITS Lab (engineering.nyu.edu/citysmart) at New York University.

Author Contributions

The authors confirm contribution to the paper as follows: study conception and design: Jingqin Gao, Kaan Ozbay, Kun Xie; data collection: Jingqin Gao, Kun Xie; analysis and interpretation of results: Jingqin Gao, Kaan Ozbay, Kun Xie; draft manuscript preparation: Jingqin Gao, Kaan Ozbay, Kun Xie. All authors reviewed the results and approved the final version of the manuscript.

References

1. Cookson, G., and B. Pishue. *The Impact of Parking Pain in the US, UK and Germany*. INRIX Research, Kirkland, WA, 2017.
2. Holguín-Veras, J., S. Hodge, J. Wojtowicz, C. Singh, C. Wang, M. Jaller, F. Aros-Vera, K. Ozbay, A. Weeks, and M. Replogle, The New York City Off-Hour Delivery Program: A Business and Community-Friendly Sustainability Program. *Interfaces*, Vol. 48, No. 1, 2018, pp. 70–86.
3. Ukkusuri, S. V., K. Ozbay, W. F. Yushimito, S. Iyer, E. F. Morgul, and J. Holguín-Veras. Assessing the Impact of Urban Off-hour Delivery Program Using City Scale Simulation Models. *EURO Journal on Transportation and Logistics*, Vol. 5, No. 2, 2016, pp. 205–230.
4. Xie, K., K. Ozbay, H. Yang, J. Holguín-Veras, and E. F. Morgul, Modeling Safety Impacts of Off-hour Delivery Programs in Urban Areas. *Transportation Research Record: Journal of the Transportation Research Board*, 2015. 2478: 19–27.
5. Morgül, E., K. Ozbay, and J. Holguín-Veras. Simulation versus Big Data: A Case Study for Traffic Impact Analysis for Off-Hour Deliveries. *Proc., 20th Conference of the International Federation of Operational Research Societies*, Barcelona, Spain, 2014.
6. Holguín-Veras, J., K. Ozbay, A. Kornhauser, M. Brom, S. Iyer, W. Yushimito, S. Ukkusuri, B. Allen, and M. Silas. Overall Impacts of Off-hour Delivery Programs in New York City Metropolitan Area. *Transportation Research Record: Journal of the Transportation Research Board*, 2011. 2238: 68–76.
7. Iyer, S., K. Ozbay, S. V. Ukkusuri, W. F. Yushimito, E. F. Morgul, and J. Holguin-Veras. Traffic Impact Analysis of an Urban Off-hour Delivery Program Using Macroscopic and Mesoscopic Simulation Models. Presented at 90th Annual Meeting of the Transportation Research Board, Washington, D.C., 2011.
8. Xie, K., K. Ozbay, and H. Yang, Spatial Analysis of Highway Incident Durations in the Context of Hurricane Sandy. *Accident Analysis & Prevention*, Vol. 74, 2015, pp. 77–86.
9. Xie, K., K. Ozbay, A. Kurkcu, and H. Yang, Analysis of Traffic Crashes Involving Pedestrians Using Big Data: Investigation of Contributing Factors and Identification of Hotspots. *Risk Analysis*, Vol. 37, No. 8, 2017, pp. 1459–1476.
10. Lum, K., and W. Isaac, To Predict and Serve? *Significance*, Vol. 13, No. 5, 2016, pp. 14–19.
11. Gao, J., and K. Ozbay, Modeling Double Parking Impacts on Urban Street. Presented at 95th Annual Meeting of the Transportation Research Board, Washington, D.C., 2016.
12. Gao, J., and K. Ozbay, A Data-Driven Approach to Estimate Double Parking Events Using Machine Learning Techniques. Presented at 96th Annual Meeting of the Transportation Research Board, Washington, D.C., 2017.
13. Wang, Q., and S. Gogineni. An Empirical Investigation of Commercial Vehicle Parking Violations in New York City. Presented at 94th Annual Meeting of the Transportation Research Board, Washington, D.C., 2015.
14. Wenneman, A., M. Roorda, and K. Habib. Illegal Commercial Vehicle Parking, Parking Demand, and the Built Environment. *Canadian Transportation Research Forum* 49th Annul Meeting, Windsor, ON, Canada, 2014.
15. Kawamura, K., P. Sriraj, H. R. Surat, and M. Menninger. Analysis of Built Environment Features and their Effects on Freight Activities. *Procedia – Social and Behavioral Sciences*, Vol. 125, 2014, pp. 28–35.
16. Smith, T., and K. Steif. *Street by Hour Probability of Getting a Parking Violation in Philadelphia*, https://imgur.com/a/TgxYw. Accessed May 15, 2017.
17. *NYC Open Data*, New York City. opendata.cityofnewyork.us, Accessed May 4, 2017.
18. U.S. Census Bureau. *Geocoding Services Web Application Programming Interface (API)*, 2017.
19. Google. *Geocoding API*, developers.google.com/maps/documentation/geocoding, Accessed July 18, 2016.
20. Xie, K., K. Ozbay, Y. Zhu, and H. Yang. Evacuation Zone Modeling under Climate Change: A Data-Driven Method. *Journal of Infrastructure Systems*, Vol. 23, No. 4, 2017, pp. 04017013.
21. ESRI. *ArcGIS Desktop Release* 10, 2011.
22. *BYTES of the BIG APPLE*. NYC Department of City Planning. www.nyc.gov/planning. Accessed June 15, 2017.
23. U.S. Census Bureau. www.census.gov. Accessed May 6, 2017.
24. *I Love NY – Places to Stay Map*. New York State. data.ny.gov/dataset/I-Love-NY-Places-to-Stay-Map/3ynz-ayvg. Accessed May 18, 2017.
25. Thompson, M. *Grade "A" Restaurants in Manhattan*, hdl.handle.net/2451/33888. Accessed June 3, 2017.
26. *Trip Record Data*. NYC Taxi& Limousine Commission. www.nyc.gov/html/tlc/html/about/trip_record_data.shtml, Accessed January 16, 2017.

27. *Data Feeds*. NYC Department of Transportation. www.nyc.gov/html/dot/html/about/datafeeds.shtml. Accessed July 2, 2017.
28. *GIS Data*. New York State Department of Transportation. gis.ny.gov/gisdata/. Accessed July, 2017.
29. *Turnstile Data*. Metropolitan Transport Authority, New York. web.mta.info/developers/turnstile.html. Accessed June 5, 2017.
30. *Precinct*. New York Police Department. www1.nyc.gov/site/nypd/bureaus/patrol/precincts-landing.page. Accessed May 4, 2017.
31. Moran, P. A. The Interpretation of Statistical Maps. *Journal of the Royal Statistical Society. Series B (Methodological)*, Vol. 10, No. 2, 1948, pp. 243–251.
32. Anselin, L. Local Indicators of Spatial Association—LISA. *Geographical Analysis*, Vol. 27, No. 2, 1995, pp. 93–115.
33. Anselin, L. *The Moran Scatterplot as an ESDA Tool to Assess Local Instability in Spatial Association*. Regional Research Institute, West Virginia University, Morgantown, WV, 1993.
34. Anselin, L., I. Syabri, and Y. Kho. GeoDa: An Introduction to Spatial Data Analysis. *Geographical Analysis*, Vol. 38, No. 1, 2006, pp. 5–22.
35. Goodchild, M. F. *Spatial Autocorrelation*. Geo Books, Norwich, UK, 1986.
36. Xie, K., X. Wang, K. Ozbay, and H. Yang. Crash Frequency Modeling for Signalized Intersections in a High-density Urban Road Network. *Analytic Methods in Accident Research*, Vol. 2, 2014, pp. 39–51.
37. Anselin, L. *GeoDa 0.9 User's Guide*, Center for Spatially Integrated Social Science, Urbana-Champaign, IL, 2003.
38. O'Brien, R. M. A Caution Regarding Rules of Thumb for Variance Inflation Factors. *Quality & Quantity*, Vol. 41, No. 5, 2007, pp. 673–690.
39. Anselin, L. *Spatial Econometrics: Methods and Models*. Kluwer Academic Publishers, Dordrecht, The Netherlands, 1988.
40. Ord, K. Estimation Methods for Models of Spatial Interaction. *Journal of the American Statistical Association*, Vol. 70, No. 349, 1975, pp. 120–126.
41. Akaike, H. A New Look at the Statistical Model Identification. *IEEE Transactions on Automatic Control*, Vol. 19, No. 6, 1974, pp. 716–723.
42. Schwarz, G. Estimating the Dimension of a Model. *The Annals of Statistics*, Vol. 6, No. 2, 1978, pp. 461–464.

The Standing Committee on Urban Transportation Data and Information Systems (ABJ30) peer-reviewed this paper (18-03899).

The opinions expressed in this paper are those of the authors, who are responsible for the facts set out in the paper. The findings presented in the paper do not represent any official views or policies of any sponsoring agencies.

Article

TRR
JOURNAL OF THE TRANSPORTATION RESEARCH BOARD

Transportation Research Record
2018, Vol. 2672(42) 170–179

DOI: 10.1177/0361198118797190
journals.sagepub.com/home/trr
SAGE

Performance Measures for Characterizing Regional Congestion using Aggregated Multi-Year Probe Vehicle Data

Thomas M. Brennan, Jr.[1], Mohan M. Venigalla[2], Ashley Hyde[1], and Anthony LaRegina[1]

Abstract

Probe vehicle speed data has become an important data source for evaluating the congestion performance of highways and arterial roads. Pre-defined spatially located segments known as traffic message channels (TMCs) are linked to commercially available, temporal anonymous probe vehicle speed data. These data have been used to develop agency-wide performance measures to better plan and manage infrastructure assets. Recent research has analyzed individual as well as aggregated TMC links on roadway systems to identify congested areas along spatially defined routes. By understanding the typical congestion of all TMCs in a region as indicated by increased travel times, a broader perspective of the congestion characteristics can be gained. This is especially important when determining the impact of such occurrences in the region as a major crash event, special events, or during extreme conditions such as a natural or human-made disaster. This paper demonstrates how aggregated probe speed data can be used to characterize regional congestion. To demonstrate the methodology, an analysis of vehicle speed data during Hurricane Sandy, the second costliest hurricane in the United States, is used to show the regional impact in 2012. Further, the analysis results are compared and contrasted with comparable periods of increased congestion in 2013, 2014, and 2016. The analysis encompasses 614 TMCs, within 10 miles of the New Jersey coast. Approximately 90 million speed records covering five counties are analyzed in the study.

By leveraging new traffic information technologies, the reliability, resiliency, and congestion of a region's roadway network can be characterized. A high-level look at a region's traffic congestion presents a means to monitor the overall health of the roadway system. The Texas A&M Transportation Institute (TTI) publishes a yearly Urban Mobility Report that ranks urban congestion levels throughout the United States (*1*). TTI's detailed methodology to rank these areas includes speed data as well as classified vehicle volumes and assumed vehicle occupancies. For this study, a direct measure of congestion based on aggregated roadway segment congestion, independent of volumes and occupancy, is performed on a study region using commercially available probe vehicle data.

Since 2010, the fidelity of probe vehicle data has improved greatly and now includes highways, arterials, and local roads, with data collected from both public and private crowd sources (*2*). The ability to obtain speeds for smaller roadway segments is fast becoming a viable means to evaluate a transportation system on a granular level, but has not been applied on a regional level. Probe vehicle data have been used in national mobility reports (*3*, *4*) and statewide mobility reports of interstates as well as pre-defined commuter corridors using different performance measures (*2*, *5–10*). These probe data have also been used in the Urban Mobility Report (*1*). In addition to the importance of taking measurements to gauge congestion in urban regions, there are several regions throughout the United States, especially along the coasts, where the collective congestion performance measure can provide insight to develop region-specific traffic management practices.

This research employed archived probe vehicle speed data to analyze the evolving congestion characteristics of the coastal region of New Jersey over the same 2-week periods of 2012, 2013, 2014, and 2016. Regional

[1]The College of New Jersey, Ewing, NJ
[2]George Mason University, Fairfax, VA

Corresponding Author:
Address correspondence to Thomas M. Brennan, Jr.: brennant@tcnj.edu

Table 1. Example TMCs Available for New Jersey

TMC	Road	Direction	Length (miles)	Start lat.	Start long.	End lat.	End long.
120-07527	NJ-138	West	0.946374	40.17121	−74.0749	40.17059	−74.0928
120N07527	NJ-138	West	0.055516	40.17059	−74.0928	40.17056	−74.0938
120-07526	NJ-138	East	0.264959	40.17056	−74.0938	40.17058	−74.0988
120N07526	NJ-138	East	0.135502	40.17058	−74.0988	40.17051	−74.1014

Table 2. Example TMC Speed Data

TMC	Date time stamp	SMS (MPH)	Confidence score	C-Value
120-07527	10/29/2012 12:30	74	30	100
120N07527	10/30/2012 12:30	71	30	100
120+07526	10/31/2012 12:30	7	30	100
120P07526	11/1/2012 12:30	13	30	100

performance measures are demonstrated, along with a data drill-down to determine the spatial relationships of areas having the highest impact on the region. The study period for 2012 specifically includes a time when a natural disaster occurs because it is definitely known to have had an impact on the region's roadway network. On October 29, 2012 Sandy, the second most expensive hurricane in United States history, made landfall on the coast of New Jersey. The total cost attributable to the storm exceeded $71 Billion (*11*). Sandy destroyed over 100,000 homes, and caused at least 132 deaths (*12*). The region being defined encompasses all monitored roadways which had archived speed data on or before October 20, 2012, and are within 10 miles of the New Jersey coast. Approximately 90 million speed records covering five counties were analyzed in the study.

The research methods incorporate previously defined performance measures (*13*) to demonstrate how aggregated probe data can be used to assess the congestion of a regional transportation system as well as the recovery from a major impact on the roadway system. The performance measures developed quantify the increase in travel time experienced by the motoring public. Although this research explores historic speed data, the methodologies presented can be applied to real-time speed data to assist first responders, residents, and freight vehicles that need to traverse a disconnected system during an emergency event.

Data

The mechanism for processing probe vehicle speed data requires a cross-reference between spatially defined traffic message channels (TMC) and commercially available temporal speed data sets. These data are collected by a commercial provider who tracks vehicle telematics over pre-defined TMCs. Each TMC is geo-located and assigned a corresponding unique roadway ID; an example of the TMC's attributes is shown in Table 1. As a probe vehicle traverses a TMC, a single or an aggregated number of vehicle speeds are associated with each TMC (Table 2). Each speed record is accompanied by a time stamp, c-value, and confidence score. A confidence of 30, along with a c-value of 100, indicates that the speed is directly based, not calculated, on probe vehicle measurements, with a high confidence that the speed represents the actual roadway conditions. For this research, only records with a score of 30 and c-value of 100 were included.

The region is defined by selecting all TMCs reporting data located within 10 miles of the coast of New Jersey (Figure 1). Other regions and the methods for selecting regions will be defined in future research, but for this study only the coast is selected to demonstrate the methodologies proposed. The study area shown in Figure 1 comprises 614, or about 3.8% of the state's 16,256 TMCs. The regional analysis encompasses all types of roadways including major highways, arterials, and local roads.

Measures of Congestion

Previous research has considered any TMC to be congested during a period of time when the calculated space mean speed (SMS) falls below a pre-defined speed, typically 45 mph for major highways (*2*, *5*, *14*, *15*). Using a fixed speed threshold does not account for variations in the road geometry, spatial location, and roadway type. This is especially important when determining the congestion thresholds along the coastal routes, where posted

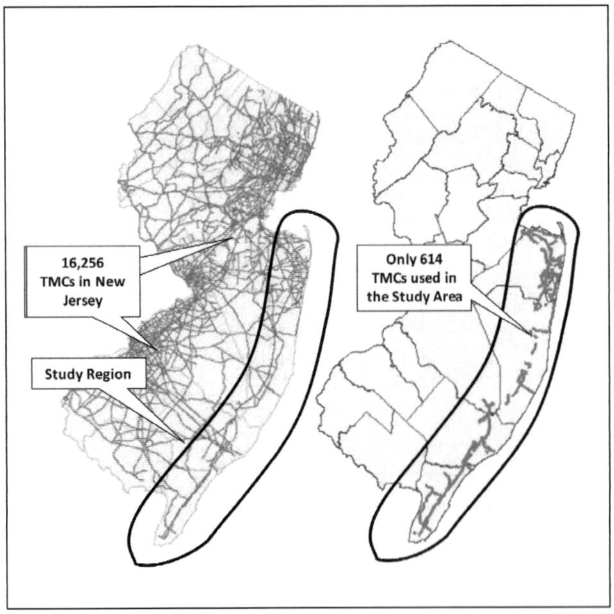

Figure 1. Study area along the Atlantic Coast of New Jersey with 614 TMC segments.

speeds are much lower than speed limits on a typical highway. Congestion is defined for each TMC as 70% of the average segment speed during time periods where congestion is unlikely (*13*). This 70% can be adjusted based on an agency's needs, but was chosen in accordance with previous studies that used 45 mph as a fixed speed threshold, which is approximately 70% of 65 mph. The variable speed threshold is 70% of the calculated base free-flow speed (BFFS), and is calculated using the following:

$$v_{ia} = 0.70 \frac{1}{n_j} \sum_{j \in F} v_{ij} \quad (1)$$

where v_{ia} is the variable speed threshold for TMC i; v_{ij} are speed records for TMC i for the 15-minute interval j; n_j is the total count of 15-minute intervals within the free-flow time F, which is defined as all of the minutes between 0200 and 0600 during June 2012. The month of June was selected as the BFFS analysis period because June was found to produce low variability in travel speeds and it occurred before any major hurricane damage would have occurred. In New Jersey there was sufficient volume to yield enough high-confidence speed records to facilitate the measurement. Other regions with lower volumes might use a wider time window. This congestion threshold was used for all time periods analyzed. Both the analysis timeframe and the percent speed threshold can be adjusted based on regional or agency requirements. Furthermore, the analysis period can also be adjusted to better understand how the regional transportation system reacts and recovers under recurring conditions as well as extreme events such as a hurricane, or even a major power outage; this had occurred in July 2017 along the Outer Banks of North Carolina, requiring a mandatory evacuation of the barrier islands (*16*).

Each of the 614 TMCs is assigned a base travel time (BTT) that accounts for the TMC's length and its observed 70% BFFS. The BTT for each TMC is computed using Equation 2 (*13*):

$$BTT_i = \frac{x_i}{v_{ia}} \quad (2)$$

where, BTT_i is the base travel time in hours for TMC i and x_i is the distance in miles for the TMC. The travel time for any TMC segment is defined as shown in Equation 3:

$$TT_{ij} = \begin{cases} \frac{x_i}{v_{ij}}, & v_{ij} < v_{ia} \\ 0, & v_{ij} \geq v_{ia} \end{cases} \quad (3)$$

where, TT_{ij} is travel time in hours of TMC i at time period j and v_{ij} and v_{ia} are as defined in Equation 1. For total travel time a travel time inflation (TI) performance measure, which accounts for TMC length and speed variability inherent with spatial location, was used (*2*). TI is the difference between the BTT and TT as shown in Equation 4:

$$TI_i = \sum_{j \in A} (TT_{ij} - BTT_i); \text{ for } TT_{ij} > 0 \quad (4)$$

where TI_i is the total travel time inflation for TMC segment i for all 15-minute bin time periods j within analysis period A. Thus, every 15-minute interval where the SMS falls beneath the segment-defined threshold leads to an accumulation of TI. This TI represents the individual TMC travel time inflation. To aggregate this on a region level, a regional performance measure is defined as shown in Equation 5:

$$RTI = \sum_{i \in R} TI_i \quad (5)$$

where, RTI is the total travel time inflation for all TMC segments i in the analysis region R within the analysis period defined in Equation 4. In summary, the RTI is the aggregation of all the increased travel time incurred on all TMCs within a region.

Regional Travel Time Inflation Applied to New Jersey Coast

The demonstration of the regional performance measure provides a high-level aggregate analysis of the regional

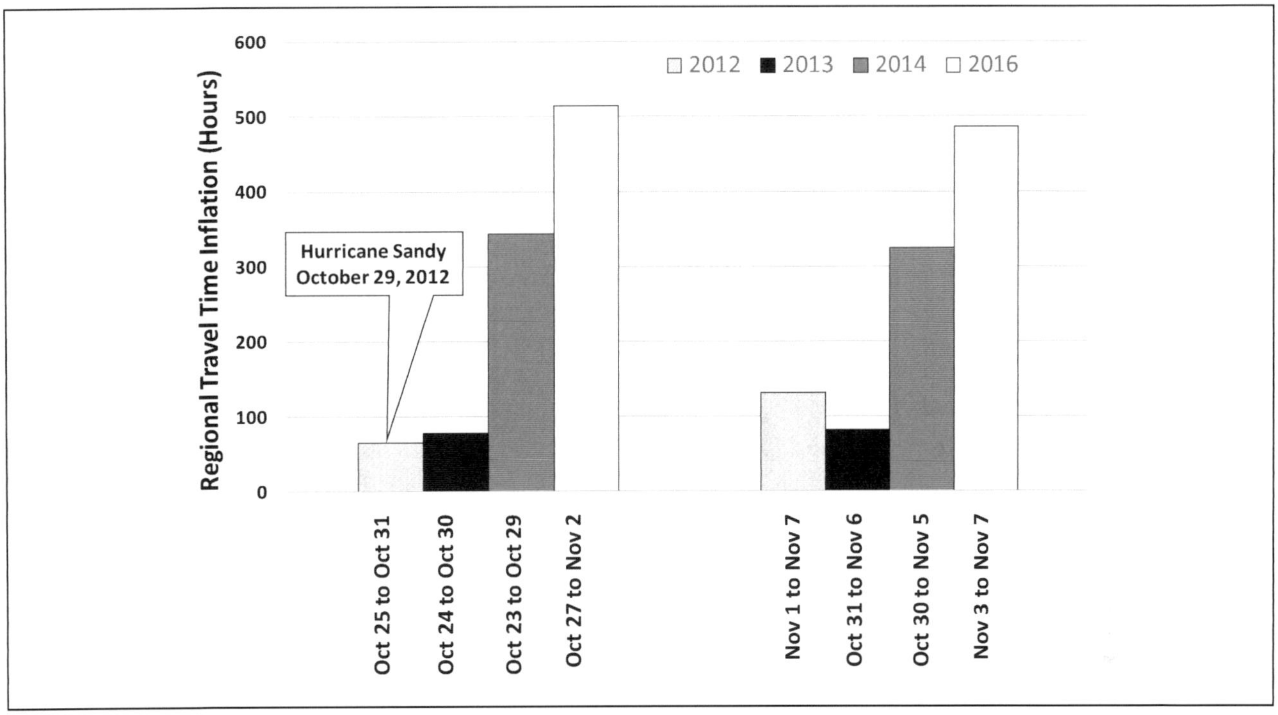

Figure 2. Total hours weekly regional travel time inflation (RTI) for study area during 2012, 2013, 2014, and 2016.

pulse of the transportation network. This aggregation begins with RTI (Equation 5), but can subsequently be disaggregated to the individual TMC's TI value (Equation 4) to better understand what is affecting the region's roadway system. The RTI provides macro-characterization of the relative congestion intensity. This is demonstrated in Figure 2, where week-to-week comparisons are made to determine fluctuations in RTI during the 2012 storm and the comparable periods in subsequent years. As seen in the figure, the storm's impact is observed during the week of October 28, 2012. There is a notable increase in RTI during the week after the hurricane, whereas in subsequent years it remained relatively steady. The increase in 2012 may be an indication that motorists are heading back to the coast to clean up after the storm and/or that there has been a decrease in the infrastructure's ability to handle capacity. Also, there was a dramatic increase in the TI from 2013 to 2014, and from 2014 to 2016.

A subsequent analysis of the daily RTI is shown in Figure 3, where traffic is lower the day of and day after Hurricane Sandy, when compared with the traffic on the same dates/day of week the following year (2013). Traffic volumes were higher on the day before the 2012 hurricane compared with the same period in 2013, potentially indicating an increase in congestion resulting from the mandatory evacuation put in place on Sunday October 28, 2012.

Data Availability during Extreme Event

The decrease in aggregated RTI for five counties on October 30, 2012 (Figure 3) may be indicative of a couple of possibilities pertaining to the data collected. The first is that no vehicles are traversing the segment, or at least the number of vehicles has decreased to a point that no travel time inflation is occurring along individual segments. This decrease was expected considering that there were mandatory evacuations beginning October 28, 2012, with forced road closures put in place. Although there was an increase in TI during the mandatory evacuation, it is minor compared with subsequent years, which appears to indicate that the evacuation was either successfully completed with minimal congestion or people did not acknowledge the evacuation. The other possibility is that the segment cannot be traversed because of road damage or some other hindrance, preventing data from being collected on a segment.

Data on all 614 segments were analyzed during a 12-h period (0700–1900) from October 28 to November 3, 2012 to determine if the infrastructure is reporting data in the study area during the storm. To quantify data availability, a binary indicator was assigned to each of the 15-minute bins during a 12-h period (48 bins). For a particular day, if a TMC was assigned 48, then it would indicate that each 15-minute period for that day had collected data. By observing these values independent of

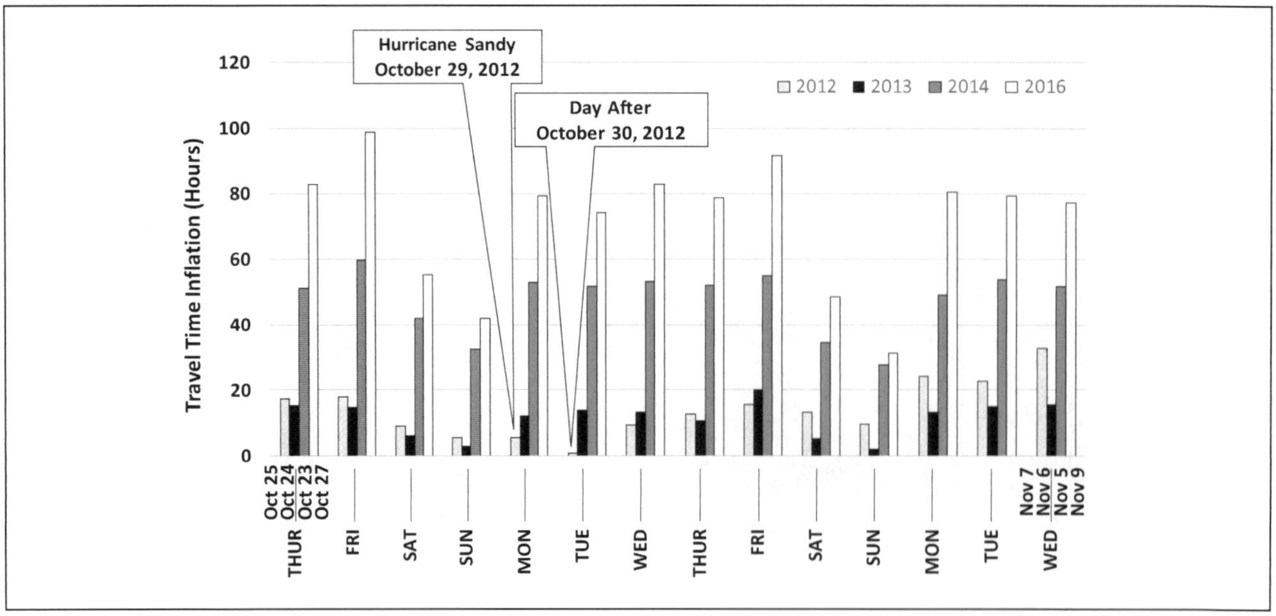

Figure 3. Total hours of daily regional travel time inflation (RTI) for study area during 2012, 2013, 2014, and 2016.

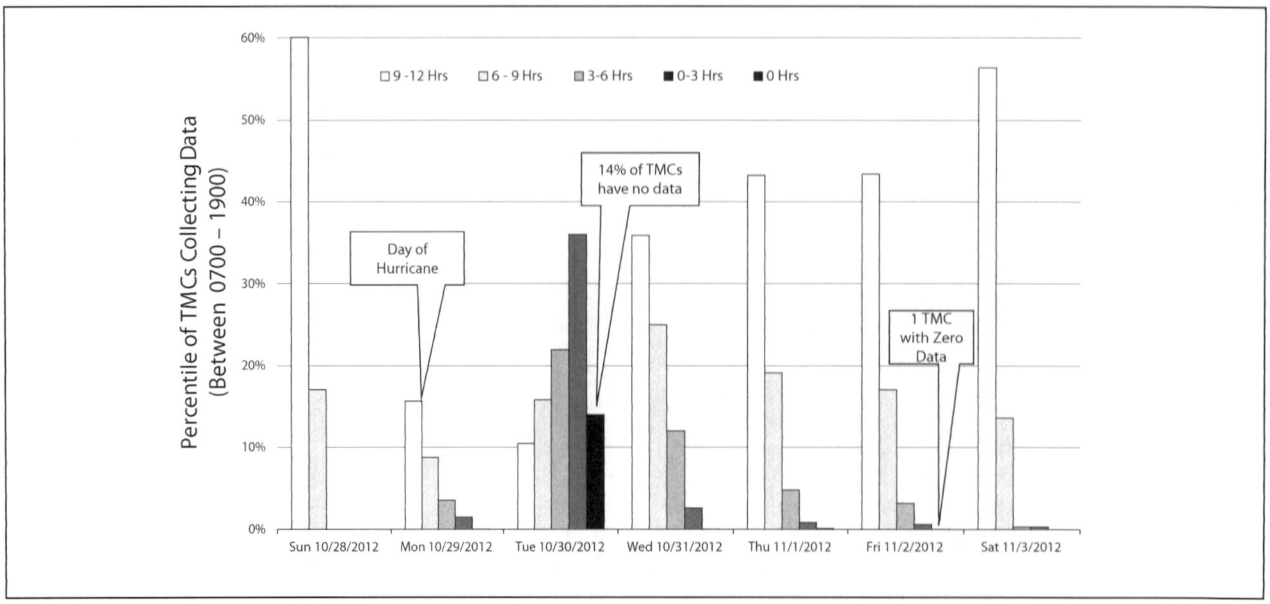

Figure 4. Percentage of segments collecting data between 0700–1900 during the week of October 28, 2012.

congestion for each day on each segment, the recovery of the system is noted as more vehicles traverse the system. This data availability indicator is shown in Figure 4, which shows that about 14% or 94 of the TMC segments did not collect data on the day following the hurricane. Although there is still one TMC with no data being collected on Friday November 2, 2012, most of the TMCs are observed to be collecting some data, which indicates vehicles are traversing the segments and the system has recovered. The figure illustrates the recovery of the system after the storm on Saturday November 3, 2012 to be about the same as the Sunday October 28, 2012.

The segment locations where data are not being collected can be spatially represented on a map (Figure 5). From the figure, there were a total of 520 segments, collecting zero and or less than 3 h of data on October 30, 2012. This decrease in data collected corresponds to the decrease in RTI shown in Figure 3. For the

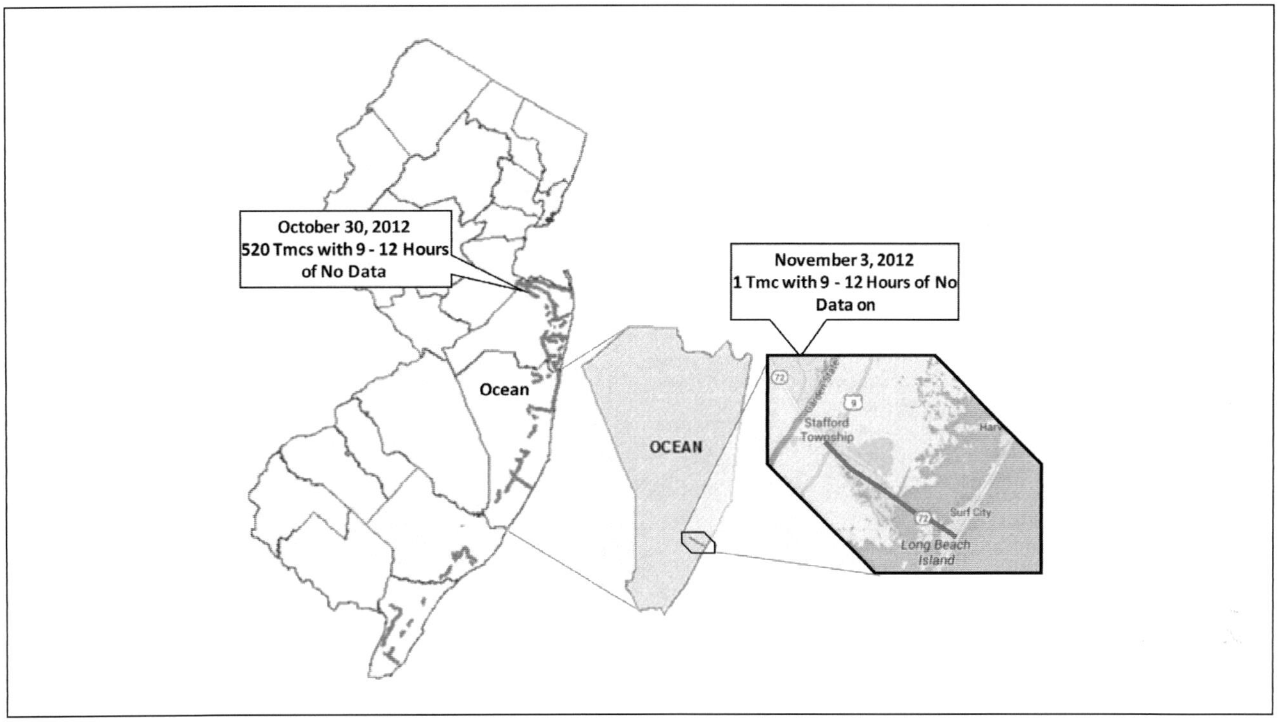

Figure 5. Spatial mapping of available probe vehicle segment data on October 30 and November 3, 2012.

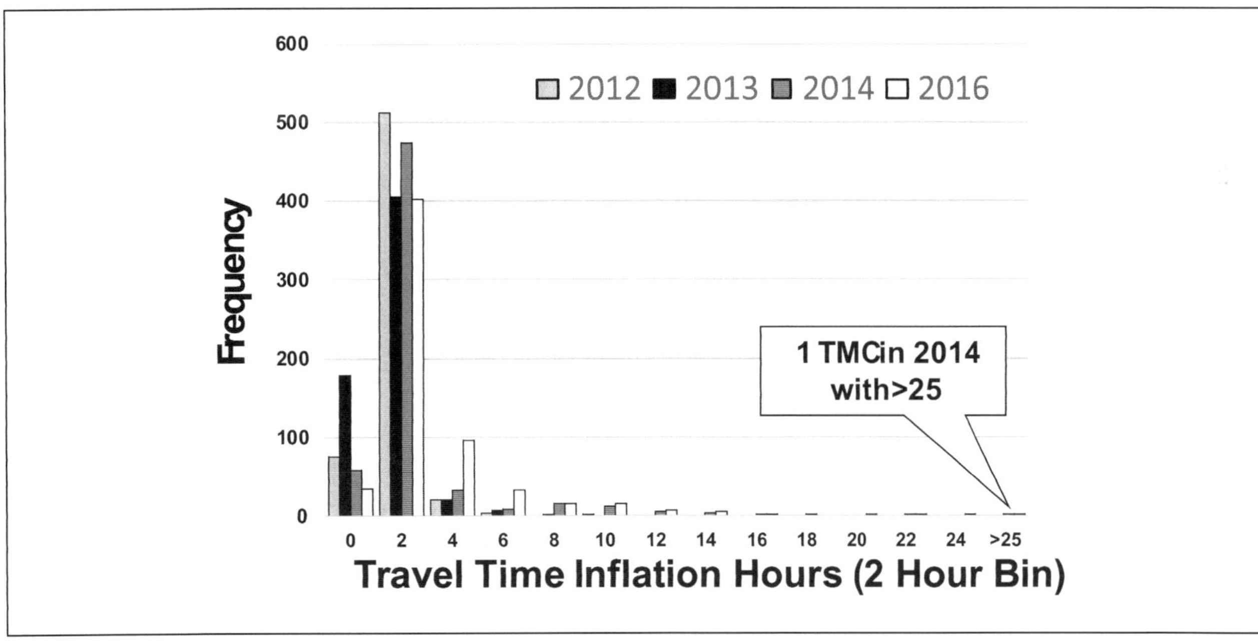

Figure 6. Number of TMCs reporting TI for 2012, 2013, 2014, 2016.

TMC in Ocean County, the Route 72 Bridge entering the barrier island was still closed as per a New Jersey Transcom report (*17*) on November 2, 2012. Ancillary data from state agencies provide insight into why some areas have no segment data being collected after the storm. Within 1 week after the storm, approximate 85% of the 614 TMCs were collecting more than 6 h of traffic data.

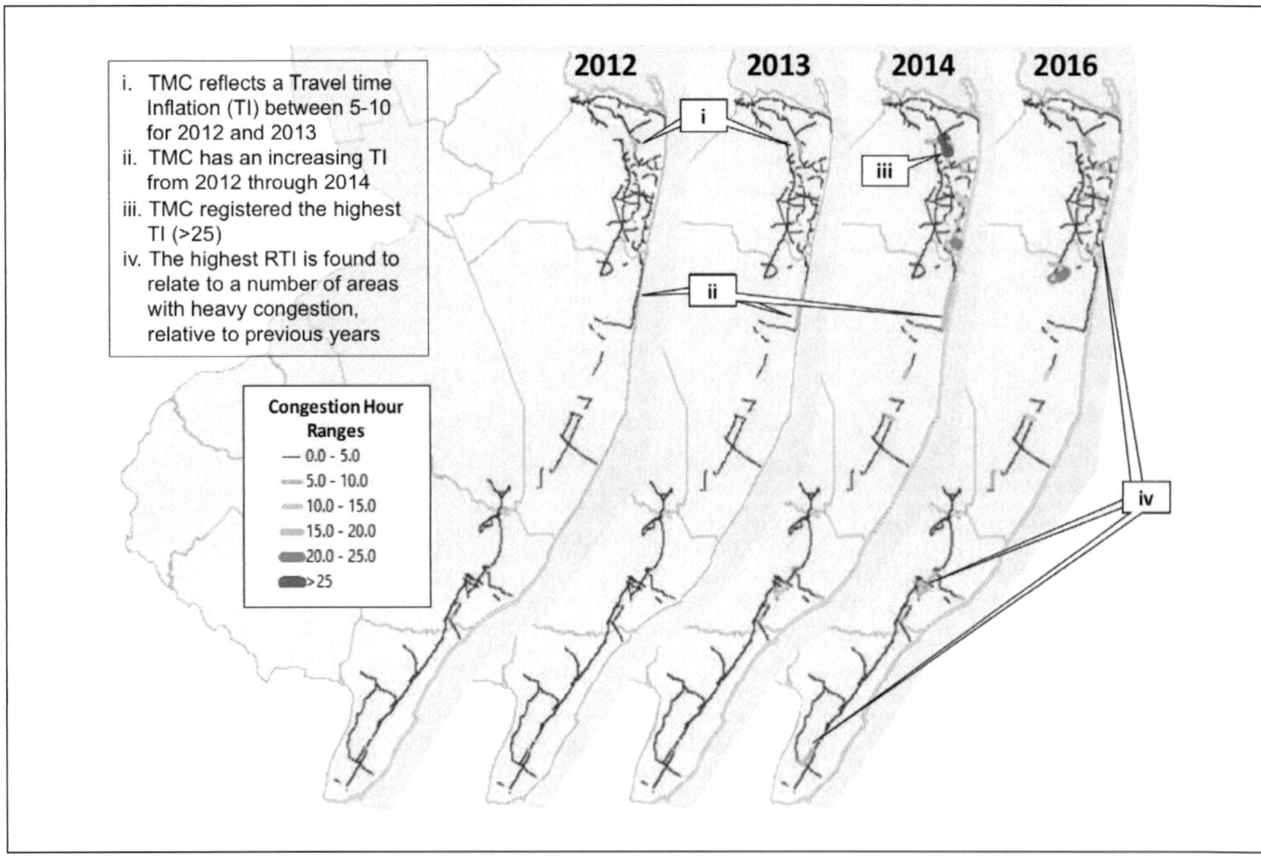

Figure 7. Two weeks of aggregated TI for 2012, 2013, 2014, 2016.

Congestion Distribution

To quantify the congestion derived from each of the TMCs, the calculated TI (Equation 4) was used to create a histogram comparing each TMC's disaggregated TI over the 2-week study period. This analysis (Figure 6) indicates that a small shift in the TI can have a dramatic impact on the RTI. For instance, in 2016, where the highest RTI is observed (Figure 2), there was a shift in the number of TMCs with a travel time inflation of 4, 6, and 8. The RTI indicated that there was an increase in the regional congestion, and the TI mapped the locations where the congestion is worse.

In the years following the hurricane, no other major events were recorded during the 2-week study periods. Therefore, in 2013, 2014, and 2016, the increase in congestion is likely caused by an increase in the recurring and construction-type congestion, as reflected by the increased TI on each TMC. To better understand the TI distribution, a spatial representation of the year-to-year change is shown in Figure 7. The observations noted on Figure 7 support the RTI analysis reflected in Figures 2 and 3. They also demonstrate how the RTI data can be disaggregated to show where areas of high congestion are located in the region.

Yearly Congestion

Although the locations of increased RTI could be qualitatively observed (as shown in Figures 2 and 3) for shorter durations, it would be difficult to represent these data over an entire year. The RTI may be used an indicator for the region which characterizes the degree of congestion experienced by drivers as a function of the increase in travel time. To view the region's data over an entire year, and compare the day-to-day congestion, the TI for each TMC is normalized with respect to the TMC's base travel time, yielding the Regional Increase in Mean Travel Time (RIMTT):

$$RIMTT_{ij} = \sum_{n} \frac{TI_i}{BTT_i} \quad (6)$$

where $RIMTT_{ij}$ is defined for a TMC segment i for a 15-minute bin time period j and TI_i is defined in Equation 4. For the RIMTT performance measure all the TI

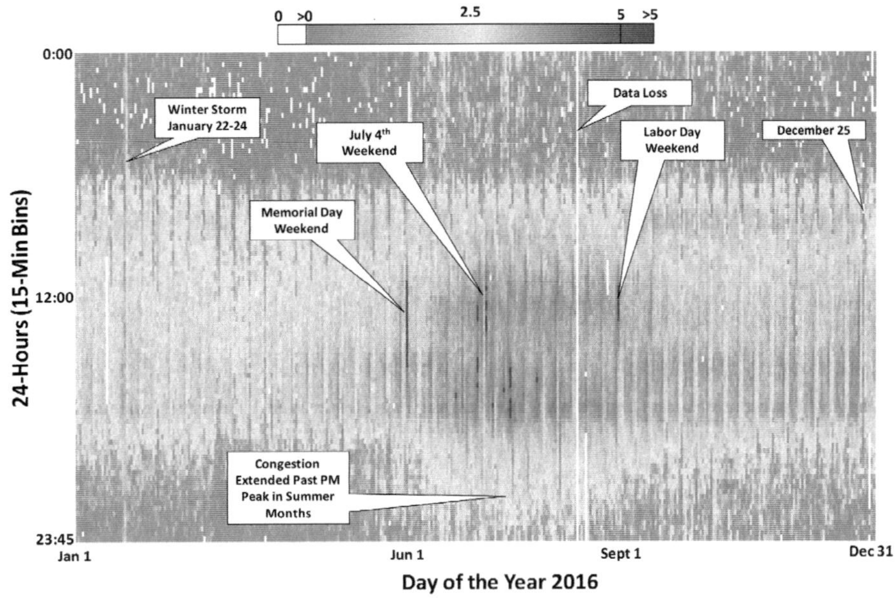

Figure 8. Daily RIMTT in 15-min bins for 2016.

values are aggregated for every TMC in 15-minute bins. A representation of this performance measure for 2016 is shown in Figure 8, in which several events occurring in the region are identifiable from the data visualization. From the figure, winter storm Jonas in January is identified, as well as the summer holidays Memorial Day, Fourth of July, and Labor Day weekend. All of these holidays were expected to increase congestion considering the region is a major summer destination in New Jersey. Other notable observations from this figure include an increase in congestion during the evening hours, data loss on August 15, and a noticeable drop in congestion on December 25. This method can be used to observe the region and characterize the congestion throughout the year. Further research is necessary to compare multiple regions and year-to-year congestion.

Conclusion

This research applied a scalable regional travel time inflation (RTI) performance measure that quantifies the increase in travel time experienced by motorists. The measure was used to aggregate probe vehicle speed data to characterize the congestion in a region within 10 miles of the New Jersey coast, thus providing a high-level characterization of the region's congestion. The RTI is derived from the travel time inflation (TI) performance measure, which was used to determine the spatial distribution of congestion throughout the region. These performance measures (RTI and TI) were derived for the study area during the same 2-week periods in 2012, 2013, 2014, and 2016. The results show that the probe vehicle speed data, when aggregated as either RTI or TI, can act as a performance measure for characterizing congestion on a dispersed transportation network.

An additional exploratory indicator, the regional increase in mean travel time (RIMTT) was also developed to characterize the yearly congestion in 2016. RIMTT, which provides a mechanism to gauge daily congestion in 15-minute intervals, is not a performance measure. Rather, it is a visualization technique that relays information that makes the probe vehicle data particularly useful. From the visualization created by the RIMTT data, specific events could be observed throughout the year as well as the changes in congestion throughout the day. This type of visualization has been applied to individual highways, but not to a region (*2*, *13*).

Creating the performance measures discussed in this study (RTI, TI) and RIMTT provides transportation agencies with the ability to determine how their system is performing year to year, as well as on a daily basis. The research specifically included the 2012 Hurricane Sandy to demonstrate how the absence or presence of data being obtained from the transportation system can be quantified to indicate where the network is being traversed by vehicles providing telematics data. For the 2012 hurricane, it was shown that there was a dramatic drop in the number of data points being collected as mandatory evacuation and road closures were put in place. The data show that the system status can be monitored just by determining data availability.

As probe vehicle data become more reliable, local, state, regional, and national standards need to be established to leverage this data to produce ongoing monitoring of the transportation system. These types of standards are especially important during a natural or manmade disaster. This is evident in North Carolina, where are a power outage caused by a construction error required the evacuation of over 10,000 tourists (*16*). The application of this type of data in real time can help improve disaster management processes by determining where and when a response should be made by an agency, and evaluating how the transportation system recovers from a disaster (*18–20*). With improved data collection, the ability to visualize traffic conditions in real time can assist first responders, residents, and freight vehicles traverse a disconnected system during an emergency event. These data can also be used to validate the need, or determine the impact, of infrastructure investment. Further research will explore the implementation of an automated system as well as comparing the performance measures in subsequent years and other regions. This type of system, once in place, will prove a more proactive means of managing the transportation system within a region.

Acknowledgments

This research is being funded by the Region 2- University Transportation Research Center (UTRC). The speed data and segment information used in this report was obtained from INRIX Inc.

Authors Contributions

The authors confirm contribution to the paper as follows: study conception and design: Brennan, Venigalla; data collection: Brennan, Hyde, LaRegina; analysis and interpretation of results: Brennan, Venigalla, Hyde, LaRegina; draft manuscript preparation: Brennan, Venigalla, Hyde, LaRegina. All authors reviewed the results and approved the final version of the manuscript.

References

1. Schrank, D., B. Eisele, T. Lomax, and J. Bak. *2015 Urban Mobility Scorecard.* Texas Transportation Institute, Texas & University System, College Station, TX, 2015.
2. Remias, S. M., T. M. Brennan, C. M. Day, H. T. Summers, D. K. Horton, E. D. Cox, and D. M. Bullock. Spatially Referenced Probe Data Performance Measures for Infrastructure Investment Decision Makers. *Transportation Research Record: Journal of the Transportation Research Board*, 2014. 2420: 33–44.
3. Lomax, T., D. Shrank, and B. Eiselee. *2012 Annual Urban Mobility Report.* The Texas A&M University System, Texas A&M Transportation Institute, College Station, TX, 2012.
4. Lomax, T., D. Shrank, and B. Eiselee. *2011 Urban Mobility Report Powered by INRIX Traffic Data.* The Texas A&M University System, Texas A&M Transportation Institute, College Station, TX, 2011.
5. Remias, S., T. Brennan, G. Grimmer, E. Cox, D. Horton, and D. Bullock. *2011 Indiana Interstate Mobility Report - Full Version.* Indiana Mobility Reports; Report No.: ISBN: 978-1-62260-209-4. Purdue University, West Lafayette, IN, 2012.
6. Maryland Department of Transportation. *2012 Maryland State Highway Mobility Report.* Maryland Department of Transportation, Hanover, MD, 2012.
7. California Department of Transportation. *Mobility Performance Report 2009.* California Department of Transportation, Sacramento, CA, 2011.
8. Washington State Department of Transportation. *The 2012 Congestion Report.* Washington State Department of Transportation, Olympia, WA, 2012.
9. Haghani, A., M. Hamdi, and K. F. Sadabadi. *I-95 Corridor Coalition Vehicle Probe Project: Validation of Inrix Data July-September 2008-Final Report.* I-95 Corridor Coalition. University of Maryland, College Park, MD, 2009.
10. Remias, S., T. Brennan, C. Day, H. Summers, E. Cox, D. Horton, and D. Bullock. *2012 Indiana Mobility Report - Full Version.* Indiana Mobility Reports. Report No.: ISBN: 978-1-62260-257-5. Purdue University, West Lafayette, Indiana, 2013.
11. National Oceananic and Atmospheric Admistration–Hurricane Research Division. Costliest Mainland United States Tropical Cyclones, 1900-2017 (Unadjusted for inflation). U.S. Department of Commerce, Washington, D.C. http://www.aoml.noaa.gov/hrd/tcfaq/costliesttable.html. Accessed August 25, 2018.
12. Reuters. Factbox: Storm Sandy Blamed for At Least 132 Deaths in U.S., Canada. Novemer 16, 2012. http://www.reuters.com/article/2012/11/16/us-storm-sandy-deaths-idusbre8af0zx20121116.
13. Brennan, T. M., S. M. Remias, and L. J. Mariani. Performance Measures to Characterize Corridor Travel Time Delay Based on Probe Vehicle Data. *Transportation Research Record: Journal of the Transportation Research Board*, 2015. 2526: 39–50.
14. Brennan, T. M., C. M. Day, G. M. Remias, D. K. Horton, E. D. Cox, and D. M. Bullock. Alternative Performance Measures and Weighting for Quantifying Spatial and Temporal Congestion using Probe Data. Presented at 21st World Congress on Intelligent Transport Systems: Reinventing Transportation in Our Connected World, Detroit, MI, 2014.
15. Brennan, T. M. Jr., S. M. Remias, D. K. Horton, and D. M. Bullock. Probe Vehicle Based Statewide Mobility Performance Measures for Decision Makers. *Transportation Research Record: Journal of the Transportation Research Board*, 2013. 2338: 78–90.

16. Sterling, J., and Schmidt, R. Thousands Evacuate North Carolina's Outer Banks after Power Outage. CNN, July 31, 2017. https://www.cnn.com/2017/07/28/us/nc-outer-banks-power-outages/index.html. Accessed August 25, 2018.
17. TRANSCOM. Regional Conditions Report, Jersey City, November 4, 2012, 9:30 a.m. http://cdn2.hubspot.net/hub/91068/file-15976096-pdf/docs/transcom_regional_conditions_report_11042012_0930.pdf?t=1411788310000. Accessed August 25, 2018
18. *Improving Disaster Management: The Role of IT in Mitigation, Preparedness, Response, and Recovery*. Committee on Using Information Technology to Enhance Disaster Management. National Research Council of the National Academies, Washington, D.C., 2007.
19. Southworth, F., B. Janson, and M. M. Venigalla. DYMOD: Towards Real Time, Dynamic Traffic Routing During Mass Evacuations. *Proc., Simulation Multi-Conference*, Orlando, FL, 1992. https://www.osti.gov/biblio/5941031
20. Southworth, F., A. K. Rathi, J. Jacobi, and M. M. Venigalla. *Database Development for Regional Evacuation Studies*. Oak Ridge National Laboratory, Prepared for the U.S. Department of Army and Federal Emergency Management Agency, Washington, D.C., 1990.

The Standing Committee on Urban Transportation Data and Information Systems (ABJ30) peer-reviewed this paper (18-04745).

The contents of this paper reflect the views of the authors, who are responsible for the facts and the accuracy of the data presented herein, and do not necessarily reflect the official views or policies of the sponsoring organizations. These contents do not constitute a standard, specification, or regulation.

TRR
JOURNAL OF THE TRANSPORTATION RESEARCH BOARD

Article

Transportation Research Record
2018, Vol. 2672(42) 180–191
© National Academy of Sciences:
Transportation Research Board 2018
Article reuse guidelines:
sagepub.com/journals-permissions
DOI: 10.1177/0361198118792768
journals.sagepub.com/home/trr
SAGE

Cross-Vendor and Cross-State Analysis of GPS Probe Data Latency

Zhongxiang Wang[1], Masoud Hamedi[2], Elham Sharifi[2], and Stanley Young[3]

Abstract

Crowd sourced GPS probe data have become a major source of real-time traffic information applications. In addition to traditional traveler advisory systems such as dynamic message signs (DMS) and 511 systems, probe data are being used for automatic incident detection, integrated corridor management (ICM), end of queue warning systems, and mobility-related smartphone applications. Several private sector vendors offer minute by minute network-wide travel time and speed probe data. The quality of such data in terms of deviation of the reported travel time and speeds from ground-truth has been extensively studied in recent years, and as a result concerns over the accuracy of probe data have mostly faded away. However, the latency of probe data—defined as the lag between the time at which disturbance in traffic speed is reported in the outsourced data feed, and the time at which the traffic is perturbed—has become a subject of interest. The extent of latency of probe data for real-time applications is critical, so it is important to have a good understanding of the amount of latency and its influencing factors. This paper uses high-quality independent Bluetooth/Wi-Fi re-identification data collected on multiple freeway segments in three different states, to measure the latency of the vehicle probe data provided by three major vendors. The statistical distribution of the latency and its sensitivity to speed slowdown and recovery periods are discussed.

In the era of Big-Data and Internet of Things (IoT), the proliferation of connected devices that continuously report their location has made it possible to capture the movement pattern of such devices. Private companies capture the location and speed samples of such devices on the transportation network, and, through proprietary algorithms and processes, convert them into real-time traffic data products. As a result, minute by minute travel time and speed data over the entire freeway and major arterial road network are available. Probe data have been gaining popularity in recent years and have quickly become a vital component of intelligent transportation systems. In addition to traditional traveler advisory systems such as dynamic message signs (DMS) and 511 systems, probe data are being used for automatic incident detection, integrated corridor management (ICM), end of queue warning systems, and mobility-related smartphone applications. The quality of such data in terms of deviation of the reported travel time and speeds from ground-truth has been extensively studied by independent researchers in recent years. The I-95 Corridor Coalition's Vehicle Probe Project (VPP) data validation program is an example, in which quality of probe data from three major vendors in participating states has been carefully analyzed periodically (http://i95coalition.org/projects/vehicle-probe-project/). As a result, concerns over the accuracy of probe data have mostly faded away. However, the latency of probe data—defined as the lag between the time at which a disturbance in traffic speed is reported in the outsourced data feed, and the time at which the traffic is perturbed—has become a subject of interest. Latency is inherent in any real-time procedure of probe data generation because of the processing, communications, and aggregation functions. Although it can be minimized, theoretically speaking it cannot be eliminated. Bluetooth travel time data are time-tagged to the nearest second based on GPS reference time from embedded GPS receivers in the equipment. So, when the travel time/harmonic reference speed is processed (not in real-time, but postprocessed) any latency in Bluetooth reference travel time/harmonic speed can be eliminated. For

[1]Department of Civil & Environmental Engineering, University of Maryland, MD
[2]Center for Advanced Transportation Technology, University of Maryland, College Park, MD
[3]Advanced Transportation and Urban Scientist, National Renewable Energy Laboratory, CO

Corresponding Author:
Address correspondence to Masoud Hamedi: masoud@umd.edu

real-time applications, it is important to have a good understanding of the magnitude of latency and its influencing factors.

Latency measurement for real-time travel time data is not well studied. Liu et al. describe the existence of latency in reporting GPS-based data (*1*). Tseng et al. discuss a method for traffic sensing with GPS-equipped probe vehicles and point out the latency as an outstanding issue (*2*). Hunter developed a low-latency state estimation algorithm trying to achieve near real-time measurement of travel time and speed using GPS trajectories on a large-scale network (*3*). Chase et al. further stated that GPS probe data has greater latency when travel speed recovers after a peak period than at the beginning of a peak period (*4*). None of these papers proposed a detailed quantitative latency measurement methodology.

Kim and Coifman measured the latency for GPS probe data compared with loop detectors (*5*). They calculated the correlation coefficient, which significantly depends on the covariance of the original time-series speed data and shifted time-series speed data. The results show that the average latency for GPS probe data is 6.8 min, and it could exceed 10 min in many situations. In a study by Iowa Department of Transportation using stationary sensor data, the latency for a probe data feed was estimated to vary from 3 to 12 min (*6*). However, loop detectors can only report spot travel speed whereas GPS probe data are reported on standardized segments, typically traffic message channel (TMC) industry standard segments. The speed reported is related to the space mean speed across a TMC segment rather than a spot speed from loop detectors. Moreover, Kim and Coifman (*5*) used 10-s aggregation intervals, whereas the granularity of commercially available GPS probe data is 1 min or more.

The authors of this paper proposed and published a methodology for calculating latency of GPS probe data compared with a secondary independent data source (*7*). The core of the methodology is a maximum pattern matching algorithm that measures the latency by finding the time shift that maximizes the overlapping of two datasets based on different objective functions. Although the methodology was proven to be effective in explaining both the significance and variation of the latency, it was only applied to a limited data set. This paper employs the same approach and uses a rich data set, consisting of high-quality independent Bluetooth/Wi-Fi re-identification data collected on multiple freeway segments in three different states, to measure the latency of the vehicle probe data provided by three major vendors. The statistical distribution of the latency and its sensitivity to speed slowdown and recovery periods are also discussed.

Methodology

This paper uses the definition and implements the methodology given by Wang et al. (*7*) to measure and analyze the latency of probe data. It includes data preparation, data filtering, interpolation, smoothing, and finally data shifting to calculate latency.

Bluetooth and Wi-Fi Data Fusion

The reference data in this paper are the fused Bluetooth/Wi-Fi re-identification data collected using portable sensors in three states. The basic idea is that two sensors are deployed at the start and end boundaries of a road segment, recording the time stamp and unique digital identifier (MACID) of in-vehicle electronic devices. By matching MACID's between two sensors, a travel time sample is generated. Details of the methodology for matching, filtering, and data aggregation can be found in Wang et al. (*7*). In total, data were collected on 41 segments for approximately 40 days. Table 1 is a summary of the data collection effort, and Figure 1 shows the location of the sensors and segments on the map.

Capturing Slowdown and Recovery Episodes

Average travel speed on freeways is subject to fluctuation. Recurring congestion is caused by increased traffic volume and insufficient capacity, whereas nonrecurring congestion often results from incidents, work zones, weather events, and special events that affect traffic flow. When traffic speeds drop significantly for a considerable

Table 1. Freeway Corridors and Bluetooth/Wi-Fi Data Collection Summary

State	Road	Start point	End point	Length (mi)	Date
South Carolina	I-85	US-276/Exit 48	SC-14/Exit 56	7.15	12/03/2015–12/17/2015
	I-26	Bush River Road/Exit 108	Harbison Boulevard/Exit 103	4.47	
New Hampshire	I-89	I-93	Stickney Hill Road/Exit 3	3.54	07/10/2016–07/24/2016
	I-93	I-393/US-202/US-4/Exit 15	Hackett Hill Road/Exit 11	15.82	
	I-93	NH-28/Rockingham Road/Exit 5	NH-102/Nashua Road/Exit 4	3.63	
North Carolina	I-240	US-70/Charlotte Street/Exit 5B	US-23/US-19/Exit 3	2.23	12/16/2016–12/26/2016
	I-40	NC-191/Exit 47	US-23/US-19/Exit 44	2.56	
	I-26	I-40/ Exit 46 A/US 74	I-26/Exit 37	14.43	

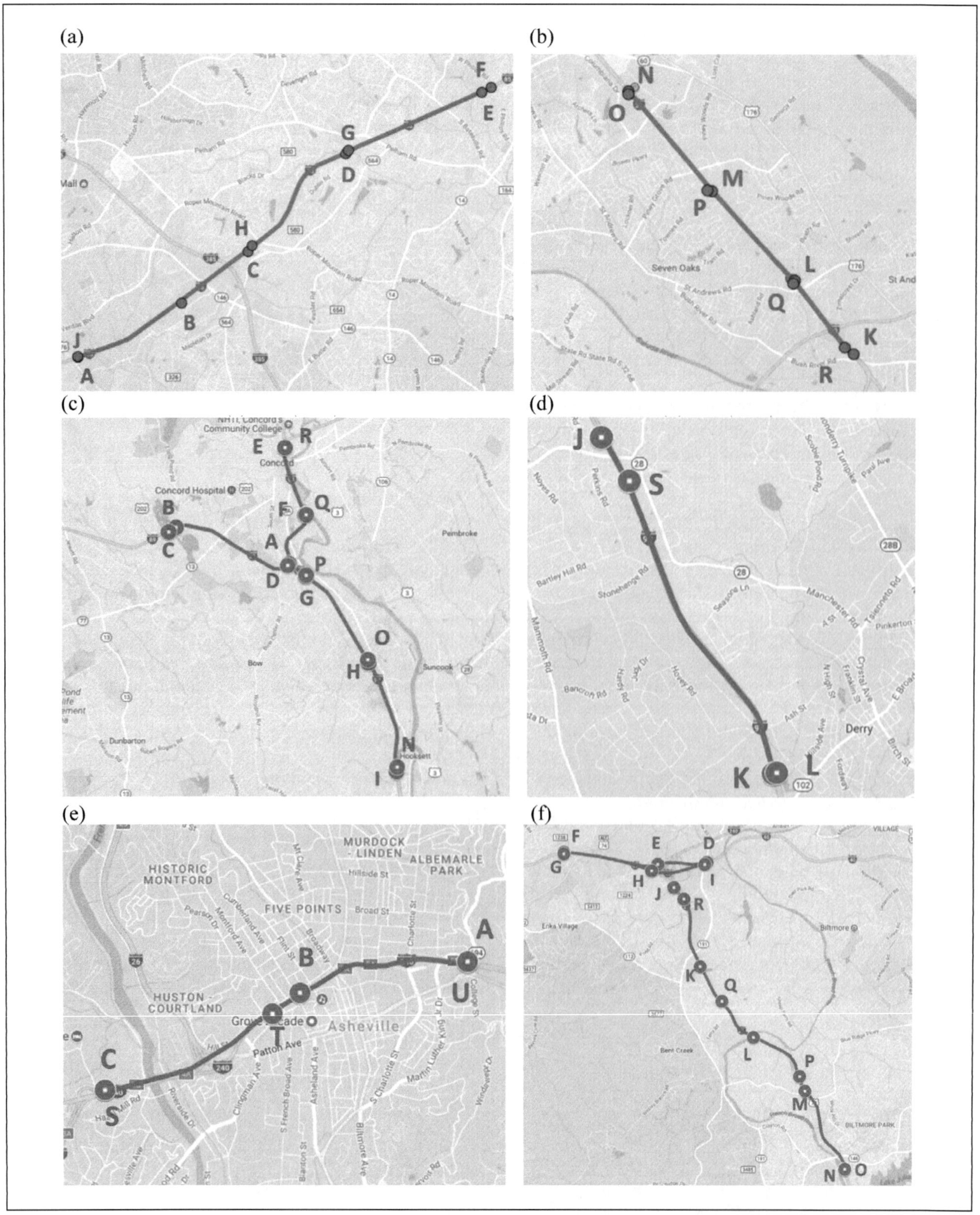

Figure 1. Freeway test segments and Bluetooth/Wi-Fi sensor locations; showing (*a*) South Carolina (I-85); (*b*) South Carolina (I-26); (*c*) New Hampshire (I-89, I-93); (*d*) New Hampshire (I-93); (e) North Carolina (I-240); and (*f*) North Carolina (I-40, I-26).

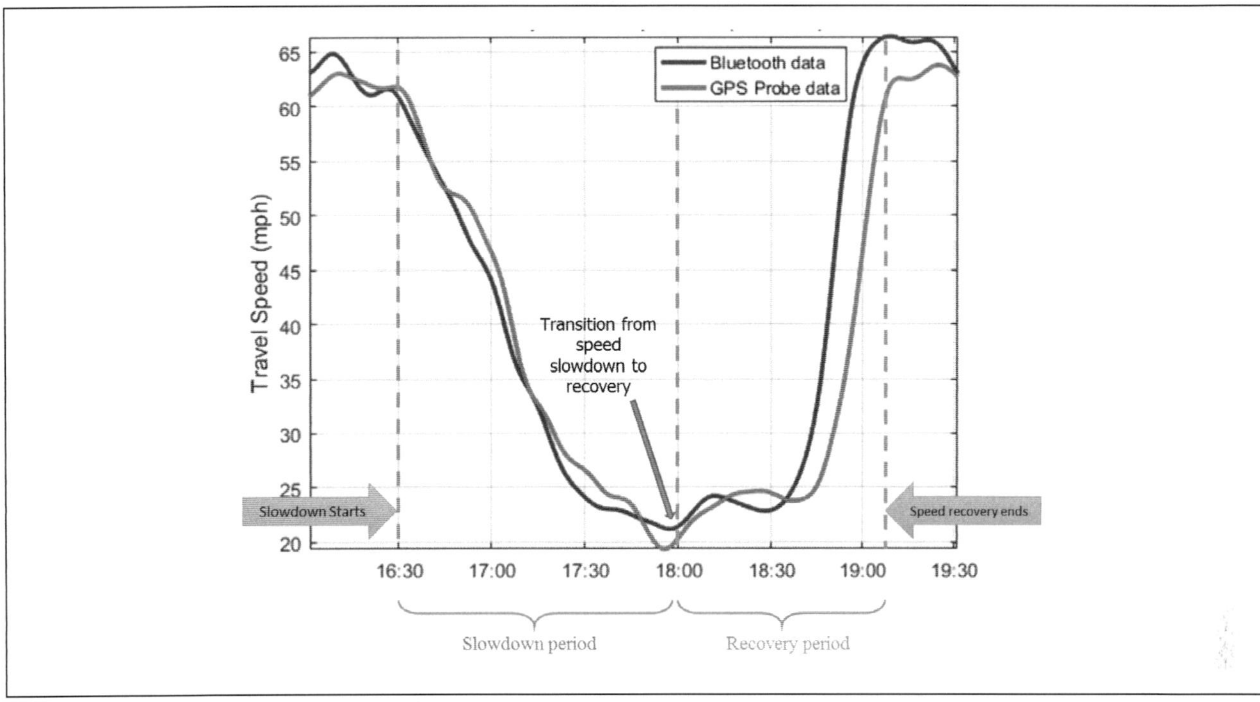

Figure 2. Illustration of traffic slowdown and recovery.

amount of time, a slowdown episode emerges. On such occasions, traffic speeds will eventually bounce back to the precongestion levels. Figure 2 shows an example of slowdown and recovery, captured by both Bluetooth and GPS probe data. To measure the latency of probe data with respect to the reference data, the start and end times of each slowdown and recovery episode (SRE), as well as the transition time between slowdown period (SP) and recovery period (RP), in the collected data must be identified. In the previous effort to measure latency, the authors visually inspected the daily speed graphs to record slowdowns (*7*). However, a manual approach is tedious and analyzing many segments for an extended time period calls for automation.

A pattern recognition algorithm was developed. The main steps of the automated speed pattern recognition filter (ASPRF) are as follows:

For every path in the data set:

For every day of the study period:

- Calculate free flow speed (FFS) as the 80th percentile of travel speed, for each specific path each day, and the minimum slowdown threshold (MST) as 40% of the free flow speed;
- Mark all data points with speed less than MST (there could exist several data groups for each day, and each group is a potential SRE);
  - For each group:
- Find the minimum speed point—its corresponding time is the transition from SP to RP;
- From the minimum speed point, search both forward and backward until the speed reaches the FFS; the point at which it does so is the start time of the SP (for backward search) and the end time of the RP (for forward search); if FFS cannot be reached, switch to alternative boundaries for SRE (start and end time of each day, the end of the previous RP, and the start of the succeeding SP);

- Rule-based SRE filtering:
  - Discard any SRE with missing speed data for five or more consecutive minutes;
  - Patch any SRE with missing speed data for less than five consecutive minutes, by interpolating neighboring data, as discussed by Wang et al. (*7*);
  - Combine two adjacent SREs if the end of the first RP is within 30 min of the start of the second SP;
  - Discard any SRE if its final duration is less than 60 min.

After implementing the ASPRF in Python, it was applied to the reference data set, and the results were visualized and inspected. Figure 3 shows an example of the algorithm output for one day on one path in South

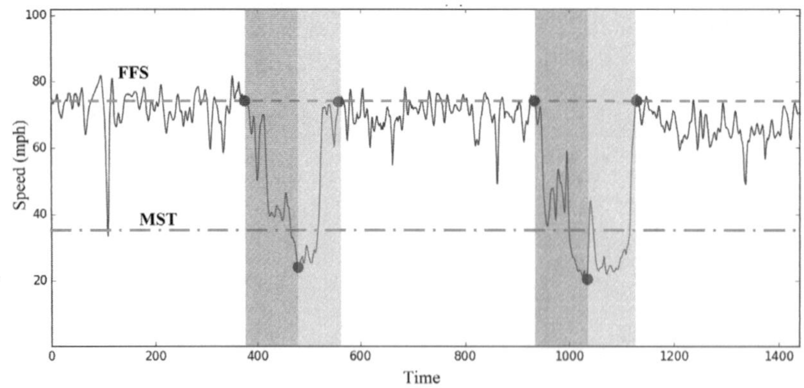

Figure 3. Sample slowdown and recovery detection results from ASPRF algorithm.

Table 2. Number of SREs per State

State	Number of paths	Number of days	Number of SREs
South Carolina	14	15	72
New Hampshire	12	15	26
North Carolina	15	11	18
Total	41	41	116

Carolina. The red dots mark the start of the SP, the transition from SP to RP, and the end of the RP. The upper dashed orange line is the FFS and the lower dash-dotted orange line is MST. Blue and green shaded areas show SP and RP respectively. Although there is a sudden speed loss in the early hours of the day, it is not detected as a slowdown because for latency measurement only significant slowdowns with at least 1-h duration are considered.

To verify the ASPRF output and calibrate the thresholds, it was applied to the same dataset used in (*7*) and results were found to be compatible with the manual process. Table 2 shows the number of SREs for each state after running the code.

Latency Measurement

As mentioned before, the same methodology used by Wang et al. (*7*) is used in this paper. The basic idea is to shift the GPS probe data from a lower bound to an upper bound minute by minute and find the time shift that yields maximum overlap between the probe and the reference data. Three different fitness functions are used to measure the overlap:

1. **AVD**: minimize sum of absolute speed difference between GPS probe data and reference data for all corresponding minute by minute data points;
2. **SVD:** minimize sum of squared speed difference between GPS probe data and reference data for all corresponding minute by minute data points;
3. **COR:** maximize correlation coefficient between GPS probe data and reference data.

Each of the above fitness functions provides a different perspective for the curve matching. AVD is the most straightforward way to measure and minimize vertical distance between the two curves. SVD puts more emphasis on higher speed differences. COR describes the statistical correlation between two curves. As discussed in (*7*), by applying all three fitness functions separately, and measuring the latency based on their average, the methodology becomes less sensitive to the irregularities in data and produces more consistent results.

Test Results

The reference data were collected on selected freeway corridors in South Carolina, New Hampshire, and North Carolina as shown in Table 1. Corridors were divided into shorter paths, with lengths varying from 1 to 3.5 mi. The paths were identical to the corresponding TMCs that GPS probe data vendors used, so path-wise comparison could be conducted without further data manipulation. Sensors were deployed at the conjunctions of the paths, recording the time that vehicles passed by the sensors. The reference traffic speed was calculated using Bluetooth re-identification technology. Approximately 2 weeks' worth of data was collected in each state. GPS probe data from three major vendors was obtained from the I-95 Corridor Coalition as part of their VPP. The purpose of this study is to investigate the latency of GPS probe data representative of the industry, and not to scrutinize any particular data vendor. Thus, the vendors

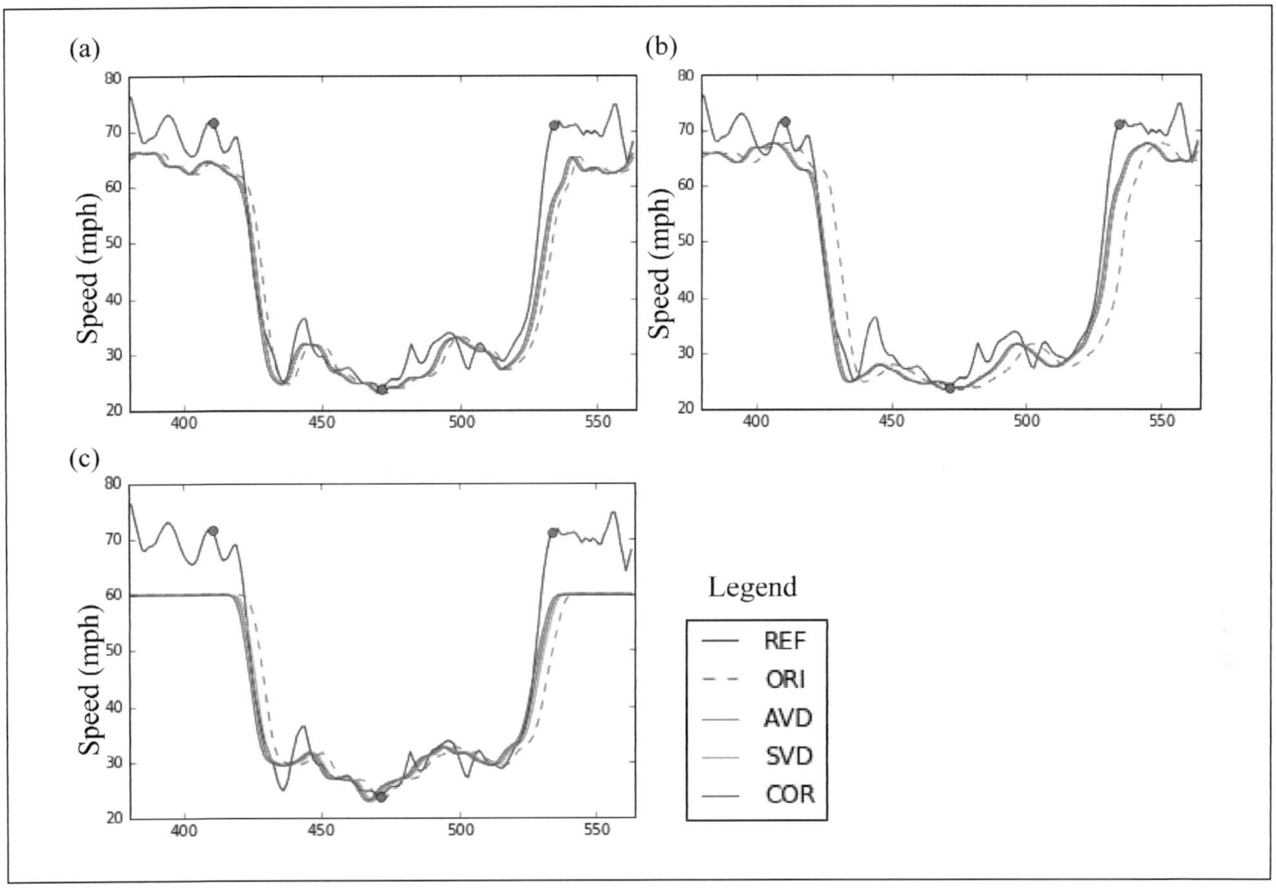

Figure 4. Latency measurement results of Segment F–G in South Carolina for 12/10/2015, showing data from (*a*) Vendor 1, (*b*) Vendor 2, and (*c*) Vendor 3. REF = reference data; ORI = original un-shifted GPS probe data; AVD = GPS probe data shifted and latency measured by AVD; SVD = GPS probe data shifted and latency measured by SVD; COR = GPS probe data shifted and latency measured by COR.

will be referred to by numbers when discussing the results and the numbers are consistent throughout the paper.

The latency measurement procedure was applied to data for all three vendors, and for all 116 SRE, SD, and RP cases identified by the ASPRF algorithm. Figure 4 shows the results for segment F to G in South Carolina on 12/10/2015 as a sample.

Figure 4 shows that all three vendors have successfully captured both speed reduction and recovery in their data feed. For each fitness function, the shifted graph corresponding to the optimal value is shown, and the shift value is reported as measured latency. Moreover, the difference among shift results based on the three applied fitness functions is marginal. The red dots on the graphs show the start time of the slowdown, start time of the recovery and the end time obtained from ASPRF.

After running the latency measurement procedure for all cases and all three vendors, graphs similar to Figure 1 were generated and carefully examined. Aggregated numerical results are presented in Table 3. In each column, average latency values obtained based on the fitness function in the column header are reported. The column marked as "mean" is the statistical mean of the latency across all three fitness functions, and is the value that is to be used.

Table 3 shows that variation of latency from vendor to vendor. For example, for all test segments in North Carolina, and for the entire SRE, Vendor 3 has the smallest average latency value of 2.5 min, whereas this number is 3.4 and 6.0 min for Vendors 1 and 2, respectively. Looking at results across states for each vendor shows that the mean SRE latency value ranges from 3.4 to 4.3, 6.0 to 7.5 and 2.5 to 3.0 for Vendors 1, 2, and 3, respectively. Vendor 3 exhibits the smallest range (in SRE and RP), suggesting that its latency is more consistent.

In their previous work (*7*), the authors had observed that latency is not symmetric for each SRE, meaning that the probe data feed might reflect the reduction in speed with less latency than for the speed recovery. To revisit that observation, all SRE cases are examined and the

Table 3. Aggregated Latency Measurement Results for All Vendors

		Vendor 1				Vendor 2				Vendor 3			
		Average latency (min)				Average latency (min)				Average latency (min)			
State	Event	AVD	SVD	COR	Mean	AVD	SVD	COR	Mean	AVD	SVD	COR	Mean
South Carolina	SRE	3.9	4.4	4.6	4.3	6.1	6.2	6.2	6.2	2.9	3.1	3.1	3.0
	SP	3.6	3.6	4.0	3.7	6.7	6.8	6.6	6.7	4.4	4.3	3.8	4.2
	RP	4.7	5.3	4.0	4.7	5.0	5.2	5.0	5.1	2.0	2.4	2.5	2.3
New Hampshire	SRE	3.3	3.4	3.6	3.4	5.6	6.0	6.4	6.0	2.3	2.5	2.7	2.5
	SP	3.3	3.2	3.2	3.2	7.0	7.0	6.7	6.9	3.5	3.0	3.2	3.2
	RP	4.0	4.2	3.9	4.0	3.6	3.7	4.7	4.0	2.9	3.5	2.6	3.0
North Carolina	SRE	3.5	4.1	4.1	3.9	7.3	7.8	7.3	7.5	3.2	3.2	2.7	3.0
	SP	4.6	4.6	3.6	4.2	7.7	7.8	7.4	7.6	3.2	2.9	2.5	2.9
	RP	3.4	3.7	4.1	3.7	5.9	6.5	6.8	6.4	3.2	3.1	3.8	3.4

Note: AVD = average latency based on absolute vertical distance fitness function; SQR = average latency based on squared vertical distance fitness function; COR = average latency based on correlation coefficient fitness function; SRE = slowdown and recovery episode; SP = slowdown period; RP = recovery period.

Table 4. Percentage of SREs with Slowdown Latency Greater Than Recovery Latency

State	Vendor 1	Vendor 2	Vendor 3
South Carolina	38%	64%	68%
New Hampshire	50%	77%	50%
North Carolina	50%	72%	56%
Total	42%	68%	62%

percentage of cases that show higher latency in SP compared with RP are reported in Table 4.

As Table 4 shows, the results vary by the vendor and state. So although there is asymmetry in the latency between slowdown and recovery periods, it cannot be said that in general SP latency is higher than RP latency or vice versa. Having said that, the last row of Table 4 shows that cases with SP latency greater than RP latency are in a minority (42%) for Vendor 1 compared with the other two vendors.

Latency Distribution

Results in Tables 3 and 4 show that the probe data latency varies from vendor to vendor, and from location to location. So instead of reducing latency to a single value, it is important to study its statistical distribution. Figure 5 shows a box plot of the latency observations created for all latency observation during the entire SRE, and during the SP and RP, separated by vendor. The box plots show that latency observations for Vendor 2 are more scattered than those of the other two vendors. RP latency for Vendor 2 is also more evenly distributed compared with SP for the same vendor. The same results are illustrated in Figure 6 from a different perspective as histogram plots.

Statistical Comparison among Vendors

As shown in Figures 5 and 6, statistical distributions for latency observations are available for each vendor. This makes it is possible to examine whether mean latency reported by a vendor is statistically equal to that of other vendors. A t-test is used to evaluate such a hypothesis. The t-test assumes the normal distribution of the mean, warranted by central limit theory. More specifically, the two-sample t-test requires independence between the two samples, which is tested using the chi-square test (*8*). The null and alternative hypothesis for t-test and chi-square tests are presented below. All the tests were conducted using latencies at SRE from all vendors across all states.

The null and alternative hypothesis for t-test:

$$\text{test 1}: \begin{cases} H0: \mu_1 - \mu_2 = 0 \\ H1: \mu_1 - \mu_2 \leq 0 \end{cases}$$

$$\text{test 2}: \begin{cases} H0: \mu_1 - \mu_3 = 0 \\ H1: \mu_1 - \mu_3 \geq 0 \end{cases}$$

$$\text{test 3}: \begin{cases} H0: \mu_2 - \mu_3 = 0 \\ H1: \mu_2 - \mu_3 \geq 0 \end{cases}$$

where μ_1, μ_2, μ_3 are the mean latencies of Vendors 1, 2, and 3, respectively.

The null and alternative hypothesis for chi-square test:

$$\begin{cases} H0: \text{vendor } a \text{ is independent of vendor } b \\ H1: \text{vendor } a \text{ is not independent of vendor } b \end{cases} \forall a, b = 1, 2, 3, a \neq b$$

The t-test and chi-square test results for each pair of vendors are shown in Table 5. The right side of this table shows chi-square test results. All P-values are above the significance level ($\alpha = 0.05$), meaning that the null hypothesis cannot be rejected. In other words,

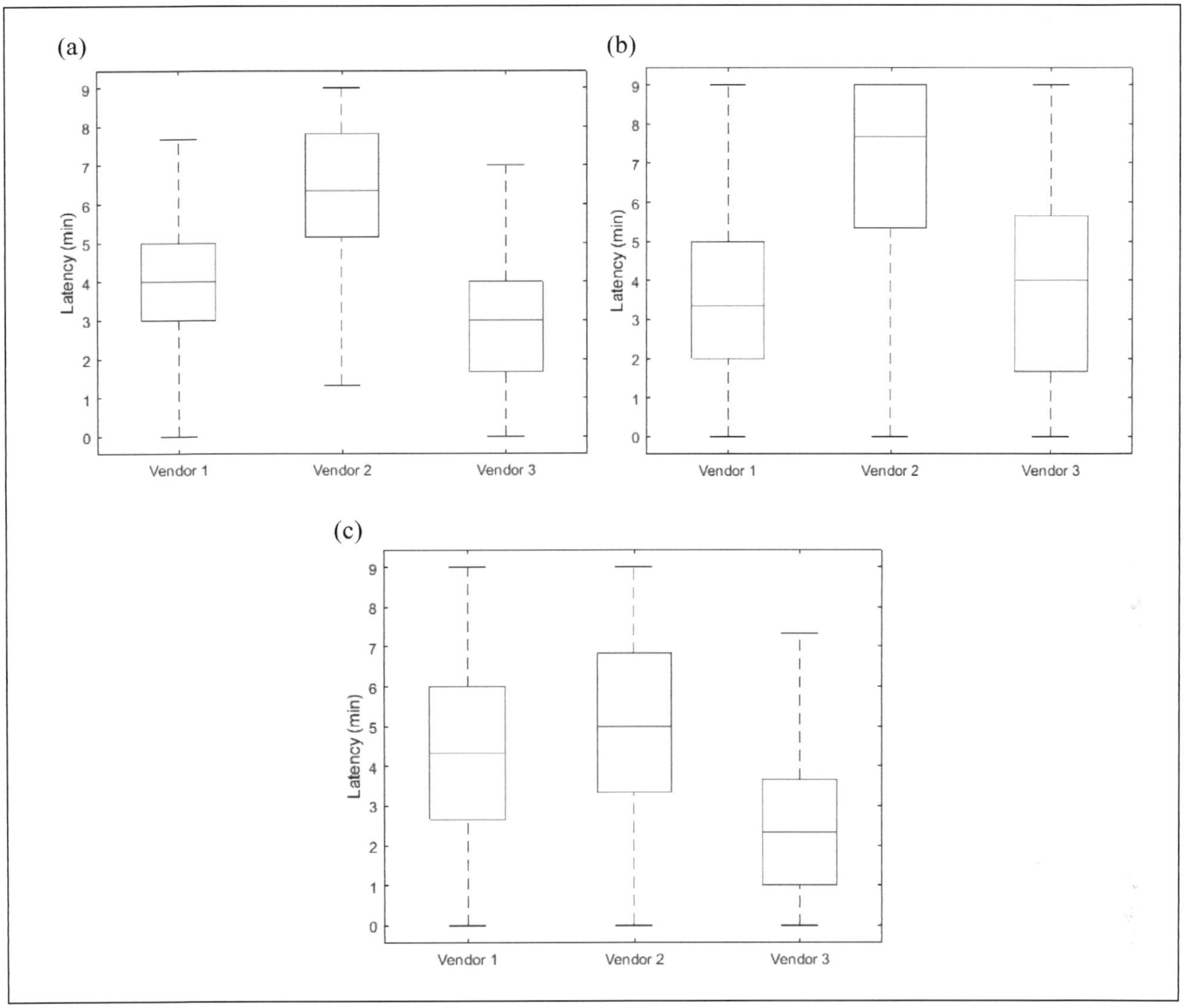

Figure 5. Latency box plot, showing (*a*) entire SRE, (*b*) SP, and (*c*) RP. Whiskers represent the 1.5 interquartile range of the lower and upper quartiles, corresponding to approximately $+/-2.7$ times standard deviation and 99.3 % coverage given that latency is normally distributed.

statistically speaking, there is not enough evidence to suggest that there is dependency among vendors which satisfies the requirement for applicability of the *t*-test. The left side of Table 5 shows *t*-stat and *P*-values for the 99.9% confidence level. Because a one-tail *t*-test is used, both *t*-statistics and *P*-value are necessary to decide whether the null hypothesis can be rejected. In test 1 (as shown above), the alternative hypothesis is "less than or equal to," so the null hypothesis should be rejected if *t*-stat is negative and the *P*-value is small. At the 99.9% confidence level, based on the values in Table 5, it is concluded that Vendor 1 has smaller mean latency than Vendor 2. For tests 2 and 3, the alternative hypothesis is "greater than or equal to," so the null hypothesis is rejected if *t*-stat is positive and the *P*-value is small. The values suggest that the mean latency of Vendor 1 is greater than that of Vendor 3, and Vendor 2 has larger mean latency than Vendor 3. The mean latency is statistically different among all vendors at the 99.9% confidence level. Results show that Vendor 2 exhibits the largest latency, followed by Vendor 1 and Vendor 3.

Sensitivity of Latency to the Time of Day

Traffic conditions vary by time of day, and the rate of speed change might be different during morning and afternoon peaks. An interesting hypothesis is that different traffic conditions at different times of day will affect

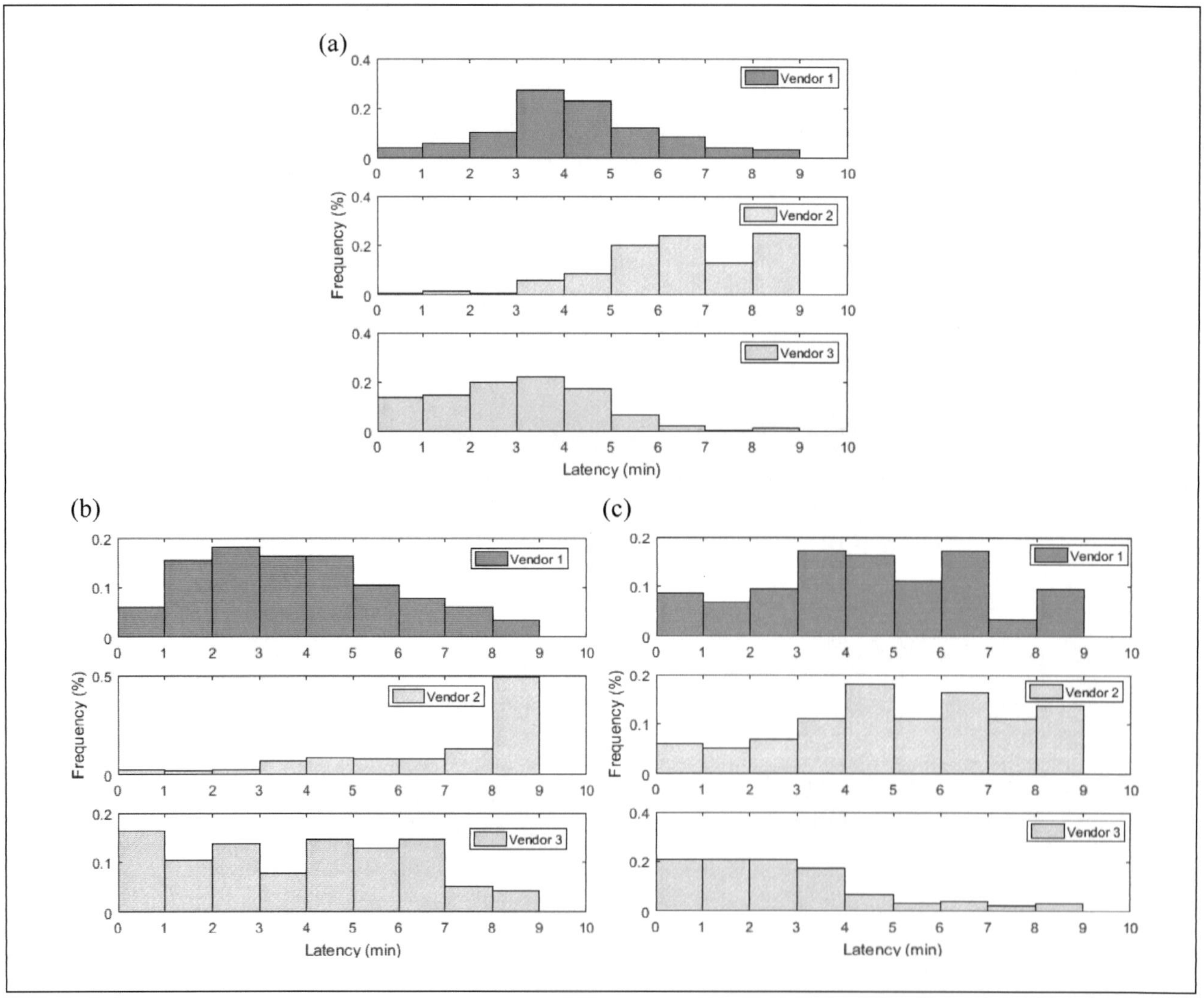

Figure 6. Latency histogram, showing (*a*) entire SRE, (*b*) SP, and (*c*) RP.

Table 5. *T*-Chi-Square Test Pairwise Comparison of Latency

	T-stat			Chi-square		
P-value	Vendor 1	Vendor 2	Vendor 3	Vendor 1	Vendor 2	Vendor 3
Vendor 1		−9.370	4.729		2.685	0.598
Vendor 2	0.000		14.185	0.975		7.454
Vendor 3	0.000	0.000		1.000	0.590	

the probe data latency. Results in (*7*) suggested that there was not a meaningful difference between latency in the morning and that in the afternoon hours. To revisit that result, latency observations were separated into morning and afternoon observations. Figure 7 shows the results for each vendor and for SRE, SP, and RP. This figure confirms the previous conclusion that there is no consistent pattern in the results across different cases and vendors. However, one interesting observation is that for SRE, Vendor 3 seems to have more observations with high latency values during afternoon hours compared with morning hours.

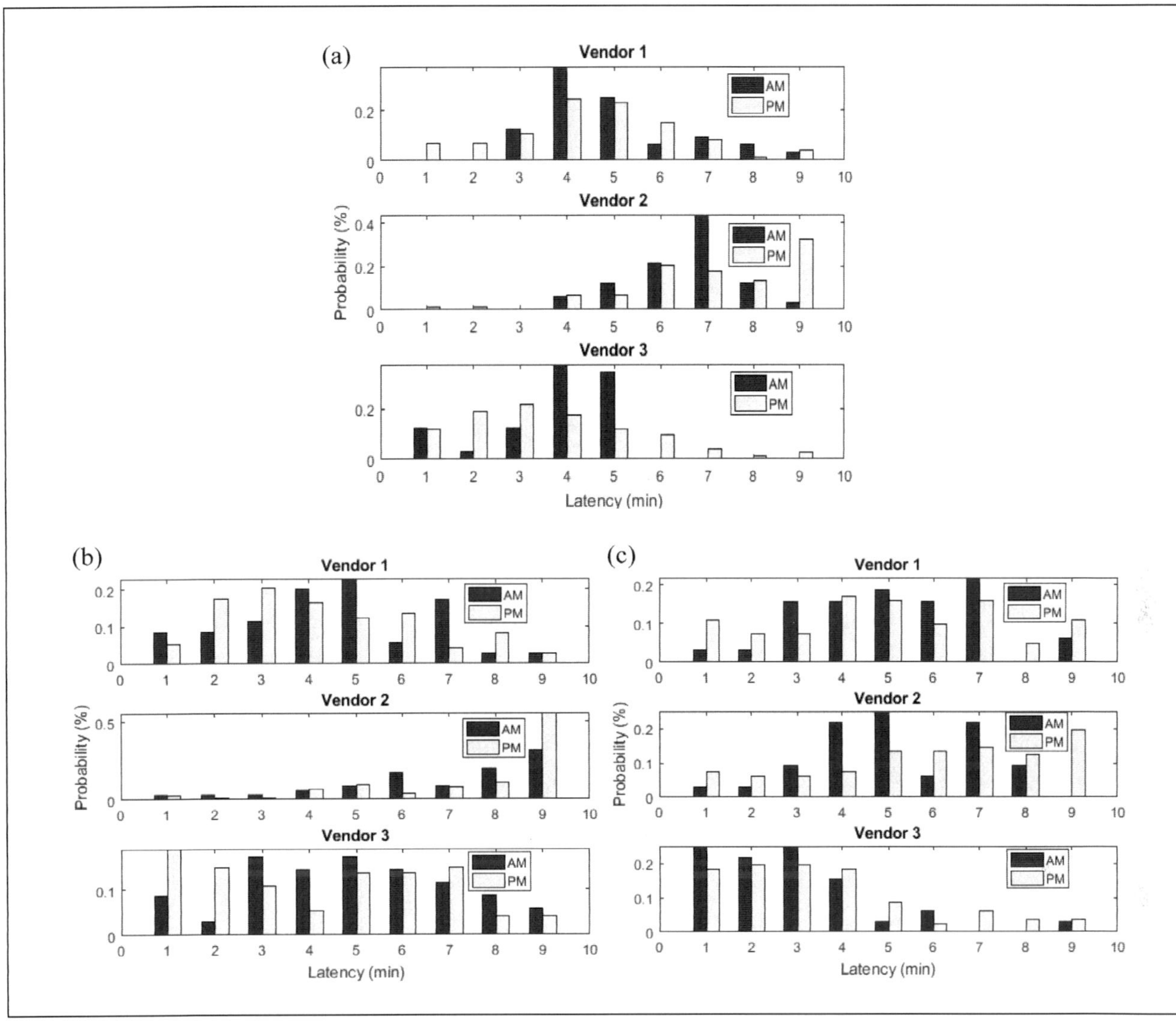

Figure 7. Morning versus afternoon latency distributions, showing (*a*) entire SRE, (*b*) SP, and (*c*) RP.

Sensitivity of Latency to the Segment Length

Correlation between latency and path length is another plausible hypothesis. Results in (*7*) show that such correlation does not exist. Results of the latency are presented as a scatter plot in Figure 8, in which the horizontal axis shows the segment length, and the vertical axis is the latency. As illustrated, there is no strong relationship between length and latency, and the observation yields for all vendors. Statistical examination of the samples confirms this conclusion.

Conclusion

This paper made an effort to measure and report latency of GPS probe data, with respect to Bluetooth/Wi-Fi reidentification data using a maximum pattern matching algorithm. Reference data were collected on several freeway segments in different states and at different times of the year. Probe data obtained from three major private sector vendors were examined. Results were presented for the entire SREs, and for SPs and RPs separately. Across all vendors on the test roads used in this research, the average latency was found to be 4.4 min for freeways (signalized arterials were not tested) with a standard deviation in latency of 2.3 min. Statistically speaking, the average latencies for different vendors are different. Results show that instead of interpreting the latency as a single number, distribution of latency should be measured and evaluated. No strong correlation between latency and time of day, or latency and segment length, was found.

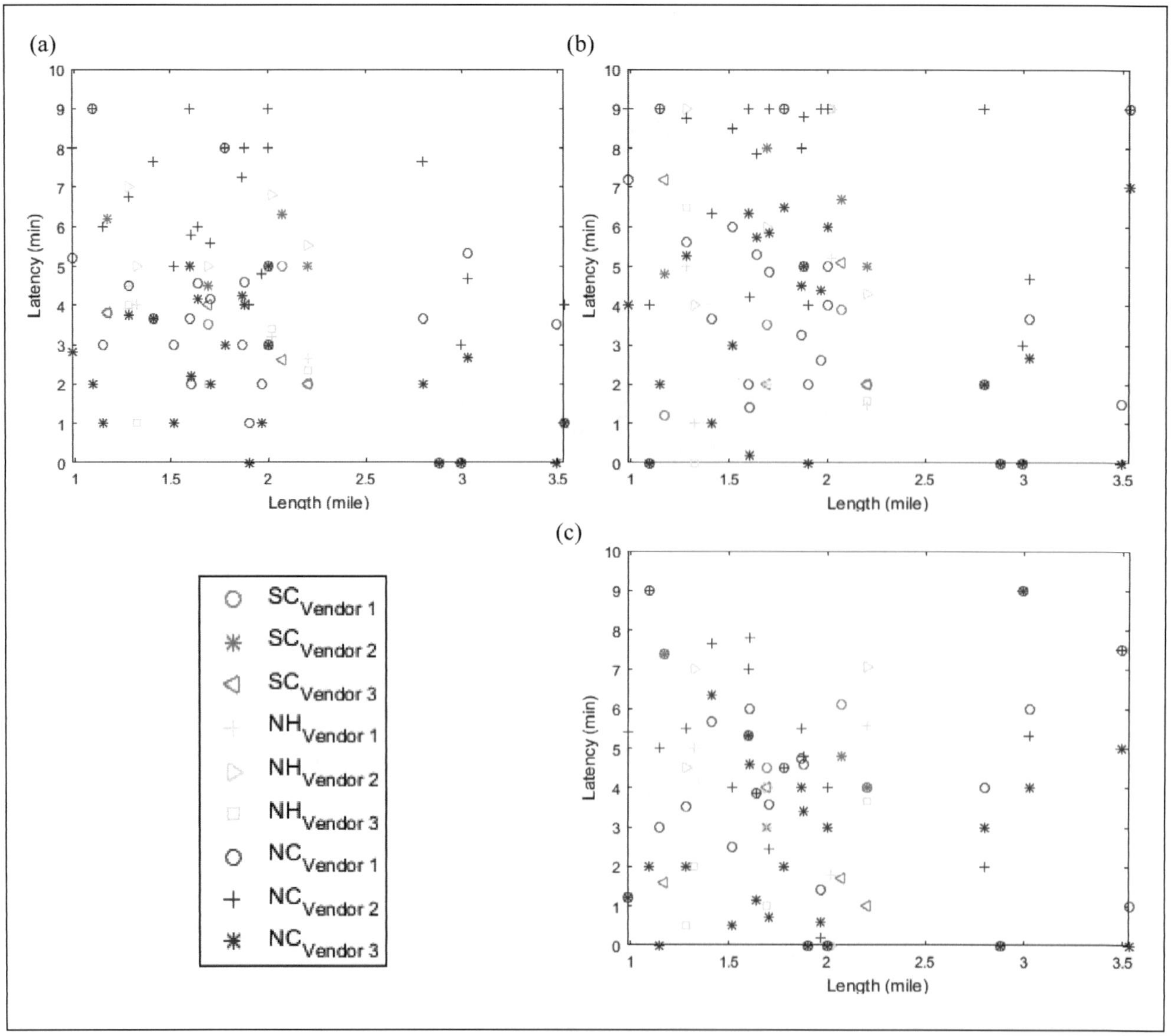

Figure 8. Latency box plot, showing (*a*) entire SRE, (*b*) SP, and (*c*) RP.

Automation of all steps in the solution approach makes it possible to apply the latency measurement procedure to future data sets. Authors plan to test more data sets as they become available, and study possible enhancements in latency over time, as both the sampling rate of probe vehicles and the performance of data fusion engines used by vendors will improve. The potential impact of traffic volume on the latency was not examined because of the lack of granular traffic volume data at the test sites. The authors will test such impact if data becomes available in the future or may use a volume surrogate measure such as factored AADT (annual average daily traffic) for the test segments. The latency of probe data on signalized arterials where speeds are subject to higher fluctuations remains unknown. Changing the methodology to accommodate arterial data is subject for future research. Probe data are used in many real-time applications. Understanding the implications of latency and its distribution for such applications and exploring solutions to compensate for latency is another area for future research.

Acknowledgments

Data used in this study were collected by the I-95 Corridor Coalition as part of their Vehicle Probe Project.

Author Contributions

The authors confirm contribution to the paper as follows: study conception and design: Zhongxiang Wang, Masoud Hamedi; data collection: Masoud Hamedi; analysis and interpretation of results: Zhongxiang Wang, Masoud Hamedi, Elham Sharifi, Stanley Young; draft manuscript preparation: Zhongxiang Wang. All authors reviewed the results and approved the final version of the manuscript.

References

1. Liu, X., S. Chien, and K. Kim. Evaluation of Floating Car Technologies for Travel Time Estimation. *Journal of Modern Transportation*, Vol. 20, No. 1, 2012, pp. 49–56.
2. Peng-Jui, T., C. C. Hung, T. H. Chang, and Y. H. Chuang. Real-Time Urban Traffic Sensing with GPS Equipped Probe Vehicles. *Proc., 2012 12th International Conference on ITS Telecommunications,* Taipei, Taiwan, IEEE, New York, 2012, pp. 306–310.
3. Hunter, T. *Large-Scale, Low-Latency State Estimation of Cyberphysical Systems with an Application to Traffic Estimation.* PhD dissertation. University of California, Berkeley, 2014.
4. Chase, R. T., B. M. Williams, N. M. Rouphail, and S. Kim. Comparative Evaluation of Reported Speeds from Corresponding Fixed-Point and Probe-Based Detection Systems. *Transportation Research Record: Journal of the Transportation Research Board*, 2012. 2308: 110–119.
5. Kim, S., and B. Coifman. Comparing INRIX Speed Data against Concurrent Loop Detector Stations over Several Months. *Transportation Research Part C: Emerging Technologies*, Vol. 49, 2014, pp. 59–72.
6. Center for Transportation Research and Education. *Evaluation of INRIX Probe Data.* Task 10 of InTrans Project 13-480. Sponsored by Iowa Department of Transportation, 2015. http://reactor.ctre.iastate.edu/publications/task10.pdf.
7. Wang, Z., M. Hamedi, and S. Young. Methodology for Calculating Latency of GPS Probe Data. *Transportation Research Record: Journal of the Transportation Research Board*, 2017. 2645: 76–85.
8. Pearson, K. On the Criterion That a Given System of Deviations from the Probable in the Case of a Correlated System of Variables Is Such That It Can Be Reasonably Supposed to Have Arisen from Random Sampling. *The London, Edinburgh, and Dublin Philosophical Magazine and Journal of Science*, Vol. 50, No. 302, 1900, pp. 157–175.

The Standing Committee on Urban Transportation Data and Information Systems (ABJ30) peer-reviewed this paper (18-05026).

The results and conclusions in this document are those of the authors and not the I-95 Corridor Coalition.

Article

TRR
JOURNAL OF THE TRANSPORTATION RESEARCH BOARD

Transportation Research Record
2018, Vol. 2672(42) 192–202

DOI: 10.1177/0361198118798459
journals.sagepub.com/home/trr
SAGE

Profiling Transport Network Company Activity using Big Data

Drew Cooper[1], Joe Castiglione[1], Alan Mislove[2], and Christo Wilson[2]

Abstract

Transportation network companies (TNCs) provide vehicle-for-hire services. They are distinguished from taxis primarily by the presumption that vehicles are privately owned by drivers. Unlike taxis, which must hold one of approximately 1,800 medallions licensed by the San Francisco Municipal Transportation Agency (SFMTA) to operate in San Francisco, there is no regulatory limit on the supply of TNCs. TNCs have an increasingly visible presence in San Francisco. However, there has been little or no objective data available on TNCs to allow planners to understand the number of trips they provide, the amount of vehicle miles traveled they generate, or their effects on congestion, transit ridership, transit operations, or safety. Without this type of data it is difficult to make informed planning and policy decisions. Discussions with Uber, Lyft, and the California Public Utilities Commission, which collects trip-level data from TNCs in California, requesting information on TNC trips have not resulted in any data being shared. Under increasing pressure from policymakers for objective data to inform policy decisions, the San Francisco County Transportation Authority (SFCTA) partnered with researchers from Northeastern University who developed a methodology for collecting data through Uber's and Lyft's application programming interfaces (APIs) with high spatial and temporal resolution. This paper provides a brief literature review on transport network company (TNC) data, and goes one to describe the methodology used to collect data, summarizes the process for converting the raw data into estimated TNC trips, and presents an analysis of the results of the TNC trip estimates. This study determined that TNCs serve a substantial number of trips in San Francisco, over 170,000 on a typical weekday, that these trips follow traditional time of day distributions, and that they tend to take place in the busiest parts of the City.

Transportation Network Companies (TNCs) provide taxi-like vehicle-for-hire services. They are defined by the California Public Utilities Commission (CPUC) as companies that "provide prearranged transportation services for compensation using an online-enabled application or platform (such as smart phone apps) to connect drivers using their personal vehicles with passengers," and are distinguished from taxis primarily by the presumption that vehicles are privately owned by drivers (*1*). TNCs are regulated at the state level by the CPUC, while taxis fall under local regulation. Unlike taxis, which must hold one of approximately 1,800 medallions licensed by the San Francisco Municipal Transportation Agency (SFMTA) to operate in San Francisco, there is no regulatory limit on the supply of TNCs.

TNCs have an increasingly visible presence in San Francisco. However, there has been little or no objective data available on TNCs that would allow planners to understand the number of trips they provide, the amount of vehicle miles traveled they generate, or their effects on congestion, transit ridership, transit operations, or safety. Without this type of data it is difficult to make informed planning and policy decisions.

Discussions with Uber, Lyft, and the CPUC, which collects trip-level data from TNCs in California, requesting information on TNC trips have not resulted in any data being shared. Under increasing pressure from policymakers for objective data to inform policy decisions, the San Francisco County Transportation Authority (SFCTA) partnered with researchers from Northeastern University who developed a methodology for collecting data through Uber's and Lyft's application programming interfaces (APIs) with high spatial and temporal resolution. The following sections of this paper will provide a brief literature review on transportation network

[1]San Francisco County Transportation Authority, San Francisco, CA
[2]Northeastern University, College of Computer and Information Science, Boston, MA

Corresponding Author:
Address correspondence to Drew Cooper: drew.cooper@sfcta.org

company (TNC) data, describe the methodology used to collect data, summarize the process for converting the raw data into estimated TNC trips, and present analysis of the results of the TNC trip estimates.

Literature Review

The San Francisco County Transportation Authority states, "Transportation network companies (TNCs) such as Uber and Lyft are an increasingly visible presence on San Francisco streets, but there has been no comprehensive data source to help the public and decision-makers understand how many TNC trips occur in San Francisco, how much vehicle travel they generate, and their potential effects on congestion, transit ridership, and other measures of system performance" (*2*). The Transportation Research Board also notes that TNC trips are a rapidly expanding market, and this growth may have implications for transportation system performance, the environment, and the economy as the demand for private vehicle, transit and non-motorized trips changes in response to the introduction of this new alternative. They observe that TNCs have resisted attempts at local regulation, including an unwillingness to share any information with public agencies. Despite the limited data, they observe that it is already apparent that the significant numbers of TNCs relative to taxis has implications for regulators (*3*). Similarly, Clewlow and Mishra identify that limited availability of data describing new mobility services impedes decision-makers and planners in making informed choices (*4*).

Rayle et al. provided some information from a limited intercept survey about the purposes, timing, potential mode shift, and induced travel associated with the introduction of TNCs in San Francisco, but notes that the survey was not representative of the ridesourcing market. They also confirm that, outside of popular media such as newspapers and websites, there has been virtually no information available to the public about the size of the TNC market and its potential impacts (*5*). Henao also notes the lack of open data on TNC services, and the resulting lack of academic literature on TNCs, and he describes that the difficulty in obtaining data for independent academic research from Lyft and Uber necessitated his primary research as a TNC driver (*6*). The City of Boston established a data sharing agreement with Uber which provided some information on the total number of rides and wait times, but this information was not useful for supporting planning efforts (*7*). The City of New York has been most assertive and successful in regulating TNCs and requiring TNC data reporting. These publicly available data were used by Schaller to reveal important trends in TNC usage in New York since 2013 (*8*). Chen demonstrated the first successful assembly of comprehensive TNC data for San Francisco's downtown core by scraping Uber information using a script that emulates the Uber app, though this effort did not cover all of San Francisco, and did not include Lyft, the other major TNC operating in San Francisco (*9*).

Methodology

Data Collection

As neither the TNCs nor the CPUC were willing to share information about TNC activity in San Francisco, it was necessary to employ atypical data collection methods. Information from San Francisco International Airport, which does track TNC activity, revealed that Uber and Lyft trips comprise nearly all of the TNC trips in the region, and thus were the focus of data collection efforts. Uber and Lyft also provide an API for developers to obtain information about price estimates and estimated wait times. However, the number of times an individual account can request information for is limited to 1,000 requests per hour. In addition, the APIs do not directly provide information on the vehicles available, as users see when they open the Uber and Lyft apps.

Researchers at Northeastern University developed a method for obtaining information on the location of vehicles in space and time. They created a Python script that emulates the requests that the Uber and Lyft client applications make. This script sends messages to the servers of Uber and Lyft reporting the location of a synthetic client, and the servers return information about the nearest vehicles by type, including a unique identifier, a vector of timestamped latitude and longitude coordinates that reflects each vehicle's recent path, estimated wait time and the presence of any peak pricing multipliers. When a vehicle driver has accepted a ride and is no longer available, or ended a shift, the vehicle no longer appears in the information returned by the servers. Similarly, when a vehicle driver drops off a passenger and is available again, or when a driver starts a shift, the vehicle appears in the information from the server. An important distinguishing difference between the data revealed by Uber and Lyft is that, while Uber appears to assign a new unique identifier to every vehicle after it has completed a trip, Lyft allows the vehicle identifiers to persist across the entire sampling period. This persistence allows us to impute origin-destination flows and to distinguish driver shift start and end times from trip start and end times. Note that each synthetic client is associated with a unique account and credit card. Figure 1 shows how telemetry data for Uber and Lyft can be associated with in-service and out-of-service travel.

Given that a single synthetic client will only receive information about the nearest eight vehicles, to develop

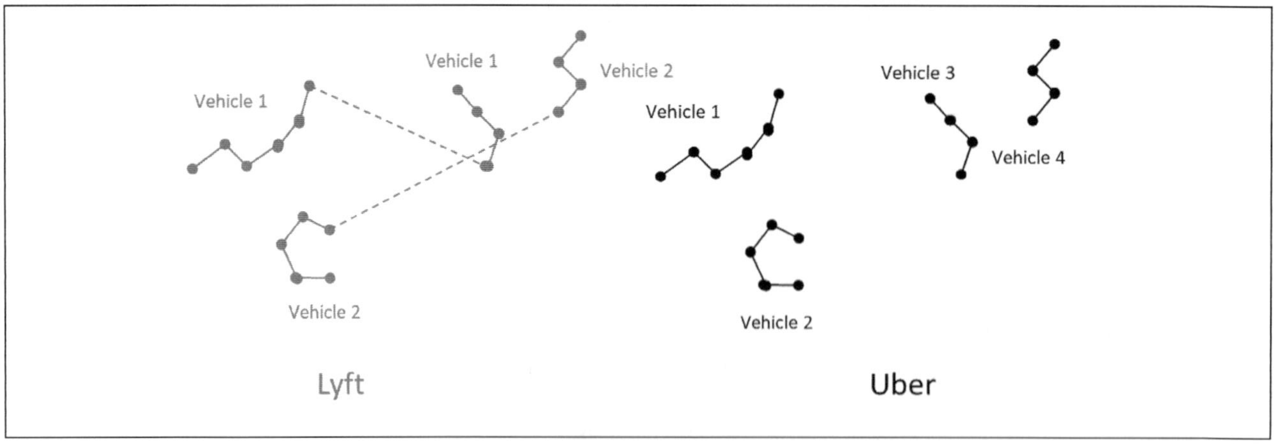

Figure 1. Lyft and Uber out-of-service and in-service segment diagram.

a dataset that comprehensively covers San Francisco it is necessary to have multiple clients requesting and receiving information simultaneously. Chen et al. developed a methodology for estimating an efficient density of clients required to provide sufficient geographic coverage while not resulting in excessive duplicative data collection (*9*). Approximately 200 synthetic client locations were used to collect data across all of San Francisco, arranged in a grid of varying density, with a higher density of locations in the northeast quadrant, which is the most developed area of the City. The script was run continuously for each of these locations, collecting information every five seconds from November 12, 2016 through to December 21, 2016. Once the data from multiple synthetic clients was received, it was then necessary to eliminate duplicate information resulting from the same vehicle information being revealed to the same synthetic client repeatedly over time (sampling every 5 seconds) as well as the same vehicle being revealed to different synthetic clients because of the clients' spatial proximity. This cleaning process is described in a subsequent section.

Validation of the data was a key concern for Chen et al. Although they were not able to validate their estimates of the number of vehicles because the TNCs and regulators would not share even aggregate data summaries, they were able to validate that the information collected is deterministic, internally coherent, and consistent with external data sources about the overall San Francisco travel market. The data revealed to the synthetic clients appears to be deterministic because the researchers performed a series of tests in which all the synthetic clients were located at the same point, and all data returned by the TNC servers to these clients was identical. The coherence of the data is also confirmed in the data cleaning process where duplicative and consistent data is returned to separate synthetic clients. Finally, the cleaned data also shows tremendous consistency in the location and timing of TNC trips across both Uber and Lyft, as well as consistency with the location and timing of trips overall in San Francisco.

Ethical concerns were considered and analyzed before collecting the data. No personal information about any TNC drivers or passengers was collected. A series of tests were also performed that confirmed that the passive data collection by the synthetic clients did not affect fares or surge pricing and thus had no effect on traveler behavior. In addition, no rides were requested, so there was no effect on drivers, either. Ultimately, the only burden on the TNCs was having a couple of hundred additional active accounts, out of tens of millions of total accounts worldwide.

Data Processing

Trip records consisting of a pre-trip (out-of-service) segment and a trip (in-service) segment were imputed from raw data through a series of data cleaning and processing techniques. The out-of-service segment is defined by the time and location a vehicle becomes available and the time and location its driver accepts a ride. It is associated with a vector of timestamped latitude and longitude data representing its path. The in-service segment is defined by the time and location the driver in the vehicle accepts a ride and the time and location that the driver and vehicle next become available. Unlike the out-of-service segment, the in-service segment does not have a vector of location data, so distance traveled is estimated. To derive these trip records, data was first restructured logically to link vehicles with associated location data, then they were cleaned to remove duplicate and anomalous data. Then, out-of-service segments were identified and attributes were imputed. Then, for Lyft, consecutive out-of-service segments were linked together into trips, and in-service attributes were imputed. Next, out-of-service segments

Table 1. Key Attributes of Vehicle, VehiclePath, and VehiclePathPoint Objects

Object	Attribute	Description
Vehicle	id	a unique identifier representing the vehicle.
Vehicle	service_types	the service types for which the vehicle is available for hire
Vehicle	path	a VehiclePath object
Vehicle	pings	the number of times this vehicle is observed by a detector
Vehicle	online_loc	A VehiclePathPoint at the location the vehicle is first observed
Vehicle	offline_loc	A VehiclePathPoint at the location the vehicle is last observe
VehiclePath	points	A collection of VehiclePathPoints representing the trajectory of a Vehicle over time, which may be comprised of one or more trips
VehiclePath	start_date	A datetime object of the time of the first point in the VehiclePath
VehiclePath	end_date	A datetime object of the time of the last point in the VehiclePath
VehiclePath	duration	A timedelta object representing the difference between start_date and end_date
VehiclePath	distance	A float value of the summation of Euclidean distance between each successive pair of points in the VehiclePath
VehiclePath	dwell_locations	A collection of VehiclePathPoints where location changed by less than 25 feet over a two-minute period.
VehiclePathPoint	native_time	The datetime representation found in the Uber or Lyft datastream
VehiclePathPoint	datetime	A datetime object and localized to the US/Pacific time zone
VehiclePathPoint	lat	The latitude in WGS84
VehiclePathPoint	lon	The longitude in WGS84

were matched between the Uber and Lyft datasets to account for drivers working simultaneously on both services. Finally, Lyft trip data was used to estimate flows, which could not be directly observed, between known Uber origins and destinations, and used to estimate vehicle miles travelled (VMT). This process is described in detail in the following sections.

Data Cleaning and Data Structure. The raw data collected through Uber's and Lyft's APIs consisted of hundreds of millions of records of vehicle location data, taking up about 12 Terabytes of hard-disk space. Much of the data is duplicated because of the spatial proximity of synthetic clients and the 5-second frequency of API requests. The datastream also contains large amounts of data not used in this analysis, including estimated wait times and price multipliers.

Data were cleaned and structured, first by reading raw data into memory and collecting related data together, and then identifying and removing duplicate information. Each time a new vehicle identifier (ID) was encountered, a new Vehicle object was created to store associated data and perform calculations. Each additional time that vehicle ID was encountered, its data was added to the Vehicle object and duplicate data removed. After all data was collected into Vehicle objects, each Vehicle's path was sorted by timestamp. Finally, this data was written out to a vehicle and a telemetry file. The vehicle file summarizes data about the entire lifetime of the vehicle in the dataset, and the telemetry file contains the detailed time and location records associated with each vehicle. The following subsections describe the in-memory data structure created to store and process vehicle information, and the cleaned, deduplicated, and formatted output data.

Intermediate In-Memory Data. This section describes the in-memory data format, implemented as a set of Python objects called Vehicle, VehiclePath, and VehiclePath Point, used to collect related data together, sort it, and remove duplicates. Table 1 identifies the key attributes of these objects.

- Vehicle: A Vehicle object is created for each distinct vehicle id in the Uber and Lyft datastreams. For Uber, this generally represents a single trip because Uber generates a new vehicle ID after each trip. In some cases, however, an Uber vehicle ID persists across multiple days or multiple trips. For Lyft, the vehicle ID is persistent, allowing the authors to identify multiple trips associated with a single vehicle.
- VehiclePath: A VehiclePath object stores locational data related to a Vehicle trajectory. It is a collection of VehiclePathPoints and additional attributes. The VehiclePath ensures that duplicate VehiclePathPoints are removed. A VehiclePath may represent one or more trips.
- VehiclePathPoint: A VehiclePathPoint represents a single point of a Vehicle's location. VehiclePath Points with missing or erroneous datetime or locational data are removed.

Cleaned, Deduplicated, and Structured Output Data. The output of this process resulted in a *vehicles* file and a *telemetry* file. The vehicles file contained summary data for the entire lifetime of the vehicle in the datastream. For Lyft, this was analogous to a vehicle, but for Uber the data in the vehicles file is more analogous to a trip. The telemetry file contained cleaned, deduplicated, and ordered telemetry records, associated with a unique vehicle identifier, with one record for each unique recorded point in the path.

Identifying Out-of-Service Segments. Vehicles only appear in datastreams when they are available to be hired, therefore the location records represent out-of-service movements only, while in-service attributes must be imputed. For all successive records spaced fewer than 10 minutes apart, 99.5% were fewer than 118 seconds apart. Out-of-service segments identified and grouped using 118 seconds as a breakpoint between segments. This break in telemetry represents one of three scenarios: a driver accepted a ride, a driver left the coverage area, a driver closed the app for another reason such as ending a shift. Each of these scenarios is addressed in later steps to avoid over-counting trips. Each out-of-service segment has a time and location at which the vehicle became available for hire, and a time and location it accepted a ride. Duration and distance were calculated for each out-of-service segment. The 118 second gap is probably too short to represent an actual trip, but may represent a break in service or in the stream of data for that vehicle. Therefore, it is useful to treat this break as a trip to impute missing attributes, and then recombine successive very-short trips into a single trip of a more realistic length.

Linking Pre-Trips into Trips. Persistent vehicle identifiers (IDs) were used to link Lyft and, where possible, Uber pre-trips together to form trips. Vehicle IDs were persistent over time for Lyft. In general, Uber vehicle IDs did not persist over time, although there were exceptions. Trip duration and distance were estimated using the location a ride was accepted as the trip origin, and the next location the vehicle becomes available for another ride as the destination. Note that the ride-acceptance location is unlikely to be the actual pickup location, so this estimate is an underestimate that will omit some amount of in-service and out-of-service VMT. Trip duration was calculated directly from the trip origin and destination times. As actual routes are unknown, trip distances were imputed using time-of-day sensitive driving distances from skim tables produced by the SF-CHAMP travel demand model.

Trip Matching. An unknown number of drivers drive for both Uber and Lyft. These drivers may drive for each service in separate sessions, or they may drive for both simultaneously, accepting a ride from whichever they are first matched with. The trip matching process described in this section was developed to identify vehicles being driven for both services simultaneously to avoid double-counting them. This process used location data from pre-trip segments to match pre-trips from the Uber dataset to the Lyft dataset. First, vehicle location points were matched to links in a network, then those links were compared to identify and link similar pre-trips. Finally, this information was used to create a master trip list with each trip assigned to either Uber or Lyft.

Map Matching. The telemetry point locations of each out-of-service segment were matched to the OpenStreetMap (OSM) network using a map-matching algorithm modified from Mapillary's map-matching open source Python software. The algorithm assigns a point to a link using the point's proximity to a link and the transition cost from the previous link in the path (*10*). The result is an OSM link associated with each point in a pre-trip trajectory.

Trip Matching. The trip matching (or more accurately, out-of-service segment matching) process was designed to filter out most of the irrelevant data using computationally inexpensive calculations so that more expensive operations would only need be performed on a small set of data. The algorithm was implemented in Python, using a distributed and parallel architecture. The process used time windows around the starting and ending times of the out-of-service segment, considered the directionality in which links were traversed, the order in which they were traversed, and the spatial proximity of the paths.

Out-of-service segments were matched if they occupied the same link(s) during the same timespan. Uber segments (candidates) were matched to Lyft segments (targets) because there were many fewer Lyft records. For each target Lyft segment, a time window of 2 minutes before its start time and 2 minutes after its end time was used to filter the universe of Uber segments to a set of candidates so that each candidate's start time was within 2 minutes of the target's start time and the candidate's end time was within 2 minutes of the target's end time. Then the candidates were filtered leaving only candidates with a link overlap of greater than or equal to 0.75. Next, the first and last observed times on that link were calculated, and only candidate trips which occupy at least one link at the same time as the target trip were kept. If multiple candidates remained, a geometric overlap between the reference trip and each candidate trip was calculated by buffering the line formed by the points in each trajectory and intersecting them. The candidate trip with the highest maximum geometric overlap was kept.

Master Trip List. Any two matching trips were assigned a common master trip ID and a master vehicle ID. Then the trip with an earlier ride-acceptance time was kept in the master trip list and the other was removed, based on the assumption that a vehicle would disappear immediately when a ride is accepted and there would be some delay as the other app is manually closed. The resulting list is a distinct collection of trips with duplicate or paired trips removed.

Imputing Missing Attributes. Trips could be imputed directly for only the subset of records where the unique vehicle ID persisted between pre-trip segments. Trip destinations, trip lengths, and durations were estimated for the remaining trips based on origin-destination distributions specific to the day-of-week and time-of-day derived from Lyft records with known origin-destination pairs.

Linking Uber Origins and Destinations and Calculating VMT. Origin and destination locations, separately, for Uber trips were revealed in the data. However, unlike for Lyft, origin-destination flows for Uber could not be estimated by linking out-of-service segments, so they were instead imputed using an iterative proportional fitting (IPF) procedure using the observed origin-destination flows from Lyft data and the total observed origins and destinations for Uber. First, a trip was assigned a destination district (collection of travel analysis zones [TAZs]), then assigned to a travel analysis zone (TAZ) within the district. This two-step process was chosen because the hourly temporal resolution combined with TAZ-level geographic resolution, even with millions of trips, resulted in sparse matrices.

IPF is an algorithm by which a known (seed) joint distribution of multidimensional data is used to estimate the joint distribution of another (target) set of data where only the independent distribution of each dimension is known. In a two-dimensional example, for each row in jointly-distributed seed matrix, each cell in that row is adjusted proportionally to the other cells in that row such that the sum of the cells equals the target row sum. Then for each column, each cell is adjusted proportionally so that the sum of the cells in the column equals the target column sum. This process is repeated iteratively until the difference in the sum of column sums and the difference in the sum of row sums between iterations falls below a certain threshold. At that point the IPF is said to be in equilibrium.

For each day of week and hour of day, a district-to-district origin-destination matrix for Uber trips was created using IPF, with the joint distribution of origins and destinations from imputed Lyft trips used as the seed matrix and the sum of origins and of destinations of Uber trips used as the row and column targets. Using these origin-destination matrices, and a given origin district, day of week, and hour of day, a destination district was chosen. Then a zone within that district was chosen based on whether the trip occurred on a weekday, Saturday, or Sunday, and the hour of day.

Imputing Sessions. As mentioned previously, the disappearance of a vehicle from the datastream may indicate that a ride has been accepted and the driver is now en-route to pick up a passenger, but it may also indicate that the driver has ended their shift or left the coverage area. As this report only considers intra-San Francisco trips, leaving San Francisco and ending a shift are treated the same, and a session in this context can be considered a San Francisco driving session.

As with imputing destinations, only the Lyft data with persistent vehicle IDs enabled direct estimation of sessions. As the actual session durations are unknown, and because it may not be appropriate to apply assumptions from the taxi industry because the taxi industry has regulations on session durations, sessions were estimated from imputed trip attributes. Trips with durations of greater than 2 hours were marked as the end of a session. These session-end trips were excluded from analysis.

Using session-end Lyft records and the number of Lyft trips by day of week and hour of day, a probability for a trip to be the last trip in a session was calculated. Using this probability, Uber trips were stochastically chosen to be the last trip in a session and excluded from further analysis.

Imputing Vehicles on the Road. The average number of vehicles on the road at a 30-minute interval for each day of week was estimated using directly-estimated session start and end times for Lyft trips, and then applying the results to Uber trips. Using estimated Lyft sessions, a vehicle was added to each 30-minute interval between the session's start and end times. Then, for each day of the week and half-hour of the day, a ratio of trips to vehicles was calculated and applied to the estimated number of Uber trips to derive an estimated number of Uber vehicles on the road.

Data Limitations and Other Considerations. The data collected by Northeastern through Uber's and Lyft's APIs has several key limitations. First, owing to the data collection methodology and the need to create hundreds of fake accounts to create synthetic apps to cover San Francisco, and the volume of data generated and amount of time needed to process it, expanding data collection to a larger geographic area to cover the entire San Francisco Bay Area was infeasible at the time. This limits the analysis to trips entirely within San Francisco. TNC operations, however, are not restricted to any city or county jurisdictions, so those trips are missing or excluded because of possible incomplete coverage. Second, true pickup locations are not known, they are based on the location a vehicle disappears from a datastream. Owing to the density of trips, we assume that the pickup location and ride-acceptance locations are geographically close, but have no means to confirm this. Regardless, both out-of-service and in-service VMT estimates are likely to be conservative (i.e., lower than true values) because they do not account for deviation from the shortest path between ride-acceptance and drop-off locations. Third, not all vehicles appear in the datastream. Both Uber and Lyft allow drivers to be matched with waiting riders while a ride is underway, so the driver will transition directly from one ride to the next without ever becoming "available." Fourth, UberPOOL and Lyft Line, like all services, do not appear in the datastream once a ride is accepted. For these shared services, it appears that only the first pickup and last drop-off would be observed, and intermediate pickups and drop-offs are omitted. Fifth, the data does not reveal any information about which product was selected, the number of passengers, fares paid and other demand related information. Finally, it has been reported that Uber creates fake data to give the appearance of greater supply of available vehicles than there are (*11*). While this last point cannot be entirely discounted, results of trip matching and the travel patterns that emerge in the following section present compelling evidence that data do reflect the real-time locations of vehicles.

Results

A key limitation of the data is that there is no independent data source to validate against. Trip matching provides a partial validation of a subset of the data, but it is also necessary to use other data on travel trends and comparisons to other modes to evaluate the reasonableness of results. We found clear TNC trends that were consistent with overall San Francisco travel patterns.

Trip Matching Results

Results from the trip matching process provide compelling evidence that the data does, in fact, represent the actual locations of real vehicles. In some instances, the unique vehicle identifiers for a small number of Uber vehicles persist for multiple trips over several hours, days, or weeks. Of the 1,278 Uber vehicles with unique identifiers that persisted for one day or longer that were matched with a Lyft vehicle, each was matched to exactly one corresponding Lyft vehicle across all matches. This implies that the trip matching algorithm was successful at correctly pairing observations from each datastream. It also implies that, if all data were fake or synthetic data, then both companies would need to be coordinating to produce the same or similar synthetic data.

Day of Week and Time of Day Patterns

The number of observed TNC trips increases throughout the week, from 130,000 trips on Monday to 220,000 on Friday and Saturday, before falling to a weekly low on Sunday. TNC trips, whose time of day distribution is shown in Figure 2, exhibit notable peaking during traditional commute hours during weekdays. The evening peak is both higher and longer than the morning peak on weekdays, and later in the week, particularly Thursday and Friday, the evening peak extends later into the evening, with a second peak later in the evening around 11 pm.

While traffic counts are not equivalent to trips, counts from the California Performance Management System (PeMS) on I-80, I-280, and US 101 have similar time-of-day distributions, but less pronounced day-over-day growth. These counts are shown in Figure 3.

Although the number of vehicles on the road and the number of trips peak during typical rush hours, the VMT per trip is the lowest during those times. Trip lengths increase late at night and in the early morning, peaking on weekdays between 3 am and 4 am, and on weekend mornings between 4 am and 6 am. Note that the underestimation of VMT is likely to be greater in the early morning and late night hours when supply is lower and available vehicle are likely to be further from their fares. Trip-length trends are shown in Figure 4.

TNC Trip Origin and Destination Locations

Origins and destinations are concentrated in the northeastern quadrant of the City, which is also the most developed part of the City. Figure 5 shows trip origins summarized to the TAZs used in the SF-CHAMP travel demand model.

Figure 6 shows the actual point locations of pickups (estimated using the location of a vehicle when a ride was accepted). These pickup locations highlight the amount of activity on the Downtown and South of Market street grid, and along major transit corridors including Geary Blvd, on which SFMTA buses carry more than 40,000 riders every day, and Mission St, served by the popular 14-Mission buses and BART rail service.

Conclusion

TNCs are an increasingly visible presence in San Francisco, putting pressure on planners to obtain objective data to understand how TNCs operate and how they interact with and impact the transportation system. So far, however, the TNCs and the CPUC have been unwilling to provide the detailed data necessary to support

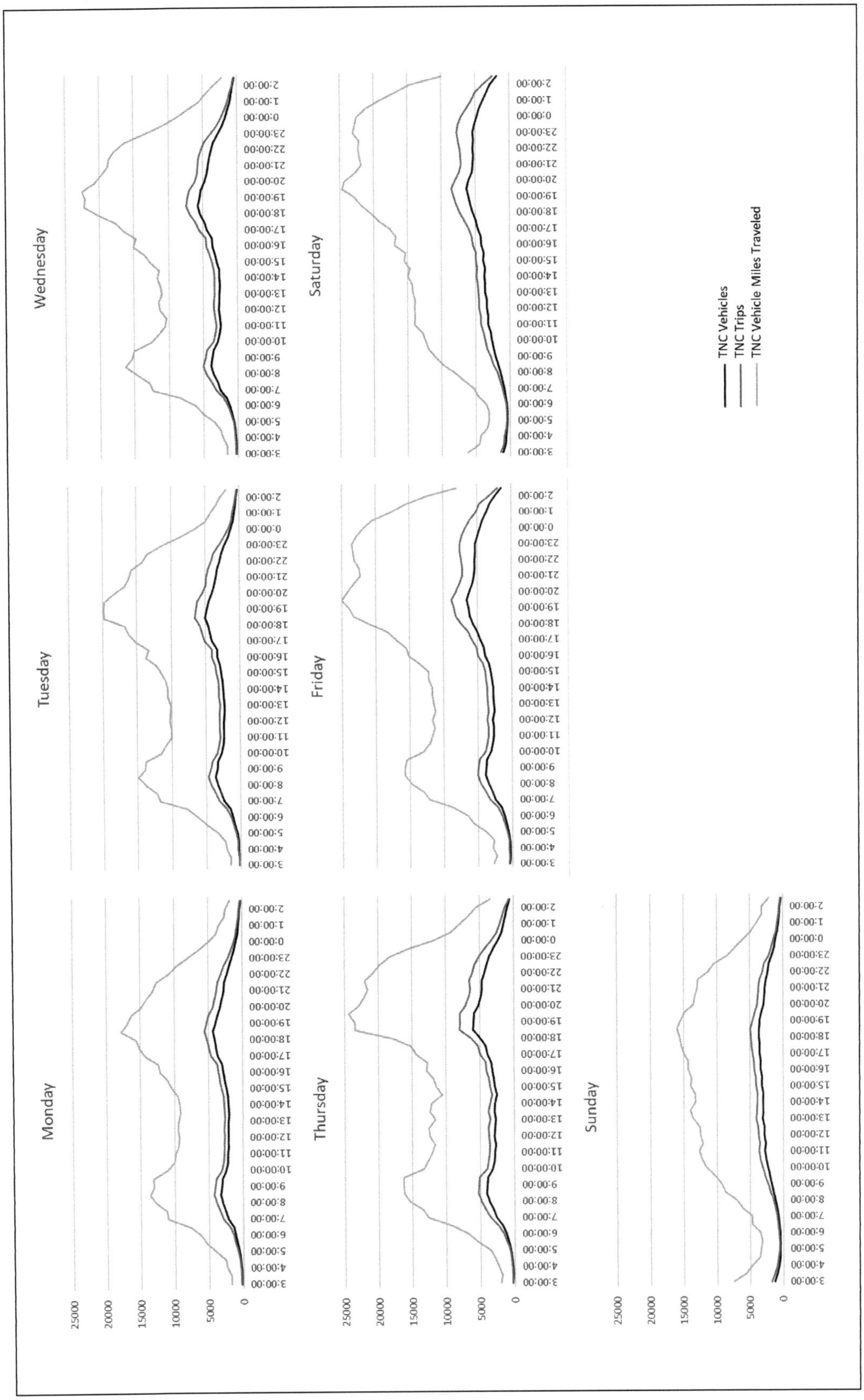

Figure 2. TNC vehicles, trips, and VMT by time of day.

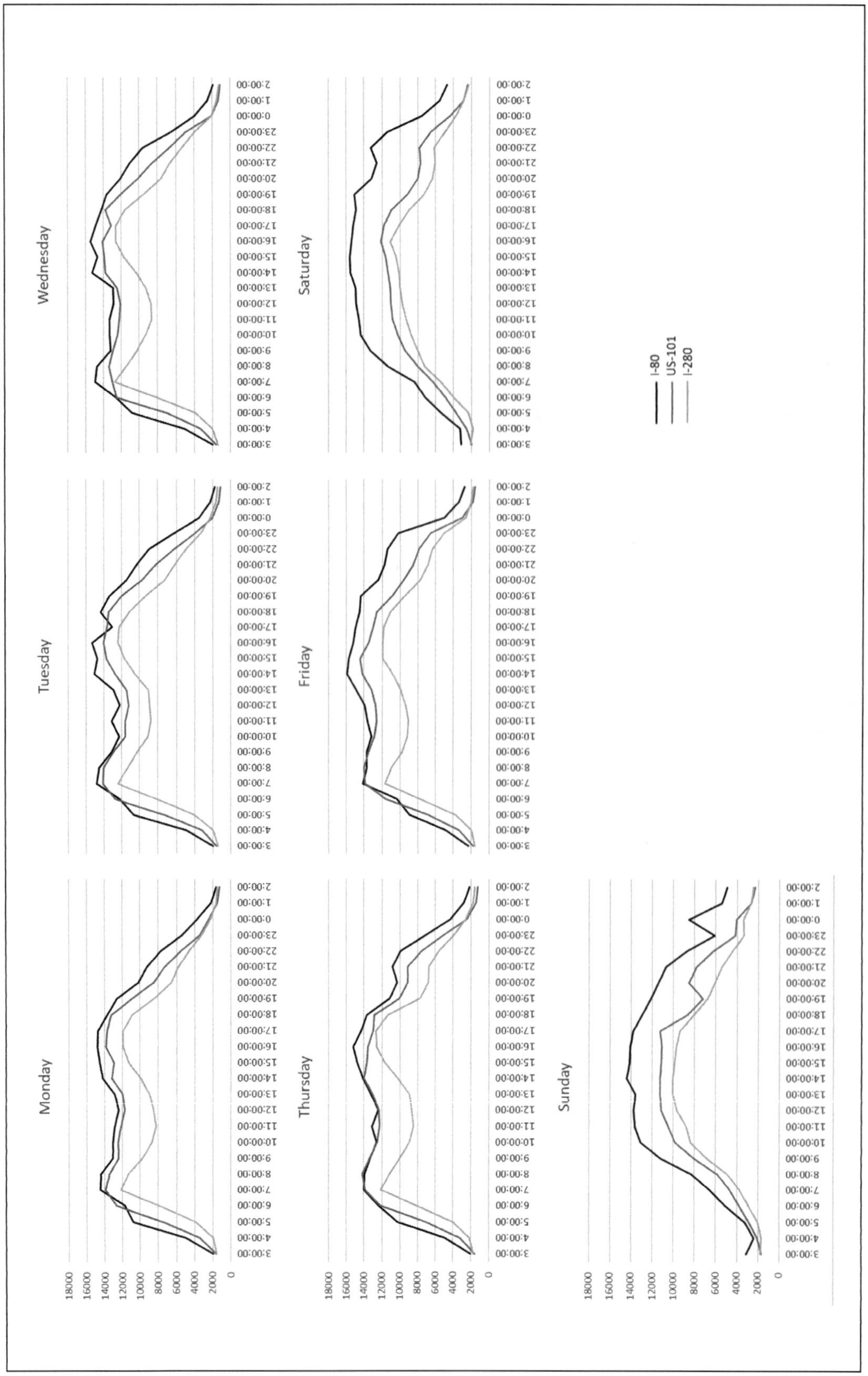

Figure 3. California performance monitoring system counts for selected locations in 2015.

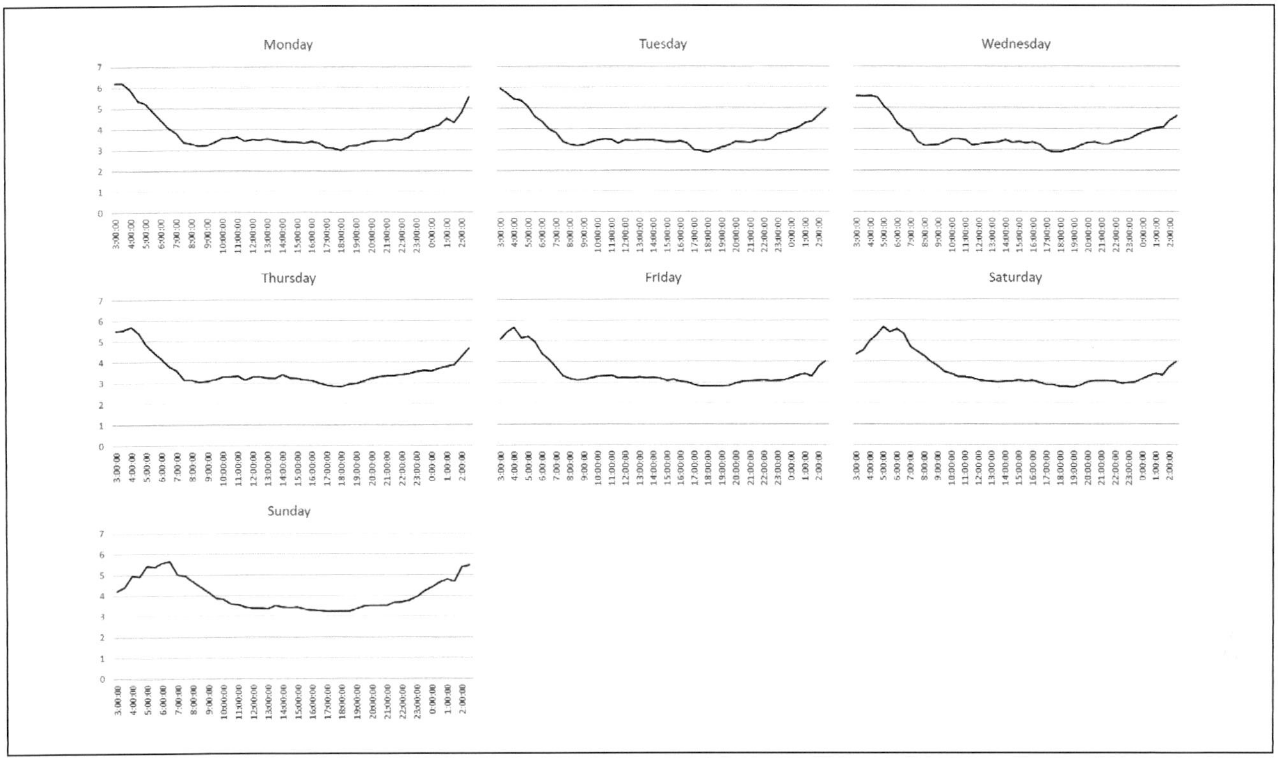

Figure 4. TNC VMT per trip by day of week and time of day.

Figure 5. TNC trip origins on a typical weekday.

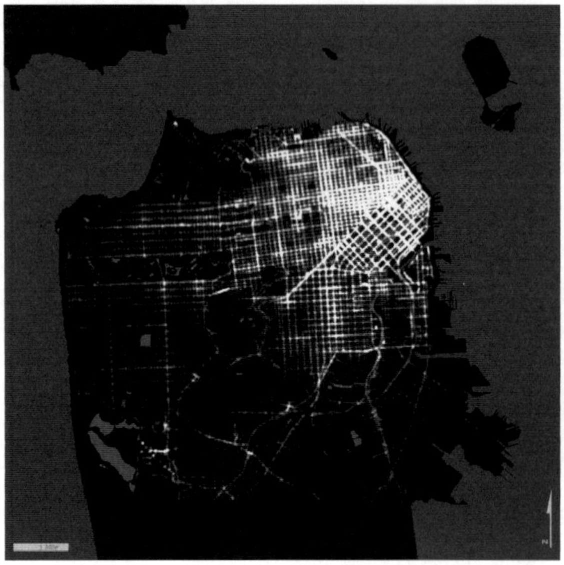

Figure 6. TNC trip pickup locations.

investment, planning, and policy decisions. This necessitated an atypical data collection through a partnership between the SFCTA and Northeastern to collect large volumes of highly detailed vehicle location data, using synthetic Uber and Lyft apps, to create a detailed trip dataset which was used to analyze TNC travel patterns.

This research produced the most detailed and comprehensive TNC data available for San Francisco, and produced results which were revealed how TNCs operate in the City. This research quantified the previously unknown size of the TNC market, where and when trips occur, how many vehicles are on the road, and how long and far drivers drive during a session. However, the data collected and used in this analysis is not the ideal data to conduct this analysis because of the limitations noted in earlier sections. There are still many remaining gaps in this data, such as mode shift, vehicle occupancy, and induced demand, that are needed to understand how these services are changing the overall transportation system. This research underscores the planning agencies' need for high-quality longitudinal data to inform investment, planning, and policy decisions.

Author Contributions

The authors confirm contribution to the paper as follows: study conception and design: Joe Castiglione, Drew Cooper, Alan Mislove, and Christo Wilson; data collection: Alan Mislove and Christo Wilson; analysis and interpretation of results: Drew Cooper and Joe Castiglione; draft manuscript preparation: Drew Cooper and Joe Castiglione. All authors reviewed the results and approved the final version of the manuscript.

References

1. California Public Utilities Commission. *Transportation Network Companies*. California Public Utilities Commission. Accessed July 31, 2017.
2. San Francisco County Transportation Authority. *TNCs Today: A Profile of San Francisco Transportation Network Company Activity*. San Francisco County Transportation Authority, San Francisco, CA, 2017. http://www.sfcta.org/sites/default/files/content/Planning/TNCs/TNCs_Today_061317.pdf
3. *TRB Special Report 319: Between Public and Private Mobility: Examining the Rise of Technology-Enabled Transportation Services*. Transportation Research Board of the National Academies, Washington, D.C., 2015.
4. Clewlow, R., and G. S. Mishra. Shared Mobility: Current Adoption, Use, and Potential Impacts on Travel Behavior. Presented at 98th Annual Meeting of the Transportation Research Board, Washington, D.C., 2017.
5. Rayle, L., D. Dai, N. Chan, R. Cervero, and S. Shaheen. Just a Better Taxi? A Survey-Based Comparison of Taxis, Transit, and Ridesourcing Services in San Francisco. *Transport Policy*, Vol. 25, 2016, pp. 168–178.
6. Henao, A. *Impacts of Ridesourcing – Lyft and Uber – on Transportation including VMT, Mode Replacement, Parking, and Travel Behavior*. Doctoral thesis. University of Colorado, Denver, CO, 2017. http://pqdtopen.proquest.com/doc/1899208739.html?FMT=ABS
7. Vaccaro, A. *Highly Touted Boston-uber Partnership has not Lived up to Hype so far*. Boston.com. 2016. http://www.boston.com/news/business/2016/06/16/bostons-uber-partnership-has-not-lived-up-to-promise.
8. Schaller Consulting. *UNSUSTAINABLE? The Growth of App-Based Ride Services and Traffic, Travel and the Future of New York City*. Schaller Consulting, 2017.
9. Chen, L., A. Mislove, and C. Wilson. Peeking Beneath the Hood of Uber. *Proc., 2015 Internet Measurement Conference*, Tokyo, Japan, 2015, pp. 495–508.
10. Mapillary. *map_matching*. GitHub Repository, 2016. https://github.com/mapillary/map_matching. Accessed August 1, 2017.
11. Clark, L. *Uber's Real Time Car Map is Fake*. Wired.com, 2015. http://www.wired.co.uk/article/uber-algorithm-fake.

The Standing Committee on Urban Transportation Data and Information Systems (ABJ30) peer-reviewed this paper (18-05899).

Article

TRR
JOURNAL OF THE TRANSPORTATION RESEARCH BOARD

Transportation Research Record
2018, Vol. 2672(42) 203–212

DOI: 10.1177/0361198118773199
journals.sagepub.com/home/trr
SAGE

A Methodology to Assess the Quality of Travel Time Estimation and Incident Detection Based on Connected Vehicle Data

Md Shahadat Iqbal[1], Samaneh Khazraeian[1], and Mohammed Hadi[1]

Abstract

Connected vehicle (CV) technologies are expected to have a significant influence on the investment decisions of transportation system management and operations (TSMO) in the near future. One of the potential applications is the use of CV data to support various TSMO processes. This study investigates the use of CV data as an alternative to existing data acquisition techniques in providing two critical functions to support TSMO: travel time estimation and incident detection. In support of this investigation, the study develops regression models to estimate the accuracy and reliability of travel time measurement and latency of incident detection as functions of the traffic demand level and the proportion of CV in the traffic stream. The developed regression models are used in conjunction with a prediction of CV proportions in future years to determine when the CV technology can provide sufficient data quality to replace existing data acquisition systems. The results can be used by TSMO programs and agencies to plan their investment in data acquisition alternatives in future years.

Connected vehicle (CV) technologies are expected to have a significant influence on transportation system management and operations (TSMO), starting in the next few years. Thus, transportation agencies need to start preparing for the next generation of advanced traffic management strategies that utilize (connected) vehicle-to-infrastructure (V2I) technologies. The draft Federal Highway Administration (FHWA) V2I (Vehicle to Infrastructure) Deployment Guidance (*1*) encourages V2I deployments, but it states that the United States Department of Transportation (USDOT) will not require public agencies to implement V2I technology or applications, and recommends that implementation should be based on agency assessments.

One of the potential applications is the use of CV data to support various TSMO processes. The National CV Field Infrastructure Footprint Analysis document produced by the American Association of State Highway and Transportation Officials (AASHTO) (*2*) states that, "Public agencies will assess and trade-off the opportunities to use connected vehicle probe data aggregation and processing versus the continued deployment, operations, and maintenance of traditional ITS vehicle detection versus purchasing commercial traffic information services." Given the accelerated advancements in technologies, there will be an increasing need to determine the quality of the data collected using CVs, in comparison to the quality of data collected from existing data sources.

Researchers have investigated utilizing CV data to assess various performance measurements (*3–19*). These performance measurements can be used to support planning, planning for operations, and operations. However, the quality of the estimation is expected to be a function of the sample size of the traffic stream including CVs. The sample size is the multiplication of the percentage of CVs in the traffic stream and the total traffic volume. The CV percentage is a function of the CV market penetration and is expected to grow with time. This study investigates the use of CV data as an alternative to existing data acquisition techniques in providing two critical functions to support TSMO: travel time estimation and incident detection. In support of this investigation, the study develops regression models to estimate the accuracy and reliability of travel time measurement, and

[1]Department of Civil and Environmental Engineering, Florida International University, Miami, FL

Corresponding Author:
Address correspondence to Md Shahadat Iqbal: miqba005@fiu.edu

latency of incident detection, as functions of the traffic demand level and the proportion of CV in the traffic stream. The developed regression models are used in conjunction with a prediction of CV proportions in future years to determine when the CV technology can provide sufficient data quality to replace existing data acquisition systems. The results can be used by TSMO programs and agencies to plan their investment in data acquisition alternatives in future years. Although a wide range of performance measurements could be derived using CV data, this study focuses on the above two performance measurements to demonstrate a methodology that can be used for other measures.

Connected Vehicle Data

The data elements generated from CVs can be used to support various mobility, safety, and environmental applications. Although some of these data elements, such as speed and acceleration and deceleration, can be obtained without the need to connect to a vehicle's onboard diagnostic port (OBD-II), obtaining many other useful data elements requires connection to the OBD-II. The data will then need to be transmitted using CV messages utilizing dedicated short-range communication (DSRC) or other communication technologies such as cellular communications.

The CV message types and components are specified in the Society of Automotive Engineers (SAE) J2735 standards (*20*). The J2735 standards specify the number of message types. The basic safety message (BSM) is one of these message types that will be used for vehicle-to-vehicle communication. The BSM contains vehicle safety-related information which is broadcast to surrounding vehicles but can also be sent to and/or captured by the infrastructure. The BSM, as defined in the J2735 standards, consists of two parts. Part 1 is sent in every BSM message broadcast 10 times per second and will be mandated to be broadcast by the National Highway Traffic Safety Administration (NHTSA) ruling. It contains core data elements, including vehicle position, heading, speed, acceleration, steering wheel angle, and vehicle size. BSM Part 2 consists of a large set of optional elements such as precipitation, air temperature, wiper status, light status, road coefficient of friction, activation of antilock brake system or traction control system, and vehicle type. BSM Part 2 elements are sent based on criteria that are not specified in the J2735 standards. However, not all of these parameters are currently available from vehicles and they will not be mandated by the USDOT. In this study, BSM Part 1 data is considered in the evaluation, as it is expected to be available in the first phase of CV implementation and is sufficient for travel time estimation and incident detection.

Findings from Previous Research

There are a number of studies in the literature focusing on travel time estimation utilizing CV data. Zou et al. (*3*) estimated travel time based on CV data using traffic simulation of a hypothetical network that simulates vehicles broadcasting probe data messages (PDM) data according to the J2735 standards (*20*). They found an average error percentage of 27.6%, 12.5%, and 8.2% for 1%, 5%, and 10% CV proportions, respectively. However, that study did not consider the demand level, which as stated earlier is one of the two factors that determine the sample size. Argote et al. (*4*) estimated measures of effectiveness based on real-world vehicle trajectories collected by the next generation simulation (NGSIM) program. These measures of effectiveness include queue length, speed, number of stops, acceleration noise, and the average delay per unit distance. A drawback of this study was that the method used in the calculation of travel time does not consider the change of vehicle ID during its course of travel, as specified in the J2735 standards to protect traveler privacy. Iqbal et al. (*5*) and Khoshmagham et al. (*21*) analyzed the BSM data to estimate the travel time utilizing the NGSIM data and simulation data respectively. Research has also investigated the optimal placement of roadside units (*22*), a snapshot generation protocol (*23*) for travel time estimation, and the robustness of connectivity in travel time estimation (*24*).

CV data has been also used in some studies to detect incidents. Utilizing microscopic simulation, Crabtree et al. (*25*) examined the use of CV data communicated using DSRC to detect freeway incidents. The utilized algorithm was based on comparing measured travel time and "normal" travel time estimated based on no-incident condition data. The results showed that the proposed algorithm can rapidly and reliably detect incidents. The study concluded that for a CV proportion of 30% (25% of the CVs were trucks and 5% were cars), the mean time to detect (MTTD) ranges from 2 minutes to 4 minutes for a DSRC roadside unit (RSU) spacing of 2.0 miles, and 2.5 minutes to 14 minutes for a reader spacing of 10 miles.

None of the above studies have investigated the effect of the demand variation on either the travel time estimation or incident detection. This is important since, as stated earlier, the accuracy of travel time estimation and the latency of incident detection are expected to be a function of the sample size, which is a function of both the percentage of CVs and the demand level. This study focuses on examining the effect of demand variation on travel time estimation and also the timeliness of incident detection under different CV proportions. The results of this examination, combined with forecasting the proportions of CV in future years, will allow agencies to determine for a given future year, the freeway and urban street segments by time-of-day, where CVs provide acceptable travel time estimation and incident detection.

Methodology

The first step to achieve the objectives of this study is to emulate the CV BSM data utilizing vehicle trajectories generated from the simulation of a freeway network and an arterial street network with different levels of traffic demand. Next, the generated BSM data is used to estimate the travel time and incident detection under the assumption of different proportions of CVs. Then, different regression equations are developed to estimate the effect of demand and the CV proportion on the performance of the travel time estimation and incident detection. These regression equations are then used to predict the performance of travel time estimation and incident detection for future years of CV implementation to determine the year when the utilization of CV data will produce acceptable accuracy and timeliness. The subsections below present more detailed discussion of the methodology.

CV Data Emulation

CV data has started to become available for research from pilots, testbeds, and initial CV deployments across the U.S.A. However, the proportions of CVs in the traffic stream at these locations are very limited. In this study, CV data is emulated using trajectories from microscopic simulation models that are used as inputs to the trajectory conversion algorithm (TCA) [Version 2.3] tool developed by the FHWA (*26*). The TCA tool then generates data that emulates CV data following the J2735 standard (*20*). The vehicle trajectories are generated utilizing the VISSIM microscopic simulation modeling tool.

The output from the TCA tool produced emulated BSM Part 1 data at 1/10th-second resolution utilizing trajectory data obtained from the simulation models. The input trajectory file to the TCA contains the identifications (ID), time stamps, speeds, position coordinates, and acceleration data of the vehicles. Using the trajectories as inputs, the TCA generates CV BSM Type 1 data following the SAE J2735 standards. The user of the TCA tool can select the CV proportion, message type (e.g., BSM or PDM), and communication type (DSRC or cellular). This study focuses on BSM Type 1 data that is expected to be captured by DSRC RSUs. The coverage area of an RSU is about 2,600 feet to 4,000 feet (*27, 28*). The TCA tool allows the specification of data transmission loss during the data transfer between the On-Board Unit (OBU) and the RSU. It has been reported that the loss rate with DSRC communications varies between 10% and 20% with an average of 12% (*29*). In this study, 12% transmission loss is specified in the input to the TCA tool. It is expected that the actual loss is a function of the availability of line of site in the coverage area and this should be considered with RSU siting. The TCA tool changes the vehicle ID every 5 minutes with a 30-second buffer window. This means that the ID will change over a period that ranges from 5 minutes to 5 minutes and 30 seconds, as is commonly used in current DSRC-based CV implementations.

There are two sources of stochasticity that are considered when assessing the utilization of the CV data. First, the accuracy of travel time estimation largely depends on the selection of vehicles on the link that are CV-equipped. Selecting different vehicles could result in variations in travel time particularly at a lower CV proportion when there is a higher variation in travel time between vehicles. The variation in speed/travel time increases due to weaving, merging, diverging, and lane drops on freeways and due to signal control on arterial streets. The second source of the stochasticity is the time at which the temporary ID for each vehicle changes within the RSU zone. As stated above, a CV vehicle ID changes every 5 minutes at different times in the simulation. A Monte Carlo analysis is used in this study to account for these two sources of stochasticity by randomly selecting the CV vehicles from the traffic stream and the time the vehicle ID changes. This requires multiple sampling of the data, each with the different random selection. As stated above, the SAE J2735 standards (*20*) specify that the BSM data contains a temporary ID which periodically changes to ensure the overall anonymity of the vehicle. One of the limitations of the TCA is in its assignment of the changeover ID, particularly if the simulation is done for a smaller segment as is the case in this study. The TCA assigns a temporary ID to each vehicle at the beginning of the simulated segment for all vehicles. Thus, the IDs for all vehicles change at the same time and if the travel time of the simulated segment is less than 5 minutes then the ID will not change during the simulation. In reality, different vehicles will come from different origins and thus the ID will change at different times within the selected link. To overcome this limitation, the TCA code is modified in this study so that the vehicle ID on the simulated link changes at random times. This is possible since the TCA is an open source software.

Travel Time Estimation

The travel time estimation is performed for each CV selection and ID changing scenario resulting from the Monte Carlo analysis and both the average and the individual selection results are used in the assessment of the accuracy of the estimation. The average of the runs represents the real-world use of data from multiple days for off-line analysis. An individual run represents using the results for the real-time operation of one day. It is expected that a lower CV proportion is acceptable when averaging across multiple days but may not be acceptable when using the data for single day real-time applications. In this study, 500 runs are performed for both the freeway and the arterial street segments.

The travel time estimation method needs to consider the vehicle ID change requirements, according to SAE J2735 standards. As mentioned earlier, the vehicle temporary ID changes periodically, so a vehicle may not pass the whole section with the same ID and a method that estimates the travel time based on full vehicle trajectories is not possible. The vehicles may change their ID once within the section. The utilized method is described next. For each of the vehicle IDs, the travel time and distance traveled by the ID are calculated. If an ID does not travel a minimum distance (e.g., 100 ft) then that ID is not considered in the travel time calculation. The summations of all the valid travel times and distances are used in the estimation of the total travel time (TT) and total distance (D). The link average travel time for each time period is calculated using

$$T = \frac{\sum_{i}^{n} t_i}{\sum_{i}^{n} d_i} D_s \tag{1}$$

where T is the average travel time of the segment which has a length of D_s; t_i, and d_i are the travel time and distance traveled by the ith ID; and n is the total number of IDs. A Monte Carlo analysis is performed with multiple runs (500 times) to obtain the distribution of travel times and the associated error at each CV proportion and demand level.

Incident Detection

This study examines detection of freeway incidents utilizing CV data. Vehicle speed, estimated based on CV BSM data, is selected as the base measurement for incident detection. The utilized method is based on hypothesis testing of the occurrence of the incident in a segment. The simulated network is decomposed to 300-foot long segments and the average speed of each segment is calculated at each time step based on CV data. This method simply takes the average of the measurements in each segment during a time period (45 seconds in this study) and if the difference between the average speeds of the downstream and upstream segments is more than a threshold, then the location is identified as an incident location. A certain percentile of speed differences distribution under no-incident conditions (99th percentile in this study) is selected as the threshold for testing the hypothesis that there is an incident on the subject segment. In other words, the probability of false alarm is set to 1% and the associated threshold is calculated. The reason for using the 99th percentile is to account for lower false alarm rate (1%). If the proposed hypothesis works for such lower false alarm rate then it could also be used for higher values.

As with travel time estimation, the method is tested using emulation of CV data generated by the TCA tool based on trajectories produced by the VISSIM microscopic simulation tool. VISSIM is used to emulate incidents occurring in a mixed traffic stream of CVs and non-CVs, as described earlier.

Regression Analysis

This study develops models of the relationship between the performance of the travel time estimation and incident detection using CV data and two independent variables: traffic demand and CV proportion. In this study, multiple linear regression is used to derive the relationships. The dependent variable in the regression is the error in travel time estimation and the independent variables are the CV proportion and the traffic flow rate. Different transformations of the independent variables are tried to identify the best fit of the regression model. It is found that the best form is

$$\begin{aligned} E_{err} = \beta_0 + \beta_1 {*} \cos(S) + \beta_2 {*} \log(CP) \\ + \beta_3 {*} S + \beta_4 {*} CP + \varepsilon \end{aligned} \tag{2}$$

where β_i are the regression coefficients, S is the flow rate measure, and CP is the CV proportion. E_{err} is the error of type "err", which is one of the two types of error considered in this study: the absolute mean percentage error (AMPE) and the standard deviation percentage error (SDPE). AMPE is the average error of all days. It is useful when assessing the average accuracy, and is possibly appropriate for the assessment when using data for planning purposes. SDPE is a measure of the reliability of travel time estimates with higher values, indicating a higher variability in the quality of the data. For the freeway, the flow rate measurement is represented by the volume/capacity (v/c) ratio. For the arterial street, the flow rate measurement is represented by the degree of saturation (volume/saturation flow ratio) of the critical intersection along the test section. The statistical software R is used in the analysis.

For incidents, MTTD is also estimated as a function of the v/c ratio and CV proportion for freeway segments. This study has found that an exponential regression equation could represent the relationship well as follows:

$$MTTD = \beta_0 + \beta_1 * e^{-(v/c)*CP} \quad (3)$$

where β are the regression coefficients, *CP* is CV proportion (%), *v* is volume (vehicles per hour) and *c* is capacity (vehicles per hour).

Test Locations

To demonstrate the proposed methodology, simulation models were developed and calibrated for a freeway network and an arterial network utilizing the VISSIM microscopic model. The details of these networks are provided below.

Freeway Segment

The freeway segment utilized in this study is a 15-mile segment of the I-95 southbound corridor in Broward County, FL, coded and calibrated in VISSIM. The segment that is selected for the assessment in this study consists of an on-ramp, an off-ramp, and a weaving segment with a total length of 1 mile. A 1-mile segment was selected considering the coverage range of one RSU. Travel time measurement and incident detection are possible at those areas where an RSU is available. However, cell-based technology could supplement the DSRC data collection and the same procedure could be used to determine the travel time and incident detection throughout the corridor.

A one-lane blockage incident was introduced based on real-world incident data to generate a bottleneck location for the purpose of assessing timeliness of incident detection. A one-lane blockage has less impact on the traffic flow, therefore it is the most challenging condition for incident detection. This study considered such a scenario to demonstrate the methodology. A better result is expected with severe incident scenarios. Since VISSIM cannot model incidents directly, the incident was modeled by creating a signal head in one lane of the freeway and the rubbernecking effect was modeled by changing the car following factors upstream of the incident location utilizing the VISSIM COM interface.

The values of the car following factors for the incident link during the incident conditions were changed using VISSIM COM based on values recommended in the literature. Knoop et al. (*30*) investigated drivers' behavior under incident conditions by analyzing the vehicles' trajectories collected by a helicopter from two actual incidents. The results showed that drivers' reaction time, headway distribution, and capacity are affected by the presence of an incident. That study observed a bimodal headway distribution, an increase in reaction time, and a 30% decrease in capacity during incident conditions. Some researchers found that the effect of an incident can be modeled by changing VISSIM car following factors. A detailed description of the VISSIM car following factors can be found in the VISSIM user manual (*31*). A study of the VISSIM calibration parameters (*32*) found that the CC0, CC1, CC2, CC4, and CC5 parameters have the most effect on the capacity of a freeway link. CC0 is the desired distance between stopped cars. CC1 is the desired time headway that the driver wants to keep to the leading vehicle. CC2 is the safety distance, which is the minimum distance a driver keeps with the leading vehicle. CC4 and CC5 are the parameters controlling the speed differences between a following and leading vehicle. Increasing CC0, CC1, and the absolute value of the CC4/CC5 ratio results in a reduction of capacity. Gomes et al. (*33*) modified three parameters: CC1, CC2, and CC4/CC5 to model the decrease in capacity resulting from curvatures. The calibrated parameters for the studied freeway segment CC1, CC2, and CC4/CC5 were 1.7, 0.9 and −2/2respectively (*33*). Rahman and Mattingly (*34*) found that the car following values proposed in prior studies for incident modeling produced better results in terms of macroscopic measures compared with the default values. Rompis et al. (*35*) developed a methodology for VISSIM calibration based on kinematic queuing theory. The incident was modeled by coding a traffic signal in VISSIM. Based on the reviewed study results, CC0, CC1, and CC2 values of 3.8 meters, 1.5 seconds and 7.5 meters, respectively, are used to replicate the actual freeway operation under incident conditions.

Arterial Street Segment

Travel time estimation is more of a challenge for urban arterial streets due to lower volumes at some times of the day and high travel time variations, which increase with the increase in the intersection density. The travel time based on CV data is also estimated for an arterial street, Glades Road, located in the City of Boca Raton, FL. The arterial network consists of three sections. The first section is 0.64 miles long between the Renaissance Way and Airport Road intersection; the second section is 0.76 of a mile long between Renaissance Way and St. Andrews Boulevard; and the third section is a 1.09-mile link between East University Drive and Airport Road. This arterial network was modeled in VISSIM and then a manual calibration process performed based on real-world measurements. The first step was to perform an initial fine-tuning of the model parameters to produce saturation flow rates that are in agreement with the Highway Capacity Manual 2010 procedures (*36*) and previous observations from the field in South Florida. The target saturation flow rate for the purpose of this calibration was set to 1,850 to 1,900 passenger cars per hour per lane. The parameters of the VISSIM model were fine-tuned to produce this value. The second part of the calibration process was to compare the traffic flow performance according to the Glades Road simulation model with the performance assessed based on the real-world traffic data. The VISSIM model parameters that were fine-tuned are the additive part (bx_add) and the multiplication part (bx_mult) of the desired safety distance in the Wideman 74 driver behavior model, which is used for urban arterial street modeling. The most appropriate combination of the two parameters was found to be 2.4 feet and 3.4 feet respectively. It was determined that the root-mean-square percentage errors for the simulated versus measured volume and speed values are below 15% for both directions of travel.

Results

The proposed methodology was applied to the test networks. The travel time estimation methodology was applied on both the freeway and arterial segments. The incident detection methodology was applied only to the freeway segment.

Travel Time Estimation Results

AMPE and SDPE of the estimated travel time were calculated for each of the 500 Monte Carlo runs performed for each CV proportion and flow rate combinations. For the arterial street segment, the investigated degrees of saturation were 0.3, 0.6, 0.7, 0.8, and 0.9. For the freeway segment, the investigated v/c ratios were 0.36, 0.5, 0.72, 0.86, and 1.01. For both freeways and arterial streets, the test CV proportions were 5, 10, 20, 50, and 80%. The results from the Monte Carlo runs were used to develop the regression equations to allow the calculation of the error percentage at a certain combination of CV proportion and demand level. The developed equations and their statistics are provided in Table 1.

Incident Detection Results

As mentioned in the methodology section, Monte Carlo simulation was performed to randomly assign connectivity to the vehicles since the position of the CVs relative to the bottleneck location affects the results. Table 2 shows the 80th and 95th percentiles of the resulted MTTD in minutes for each of the scenarios. The result shows that the MTTD decreases as the v/c ratio and CV proportions (CP) increase. It should be noted that the equations in Table 2 are applicable to one-lane blockage incidents on four-lane freeways. The equations are also applicable when the v/c ratio is greater than 0.4. This is because the incident is not detectable based on speed for v/c less than 0.4, as no queue is formed.

Quality of CV Utilization for Different Future Years

This section discusses the application of the models developed, as discussed earlier in this paper, to determine the quality of CV-based data in terms of travel time measurement accuracy (in terms of AMPE), travel time measurement reliability (in terms of SDPE), and incident detection latency (in terms of MTTD) for different years in the future. Such analysis is critical for agencies to make investments in CV versus another mode of data acquisition in future years. The proportion of CVs in the traffic stream will increase as new vehicles join the fleet every year. Different studies have predicted the future CV proportion after the CV technology mandate by the USDOT becomes effective. In this study, the prediction model developed by Iqbal et al. (*37*) was used to determine the accuracy of performance measurement at different demand levels and at different years after the CV mandate becomes effective. Iqbal et al. (*37*) has predicted the future market penetration of CVs considering the socio-economic variance between different areas. Therefore the predicted market penetration has a range instead of a fixed value for each year after the CV is mandated. In this study the lower market penetration is considered to calculate the accuracy and incident detection timeliness to consider the worst condition.

The results of the analysis are shown in Table 3. Previous guidelines and practices for acceptable error of travel time point to an AMPE of less than 10% absolute (*38, 39*). Toppen et al. (*40*) reported that if the SDPE is greater than 12%, then there are no user benefits when using the travel time estimates in advanced traveler information systems. Following those guidelines, Table 3 shows that the travel time data estimation on freeways starting the second after the CV is mandated. For the arterial facility, it will take four to five years to get an acceptable accuracy of travel time estimation. The accuracy of the travel time is also affected by the v/c ratio. At lower v/c ratio, the accuracy is lower because of the smaller sample size. The accuracy increases with the increase of the sample size up to a certain point and then it starts to decrease. At higher v/c ratios, there are higher variations in travel time between vehicles due to congestion, which results in a lower accuracy in the estimated travel time. CV data could be used for incident detection from the very first year of CV implementation on freeways with a latency of less than 5 minutes. Martin et al. (*41*) has reviewed a wide range of incident detection algorithms that are implemented in different location or simulation environments. The review identified the detection latency to be between 0.85 to 4 minutes. Therefore, the existing detectors could be completely replaced by the CV technology for the purpose of incident detection at year 4. At higher market penetration, the MTTD is

Table 1. Estimated Travel Time Error Equation

Arterial street	AMPE				SDPE			
Equation	$E_{AMEP} = 68.79 - 52.9^{*}\cos(S) - 0.996^{*}\log(CP) - 27.3^{*}S$ (5)				$E_{SDPE} = 9.6 + 3.04^{*}\cos(S) - 3.14^{*}\log(CP) + 0.038^{*}CP$ (6)			
Coefficients statistics	Coefficients	Std Error	*t* value	*Pr(>|t|)*	Coefficients	Std Error	*t* value	*Pr(>|t|)*
	β_0	11.55	5.96	6.54e-06	β_0	1.09	8.8	1.72e-08
	β_1	5.62	-4.86	8.41e-05	β_1	1.05	2.88	0.0089
	β_2	10.46	-5.2	3.71e-05	β_2	0.36	-8.62	2.45e-08
	β_3	0.15	-6.46	2.12e-06	β_3	0.013	2.88	0.009
	Multiple R-squared: 0.78				Multiple R-squared: 0.94			
	Adjusted R-squared: 0.75				Adjusted R-squared: 0.93			
Freeway								
Equation	$E_{AMEP} = 12.05 - 9.18^{*}\cos(S) - 0.52^{*}\log(CP) - 4.81^{*}S + 0.0085^{*}CP$ (7)				$E_{SDPE} = 25.15 - 19.81^{*}\cos(S) - 0.75^{*}\log(CP) - 11.56^{*}S + 0.02^{*}CP$ (8)			
Coefficients statistics	Coefficients	Std Error	*t* value	*Pr(>|t|)*	Coefficients	Std Error	*t* value	*Pr(>|t|)*
	β_0	3.42	3.56	0.001	β_0	4.66	5.4	3.27e-068
	β_1	2.9	−3.164	0.003	β_1	3.95	−5.02	1.12e-05
	β_2	0.075	−6.95	2.2e-08	β_2	0.1	−7.33	6.5e-09
	β_3	1.81	−2.66	0.01	β_3	2.46	−4.69	3.13e-05
	β_4	0.0085	0.004	0.04	β_4	0.01	3.09	0.00359
	Multiple R-squared: 0.8				Multiple R-squared: 0.78			
	Adjusted R-squared: 0.78				Adjusted R-squared: 0.76			

Table 2. MTTD for Different CV Proportions and Different V/C Ratios

	Mean time to detect (MTTD) (80%)			
Equation	$MTTD = \begin{cases} 0.75 & CP>6\% \\ -51.649 + 54.563^{*}e^{-(v/c)^{*}CP} & CP\leq 6\% \end{cases}$ (9)			
Coefficients statistics	Coefficients	Std Error	t value	Pr(>|t|)
	β_0	7.447	−6.936	0.000
	β_1	7.607	7.173	0.000
	Multiple R-squared: 0.798			
	Adjusted R-squared: 0.783			
Equation	$MTTD = \begin{cases} 0.75 & CP>6\% \\ -102.892 + 108.192^{*}e^{-(v/c)^{*}CP} & CP\leq 6\% \end{cases}$ (10)			
Coefficients statistics	Coefficients	Std Error	t value	Pr(>|t|)
	β_0	20.864	−4.931	0.000
	β_1	21.313	5.076	0.000
	Multiple R-squared: 0.665			
	Adjusted R-squared: 0.639			

less than a minute. Therefore, the algorithm provides the minimum value of 0.75 minutes which is equal to the time resolution (45 seconds or 0.75 minutes) of the incident detection algorithm.

Conclusion

This study investigated the use of CV data as an alternative to existing data acquisition techniques in providing two critical functions to support TSMO: travel time estimation and incident detection. In support of this investigation, the study examined the accuracy and reliability of measurement of the estimated travel time and the latency of incident detection under different demands and proportions of CV in the traffic stream. Regression models were developed to estimate measures of travel time accuracy (AMPE), reliability (SDPE), and latency (MTTD). Significant relationships were found between

Table 3. Minimum Accuracy at Different Year after the CV Mandate Becomes Effective

		Travel time accuracy (%)				Incident detection time (minutes)	
		Freeway		Arterial		Freeway	
v/c	Year (Iqbal et al. (*33*))	AMPE	SDPE	AMPE	SDPE	80% MTTD	95% MTTD
0.1	1 (0.80)	2.66	13.2	13.60	13.2	Accident cannot be detected	
	2 (2.49)	2.07	9.64	12.45	9.64		
	3 (4.61)	1.82	8.1	11.94	8.09		
	4 (9.39)	1.46	5.86	11.16	5.84		
	5 (13.79)	1.30	4.89	10.80	4.87		
	10 (18.80)	0.95	2.47	9.70	2.45		
	15 (73.91)	0.91	1.92	9.14	1.88		
	20 (89.13)	0.94	1.91	8.95	1.87		
	25 (94.76)	0.95	1.93	8.89	1.89		
0.5	1 (0.80)	1.81	12.84	8.89	12.86	2.94	4.76
	2 (2.49)	1.23	9.28	7.74	9.29	2.19	3.86
	3 (4.61)	0.97	7.74	7.23	7.75	1.71	2.91
	4 (9.39)	0.61	5.5	6.45	5.5	0.75	0.75
	5 (13.79)	0.46	4.53	6.10	4.53	0.75	0.75
	10 (18.80)	0.11	2.12	4.99	2.11	0.75	0.75
	15 (73.91)	0.07	1.56	4.43	1.54	0.75	0.75
	20 (89.13)	0.09	1.55	4.24	1.53	0.75	0.75
	25 (94.76)	0.11	1.57	4.18	1.55	0.75	0.75
1	1 (0.80)	2.5	11.82	13.08	11.85	2.37	4.22
	2 (2.49)	1.92	8.26	11.93	8.29	1.47	2.45
	3 (4.61)	1.66	6.72	11.42	6.74	0.75	0.75
	4 (9.39)	1.3	4.47	10.65	4.5	0.75	0.75
	5 (13.79)	1.15	3.50	10.29	3.52	0.75	0.75
	10 (18.80)	0.8	1.09	9.18	1.1	0.75	0.75
	15 (73.91)	0.76	0.53	8.62	0.54	0.75	0.75
	20 (89.13)	0.78	0.53	8.43	0.52	0.75	0.75
	25 (94.76)	0.8	0.55	8.37	0.54	0.75	0.75

Error more than the minimum acceptable threshold

these measures and the demand level and CV proportions. The developed regression models were used in conjunction with a prediction of CV proportion in future years to estimate the accuracy, reliability, and latency measures for each future year after the USDOT CV mandate becomes effective. The results show that the CV data can be used for freeway travel time estimation from the very first year of the CV implementation. For arterials, it may take four to five years to reach an acceptable accuracy. The use of CV data for incident detection on freeways will provide a latency of 5 minutes or less for the first year after implementation when the v/c ratio is higher than 0.4. The latency is expected to be lower than the existing detectors for all incidents four years after the CV mandate becomes effective.

The results of this study can be used by TSMO programs and agencies to plan their investment in alternative means of data collection. They can be used to support the preparation of a timeline to replace the existing data collection systems with the CV technology, depending on the traffic demand of a specific roadway. This should be done, however, with the consideration that some point detectors may still be necessary to provide other parameters such as traffic volume.

Acknowledgments

This paper is based on work supported by the Florida Department of Transportation under grant BDV29 TWO 977-21.

Author Contributions

The authors confirm contributions to the paper as follows: study conception and design: Mohammed Hadi, Md Shahadat Iqbal, Samaneh Khazraeian; data collection: Md Shahadat Iqbal, Samaneh Khazraeian; analysis and interpretation of results: Md Shahadat Iqbal, Samaneh Khazraeian; draft manuscript preparation: Md Shahadat Iqbal. All authors reviewed the results and approved the final version of the manuscript.

References

1. Federal Highway Administration. *Vehicle-to-Infrastructure Deployment Guidance and Products*. FHWA-HOP-15-015. U.S. Department of Transportation, Washington, D.C., 2015.

2. Wright, J., J. K. Garrett, C. J. Hill, G. D. Krueger, J. H. Evans, S. Andrews, C. K. Wilson, R. Rajbhandari, and B. Burkhard. *National Connected Vehicle Field Infrastructure Footprint Analysis*. Final Report, FHWA-JPO-14-125, U.S. Department of Transportation Research and Innovation Technology, Washington, D.C., 2014.
3. Zou, Z., M. Li, and F. Bu. Link Travel Time Estimation Based on Vehicle Infrastructure Integration Probe Data. *ICCTP 2010: Integrated Transportation Systems: Green, Intelligent, Reliable*. 2010, pp. 2266–2276.
4. Argote, J., E. Christofa, Y. Xuan, and A. Skabardonis. Estimation of Arterial Measures of Effectiveness with Connected Vehicle Data. Presented at 91st Annual Meeting of Transportation Research Board, Washington, D.C., 2012.
5. Iqbal, M. S., M. Hadi, and Y. Xiao. Accuracy, and Reliability of Estimated Travel Time Using Basic Safety Message (BSM) Data Collected from Connected Vehicles. Presented at 96th Annual Meeting of the Transportation Research Board, January, Washington, D.C., 2017.
6. Lee, J., B. Park, and I. Yun. Cumulative Travel-Time Responsive Real-Time Intersection Control Algorithm in the Connected Vehicle Environment. *Journal of Transportation Engineering*, Vol. 139, No. 10, 2013, pp. 1020–1029.
7. Garelli, L., C. Casetti, C. F. Chiasserini, and M. Fiore. Mobsampling: V2V Communications for Traffic Density Estimation. *Proc., Vehicular Technology Conference (VTC Spring)*, 2011 IEEE 73rd, Yokohama, Japan, 2011, pp. 1–5.
8. Barrachina, J., P. Garrido, M. Fogue, F. J. Martinez, J. C. Cano, C. T. Calafate, and P. Manzoni. A V2I-based Real-Time Traffic Density Estimation System. *Wireless Personal Communications*, Vol. 83, No. 1, 2015, pp. 259–280.
9. Mao, R., and G. Mao. Road Traffic Density Estimation in Vehicular Networks. *Proc., Wireless Communications and Networking Conference (WCNC)*, Shanghai, China, IEEE, New. York, USA, 2013, pp. 4653–4658.
10. Qiu, T. Z., X. Y. Lu, and A. H. Chow. Real-Time Density Estimation on Freeways with Loop Detector and Probe Data. *IFAC Proceedings*, Vol. 42, No. 15, 2009, pp. 298–303.
11. Khan, S. M. *Real-time Traffic Condition Assessment with Connected Vehicles*. Masters dissertation. Clemson University, 2015.
12. Barrachina, J., J. A. Sanguesa, M. Fogue, P. Garrido, F. J. Martinez, J. C. Cano, C. T. Calafate, and P. Manzoni. V2X-d: A Vehicular Density Estimation System that Combines V2V and V2I Communications. *Proc., Wireless Days (WD)*, IEEE, New. York, USA, 2013, pp. 1–6.
13. Sanguesa, J. A., J. Barrachina, M. Fogue, P. Garrido, F. J. Martinez, J. C. Cano, and P. Manzoni. Sensing Traffic Density Combining V2V and V2I Wireless Communications. *Sensors*, Vol. 15, No. 12, 2015, pp. 31794–31810.
14. Doan, Q. C., T. Berradia, and J. Mouzna. Vehicle Speed and Volume Measurement Using Vehicle-to-Infrastructure Communication. *WSEAS Transactions on Information Science and Applications*, Vol. 6, No. 9, 2009, pp. 1468–1477.
15. Zhou, Y., S. Chen, Y. Zhou, M. Chen, and Q. Xiao. Privacy-Preserving Multi-Point Traffic Volume Measurement through Vehicle-to-Infrastructure Communications. *IEEE Transactions on Vehicular Technology*, Vol. 64, No. 12, 2015, pp. 5619–5630.
16. Zheng, J., and H. X. Liu. Estimating Traffic Volumes for Signalized Intersections Using Connected Vehicle Data. *Transportation Research Part C: Emerging Technologies*, Vol. 79, 2017, pp. 347–362.
17. Zou, Z., M. Li, and F. Bu. Link Travel Time Estimation Based on Vehicle Infrastructure Integration Probe Data. *ICCTP 2010: Integrated Transportation Systems: Green, Intelligent, Reliable*, 2010, pp. 2266–2276.
18. Khondaker, B., and L. Kattan. Variable Speed Limit: A Microscopic Analysis in a Connected Vehicle Environment. *Transportation Research Part C: Emerging Technologies*, Vol. 58, 2015, pp. 146–159.
19. Minelli, S., P. Izadpanah, and S. Razavi. Evaluation of Connected Vehicle Impact on Mobility and Mode Choice. *Journal of Traffic and Transportation Engineering*, Vol. 2, No. 5, 2015, pp. 301–312.
20. SAE International. *Dedicated Short Range Communications (DSRC) Message Set Dictionary*, SAE J2735, DSRC Technical Committee, PA, USA, 2016.
21. Khoshmagham, S., K. L. Head, Y. Feng, M. Zamanipour, and Y. Yuan. Travel Time Observation in Privacy Ensured Connected Vehicle Environment Using Partial Vehicle Trajectories and Extended Tardity. Presented at 96th Annual Meeting of the Transportation Research Board, Washington D.C., 2017.
22. Kianfar, J., and P. Edara. Placement of Roadside Equipment in Connected Vehicle Environment for Travel Time Estimation. *Transportation Research Record: Journal of the Transportation Research Board*, 2013. 2381: 20–27.
23. Chen, C., J. Kianfar, and P. Edara. New Snapshot Generation Protocol for Travel Time Estimation in a Connected Vehicle Environment. *Transportation Research Record: Journal of the Transportation Research Board*, 2014. 2424: 1–10.
24. Osman, O. A., and Ishak, S. Travel Time Estimation and Connectivity Robustness for Connected Vehicle Environments. Presented at 94th Annual Meeting of the Transportation Research Board, Washington D.C., 2015.
25. Crabtree, J., and N. Stamatiadis. Dedicated Short-Range Communications Technology for Freeway Incident Detection. *Transportation Research Record: Journal of the Transportation Research Board*, 2007. 2000: 59–69.
26. Open Source Application Development Portal. *A Channel for Distributing and Collaborating on Transportation Related Open Source Applications*. U.S. Department of Transportation, Federal Highway Administration, Washington, D.C.http://www.itsforge.net/78-news/140-tca-2-3-release. Accessed July 13, 2015.
27. Andrews, S., and M. Cops. *Vehicle Infrastructure Integration Proof of Concept Technical Description – Vehicle*. Final Report, FHWA - JPO-09-043. U.S. Department of Transportation, Federal Highway Administration, Washington, D.C., 2009.
28. McGurrin, M. *Vehicle Information Exchange Needs for Mobility Applications*. FHWA-JPO-12-021. U.S. Department of Transportation, Federal Highway Administration, Washington, D.C., 2012.

29. Kandarpa, R., M. Chenzaie, J. Anderson, J. Marousek, T. Weil, F. Perry, L. Schworer, J. Beal, and C. Anderson. *Vehicle Infrastructure Integration Proof-of-Concept Results and Findings - Infrastructure, Volume 3B*, Final Report, FHWA-JPO-09-057. Research and Innovative Technology Administration (RITA), U.S. Department of Transportation, Washington, D.C., 2009.
30. Knoop, V. L., H. J. Van Zuylen, and S.P. Hoogendoorn. *Microscopic Traffic Behavior near Incidents. Transportation and Traffic Theory 2009: Golden Jubilee*. Springer, Boston, MA, 2009, pp. 75–97.
31. PTV AG. *VISSIM 5.4 - User Manual*. Karlsruhe, Germany, 2012.
32. Woody T. *Calibrating Freeway Simulation Models in VISSIM*. University of Washington, Seattle, WA, 2006.
33. Gomez, G., A. May, and R. Horowitz. *Calibration of VISSIM for a Congested Freeway*. California PATH Program, Institute of Transportation Studies, University of California, Berkeley, California, 2004.
34. Rahman Z., and S. P. Mattingly. Macroscopic Evaluation of Incident-Induction Driver Behavior Changes. *Proc., Symposium Celebrating 50 Years of Traffic Flow Theory*, Portland, Oregon, 2014.
35. Rompis, S. Y., F.G. Habtemichael, and M. Cetin. A Methodology for Calibrating Microscopic Simulation for Modeling Traffic Flow under Incidents. *Proc., 17th International Conference on Intelligent Transportation Systems (ITSC)*, New. York, USA, IEEE, 2014.
36. *Highway Capacity Manual 2010*. Transportation Research Board of the National Academies, Washington, D.C., 2010.
37. Iqbal M. S., M. Hadi, and Y. Xiao. Assessment of Link Level Variation of Connected Vehicle Market Penetration. Presented at 96th Annual Meeting of the Transportation Research Board, Washington, D.C., 2017.
38. Turner, S. M., W. L. Eisele, R. J. Benz, and D. J. Holdener. *Travel Time Data Collection Handbook*. FHWA-PL-98-035. Federal Highway Administration, U.S. Department of Transportation, Washington, D.C., 1998.
39. Fries, R., P. Yousefzadehfard, K. Bahaaldin, A. Fadoul, K. Ghale, M. Atiquzzaman, M. A. Chowdhury, S. Petersen, and E. Minge. *Real-Time Information Dissemination Requirements for Illinois per New Federal Rule*. Produced for Illinois Department of Transportation, Springfield, IL, 2015.
40. Toppen, A., and K. Wunderlich. *Travel Time Data Collection for Measurement of Advanced Traveler Information Systems Accuracy*. Presented at the 14th Annual Meeting and Exposition of ITS America: At the Crossroads: Integrating Mobility Safety and Security, San Antonio, Texas, 2004.
41. Martin, P., J. Perrin, B. Hansen, R. Kump, and D. Moore. *Incident Detection Algorithm Evaluation*. Utah Department of Transportation, Utah, 2009.

The Standing Committee on Urban Transportation Data and Information Systems (ABJ30) peer-reviewed this paper (18-05941).

The opinions, findings, and conclusions expressed in this publication are those of the author(s) and not necessarily those of the Florida Department of Transportation or the U.S. Department of Transportation.

Article

Transportation Research Record
2018, Vol. 2672(42) 213–223

DOI: 10.1177/0361198118798720
journals.sagepub.com/home/trr
SAGE

Discovering Regularity in Mobility Patterns to Identify Predictable Aggregate Supply for Ridesharing

Ivan Mendoza[1,2], Clas Rydergren[3], and Chris M. J. Tampère[1]

Abstract

Heterogeneous data collected by smartphone sensors offer new opportunities to study a person's mobility behavior. The mobility patterns extracted from the travel histories found in these data enable agents residing in mobile devices to model transitions between visited locations, so that upcoming trips can be predicted after observing a set of events, and assistance can be planned in advance. When several agents cooperate, the forecasted trips made by multiple users can provide a potential supply for shared mobility systems such as dynamic ridesharing. These trips must be sufficiently regular and frequent to be reliably announced as shareable trips. This paper describes a methodology to identify a predictable aggregate supply for ridesharing via mobility patterns discovered in users' travel histories. The methodology empirically quantifies measures like the regularity and frequency of these patterns on a dataset consisting of 967 users scattered across different geographical areas. The sample exhibits high heterogeneity with respect to these measures (hence, of predictability, regardless of the prediction method). This paper shows how frequency of trip patterns decreases, while regularity increases, when additional dimensions such as departure times are added to the analysis. It was concluded that the traveler flexibility with regard to accepting less regular trips is vital to discover a larger supply. These results provide insights to develop future applications able to take advantage of this approach, to increase ridesharing rates, allowing a critical mass to be more easily attained.

Today, some mobile apps can track a user's activities, producing logs for which the statistics are mainly used for sporting and health purposes; other apps in the "life-logging" category use sensors that allow users to track themselves during an entire day, logging the complete chain of events. Once processed, the heterogeneous data collected by these apps through the sensors of a mobile device, including but not limited to global positioning systems (GPS), accelerometers, and gyroscopes, can also provide insights about the person's mobility behavior. The distinctive information found in these data is the chain of activities carried out during the day, including visited locations, time spent at these places, and additional characteristics depending on the complexity of the algorithms.

Data-mining techniques allow extraction of mobility patterns from travel histories constructed from these logs, making it possible to identify regular transitions between repeatedly visited locations. These transitions can be conditioned to a particular detected state or context defined by attributes such as current location, time of day, or day of the week. The regularity of these patterns enables the prediction of future trips, so that software agents residing in a user's mobile device can autonomously decide about planning service assistance in advance.

One of the transport services that could potentially benefit from this approach is dynamic ridesharing: when multiple agents cooperate within a region. If everyday car trips with available seats made by multiple users could be predicted for a certain time interval, day, or destination, information about a potential supply for ridesharing, matching an upcoming ride request, could be inferred and announced earlier. Ridesharing passengers could for instance visualize the expected supply, consisting of potential shareable trips to a destination of interest compatible with their own schedules. This visual

[1]KU Leuven, L-Mob, Leuven Mobility Research Center, CIB, Leuven, Belgium
[2]Universidad del Azuay, Cuenca, Ecuador
[3]Department of Science and Technology, Linköping University, Norrköping, Sweden

Corresponding Author:
Address correspondence to Ivan Mendoza: ivan.mendoza@kuleuven.be

information could display hotspots where the expected trips will occur, so that passengers can find suitable pickup points to enable ridesharing. The possible software applications using this approach could provide an instrument to attain a critical mass for ridesharing services by promoting an increase in the number of participants. In this context, a critical mass refers to having a minimum number of drivers, so that passengers can find supply for a ride requirement; but at the same time, having a minimum number of passengers, enabling drivers that want to share their trip costs to find sufficient demand for requests. This paper proposes a solution from the supply perspective: assuming that without a supply, no passengers can choose to share rides, but without passengers, drivers could still offer rides and hence create a basis for a community of ridesharing users to grow steadily.

The objective of this paper is to define a methodology to estimate a supply of shareable trips for ridesharing that can be conditioned to a context, based on the analysis of trip frequency and regularity in travel histories. Then, a subset of these trips with a minimum level of regularity is used to produce visualizations of the potential supply. The key concept of trip regularity is explored empirically in a dataset of travel histories of 967 users scattered across different geographical areas. The present paper is organized as follows: the first section introduces the context of the research and states the paper's objective; a literature review of related works is then presented. The next section describes the proposed methodology, followed by a discussion of the results obtained from the empirical data, and finally the conclusions and potential future works are presented.

Literature Review

Ridesharing has received considerable attention in recent transport-related research; this alternative travel mode allows traffic congestion to be mitigated as the number of vehicles is reduced, bringing environmental and societal benefits. A review of some recent research and possible future directions on ridesharing can be found in Furuhata et al., in which the authors also provide the following unified definition: "ridesharing is a transportation mode in which individual travelers share a vehicle for a trip, with the purpose of splitting travel costs among users with similar itineraries and time schedules" (*1*). Some of the suggested improvements included better ride-matching strategies, pricing systems, and multimodal integration so that ridesharing is combined with another travel mode to complete a trip. In this same article (1), it is also stated that: "A great number of choices for participants increases the importance of assistance by software agents that enable personalized travel planning and execution."; so that our research uses this suggestion to propose an improvement for ridesharing. Selected relevant literature related to this research is studied in the following paragraphs.

An attempt to enable ridesharing through a recommender system, implemented as a web-based platform that uses large-scale smartphone mobility data is presented in Bicocchi and Mamei (*2*). Here, the authors extracted information from two datasets containing mobility traces to identify potential rides. Routines consisting of repeated transitions on certain days of the week between a user's frequent locations were found when mining trips endpoints, then rides were matched based on similarities between the origins and destinations of different users' routines. Recently, as detailed by Bicocchi et al. (*3*), the previous work was extended through a set of methods, which analyzed urban mobility traces to recognize matching rides along similar routes. A list of optimization alternatives for ride-matching can be found in Agatz et al. (*4*) and a review of techniques to extract different mobility patterns can be found in Lin and Hsu (*5*).

Cici et al. provided an upper bound for the potential reduction of traffic in three different cities through ridesharing (*6*). Mobility patterns concerning home and work locations found in data from different heterogeneous sources were extracted, then an algorithm for matching users with similar patterns considering additional constraints such as social distance was proposed. Research exploring the impact of ridesharing on congestion using mobile phone data is presented in Alexander and González (*7*). Here, the authors extracted the average daily origin–destination (OD) matrix per travel mode from mobile phone data to match trips with spatiotemporal similarities; the impacts on congestion were evaluated by considering different adoption rates.

The potential benefits of introducing meeting points in a ridesharing system to attain a critical mass are evaluated in Stiglic et al. (*8*). The authors used simulation to measure the impact of picking up and dropping off passengers at locations different than the actual origins or destinations, obtaining a significant increase in the number of matched trips. Later research by Stiglic et al. was extended by adding flexibility in departure and detour times (*9*). Finally, research performed in Goel et al. provides a method for choosing the best locations for these pickup points based on Voronoi diagrams (*10*).

Research evaluating regularity in mobility patterns can be found in Williams et al. (*11*), Wang et al. (*12*), and Zhong et al. (*13*). High regularity of trips leads to a low randomness or entropy in a choice set of destinations; these metrics are extensively used in literature for attribute selection, particularly in classification models such as decision trees (*14*).

The present paper extends these works by using travel data acquired from modern smartphone tracking apps: after a multi-step data-mining process, spatial characteristics of potential "sharable" trips are inferred and announced as hotspots to find ridesharing opportunities if patterns are sufficiently regular.

Methodology

In this section, the required methods to attain the paper's objective are elaborated. These methods include the extraction of personal points of interest (POIs) and mobility patterns, when trips occur under certain conditions. Frequency and regularity measures are subsequently formulated so that a subset of candidate trips can be proposed as the potential aggregate supply for ridesharing.

Extracting Personal Points of Interest

To recognize recurrent visits to the same location in a travel history, the captured spatial coordinates of the trip's endpoints cannot be used directly. Given that coordinates are estimates of a user's position, they can be different on every new visit to the same location. Methods for detecting a trip's endpoints can be found in Zheng et al. (*15*) and Mendoza et al. (*16*), and the problem of labeling endpoints as visits to known or new locations are detailed in Lin and Hsu (*5*) and Zheng et al. (*15*) among others (*17–19*). In this paper, POI refers to a frequently visited location in a user's travel history. Typical approaches in literature to identify these spatial patterns use density-based clustering techniques such as DBSCAN (*20*) or OPTICS (*21*). The latter also allows a hierarchy to be obtained, facilitating the extraction of POIs at a desired scale (i.e., buildings, neighborhoods, regions, etc.). A general density-based procedure used to extract POIs will now be explained.

Let us assume a dataset H consisting of travel histories from multiple users with data collected by a smartphone app. Then, a travel history H^v, defined as a chain of trips $b_1, b_2, \ldots b_N$ between the endpoints where a user v stops during the day to start a new activity, is created. The following notation describes each trip i's characteristics:

$b_i.o$ origin coordinates {latitude, longitude}, $\forall b_i \in H^v$
$b_i.d$ destination coordinates {latitude, longitude}, $\forall b_i \in H^v$
$b_i.m$ travel mode, $\forall b_i \in H^v$
$b_i.s$ departure time in 24-h notation, $\forall b_i \in H^v$
$b_i.e$ arrival time in 24-h notation, $\forall b_i \in H^v$
$b_i.w$ day type, $w \in$ {weekday, weekend}, $\forall b_i \in H^v$

Then, the spatial dataset containing coordinates of trips' endpoints to be clustered is

$$S = \{x_1, x_2, \ldots, x_{2N} \mid \forall b_i \in H^v : x_i \in \{b_i.o, b_i.d\}\}$$

The distance between two points, which is typically Euclidean after transforming coordinates to a Cartesian system (*22*), is denoted by $dist(x_1, x_2)$, then the neighborhood of a point x_i as shown in Figure 1*a* with points directly reachable at a radius $\varepsilon \geqslant 0$ is

$$N_\varepsilon(x_i) = \{x_j \in S \backslash \{x_i\} \mid dist(x_i, x_j) \leqslant \varepsilon\}$$

The complete neighborhood $N'_\varepsilon(x_i)$ shown in Figure 1*b*, consists of all merged points in neighborhoods of those previously discovered, that is,

$$N'_\varepsilon(x_i) = N_\varepsilon(x_i) \cup N_\varepsilon(x_j) \cup N_\varepsilon(x_k) \cup .. \quad \forall x_j \in N_\varepsilon(x_i), \forall x_k \in N_\varepsilon(x_j), ..$$

Algorithm 1 Find Complete Neighborhood

```
Function: Neighborhood (x_i, N_ε(x_i), dataset S)
For each unvisited x_j ∈ S
  Label x_j as visited
  If dist(x_i, x_j) ≤ ε
    Set N_ε(x_i) = Neighborhood (x_j, N_ε(x_i) ∪ {x_j}, S\{x_j})
Return N_ε(x_i)
```

Algorithm 2 Density-Based Clustering Structure

```
Function: DB-Clustering (dataset S)
Set C = ∅, k = 0, ε≥0, minPts ∈ ℤ
For each unvisited x_i ∈ S
  Label x_i as visited
  Set N'_ε(x_i) = Neighborhood(x_i, {x_i}, S \ {x_i})
  Set k = k + 1
  Set C = C ∪ newcluster(N'_ε(x_i))
Return C
```

The process to obtain $N'_\varepsilon(x_i)$ is presented in Algorithm 1, and the final density-based procedure to find the set C of repeated locations in Algorithm 2, in which complete neighborhoods are included in the final result only if a minimum number of points (visits) specified by parameter $minPts \in \mathbb{Z}$ is reached, otherwise they are tagged as noise (see Figure 1, *c* and *d*). Then, the set of POIs in user v's travel history is

$$C^v minPts, \varepsilon = \{c_i \in C \mid minPts \leqslant |c_i|\}$$

Identifying Transitions between POIs

Following this process, the trip endpoints are tagged to identify the cluster they belong to, so that all visits to the same location have the same label. That is, assuming u_k is a unique identifier for cluster c_k and x'_i the label on point x_i is

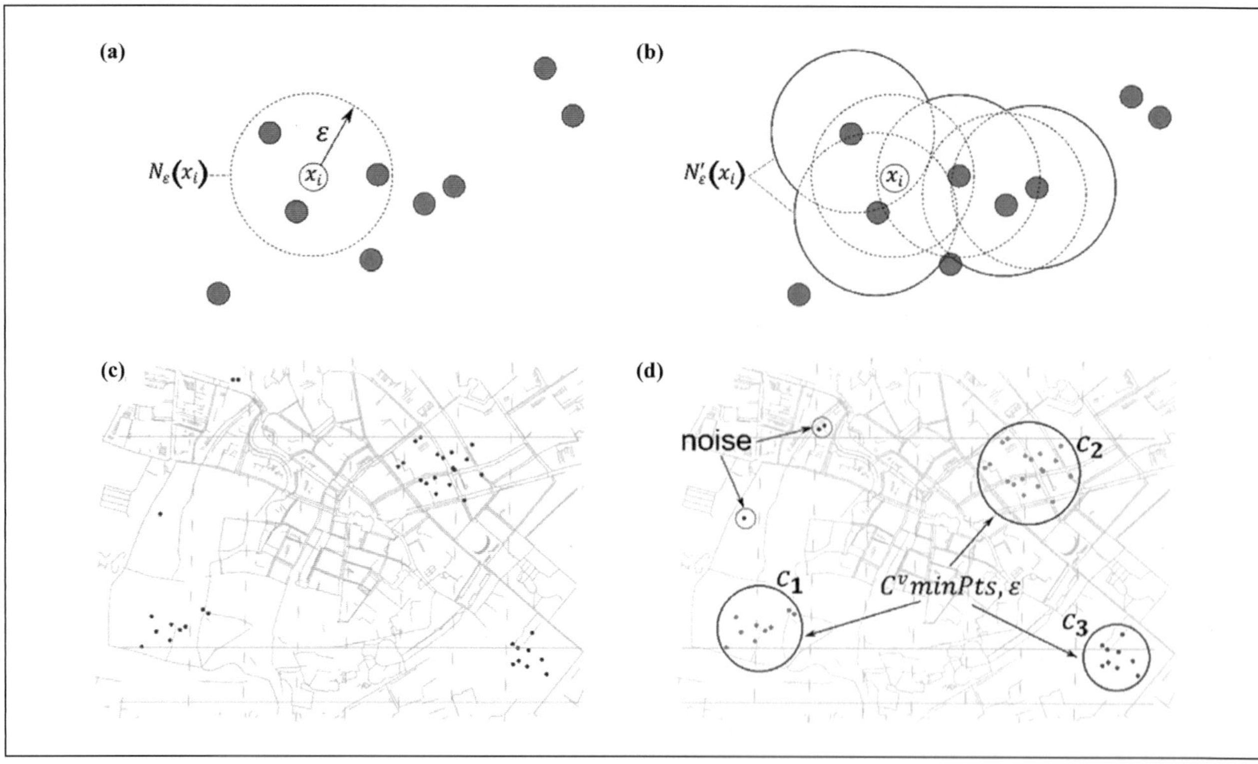

Figure 1. (*a*) Neighborhood of a point; (*b*) complete neighborhood of the same point; (*c*) trip endpoints in a travel history; and (*d*) their clustering structure when ***minPts*** = 5.

$$x'_i = \begin{cases} u_k & \text{if } x_i \in c_k \\ \text{undefined} & \text{otherwise} \end{cases}$$

Let u_0 be a label to denote endpoints at non-frequent locations, that is, those not included in a cluster. The following new notation is required for the updated trip i's characteristics:

$b_i.o'$ trip origin's label, $\forall b_i \in H^v$
$b_i.d'$ trip destination's label, $\forall b_i \in H^v$
u_0 label for "noise" points, in which $|c_0| < minPts$

Two trips b_i and b_j by the same person are then assumed to have the same OD pair if $b_i.o' = b_j.o' \wedge b_i.d' = b_j.d'$, even though their actual endpoints' coordinates could be different. If the travel history is consistent it should be expected that $b_i.d' = b_{i+1}.o', \forall b_i \in H^v$, producing a reliable chain of trips. After transitions between pairs of POIs are identified, regularity and frequency of these transitions with respect to their travel histories can be quantified.

Regularity of Transitions between POIs

In this paper, regularity of a certain trip characteristic (i.e., destination) is defined as the relative frequency of that characteristic in a user's travel history, conditioned to a state defined by other characteristics; this relative aspect distinguishes regularity from frequency. Frequency is the number of events per unit of time; it denotes the probability that something happens, such that the expected number of events in a period is found as frequency × length of period. On the other hand, regularity is the relative frequency of an event i within a user's pattern: it denotes the probability that among all possible events in the given conditions, i is the current one. Values closer to 1 are more regular and hence easier to predict.

A high frequency does not necessarily imply high regularity or vice versa. For instance, an event (e.g., to travel from A to B) may be rare (e.g., only once per month), yet any time the person resides in A his next move may be to go to B, hence his move is 100% regular, conditional on residing in A. An agent observing this person for a while and identifying its current location to be A, may therefore have an easy job predicting B as the next destination. Inversely, a person might frequently travel between C and D (e.g., twice a day); nevertheless, when residing in C, there might be five more very frequent destinations other than D, which makes the trip C to D frequent but not very regular when only conditioned on residing in C.

Therefore, an agent that knows its current position in C cannot predict the next destination with confidence. The following functions provide different measures of regularity of a destination in a user v's travel history.

Let $y_{i,j}$ be the number of trips in H^v between POIs u_i and u_j :

$$y_{i,j} = \left|\{b \in H^v | b.o' = u_i \wedge b.d' = u_j\}\right|$$

Let V be the set of all users with consistent travel histories. A passenger requesting a ride to a location q could potentially use trips toward its neighborhood:

$$N_\varphi(q) = \{b.d' \in H | dist(b.d', q) \leq \varphi\}, \text{ where } H = \bigcup_{v \in V} H^v$$

That is, $N_\varphi(q)$ contains all POIs destinations in every user's travel history near location q, where φ could denote the maximum distance a passenger is willing to walk from the drop-off location. Let P be the number of unique POIs in a user's travel history, the total number of inter-cluster trips to destination $u_j \in N_\varphi(q)$, including origins in u_0 is

$$y_{*,j} = \sum_{\substack{p=0 \\ j \neq p}}^{P} y_{p,j}$$

The regularity of a u_j with respect to other destinations in H^v is defined by

$$R(u_j) = P(u_j|v) = \frac{y_{*,j}}{|H^v|}$$

That is, the probability that user v visits location u_j without being conditioned to any current state. On the other hand, the frequency, measured in number of visits per day to u_j is

$$f(u_j) = \frac{y_{*,j}}{T^v}$$

where T^v, is the size of user v's travel history in days. The total number of trips from an origin POI u_i, including destinations in u_0 is

$$y_{i,*} = \sum_{\substack{p=0 \\ i \neq p}}^{P} y_{i,p}$$

The regularity of transitions to u_j (OD regularity) from an origin u_i is then

$$R(u_j|u_i) = \frac{y_{i,j}}{y_{i,*}}$$

Furthermore, transitions conditioned to a certain departure time interval can be evaluated. First, the 24-h format of departure times used in the dataset is transformed to a discrete value, so that trips with similar times can be clustered in the same group. Consider a finite number of time segments ω during the day. For instance, with a time interval $\Delta t = 15$ min, a day consists of $\omega = 96$ segments. The resulting departure and arrival time periods are $b.s' = \frac{b.s}{\Delta t} + 1$ and $b.e' = \frac{b.e}{\Delta t} + 1$ respectively, with values $\{x \in \mathbb{Z} | 1 \leq x \leq 96\}$. Then the regularity of destination u_j conditioned to an origin and departure time is

$$R(u_j|u_i, s_k) = \frac{y_{i,j,k}}{y_{i,*,k}}$$

where $y_{i,*,k}$, represents the number of all trips starting from u_i at departure period s_k, defined as

$$y_{i,*,k} = |\{b \in H^v | b.o' = u_i \wedge b.s' = s_k\}|$$

whereas $y_{i,j,k}$, considers only those to destination u_j, that is,

$$y_{i,j,k} = \left|\{b \in H^v | b.o' = u_i \wedge b.d' = u_j \wedge b.s' = s_k\}\right|$$

Finally, the frequency of these trips would be

$$f(u_j|u_i, s_k) = \frac{y_{i,j,k}}{T^v}$$

More dimensions can be added, although the number of trips constrained to the new conditions is significantly reduced, as well as their frequency. These new conditions can include day type: $b_i.w \in \{\text{weekday, weekend}\}$ and travel mode: $b_i.m \in \{\text{car, public transport}\}$.

The overall regularity $R(u_j|\theta)$ of destination u_j in travel history of user v, conditioned to a multidimensional state $\theta = (\theta_1, \theta_2, ..)$, consists of any information sensed by an agent to be used when deciding to announce a potential trip.

Identifying the Aggregate Supply

Contrary to selecting the location with the highest probability to be the next trip's destination, as done by a typical predictor (i.e., a classifier), all patterns above a minimum regularity value are considered by the agent. Therefore, when a threshold ρ is applied to filter out unreliable trip predictions, the expected trips of user v conforming conditions stated in θ are

$$H^{v,\rho}(\theta) = \{b \in H^v | \forall i \in \theta : b.i = \theta_i \wedge R(b.d'|\theta) \geq \rho\}$$

For a specific passenger's request to location q, the fraction of the aggregate supply, that is, the car trips subset that could potentially be used for ridesharing is

$H^p(\theta, q) = \{b \in H^p(\theta) | b.d' \in N_\varphi(q) \wedge b.m = 'car'\}$, where $H^p(\theta) = \bigcup_{v \in V} H^{v,p}(\theta)$

The relevance of a trip pattern, when presented to a passenger, depends on how accurately it can be predicted and how frequent it is. Then the relevance of trip b_i can be quantified by predictability × frequency. Regularity denotes the probability of making a trip conditioned to some known state, and allows identifying which characteristics are relevant to correctly predict a specific destination; then predictability is a function of regularity and correlates positively (the more regular the more predictable), meaning they can be used interchangeably.

A first visualization highlights the origins of this predicted aggregate supply, so that these hotspots can be used by passengers as pickup points. In Figure 2*a*, origins of trips in $H^p(\theta, q)$ denoted by red circles, have different relevance for passengers, which is indicated by their radius size.

Let r_i be the circle's radius for trip i's origin, then

$$r_i = \lambda R(b_i.d'|\theta) \times f(b_i.d'|\theta), \ \forall b_i \in H^p(\theta, q)$$

where λ, is a configurable scaling parameter to correctly visualize the plot. Another useful piece of information is the predicted paths used by those trips so that passengers can find a ride on the way as shown in Figure 2*b*. The line weight of links in a network $G = (V, E)$ can represent the link relevance, calculated by

$$w_e = \lambda \sum_{b_i \in H^p(\theta, q)} R(b_i.d'|\theta) \times f(b_i.d'|\theta) z_{l,e}$$

where w_e is the line weight for link $e \in E$ for θ, and $z_{l,i} \in \{0, 1\}$ is a binary variable indicating whether link e is included in the predicted path of trip b_i. The line weight is then determined by the number of trips using that link, and also by each trip's predictability and frequency.

Results and Discussion

The previous methods are now applied to a dataset with multiple users' travel histories collected via a smartphone tracking app called MOVES; the dataset is briefly described in the following.

Dataset Description

The MOVES dataset contains travel histories of users in different geographical areas represented by unimodal trip chains. The variation in size of travel histories with respect to the number of days and trips is presented in Figure 3, *a* and *b*.

The travel modes automatically inferred by the app are: walking, running, cycling, and motorized vehicles; there is no distinction, for instance, between car or public transport. The distributions of travel distance and time per travel mode are plotted in Figure 3, *c* and *d*; outliers corresponding to very infrequent trips' characteristics have been filtered out using Tukey's method (*23*). These plots exhibit high heterogeneity in data, nevertheless most trips are short both in time and distance.

To evidence the need of further filtering, as expected when treating big data, a summary of the main characteristics of this dataset is given:

- 967 unique registered users as well as travel histories;
- 692,306 registered trips in 1,070 days;

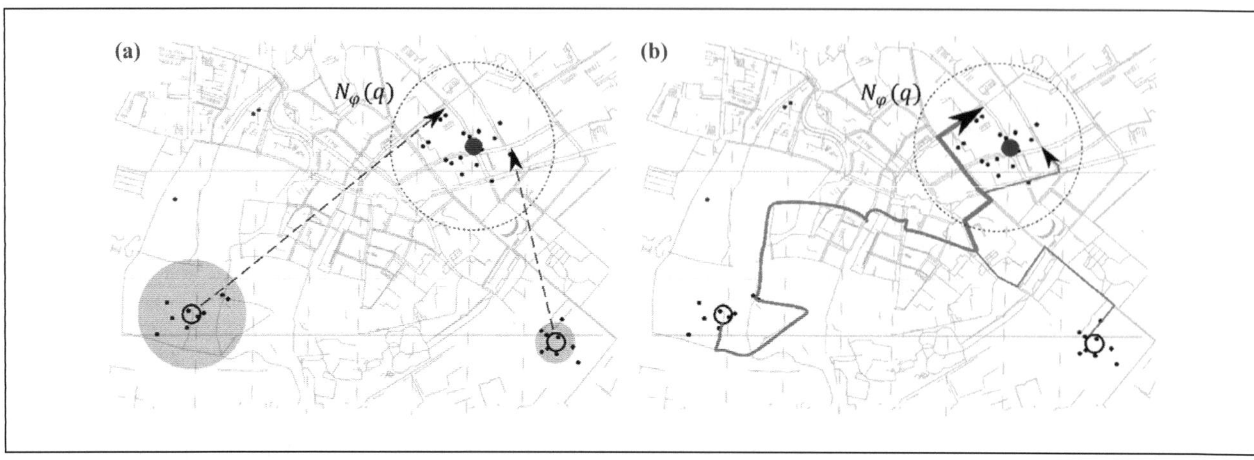

Figure 2. (*a*) Origins (red circles) of trips neighboring a location (blue circle), acting as ridesharing hotspots for which relevance is denoted by the radius size; and (*b*) trip's paths, where relevance is specified by line weight.

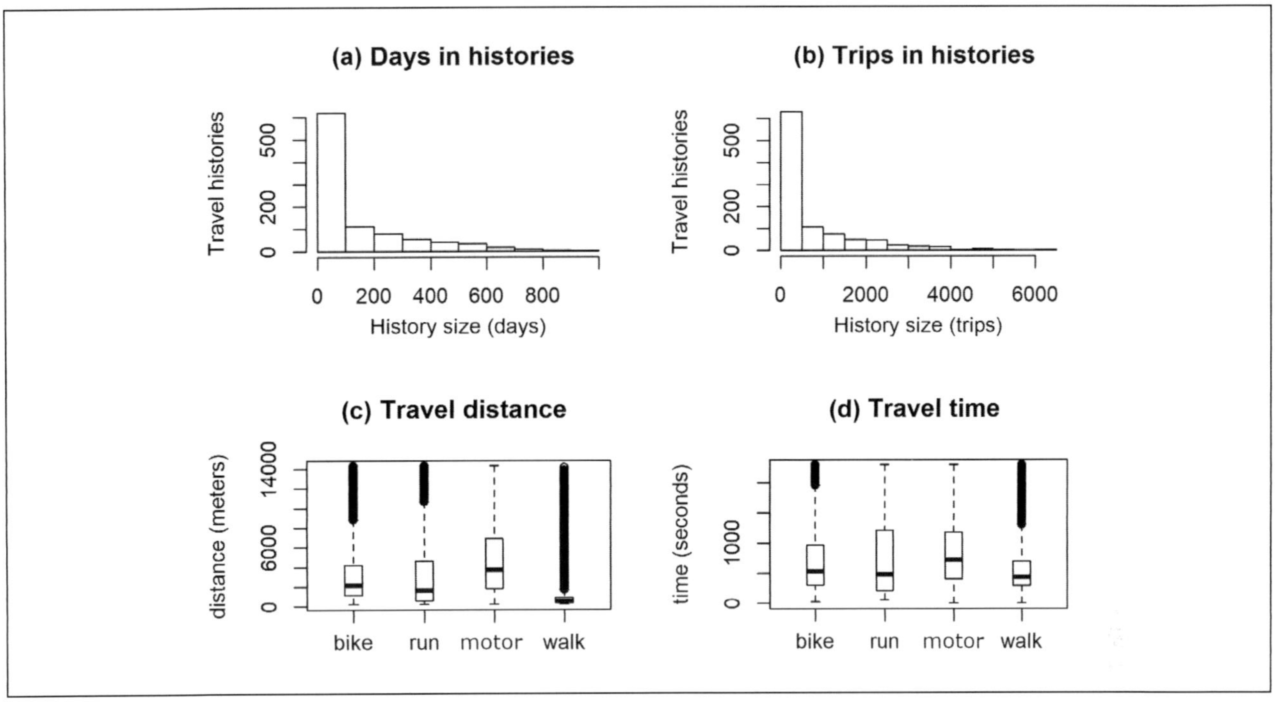

Figure 3. Heterogeneity in travel histories in MOVES dataset: (*a*) size in days; (*b*) in number of trips; (*c*) distribution of travel distance; and (*d*) time in the dataset.

- The time frame of travel histories ranges from 2 to 928 days.

The patterns studied in the next section were extracted from histories with at least 30 days of tracking, reducing the number of users to 573. The travel mode of trips was not considered in the study.

Extraction of Mobility Patterns

As parameter ε, corresponding to the search radius for the data-mining procedure, must be previously calibrated, trips with lengths smaller than this value were filtered out; this avoids trips with origins and destinations ending in the same POI. The heterogeneity in number of POIs per user, when $\varepsilon = 200$ m and $minPts = 5$ is presented in Figure 4*a*. Here, several POIs per user were found due to the low number of required visits specified by parameter *minPts*; nevertheless, the number of actual points of interest a user regularly visits is much smaller, as seen in Figure 4*b*, when only those with at least two visits per week were considered.

The following patterns correspond to repeated trips between the same pair of OD patterns after removing intra-cluster trips, when constrained to a minimum of observations denoted by parameter *minTrips*. The pattern's regularity, indicating the probability of the trip's destination conditioned to an origin, and the frequency denoting the average number of times the pattern was observed per day are presented in Figure 4, *c* and *d*.

The most regular users ranked by the largest number of OD patterns are presented in Table 1, when a minimum regularity level of 0.5 and a minimum frequency of two trips per week are applied.

With respect to spatiotemporal regularity, the most regular users are displayed in Table 2 when adding departure time intervals, with parameters $minTrips = 5$ and $\Delta t = 15$ min. The corresponding regularity and frequency of these patterns can be seen in Figure 5, *a* and *b*. As noted, when more dimensions are added to the analysis, the number of trips and their frequency are reduced, therefore, the supply for more specific requests (e.g., those also including a desired arrival time) will be reduced; nevertheless, regularity and therefore predictability of patterns rises due to the increased availability of context information.

The most regular users with respect to OD-arrival patterns, which includes trips between repeated OD pairs arriving at regular times are displayed in Table 3; Figure 5, *c* and *d* show the corresponding frequency and regularity of these patterns. Accuracy in predictions is always wanted; nevertheless, if agents are less demanding when announcing upcoming trips, the potential aggregate supply would be sufficiently large to encourage passengers to participate. The findings support the idea that stronger trip patterns only involve a few POIs per travel

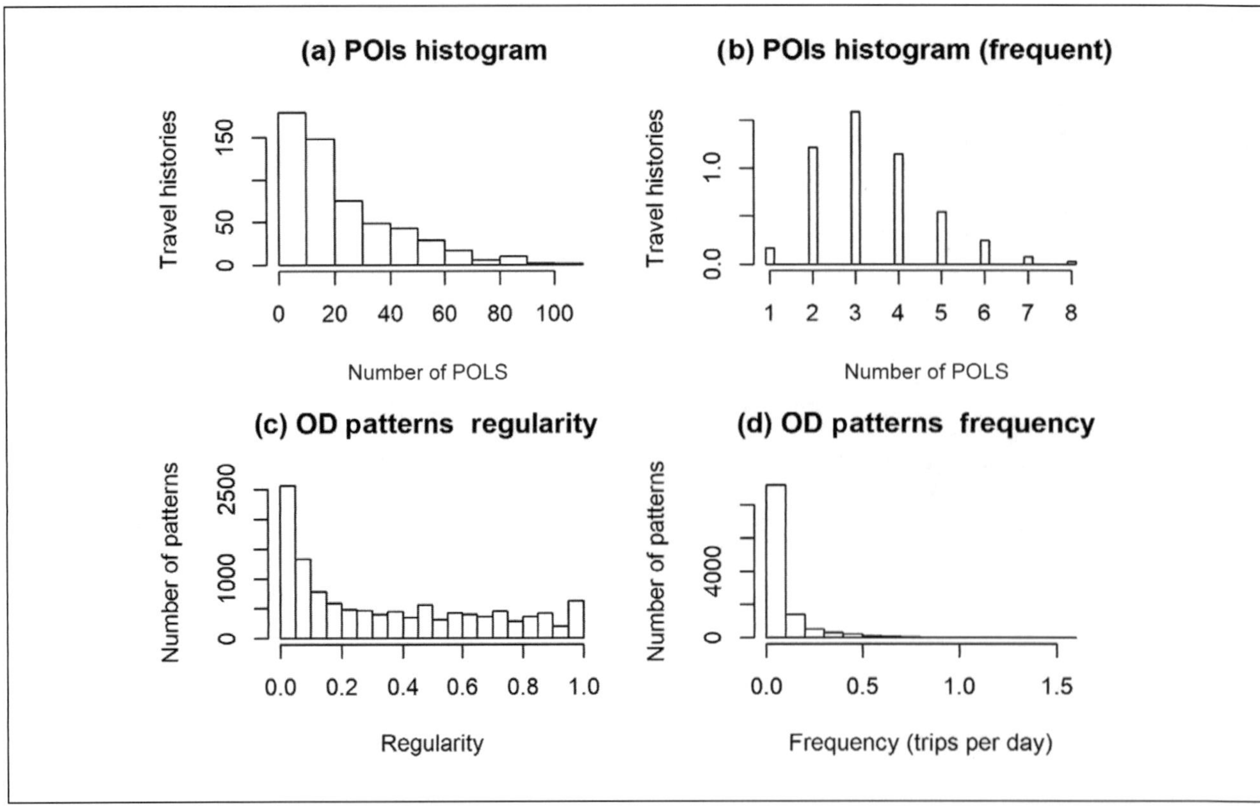

Figure 4. Variation in number of POIs per travel history: (*a*) those with a minimum of five occurrences in the entire history; (*b*) those with at least two visits per week; (*c*) regularity; and (*d*) frequency of the OD patterns.

Table 1. List of Most Regular Users with Respect to Their OD Patterns

User ID	OD patterns	Trips in patterns	Average frequency	Average regularity	POIs
U777	3	67	0.35	0.81	2
U992	3	109	0.54	0.71	3
U603	3	562	0.39	0.68	3
U000	3	236	0.3	0.65	2
U873	3	142	0.68	0.61	3
U211	3	166	0.32	0.6	3
U980	3	43	0.41	0.58	4
U552	2	87	0.29	0.93	2
U589	2	159	0.4	0.9	3
U519	2	176	0.61	0.85	3

Note: User ID = a unique user pseudo-identifier; OD patterns = the number of discovered patterns; Trips in patterns = the number of trips in history contained in patterns; Average frequency = the average daily frequency; Average regularity = the average regularity; POIs = the number of unique points of interest observed in the patterns.

history, as observed in the last two tables and as previously suggested by Figure 4*b*.

As the pickup hotspots of the potential aggregate supply conditioned to a hypothetical ride request include all patterns with minimum regularity, collecting several data about the current context is necessary. However, an agent should select only a few characteristics to announce the predicted trips, as looking for too-specific and predictable patterns would produce a short or even empty supply.

Table 2. List of Most Regular Users with Respect to Their OD-Departure Patterns

User ID	ODD patterns	Trips in patterns	Average frequency	Average regularity	POIs
U516	4	53	0.44	0.99	3
U254	2	44	0.33	0.98	3
U925	2	21	0.31	0.96	3
U992	2	74	0.55	0.95	3
U098	2	33	0.34	0.94	3
U777	2	50	0.39	0.88	3
U040	2	93	0.37	0.87	3
U730	2	42	0.34	0.86	3
U282	2	40	0.43	0.85	3
U661	2	67	0.47	0.83	3

Note: User ID = a unique user pseudo-identifier; ODD patterns = the number of OD patterns within specific departure time periods; Trips in patterns = the number of trips in history contained in patterns; Average frequency = the average daily frequency; Average regularity = the average regularity; POIs = the number of unique points of interest observed in the patterns.

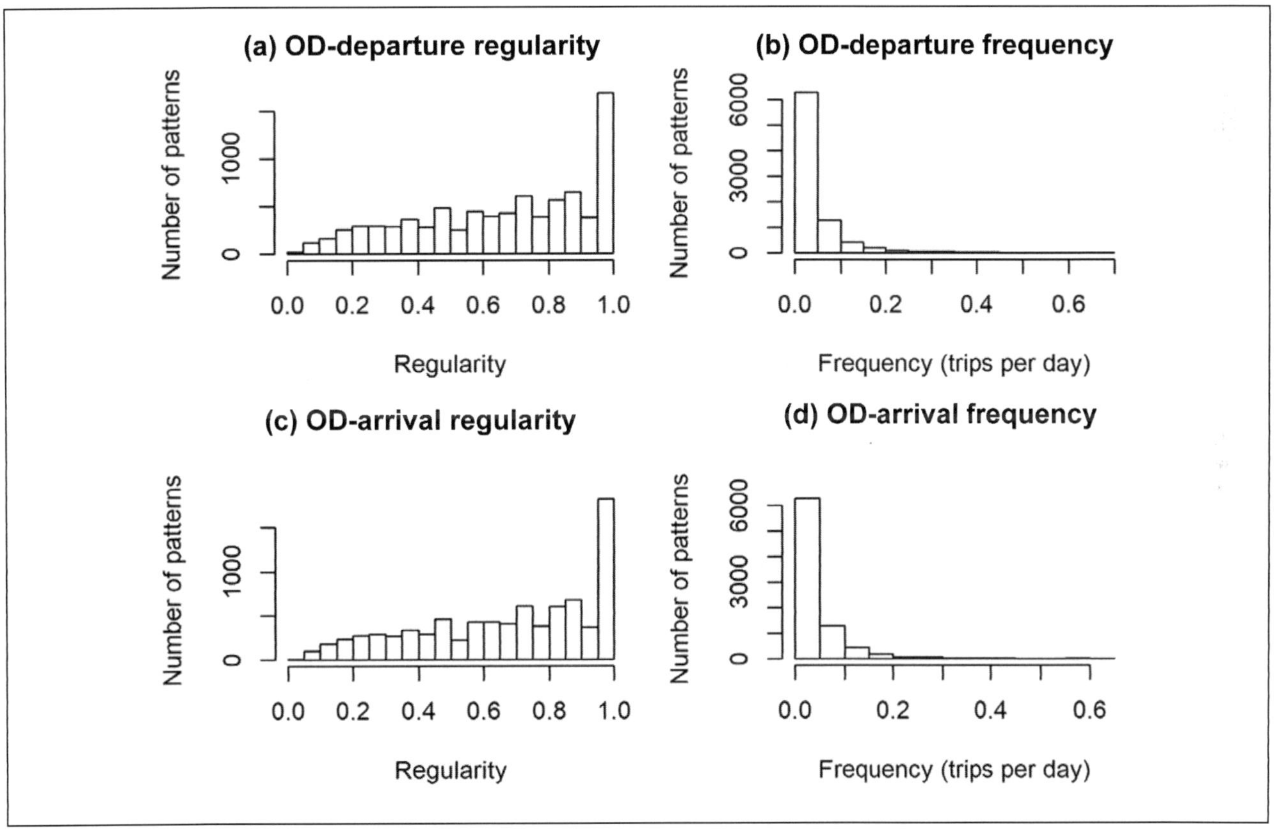

Figure 5. (*a*) Variation in regularity when adding departure period; (*b*) frequency of the OD-departure patterns; (*c*) variation in regularity when adding arrival period; and (*d*) frequency of the OD-arrival patterns.

Dealing with Other Datasets

The methods used in this paper can be used with datasets from other existing tracking apps, as long as trip chains can be retrieved containing spatiotemporal characteristics. The accuracy with respect to the collection of spatial data may differ for each data set, because even though sensors and operating system programming interfaces are shared among mobile apps, each app may have its own inference and aggregation methods. Another required

Table 3. List of Most Regular Users with Respect to Their OD-Arrival Patterns

User ID	ODA patterns	Trips in patterns	Average frequency	Average regularity	POIs
U516	4	57	0.48	1	3
U966	3	80	0.33	1	3
U873	3	65	0.31	0.96	3
U867	2	38	0.35	1	3
U992	2	77	0.57	0.97	3
U254	2	43	0.33	0.96	3
U980	2	22	0.31	0.92	3
U282	2	37	0.39	0.89	3
U777	2	48	0.38	0.88	3
U040	2	91	0.36	0.86	3

Note: User ID = a unique user pseudo-identifier; ODA patterns = the number of OD patterns within specific arrival time periods; Trips in patterns = the number of trips in history contained in patterns; Average frequency = the average daily frequency; Average regularity = the average regularity; POIs = the number of unique points of interest observed in the patterns.

item of information not included in the previous analysis is used travel modes. MOVES, unfortunately, does not differentiate motorized vehicles, though in some cases, modes can be inferred from existing data such as waypoints and instant speeds. It must be noted that databases consisting of raw data directly retrieved from sensors, should first be converted to a trip-chain format; this means that trip endpoints (stops) must be identified prior to data-mining procedures.

The main concern for research using user data is privacy. An appropriate data collection mechanism to certify that information is stored anonymously and with the user's explicit consent is compulsory. The methodology described in this paper does not require tracks or patterns to be stored in a remote server; instead these can reside in the smartphone for further use by software agents, avoiding the needless flow of data through insecure channels. Finally, the predicted aggregate supply consisting of the announced trips by other agents does not need to include any driver contact information.

Conclusions and Future Work

A methodology to identify mobility patterns at different levels from smartphone data collected by apps tracking a user's daily activities has been presented, together with measures to evaluate their relevance for ridesharing. The origins and trajectories of the predicted trips through these patterns represent hotspots for ridesharing in which passengers may find a potential aggregate supply conditioned to their ride requirements. For ridesharing, regularity that allows accurate prediction of trip patterns and frequency is important to determine realistic supply. It has been shown that mobility regularity can be identified from smartphone data via data-mining procedures, although heterogeneity in the travel histories is high. The number of patterns and their frequency has been found to decrease when more trip characteristics are considered, although regularity and prediction accuracy with respect to a set of conditions will increase. Flexibility on accepting trips with lower regularity, as well as different pickup windows or locations from their actual origins, is important to discover a larger aggregate supply. The method described in this paper is believed to facilitate the design of future apps that will help increase ridesharing rates.

Some future directions might include adding travel mode to the mobility patterns, as the user's role in ridesharing depends on this characteristic. The potential of ridesharing could be evaluated in specific regions via other datasets or activity-based simulators. The latter approach could be attained through the generation of a synthetic population consisting of users with heterogenous mobility behaviors, for instance, by transferring the distributions found in patterns extracted from datasets generated by tracking apps to an alternative dataset containing demographic data. Each possible scenario for the simulation would involve distinct ridesharing penetration rates, as well as thresholds of regularity and frequency, so that in each case the supply announced by an agent would be different.

Acknowledgments

This research project was financially supported by the National Secretariat of Higher Education, Science, Technology and Innovation of Ecuador (SENESCYT).

Author Contributions

The authors confirm contribution to the paper as follows: study conception and design: IM, CT; data collection: CR; analysis and interpretation of results: IM, CR, CT; draft manuscript preparation: IM. All authors reviewed the results and approved the final version of the manuscript.

References

1. Furuhata, M., M. Dessouky, F. Ordóñez, M. Brunet, X. Wang, and S. Koenig. Ridesharing: The State-of-the-Art and Future Directions. *Transportation Research Part B*, No. 57, 2013, pp. 28–46.
2. Bicocchi, N., and M. Mamei. Investigating Ride Sharing Opportunities through Mobility Data Analysis. *Pervasive and Mobile Computing*, Vol. 14, 2014, pp. 83–94.
3. Bicocchi, N., M. Mamei, A. Sassi, and F. Zambonelli. On Recommending Opportunistic Rides. *IEEE Transactions on Intelligent Transportation Systems*, 2017.
4. Agatz, N., A. Erera, M. Savelsbergh, and X. Wang. *Optimization for Dynamic Ride-Sharing: A Review*. North-Holland, 2012.
5. Lin, M., and W.-J. Hsu. Mining GPS Data for Mobility Patterns: A Survey. *Pervasive and Mobile Computing*, Vol. 12, 2014, pp. 1–16.
6. Cici, B., A. Markopoulou, E. Frias-Martinez, and N. Laoutaris. Assessing the Potential of Ride-Sharing Using Mobile and Social Data: A Tale of Four Cities. *Presented at the International Joint Conference on Pervasive and Ubiquitous Computing*, 2014, pp. 201–211.
7. Alexander, L. P., and M. C. González. Assessing the Impact of Real-Time Ridesharing on Urban Traffic Using Mobile Phone Data. *Proc., UrbComp*, 2015, pp. 1–9.
8. Stiglic, M., N. Agatz, M. Savelsbergh, and M. Gradisar. The Benefits of Meeting Points in Ride-Sharing Systems. *Transportation Research Part B: Methodological*, Vol. 82, 2015, pp. 36–53.
9. Stiglic, M., N. Agatz, M. Savelsbergh, and M. Gradisar. Making Dynamic Ride-Sharing Work: The Impact of Driver and Rider Flexibility. *Transportation Research Part E: Logistics and Transportation Review*, Vol. 91, 2016, pp. 190–207.
10. Goel, P., L. Kulik, and K. Ramamohanarao. Optimal Pick up Point Selection for Effective Ride Sharing. *IEEE Transactions on Big Data*, Vol. 3, No. 2, 2017, pp. 154–168.
11. Williams, M. J., R. M. Whitaker, and S. M. Allen. *Measuring Individual Regularity in Human Visiting Patterns*. 2012, pp. 117–122.
12. Wang, Y., N. J. Yuan, D. Lian, L. Xu, X. Xie, E. Chen, and Y. Rui. *Regularity and Conformity: Location Prediction Using Heterogeneous Mobility Data*. New York, NY, 2015, pp. 1275–1284.
13. Zhong, C., M. Batty, E. Manley, J. Wang, Z. Wang, F. Chen, and G. Schmitt. Variability in Regularity: Mining Temporal Mobility Patterns in London, Singapore and Beijing Using Smart-Card Data. *PLoS One*, Vol. 11, No. 2, 2016, pp. 1–17.
14. Caruana, R., and D. Freitag. Greedy Attribute Selection. In *Machine Learning Proceedings* (Cohen, W. W., and H. Hirsh, eds.), Morgan Kaufmann, San Francisco, CA, 1994, pp. 28–36.
15. Zheng, Y., L. Zhang, X. Xie, and W.-Y. Ma. *Mining Interesting Locations and Travel Sequences from GPS Trajectories*. ACM, New York, NY, 2009, pp. 791–800.
16. Mendoza, I., C. Tampère, and P. Mekerlé. Anticipatory Assistance for Real Time Ride-Sharing in Environments of Pervasive Computing. Presented at the 95th Annual Meeting of the Transportation Research Board, Washington, D.C., 2016.
17. Ashbrook, D., and T. Starner. Learning Significant Locations and Predicting User Movement with GPS. *Proc., Sixth International Symposium on Wearable Computers*, 2002, pp. 101–108.
18. Ashbrook, D., and T. Starner. Using GPS to Learn Significant Locations and Predict Movement across Multiple Users. *Personal and Ubiquitous Computing*, Vol. 7, No. 5, 2003, pp. 275–286.
19. Nurmi, P., and S. Bhattacharya. Identifying Meaningful Places: The Non-Parametric Way. In *Pervasive Computing* (Indulska, J., D. Patterson, T. Rodden, and M. Ott, eds.), Springer, Berlin Heidelberg, 2008, pp. 111–127.
20. Ester, M., H.-P. Kriegel, J. Sander, and X. Xu. A Density-Based Algorithm for Discovering Clusters in Large Spatial Databases with Noise. *Proc., 2nd Conference of ACM SIGKDD*, ACM Press, New York, 1996, pp. 226–231.
21. Ankerst, M., M. M. Breunig, H.-P. Kriegel, and J. Sander. OPTICS: Ordering Points to Identify the Clustering Structure. No. 28, 1999, pp. 49–60.
22. Butler, H., C. Schmid, D. Springmeyer, and J. Livni. *Spatial Reference*. EPSG Projection 31370 - Belgian Lambert 72. 2016.
23. Tukey, J. W. *Exploratory Data Analysis*. Addison-Wesley Publishing Company, 1977.

The Standing Committee on Urban Transportation Data and Information Systems (ABJ30) peer-reviewed this paper (18-06037).

TRR
JOURNAL OF THE TRANSPORTATION RESEARCH BOARD

Article

Transportation Research Record
2018, Vol. 2672(42) 224–234

DOI: 10.1177/0361198118791388
journals.sagepub.com/home/trr
SAGE

Closing the Loop: Improving Transit through Crowdsourced Information

Sean J. Barbeau[1]

Abstract

Offering real-time arrival information to riders via mobile applications has been shown to improve the rider's perception of transit, and even increase ridership. This direct connection to riders also offers the transit agency an opportunity to collect feedback on how transit service and infrastructure can be improved, including pedestrian and bike access to transit. These improvements will lead to an enhanced customer experience and can potentially help address Title VI access equity concerns. However, managing the sheer volume of this rider feedback can be very challenging, especially when various departments and agencies (e.g., city/county government) are involved (e.g., who owns the bench by the bus stop?). This paper discusses the design and deployment of a pilot project in Tampa, Florida, which focused on the improvement of the feedback loop from riders back to transit agencies, local government, and departments of transportation. This project made enhancements to the open-source OneBusAway mobile app, originally deployed in Tampa in 2013, to include support for the Open311 standard for issue reporting. Open311 support gives agencies the option of selecting a hosted issue management solution such as SeeClickFix.com and PublicStuff.com, or the option to utilize existing open-source Open311-compliant software. Lessons learned from regional collaboration surrounding issue reporting and infrastructure improvements are discussed, as are the technical design and challenges behind implementing such a system. The results of the first 6 months of system deployment covering 677 issue reports are presented, including specific examples of cross-jurisdictional and multimodal issues reported by the public.

Offering real-time arrival information to riders via mobile applications has been shown to improve the rider's perception of transit, and even increase ridership (*1–3*). This direct connection to riders also offers the transit agency an opportunity to collect feedback on how transit service and infrastructure can be improved, including pedestrian and bike access to transit. Directly addressing rider concerns will lead to an enhanced customer experience and potentially help address Title VI access equity concerns. However, managing the sheer volume of feedback from thousands of riders can be very challenging, especially when various departments and agencies (e.g., city/county government, departments of transportation) are involved (e.g., does this broken bench belong to the transit agency or the county?). If this feedback can be appropriately triaged and turned into actionable information, it can assist transit agencies, local governments, and departments of transportation in prioritizing costly infrastructure improvements. This data-rich approach is a significant improvement to the traditional prioritization process, which is often based on a simple "buffer" surrounding the planned transit route and calculations of the number of people in an area.

This paper discusses the design and deployment of a pilot project in Tampa, Florida, funded by the Florida Department of Transportation (FDOT) and the National Center for Transit Research, which focused on the improvement of the feedback loop from riders back to transit agencies, local government, and departments of transportation. This project made improvements to the OneBusAway open-source mobile app, deployed in Tampa in 2013, to include support for the Open311 standard for issue reporting, and expanded the region to include the two largest regional transit agencies, Hillsborough Area Regional Transit (HART) and Pinellas Suncoast Transit Authority (PSTA) (*4*). Open311 support gives agencies in all OneBusAway regions (e.g., Seattle, Washington, D.C., San Diego) the option of selecting a hosted issue management solution that supports Open311 (e.g., SeeClickFix, Connected

[1]Center for Urban Transportation Research, University of South Florida, Tampa, FL

Corresponding Author:
Address correspondence to Sean J. Barbeau: barbeau@cutr.usf.edu

Bits, PublicStuff (*5–7*)) or the option to utilize existing open-source Open311-compliant software. The following sections discuss the technical infrastructure developed to deploy a standard-based system for collecting user feedback, design and deployment challenges, as well as the results from the deployment, including lessons learned and example user feedback.

Challenges with Managing User Feedback

OneBusAway is a mobile application for real-time multimodal information that allows transit riders to see predicted arrival and departure times for each bus stop. Unlike other mobile transit apps, OneBusAway is open source, which means that anyone can download and deploy the software source code in their region. OneBusAway started as a project created by graduate students at the University of Washington, and has since spread to more than 10 cities with approximately 340,000 active users. The open-source nature of OneBusAway differentiates it from all other transit applications (e.g., Transit App, Moovit), as it allows researchers to deploy and evaluate solutions designed to enhance and evaluate both the rider and agency experience (*1*, *3*, *8–12*). This research provides decision makers with objective evidence to help evaluate and justify investments in new transit technologies.

OneBusAway was officially launched in Tampa, Florida, in August 2013, and continues to be the main source of real-time public transportation information for HART for over 35,000 active users in Tampa. However, one of the significant challenges in deploying such an app is the large amount of feedback provided by users. The initial design of OneBusAway provided a single regional email contact point for feedback—tapping on "Contact Us" in the OneBusAway app would open an email directed to the regional contact point (initially HART). Yet this creates several challenges:

1. Managing the issue lifecycle via email is inefficient. Email is not a good medium for tracking individual issues, including who they are assigned to and when a specific issue is resolved.
2. OneBusAway regions can contain multiple agencies. If there is only a single email contact point for a region, multiple transit agencies must share that email address. This causes additional confusion and issue lifecycle management difficulties.
3. Intermodal issues are difficult to manage. Because the transit agency is the lead organization responding to issues, any issues related to sidewalks, bicycle lanes, parking, right of way, roads, or other infrastructure not managed by the transit agency must be shared with other organizations.

In 2011, Ferris et al implemented a proof-of-concept crowdsourcing issue reporting system within OneBusAway that focused on arrival times (i.e., trip) problem reports, and stop information problem reports (*8*). These improvements gave users a way to categorize problems falling into these two categories and submit them to a OneBusAway server database, although additional challenges were encountered:

1. Processing and analyzing the volume of reports is burdensome. The issue management interface within the OneBusAway administrative website does not allow the transit agency to search, analyze, and aggregate issues. As a result, dealing with a large volume of reports is inefficient and costly, leading to concerns of delays in response.
2. Arrival times and stop information does not sufficiently cover problems reported by users. HART wanted to add more categories for which users could report problems (e.g., Wi-Fi on bus not working, broken stop benches).
3. The problem-reporting categories were rigid and hard-coded into the mobile app. HART needed the ongoing ability to add and update problem-reporting categories within the app without requiring software code changes.
4. The administrative interface within OneBusAway to manage the issue lifecycle was very limiting. All reports were anonymous, which did not give HART the ability to respond to users to let them know that an issue was resolved. This also creates the rider perception that the transit agency is not receiving or reacting to any of the feedback. Additionally, multiple tiers of user permissions were not supported in the administrative interface, which did not give the ability to assign restricted permissions to customer service staff, who should be able to manage reported issues, but not have access to other OneBusAway administrative tools (e.g., updating system schedule data).

Improving the Issue Reporting Experience

Based on the above challenges encountered with the existing email and arrival time and stop information reporting mechanisms, the research team developed the following criteria for creating a new issue reporting solution within OneBusAway:

- Interoperability: Issue reporting management systems need to allow interoperability with external systems to be successfully integrated with OneBusAway. This type of integration is typically

accomplished via a standardized open application programming interface (API) that would allow OneBusAway to interact and submit issue reports to the system.
- User feedback and interaction: After issue submission, the data collection system should return useful feedback to the application, and allow additional communication between the user and managing agency if needed. Anonymous reporting of issues is also desired.
- Multi-jurisdictional issue notification and management: Transit service often occurs at the boundary of service jurisdictions where infrastructure may be owned and maintained by several different parties. For example, a traveler walking to a bus stop may encounter a broken sidewalk that is maintained by the county. When they reach their bus stop, if the bench is broken, it may be maintained by the transit agency. If a bus consistently gets stuck at a red traffic light, that traffic signal may be owned by the city. And a pothole on the road, which the traveler must step around to board the bus, may be owned by the department of transportation district. The public should be shielded from these complexities and allowed to report an issue to a single contact point, from which it is automatically directed to the correct agency.
- Standardization of data elements: If possible, standardized reporting protocols and data elements should be used so that a common set of information is reported across all OneBusAway deployments.

The research team worked closely with FDOT District 7 as well as HART, PSTA, and Pinellas County to ensure that a solution could meet the needs of each of these agencies.

Open311: A Standardized Issue Reporting Protocol

The research team reviewed several major civic issue management products (e.g., SeeClickFix, Connected Bits, Public Stuff) and found that many supported the Open311 protocol. Open311 is an open API specification for reporting non-emergency issues deployed in over 35 cities around the world, and was selected by the research team as the issue reporting protocol for OneBusAway (*4*). To the research team's knowledge, this is the first time the Open311 format has been used for issue reporting for public transportation. Open311 support in OneBusAway would allow a region to choose the issue reporting tool/service that they want to receive and store information in, and if that tool/service supports the Open311 format, OneBusAway and the tool should work together.

Given that SeeClickFix was actively used by Pinellas County, FL, and the City of St Petersburg, and that OneBusAway was deployed in Tampa with HART, the greater Tampa Bay area was a model region for a pilot of enhanced issue reporting within OneBusAway. The research team added PSTA, which provides transit service to Pinellas County, to OneBusAway in Tampa to demonstrate the ability of more than one agency to triage and manage issues reported via the Open311 interface. The research team carefully analyzed PSTA's real-time arrival data as it was added to OneBusAway to ensure high-quality information was shown to riders (challenges and lessons learned during this process are outside the scope of this paper but discussed extensively in Barbeau (*13*)). FDOT D7 participated in the project to give a perspective on intermodal issue management and streamline issue-related communications with HART and PSTA.

The following sections discuss the technical design of the system that adds Open311 support to OneBusAway. Because all enhancements to OneBusAway are open-source software, other apps can use the source code developed by the research team to implement the same issue reporting improvements.

Integration of Open311 Support in OneBusAway

System Architecture. OneBusAway uses a regional model, in which the mobile applications contact a centralized Regions API to discover the available OneBusAway deployments in each city. The Regions API contains information for each region, including the region name (e.g., "Tampa"), region OneBusAway server URL (e.g., http://api.tampa.onebusaway.org), and the geographic bounds for each region. Regional OneBusAway servers are typically maintained by a transit agency in that region.

To add support for enhanced issue reporting, a set of new fields for each region was created in the OneBusAway Server Directory to allow the designation of Open311-compliant servers for each region.

The new fields in the Regions API for each region are:

- Open311_Base_URLs: A pipe (|) separated list of Open311 endpoints in the region (e.g., http://api.seeclickfix.com/|http://api.myserver.com/). If no Open311 server is available for a region, this field should be left blank.
- Open311_API_Keys: A pipe (|) separated list of API keys for the Open311 endpoints in the region. If no Open311 server is available, this field should be left blank.

- Open311_Jurisdiction_ID: The default jurisdiction identity (ID) for the Open311 endpoint in the region (e.g., tampa-fl). This field is only populated if there is a single jurisdiction for the entire OneBusAway region and that area has been assigned a jurisdiction ID by the Open311 provider, and if the Open311 provider doesn't support a server discovery protocol (discussed later in this paper). If no Open311 server is available, or if the Open311 provider(s) support(s) a server discovery protocol, this field should be left blank. This field is not used in the Tampa deployment of Open311 but is provided for future compatibility with other Open311 providers.

This information allows the mobile apps to discover the URLs of the local Open311-compliant issue reporting servers so the Open311 client software knows where to submit issue report data.

For regions without Open311 servers, the same issue reporting process that is currently supported within the mobile apps (i.e., directly sending feedback to the OneBusAway servers or via a single regional email contact point) is used. If the region provides an Open311 server, then the enhanced issue reporting processes created as part of this project are made available to that region. Therefore, the results of this research project are instantly available to any OneBusAway region that wants to implement an Open311-based issue management system.

The project team reviewed the Open311 GeoReport v2 specification and designed and implemented the software library necessary to submit issue reports to an Open311-compliant server (*4,14*). A diagram showing the protocol used to communicate between an application using the Open311 library and an Open311 server is shown in Figure 1.

A new issue reporting user interface for OneBusAway was also designed and implemented in the OneBusAway Android app (Figure 2). As discussed below, this design evolved throughout the project as the research team received feedback from agencies—the final design is shown in Figure 2.

After tapping on the new "Send feedback" link in the main navigation drawer (Figure 2, first screen) for the first time, the user is prompted to confirm that they are connected to the proper regional OneBusAway server (Figure 2, second screen). If the user indicates they are not in the region the app has auto-selected (e.g., if there is a GPS error with their device), they are prompted to enter the correct region. This screen helps users resolve location errors from their device without needing to contact customer service.

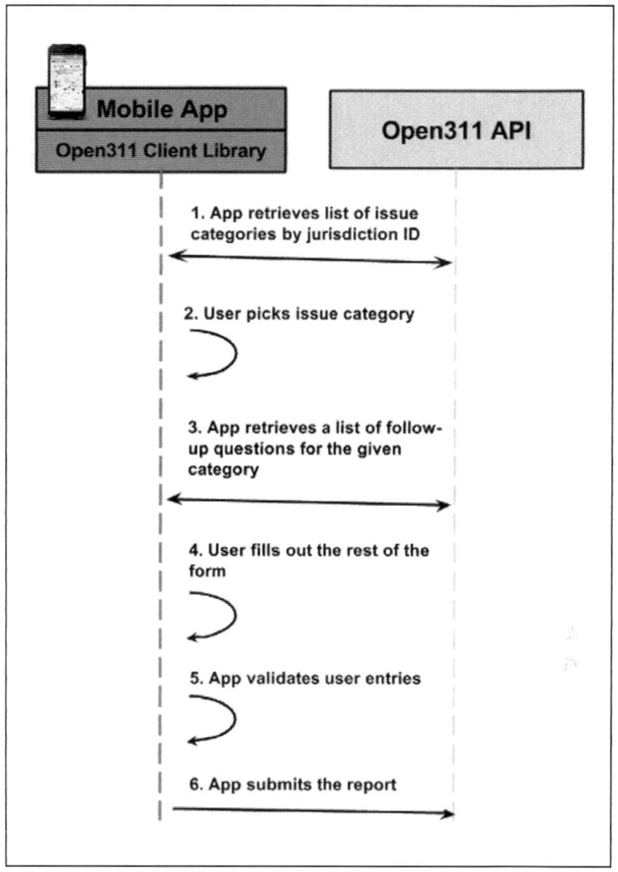

Figure 1. The Open311 protocol for the communication between an app and an Open311 API server.

Based on discussions with participating agencies, early beta users of the feedback system and the OneBusAway community, the organization of the general feedback categories visible to the user within the app (Figure 2, third screen) evolved to facilitate the flow of issues to the correct contact point at the agency. Each category shown on the third screen of Figure 2 channels user feedback to the most appropriate entity based on specific examples shown for each category.

The "Contact Customer Service" category was added to capture time-sensitive issues that should be directed straight to the agency's customer service department. Examples best addressed by customer service directly include safety or security situations, bus driver compliments and complaints, requests for directions, and lost and found requests that don't have a specific geographic component. When users select this option they typically know which agency they wish to reach (e.g., if the rider is on a HART bus and there is an escalating dispute between a driver and passenger). HART requested the ability to send feedback via email directly to their customer service department, and this feature was

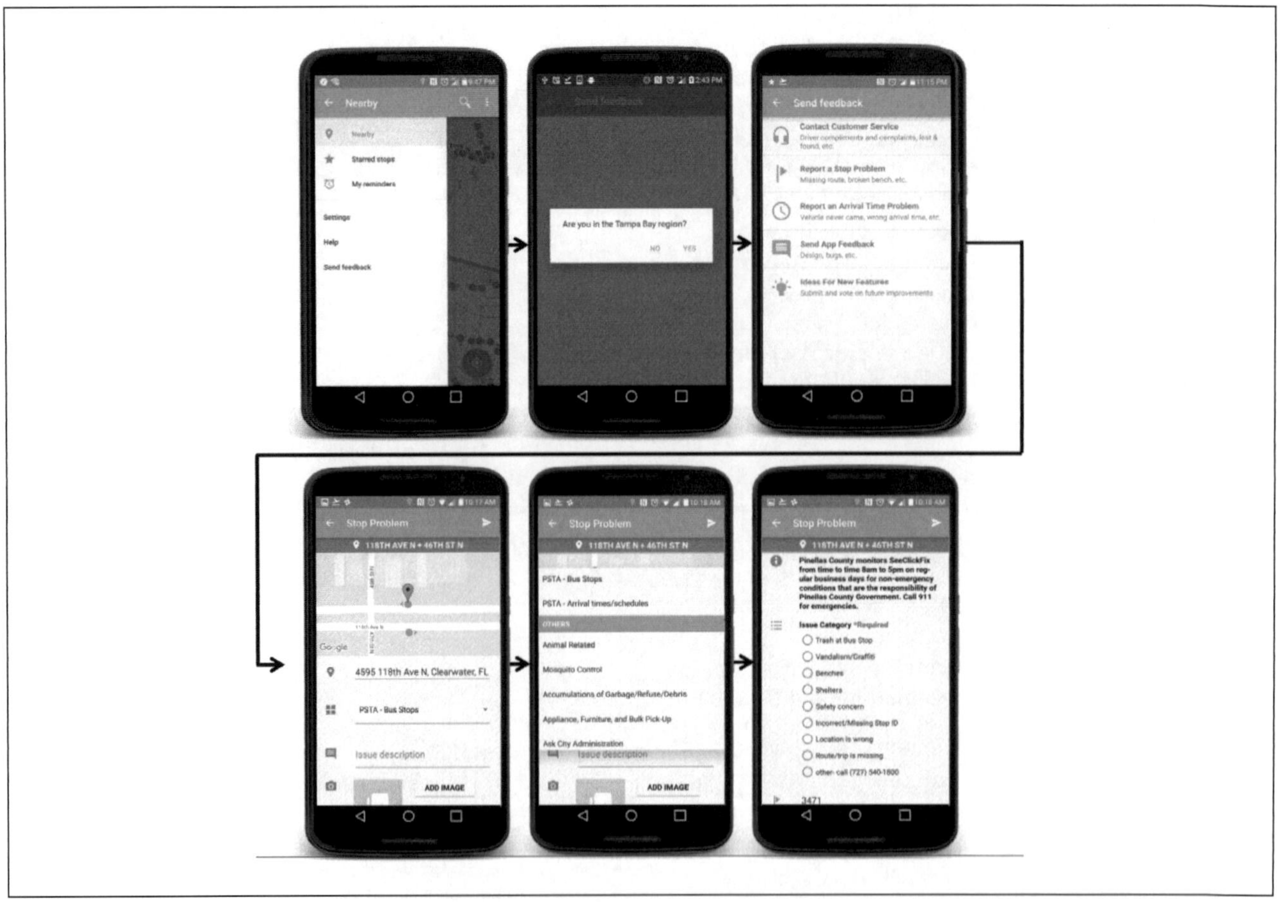

Figure 2. The new issue reporting process for the OneBusAway mobile app, including directing issues to regional Open311-compliant servers.

accomplished via a proposal for adding agency-specific email addresses to the General Transit Feed Specification (GTFS) format used by OneBusAway and many other apps (*15*).

The "Ideas For New Features" category was added to direct community-generated ideas directly to the main OneBusAway project IdeaScale site (http://onebusaway.ideascale.com/). The OneBusAway IdeaScale site allows users from all regions to submit and vote on ideas for new features in the OneBusAway open-source project, which will help transit agencies from around the world to prioritize and collaboratively fund enhancements to the OneBusAway open-source project. Agencies participating in the OneBusAway project are already reaping these cross-regional benefits, with improvements from New York City being deployed in Tampa, and improvements from Tampa being deployed in Atlanta and Seattle.

The fifth screen in Figure 2 shows the available top-level categories for which an issue can be reported for a given geographic location. One advantage of the Open311 standard is that it allows each local jurisdiction to define the reporting issue categories. The Open311 client retrieves a list of these categories from the local jurisdiction, and displays these to the user. The issue categories used by Pinellas County and the City of St Petersburg in SeeClickFix included a variety of non-transit categories, such as animal-related ("barking dog"), code compliance ("noise," "trash"), ditch obstruction, flooding, graffiti, and mosquito control. PSTA added the categories of "Bus Stops" and "Arrival Times" to this list. HART defined many additional top-level issue categories specific to issues reported in their service, including "ADA Complaints," "Commendation," "Environmental Issues (recycling, fuel spills...)," "Operator/Passenger Behavior," "Vehicle Maintenance," and more. Additionally, each jurisdiction can define sub-categories and follow-up questions for each top-level category that the user must answer before submitting the issue report (Figure 2, sixth screen). This enables the jurisdiction to ensure they have all information needed for the report to be actionable.

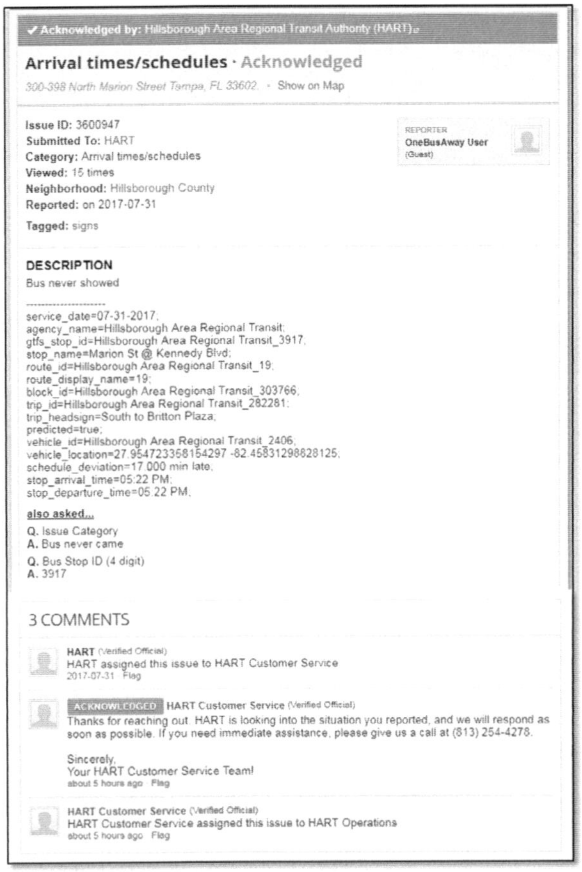

Figure 3. Metadata for the vehicle included in each issue report.

An example issue reported by a OneBusAway user via the Open311 protocol to the SeeClickFix system is shown in Figure 3. In this example, a user reports that the app showed a predicted arrival for a bus, but no bus arrived (i.e., a "ghost" bus). When the user taps on the arrival time in the app to report a problem for that arrival, OneBusAway automatically captures metadata for that vehicle (e.g., user location, stop_id, tripoidal, route_id, block_id, trip_headsign, predicted and scheduled arrival, and departure time) and attaches this to the issue report. This information, visible in the "Description" section of Figure 3, enables the transit agency to quickly triage the issue report and respond with preset answers. In this case, HART customer service has assigned the issue to HART operations so that operations can evaluate if a problem with the automatic vehicle location (AVL) system is causing these "ghost" buses. After operations reviews and responds to the issue, HART can close the issue and provide a resolution to the user. Issue reporting products such as SeeClickFix, PublicStuff, and Connected Bits track these actions and can provide detailed reports to the agency, so they can track how long it takes various departments within the agency to acknowledge, assign, and close out issues.

Deployment and Lessons Learned

The Open311-based issue reporting feature within OneBusAway was deployed throughout 2016 as part of beta testing with HART and PSTA transit riders in Tampa Bay, using SeeClickFix as the Open311-compliant issue management system. HART announced an official launch of SeeClickFix as their issue reporting management tool on January 30, 2017. The following sections discuss lessons learned from the first 6 months of deployment through August 1, 2017.

Cross-Jurisdictional Issues

One goal of using an Open311-compliant issue reporting platform was to enable the app to capture diverse types of issues beyond simply reporting errors in arrival times. Figure 4 shows that of the 677 issue reports in Hillsborough County, approximately 87% were related to arrival times and schedules, which means that approximately 13% were related to other categories.

Figure 5 shows photos from two different safety-related issues reported by OneBusAway users. On the left, an electrical utility box was left open near a bus stop (*16*). On the right, a storm drain has been damaged with the debris extending into the pedestrian and bike travel paths (*17*). These issues demonstrate the cross-jurisdictional nature of problems reported by the public. In the case of the electrical box, HART shared a link to this issue, including the exact geographic latitude and longitude and photo, with Frontier Communications, a private telecommunications company, who was able to quickly resolve the problem. HART shared the information about the storm drain, again including the exact latitude, longitude, and photo, to the City of Tampa, who responded and repaired the drain the following week. The photos and exact geographic location help users communicate a first-person perspective to HART, who can in turn share it with other agencies as needed, which facilities a rapid response for truly hazardous situations.

In the above cases, Frontier Communications and the City of Tampa didn't have existing SeeClickFix accounts (either via their own organization or via sharing with HART) and therefore they were not automatically notified of these issues. Instead, HART shared URLs to these issues via email, and the organizations then used their own internal management tools to log and resolve the issue, reporting back to HART when they were resolved. Pinellas County reported an analogous situation when they first started using SeeClickFix—at first,

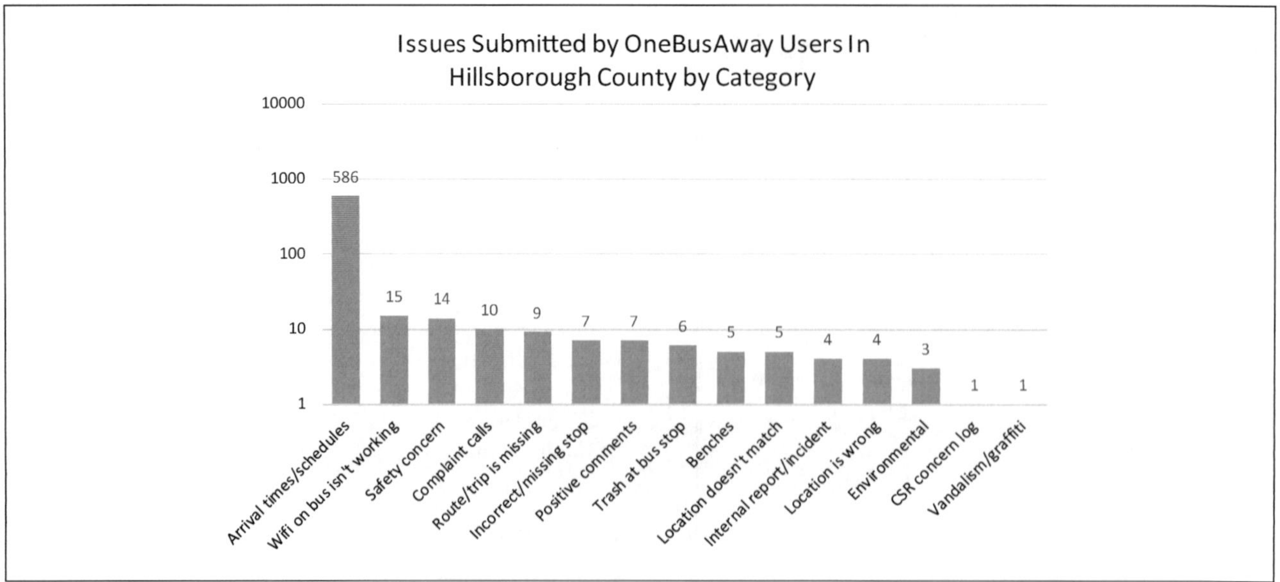

Figure 4. Arrival times and schedules issues comprise the bulk of reports, but other issue types are also captured.

Figure 5. Two cross-jurisdictional safety issues reported by OneBusAway users: left, electrical box open; right, broken storm drain.

Pinellas County was the only account holder and shared issue URLs with other organizations. However, as other organizations began working with Pinellas County on resolving issues, they saw the benefit of using a shared, dedicated issue management tool and began to adopt the technology. The City of St Petersburg, City of Pinellas Park, City of Clearwater, City of St Pete Beach, City of Tarpon Springs, City of Treasure Island, and Belleair are all independent jurisdictions contained within Pinellas County that have gradually adopted the shared issue management platform because they saw the benefits. The ability to either share issue links (when both organizations do not have an account on the same issue management system) or directly auto-assign issues (for shared issue management systems) illustrates the flexibility of these systems. Pinellas County cited this flexibility as key to the gradual adoption of a shared issue management platform by the region—wholesale adoption and

Figure 6. Dangerous bus stop location reported by a OneBusAway user.

purchase of licenses for all jurisdictions at once would have likely been politically prohibitive. Nonetheless, gradual adoption allows each jurisdiction to try lower-priced pilots at their own pace before adopting the technology for the entire organization. And, as each new organization comes on board, the issue management process for the nearby organizations becomes simpler as they can use the automated tools.

In another example, a user reported a bus stop in dangerous locations via OneBusAway (*18*):

> This stop is dangerous and should be moved. It is at MacDill and Granada serving Northbound Route 574 (South Tampa Flex). There is on-street parking directly in front of it. If someone is parked there bus patrons must step partially into the street to be seen by the driver, which puts them at risk of being hit by a passing motorist. Better locations for this stop would be either north of Barcelona, in front of Datz Deli, which has a parking lot, or MacDill and Santiago, under the Lee Roy Selmon Expressway and just before the right turn lane for traffic going onto northbound Bay To Bay starts. There is a bus bench and a sign stanchion at this intersection but no sign and according to the driver it is not a stop.

In this case, the user didn't attach a picture to the report, but given the stop ID and exact location from the metadata, a quick review of Google StreetView shows that there is indeed unrestricted on-street parking directly in front of the bus stop (Figure 6), creating a conflict between riders and vehicular traffic.

HART's service planning and safety and security teams reviewed the location of this stop based on this feedback and relocated it.

HART has also seen positive feedback via the app. One user reported (*19*):

> *Bus driver was kind enough to wait for everyone that got there by departure time. My husband and I are disabled so we can't run for buses anymore. This driver was patient while we walked to the stop. Just polite and kind. Bus 1510 not sure of operator number.*
>
> Please thank him for his good attitude from 2 loyal patrons.

HART has intentionally provided an issue category for positive comments to encourage patrons to report good service when they receive it, so they can pass positive feedback back to operators.

Potential Areas of Improvement in the Open311 Specification

During the design and deployment of the system, the research team learned that there are certain aspects of issue reporting that are not covered by the Open311 standard, which are discussed below.

The discovery protocol of determining Open311 system coverage has emerged as an important feature for regional deployments. Initially it was thought that a single Open311 server would be used for an entire OneBusAway region. However, after further discussions, it is apparent that different agencies within the same large OneBusAway region may use different issue reporting vendors, and some areas may not be monitored by an Open311-compliant system at all.

To ensure that issues for a geographic area are submitted to the correct issue reporting server, the project team designed and implemented a discovery protocol for Open311 servers within a OneBusAway region. When a user reports an issue, each of the Open311 servers for the region are queried to determine which issue categories exist for that geographic location. If an Open311 server returns more than one category, then the issue is sent to that Open311 server. If only one "Other" category is reported for a specific Open311 server, or if no Open311 servers exist for a region, the issue is sent to the OneBusAway server.

The project team worked with the SeeClickFix engineering team to design, implement, and test software that executes this new Open311 server discovery protocol. It was successfully deployed during this project, with Pinellas County being monitored by the SeeClickFix server, and Hillsborough County initially being unmonitored (and therefore issues were still reported to the OneBusAway server for Hillsborough County). HART started monitoring Hillsborough County in SeeClickFix in January 2017, at which point issues reported in Hillsborough County were instantly directed to SeeClickFix and successfully received by HART. Future work could examine proposing this design as part of the Open311 standard.

Similarly, the project team also worked with SeeClickFix to enable the direct submission of images via their Open311 API, although this design currently falls outside the exact Open311 specification. The current Open311 specification requires a link to an image, which then requires a secondary system to be set up to store these images. Future work could propose the addition of direct image submissions to the Open311 standard.

As mentioned earlier, Open311 allows local jurisdictions to define the issue reporting categories for their location. While this flexibility is needed, general guidance for the issue category types related to public transportation would help agencies standardize some categories and enable easier cross-agency coordination and reporting. TCRP Report 179 "Use of Web-Based Rider Feedback to Improve Public Transit Services" came to the same conclusion and identified category standardization as a future research need (*20*).

Currently, all vehicle and stop metadata are inserted into the body of the description, which makes it difficult to produce reports for trip_ids, route_ids, or schedule deviations (e.g., show me all issues for Route 5 when it was running more than 10 min behind schedule). At present, Open311 does not support "hidden" fields—this means that if these fields are added individually, they will be shown to the user when they submit the issue. Ideally, these fields would be hidden from the user's view, but the OneBusAway app would still have a way to discover these field names and submit discrete data for these fields. Future work could examine expanding the Open311 specification to include this use case.

Potential Areas of Improvement for the OneBusAway Mobile App

HART requested that when a user is preparing to submit a new issue report, the OneBusAway app show existing issues that are nearby the user's location, so the user can determine if the problem has already been reported. This could be implemented as part of future work.

HART was interested in capturing more information about the transit rider, including when they arrived at a bus stop (i.e., how long they waited for the bus), as well as travel path and origin/destination information. This feature would need to be balanced with privacy protections and transparency for the user.

Potential Areas of Improvement within HART's Issue Management Process

When reviewing issue reports, HART discovered that the block_id values publicly shared in their GTFS data do not match the block_ids that are visible to customer service representatives in the internal AVL system interface. This is relevant for issues related to bus driver behavior and potential disciplinary issues, as the customer service representatives need to look up the driver's ID based on block_id. Future work could examine transforming block_ids so customer service representatives don't have to go through a manual lookup process to map the GTFS block_ids to internal block_ids.

Overall, HART had a very positive experience with the new issue management platform. HART plans to eventually move their entire customer service department to using SeeClickFix instead of their legacy customer service software. However, to satisfy their customer service requirements, HART would need to be able to automatically generate mailers' responses to people that request them, as well as generate monthly letters to the driver union for any actionable items. HART is currently investigating to determine if SeeClickFix can meet these requirements as well.

There have been two types of "arrival time problems" reported via the app—errors in predictions (e.g., the app says the bus is still 5 min away but it has just passed the rider) and complaints about the bus running early or late (but the prediction itself is accurate). The issue report is timestamped and the "schedule_deviation" shown to the rider when they report the problem is included in the issue metadata, so it is possible for customer service staff to distinguish between these two scenarios without having to go through the process of pulling logs from the AVL system. Discussions with customer service staff determined that the research team should not attempt to break the "arrival time problems" down into further subcategories so the riders could label the issue categories themselves—this would likely only cause more confusion.

Instead, future work can focus on establishing internal categories that customer service staff can use to label each type of category as they are reported. This will allow HART to generate better reports, so they can share prediction errors with AVL vendors and complaints about early and late buses with operations. This process will require additional staff training so that all customer service team members can properly identify the schedule deviation in the metadata and categorize the issue correctly. Providing hidden fields via the Open311 specification as discussed earlier would also assist in this effort, as schedule deviation metadata elements could be pulled out separately and displayed more prominently to customer service staff in the administrative user interface.

Transit Agency Lessons Learned

HART has provided the following suggestions to agencies considering moving to a dedicated issue management platform like SeeClickFix:

1. Have dedicated resources ready and trained before making the system available to the public. HART opened issue submissions to the public before having all departments trained on the tool, as the assumption was that customer service would train other departments (e.g., operations, safety and security) as needed as issues came in related to those departments. If they were to repeat the process again, they would train the other departments before opening issue submissions to the public.
2. Be prepared to manage two systems during a transition period. There are a few remaining tasks that HART's legacy issue management system can perform that are required functions that HART cannot yet perform in the new software. HART advised new agencies to be prepared to manage more than one system during this transition period until the legacy system can be completely replaced.
3. Put thought into secondary questions for users. When defining issue categories, give a lot of thought to the secondary required fields that a user must fill out before submitting the issue report. Make sure you include all information needed to resolve the problem to help minimize unnecessary back-and-forth communication with the user. If all information is gathered in the initial submission, it makes it much simpler to use "canned" responses to quickly triage reports.
4. Be careful with private/public comments. Ensure that staff understand that some comments via the administrative portal are private among team members and others are publicly shown on the site. Be aware that even internal comments may be subject to public records requests depending on the agency and local regulations.

Conclusion

A pilot of the Open311 issue reporting specification as a method for capturing multimodal issue reports via the open-source OneBusAway mobile app was successfully deployed in Tampa, Florida. HART plans to continue using the enhanced customer service features provided by the improvements to the OneBusAway app beyond the end of the research project as part of their normal customer service operations. The open-source nature of OneBusAway allows the integration of such tools as part of research designed to enhance both the transit agency and riders' experience—this is a significant differentiation of OneBusAway from all other transit applications (e.g., Transit App, Moovit), as none of them provide direct, structured feedback channels to the transit agency that are directly under the transit agency's control (e.g., custom categories defined via Open311). Future research efforts in this area should focus on identifying organizations that track issue management metrics before adopting the technology presented in this paper, and then comparing these "before" metrics to metrics after the technology has been adopted. Such research will be able to provide quantitative measures of change for important items such as the number of issues processed per hour and per person, the average time to acknowledge, triage, and close an issue, and more valuable metrics. Future research could also examine the impact of this technology on existing issue reporting channels (e.g., do the number of calls to customer service drop significantly?). Anecdotal evidence from discussions with public agencies suggest this is true although quantitative evidence should be pursued. Readers interested in additional details of the project discussed in this paper are directed to the project final report (*21*).

Acknowledgments

The project deploying OneBusAway in multiple agencies in Tampa Bay as part of an improved issue reporting infrastructure was funded by the National Center for Transit Research and the Florida Department of Transportation. The author would also like to thank the Florida Department of Transportation District 7 (including Project Manager Elba Lopez), Hillsborough Area Regional Transit, and Pinellas Suncoast Transit Authority for their participation in the project, as well as Cambridge Systematics that supports OneBusAway for HART, as well as Clever Devices, PSTA's AVL vendor, and SeeClickFix for their implementation of the Open311 API. Thanks also to Cagri Cetin for his invaluable work implementing many software components critical to the success of this project, and troubleshooting and documenting data issues.

Author Contributions

The author confirms contribution to the paper as follows: study conception and design: Sean J. Barbeau; data collection: Sean J. Barbeau; analysis and interpretation of results: Sean J. Barbeau; draft manuscript preparation: Sean J. Barbeau. The author reviewed the results and approved the final version of the manuscript.

References

1. Brakewood, C., S. Barbeau, and K. Watkins. An Experiment Evaluating the Impacts of Real-Time Transit Information on Bus Riders in Tampa, FL. *Transportation Research Part A: Policy and Practice*, Vol. 69, 2014, pp. 409–422.
2. Lei, T., and P. Thakuriah. Ridership Effects of Real-Time Bus Information System: A Case Study in the City of Chicago. *Transportation Research Part C: Emerging Technologies*, Vol. 22, 2012, pp. 146–161.
3. Brakewood, C., G. Macfarlane, and K. Watkins. The Impact of Real-Time Information on Bus Ridership in New York City. *Transportation Research Part C: Emerging Technologies*, Vol. 53, 2015, pp. 59–75.
4. Open311 Community. *Open311: A Collaborative Model and Open Standard for Civic Issue Tracking*. http://www.open311.org/. Accessed August 1, 2017.
5. SeeClickFix. *Strong Communities, More Efficient Government*. https://seeclickfix.com/. Accessed August 1, 2017.
6. Connected Bits. *What Could a Team of Mobile Reporters Do For You?* http://www.spotreporters.com/. Accessed August 1, 2017.
7. PublicStuff. *Drive Positive Change in Your Community*. http://www.publicstuff.com/. Accessed August 1, 2017.
8. Ferris, B. *OneBusAway: Improving the Usability of Public Transit*. Doctor of Philosophy Thesis. Computer Science and Engineering, University of Washington, Seattle, WA, 2011.
9. Ferris, B., K. Watkins, and A. Borning. OneBusAway: Results from Providing Real-Time Arrival Information for Public Transit. *Proc., 28th International Conference on Human Factors in Computing Systems*, Association for Computing Machinery, Atlanta, GA, 2010, pp. 1–10.
10. Watkins, K. E., B. Ferris, A. Borning, G. S. Rutherford, and D. Layton. Where Is My Bus? Impact of Mobile Real-Time Information on the Perceived and Actual Wait Time of Transit Riders. *Transportation Research Part A: Policy and Practice*, Vol. 45, No. 8, 2011, pp. 839–848.
11. Brakewood, C. *Quantifying the Impact of Real-Time Information on Transit Ridership*. Doctor of Philosophy Thesis, Department of Civil and Environmental Engineering, Georgia Institute of Technology, Atlanta, GA, 2014.
12. Gooze, A., K. Watkins, and A. Borning. Benefits of Real-Time Information and the Impacts of Data Accuracy on the Rider Experience. *Transportation Research Record: Journal of the Transportation Research Board*, 2013. 2351: 95–103.
13. Barbeau, S. J. Quality Control - Lessons Learned from the Deployment and Evaluation of GTFS-Realtime Feeds. Presented at 97th Annual Meeting of the Transportation Research Board, Washington, D.C., 2018.
14. Center for Urban Transportation Research at University of South Florida. open311-client. https://github.com/CUTR-at-USF/open311-client. Accessed August 1, 2017.
15. Barbeau, S. J. GTFS Changes - agency_email proposal. https://groups.google.com/forum/#!topic/gtfs-changes/aezjQsriLYA. Accessed November 15, 2017.
16. SeeClickFix. *Issue ID: 3200286: Open Electrical Box*. https://seeclickfix.com/issues/3200286. Accessed August 1, 2017.
17. SeeClickFix. *Issue ID: 3339143: Broken Storm Drain*. https://seeclickfix.com/issues/3339143. Accessed August 1, 2017.
18. SeeClickFix. *Issue ID: 3511434: Dangerous Bus Stop Location*. https://seeclickfix.com/issues/3511434. Accessed August 1, 2017.
19. SeeClickFix. *Issue ID: 3128306: Bus Driver Compliment*. https://seeclickfix.com/issues/3128306. Accessed August 1, 2017.
20. Watkins, K., Y. Xu, S. Bregman, and K. Coffel. *TCRP REPORT 179: Use of Web-Based Rider Feedback to Improve Public Transit Services*. Transportation Research Board of the National Academies, Washington, D.C., 2015.
21. Barbeau, S. J. *Improving Access to Transit through Crowd-sourced Information*. University of South Florida, Florida Department of Transportation, Tallahassee, FL, 2017.

The Standing Committee on Urban Transportation Data and Information Systems (ABJ30) peer-reviewed this paper (18-06225).

The opinions, findings, and conclusions expressed in this paper are those of the author's and not necessarily those of the Florida Department of Transportation or the U.S. Department of Transportation.

Article

TRR
JOURNAL OF THE TRANSPORTATION RESEARCH BOARD

Transportation Research Record
2018, Vol. 2672(42) 235–246

DOI: 10.1177/0361198118798991
journals.sagepub.com/home/trr
SAGE

Characteristics and Temporal Stability of Recurring Bottlenecks

Ishtiak Ahmed[1], Nagui M. Rouphail[2], and Shams Tanvir[1]

Abstract

This study applies and updates a method which identifies and quantifies the extent of traffic congestion from recurring freeway bottlenecks. Additionally, the spatiotemporal stability of bottlenecks over an extended period was tested. Over time congestion at bottlenecks may increase, may decrease, or may migrate to other nearby locations. Analysis of stability is important since prioritizing and applying treatments at bottlenecks is a multiyear process. In addition, a robust method for selecting sensitivity based parameters to identify and quantify bottleneck effects is presented. Subsequently, a systematic framework is developed for tracking and archiving the spatiotemporal changes in the recurring bottlenecks. The proposed method is demonstrated on a case study on Interstate 40 in North Carolina using three years of probe data. A congestion speed ratio detection threshold of 0.7 and a probability of activation threshold of 33% for the study area were determined from a sensitivity test to ascertain their recurrence. The method identified 13 bottlenecks with their impacts ranging from 35 to 3,278 mi-hours of congestion per year. Eight bottlenecks either newly emerged or had their queues merged or shifted between successive years. Even spatially stable bottlenecks had significant variation in their activation frequency and queue length. Exploration of the changes in bottleneck severity and locations revealed the influence of a long-term work zone in the area and the effect of the rapid growth in traffic demand. Local agencies can use this method to shortlist recurring bottlenecks and track changes to plan mitigation strategies.

Increasing traffic congestion is an inevitable predicament in large and growing cities across the world. Many techniques have been proposed and applied for mitigating traffic congestion (*1*). The foremost task for congestion mitigation is locating its origin, also known as the choke point or bottleneck. A bottleneck is a location where demand for the usage of a highway section exceeds the section's capacity to handle it (*2*). Several studies related to freeway bottleneck identification and ranking have been conducted in the past few years (*3–7*). However, few focused exclusively on the recurrence of bottlenecks and their varying configuration over time.

In the U.S., about 40% of congestion nationwide has been attributed to recurring bottlenecks (*2*). To mitigate these bottlenecks, it is important to generate a shortlist of only the truly recurring ones since bottleneck treatment resources of public agencies are limited. However, the shortlisted bottlenecks can migrate or transform over an extended period due to the dynamic and interactive nature of recurring bottlenecks. Hence, it is imperative to monitor the pattern of spatiotemporal changes of recurring bottlenecks to enable judicious investments. Such spatiotemporal changes are more frequent and complex in fast-growing urban roadway networks which makes the monitoring task challenging. However, with the advent of new data sources relevant to traffic monitoring, it is possible to track these dynamics and document them in a consistent manner. In this study, a method that can identify recurring freeway bottlenecks, characterize them using readily understood performance measures, and monitor their changes over a long span of time is proposed. The recurring bottleneck identification method incorporates the frequency of bottleneck activation along with a robust definition of congestion onset. The performance measures provide a foundation for the characterization and prioritization of bottlenecks. Classification of different transitions of bottleneck impact and location contributes to the research need for developing a long-term monitoring technique for bottlenecks.

A review of past studies is presented in the next section. Following that, methodological tools for recurring

[1]Institute for Transportation Research and Education (ITRE), North Carolina State University, Raleigh, NC
[2]North Carolina State University, Raleigh, NC

Corresponding Author:
Address correspondence to Ishtiak Ahmed: iahmed2@ncsu.edu

bottleneck identification, quantifying their impact, and monitoring changes are introduced. A robust technique for the selection of thresholds for congestion detection and for filtering recurring bottlenecks is described as well. The proposed method is demonstrated through a case study on an urban freeway network. The identified recurring bottlenecks and changes over a three-year time span are discussed in detail. Key findings from the case study results are summarized in the last section.

Background

The common approach of bottleneck identification is the detection of congestion by imposing a threshold on the observed speed (*3–7*). The Florida Department of Transportation (*5*) gathered probe vehicle speed data from INRIX (*8*) and used the combination of frequency of congestion (FOC) and planning time index (PTI) to identify recurrent congestion. FOC is the fraction of time when travel speeds fall below a certain threshold, in this case, 75% of the free-flow speed. PTI is defined as the 95th percentile travel time index. Threshold values of 40% for FOC and three for PTI revealed the highly congested freeway segments. A combination of these two parameters identified the recurring bottlenecks and ranked them.

RITIS (*7*) established a method for bottleneck identification and ranking, also based on archived speed data delivered by INRIX (*8*). This approach used 60% of the free-flow speed as the cut-off threshold at which the free-flow speed was estimated as the 85th percentile observed speed for all time periods. Additionally, it identified the most downstream road segment in a series of congested road segments to locate the physical bottleneck, which is typically the "head" or most downstream location of congestion. Each identified bottleneck can be ranked by an impact factor, which is estimated by combining the resulting queue length and duration of congestion during each activation. When this ranking algorithm is applied over a user-specified time period, all occurrences of activation that share an identical "head" location are combined into an "Element". The total impact of an Element is estimated by summing up the impacts of the associated "Occurrences", based on which the ranking is provided.

Song (*6*) used the ratio of speed at capacity to free-flow speed to estimate the cut-off threshold for congestion identification. This ratio was calculated using the speed-flow curves described in the *Highway Capacity Manual* (*9*). It also distinguished between recurring and nonrecurring bottlenecks by setting up a minimum threshold for the probability of congestion occurrence and developed a congestion extent-based measure. Details of this method are given in the Methodology section.

Chen et al. (*4*) used data from loop detectors to identify freeway bottlenecks. Their algorithm stipulates that an active bottleneck exists between any loop detector pair during period t if the speed at the upstream detector is at least 20 mph lower than that at the downstream detector at time t. Also, speed at the upstream detector must be less than 40 mph to declare the activation. Other speed conditions at intermediate sensor locations that may be present between the first sensor pair must also be met to declare a bottleneck activation.

Ban et al. (*3*) identified bottlenecks using a plot of a speed contour map (SCM) in a time–speed domain. A simplified algorithm was proposed based on a binary SCM (BSCM). BCSM convert SCM's values to either 0 (uncongested) or 1 (congested) based on the congested threshold speed (35 mph). The length and duration of each bottleneck were visually estimated by the number of contiguous cells containing a "1" index. The identified bottlenecks were calibrated using a microscopic simulation model.

Although a number of studies dealt with the systematic tracking of bottlenecks on a daily basis (*10*, *11*), only a few focused on their long-term changing pattern. Identification of different congestion pattern by monitoring loop detector data over a long period (1995–2001) was conducted on four German highways (*12*). The study categorized two congestion patterns for isolated bottlenecks: the general pattern (GP) which consists of both synchronized flow and wide moving jam, and the synchronized flow pattern (SP) which consists of synchronized flow upstream of the bottleneck only. Findings from their study showed that the spatiotemporal structure of GP possesses reproducible and predictable features.

The online tools provided by RITIS (*7*) can be used to track the progression of bottlenecks over time. The freeway performance measurement system (PeMS) operated by Caltrans (*13*) provides a suitable platform for monitoring operating conditions of the freeways in California. Several studies used PeMs for bottleneck identification and congestion impact quantification (*3*, *4*). The advantage of this platform is that it provides the necessary tools to evaluate link performance measures in terms of vehicle miles traveled, vehicle hours traveled, and delays. Additionally, all these analyses can be conducted using both real-time and historical data.

From the review of literature, it is evident that methods to distinguish between recurring and nonrecurring bottlenecks such as FDOT (*5*) and Song et al. (*6*) while sensible, did not validate the selection of the thresholds used for congestion detection and recurring bottleneck identification. Moreover, prior research on developing congestion extent-based performance measure is limited in number. Consequently, only a few studies were found that focused on monitoring and documenting the long-term changes in the spatiotemporal characteristics of traffic bottlenecks.

Methodology

This paper presents a systematic approach for (a) the identification of recurring freeway bottlenecks, (b) the development of robust performance measures associated with them, and (c) the analysis of recurring bottleneck variations over time, using probe data. Data were downloaded from INRIX (*8*) via RITIS (*7*). The overall process is described step-by-step in the following subsections.

Congestion Detection and Bottleneck Activation Identification

Song (*6*) identified a congested link or traffic message channel (TMC) and congested clock time by ascertaining whether the ratio of reported to free-flow speed exceeds a certain threshold. The method introduces two parameters, congestion value (CV) and congestion index (CI), as expressed by Equations 1 and 2 respectively:

$$\text{Congestion Value, CV}(i,t,m) = \frac{\text{RS}(i,t,m)}{\text{FFS}(i)} \quad (1)$$

$$\text{Congestion Index, CI}(i,t,m) = \begin{cases} 1 \; if \; \text{CV}(i,t,m) < \text{CI} \; threshold \\ 0 \quad otherwise \end{cases} \quad (2)$$

where

RS (i, t, m) = reported speed (mph) for TMC i during time interval t and weekday m, and

FFS (i) = free-flow speed for TMC i (mph).

The reference speed of a TMC adopted in RITIS (*7*) is the 85th percentile observed speed for all time periods. In this study, the threshold is selected upon investigating the sensitivity of the proposed performance measure of the bottlenecks to different threshold values. The performance measure of the bottlenecks discussed later is estimated for a series of threshold values. The appropriate threshold is the one beyond which a small change in the threshold value results in a drastic shift in the performance measure.

Identification of Recurring Bottlenecks

To declare a bottleneck as recurring, it must activate a minimum number of days within a given study period. The parameter, average historic congestion index (AHCI) was proposed by Song (*6*). AHCI represents the percent of days a road segment during a specific clock time interval is congested, relative to the total number of days observed (see Equation 3):

$$\text{AHCI}(i,t) = \sum_{m=1}^{M} \text{CI}(i,t,m)/\text{M}*100 \quad (3)$$

where M is the number of days included in the study period.

If for a TMC segment the AHCI over a time interval is greater than or equal to a threshold value, that segment during that time can be considered to be experiencing recurrent congestion. Like the CI threshold, the threshold for AHCI needs to be selected after observing the sensitivity of the proposed performance measure to it. A recurring bottleneck and its spatiotemporal extent can be mapped when the AHCI values are plotted in a space–time region (see Table 1). In this plot, each cell contains the AHCI value of a TMC for each 15-minute duration. Cells meeting the threshold criterion (in this case, 33%) are shaded in yellow, and a continuous cluster of these colored cells make up a recurring bottleneck region. Some bottleneck regions may contain multiple bottleneck activations, where the queue from the downstream activation may have merged with an upstream bottleneck activation.

In Table 1, TMC 1 is the furthest downstream TMC that meets the AHCI threshold criterion. Therefore, the

Table 1. Recurring Bottleneck Region using an AHCI = 33% Criterion

TMC ID	Length (mi)	TMC tag	16:30	16:45	17:00	17:15	17:30	17:45	18:00	18:15	18:30
125-17000	3.29	0	4.87	4.87	5.31	6.19	6.67	12.89	12.89	8.93	7.14
125N04836	0.75	1	14.67	43.3	47.13	55.94	65.38	63.08	50	39	10.77
125-04836	1.06	2	11.97	31.42	32.95	45.21	48.08	44.62	36.54	25.87	15.44
124N04837	0.85	3	11.97	31.42	31.03	43.68	50.38	44.23	37.31	23.17	15.06
125-04837	1.31	4	13.41	24.52	31.03	35.63	44.62	38.85	33.46	21.62	15.44
125N04838	0.86	5	14.67	27.88	32.3	38.05	43.11	41.78	34.67	23.21	16.07
125-04960	0.6	6	14.67	21.07	24.52	28.74	33.46	34.23	25.77	16.99	11.97
125N04961	0.37	7	13.41	12.89	12.89	12.89	15.44	15.06	15.06	14.67	10.04

boundary between TMC 1 and TMC 0 is tagged as the location of the recurring bottleneck. The region of the recurring bottleneck is extended from clock time 16:45 to 18:15 and from TMC 1 to TMC 6.

Development of Performance Measures

To characterize the identified recurring bottlenecks, two congestion extent-based measures are developed. These measures reflect the combined effect of the queue length extent, the duration of congestion, the probability of congestion, and the number of activations of a bottleneck within a specified time duration. Two measures are proposed as follows.

The first measure is termed as the recurring bottleneck impact factor (RBIF). It represents the expected extent of the impact for each activation of a bottleneck over multiple days. It implies that some bottlenecks may activate more than once a day. It is estimated as the spatiotemporal impact area of the bottleneck across weekdays and peaks in miles-hours of congestion per activation (see Equation 4):

$$\text{RBIF per activation} = \left(\frac{\Delta T}{60}\right) * \sum_{i=1}^{I} \sum_{t=1}^{N(i)} L_i * \text{AHCI}(i, t)) \text{ in miles.hrs} \quad (4)$$

where

ΔT = time aggregation (in minutes), and

L_i = length of TMC segment i *(miles)*.

This measure is similar to what Song (*6*) termed the recurrent bottleneck spatiotemporal impact index (RBSII). However, the estimation of RBIF is more robust in this study since it uses a sensitivity analysis to determine the appropriate thresholds for CI and AHCI. Additionally, Song used RBSII to rank the identified bottlenecks without incorporating the number of activations of a bottleneck, while this study uses RBIF to analyze the sensitivity of the thresholds only. RBIF is estimated for different values of CI and AHCI thresholds and the sensitivity of RBIF to the thresholds are investigated. A critical point at which a sharp change (or "knee") in RBIF occurs for a slight change in the threshold can be indicative of an appropriate threshold.

The second measure, Overall RBIF takes into account the number of activations of a recurring bottleneck over the study period, in addition to the parameters included in RBIF. It is estimated in miles-hours of congestion according to Equation 5:

$$\text{Overall RBIF} = \text{RBIF per activation} * \text{Number of activations} \quad (5)$$

Since Overall RBIF includes all the parameters related to the impact of a recurring bottleneck, it is used to quantify and investigate the changes in the bottleneck impact over time.

Tracking Changes in Recurring Bottleneck Characteristics

Although the purpose of using the AHCI threshold is to identify bottlenecks that are recurring, even a truly recurring bottleneck can be very dynamic in nature. A long-term recurring bottleneck may progress over time in many ways. Changes to bottlenecks that are monitored and considered in this study include:

- Emergence *and disappearance of bottlenecks*: The emergence of bottleneck at a TMC is signalled when the TMC was uncongested during year "y" and is the front of a recurring bottleneck (with newly congested TMCs upstream of it) during year "y + 1". The reverse phenomenon is referred to as the disappearance of the bottleneck.
- *Merging and separation of bottleneck queues*: A merging event occurs when the queues of two bottlenecks separated by at least one TMC during year "y" and the TMC gap fills up during year "y + 1". Conversely, a separation is declared when a congested existing TMC in year y becomes uncongested in year y + 1 and hence, creates two separate bottlenecks.
- *Shift in congestion*: This phenomena occurs if the resulting congestion migrates over time to a nearby upstream or downstream location. The shift of the most downstream TMC is used as the spatial unit to signal a shift in congestion. The shift in congestion can be viewed as a disappearance of a new bottleneck and emergence of a new one at a nearby location.
- *Change* in *performance measures*: It occurs if the bottleneck location is stable but the Overall RBIF increased or decreased

Thus, the identification of recurring bottlenecks is carried out over a multiyear period, and each identified location is tagged by one or more changes shown in the above list. Here, a calendar year is selected as the period based on which the bottlenecks are identified and monitored. Such use of calendar year is a common practice in transportation planning and in other research fields (*14*).

The end product of this monitoring task is a multiyear database of the system of facilities under consideration with fields corresponding to each feature identified. The developed database is spatiotemporal in nature with space discretized to the level of TMC and time discretized

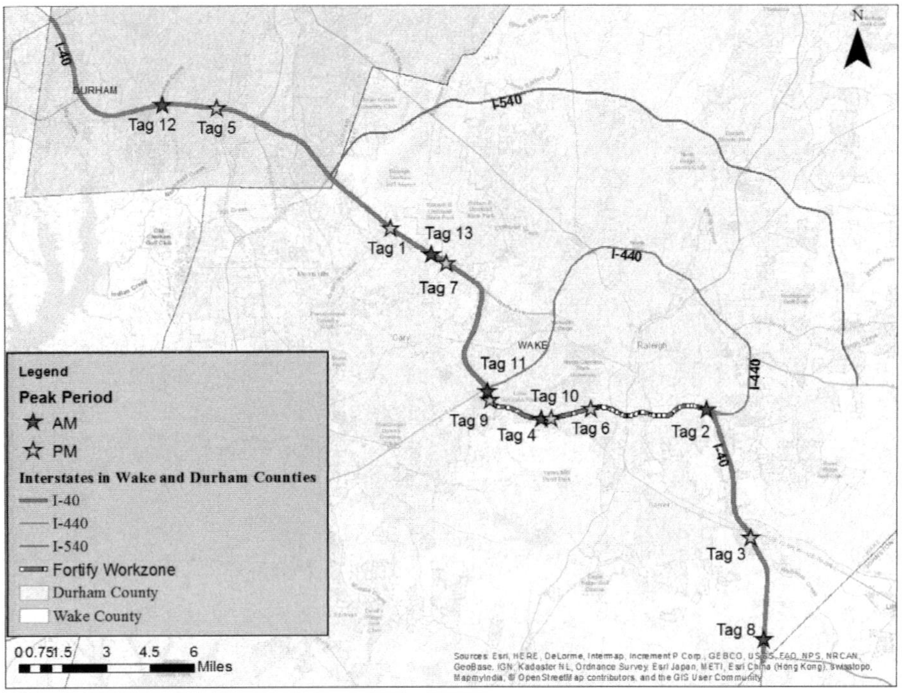

Figure 1. Study section on I-40 in Wake and Durham Counties, NC.

to 15-minute periods. The database can be packaged as a tool with a visual representation of the estimated features for operation managers and planners. It also allows developing a framework for quantitative multiyear bottleneck analysis.

Case Study and Results

This section presents an application of the methodology described in the preceding section. The proposed method was applied to Interstate 40 corridor (both east and westbound directions) within Wake and Durham Counties in North Carolina. This facility spans a distance of about 42 centerline miles. All weekdays in three successive calendar years, namely 2014, 2015, and 2016 constitute the study period for which probe vehicle speed data was downloaded at a 15-minute time aggregate.

The selected I-40 corridor is a part of an urban network within the research-triangle park (RTP) area. This area has experienced increased traffic growth in recent years. Fixed sensors along this corridor maintained by North Carolina Department of Transportation (*15*) recorded upwards of a 5% increase in AADT between 2013 and 2015 on this corridor. Figure 1 shows the study corridor marked by a thick line. The tags in Figure 1 represent the location of recurring bottlenecks that are described in a later section. It should be noted that a work zone (termed Fortify) involving major pavement reconstruction on sections of I-40 (*16*) began on this corridor from mid-2015 to date. The affected area is indicated by white spots on the thick facility line.

Selection of CI and AHCI Thresholds

To appropriately select the threshold values for CI and AHCI, one might consider estimating the number of recurring bottlenecks obtained using different thresholds and plotting the number against the thresholds to study the sensitivity. However, once the congestion detection (CI) threshold is raised significantly, recurring bottleneck regions at different locations of a route start to merge with each other. Similar phenomena are observed when the threshold for recurring bottleneck identification (AHCI threshold) is tested against the number of bottlenecks. To tackle this challenge, rather than considering the number of bottlenecks along a route, the RBIF value per activation for the entire length of each direction of I-40 is evaluated acorss various thresholds for CI and AHCI.

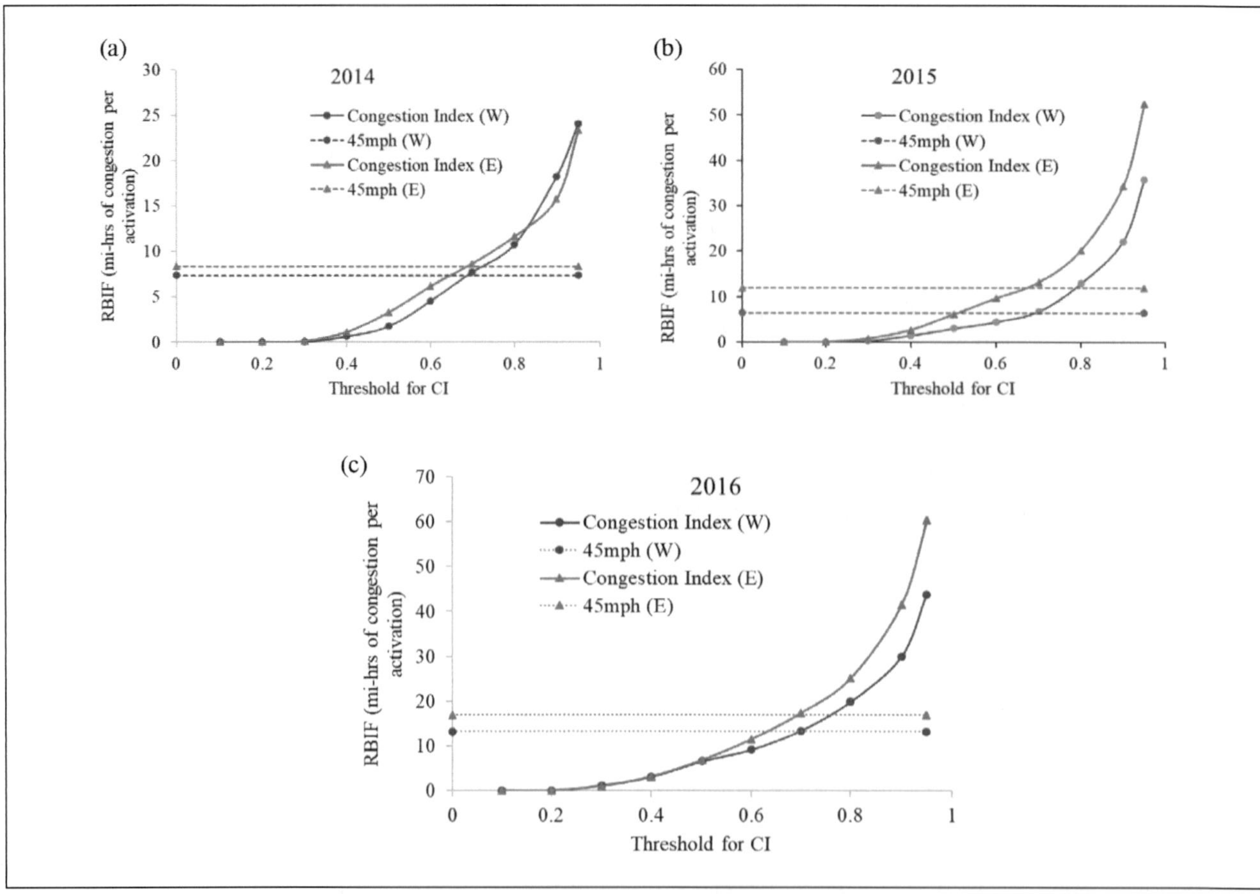

Figure 2. Sensitivity of *RBIF* per activation to CI thresholds in calendar years: (*a*) 2014; (*b*) 2015; and (*c*) 2016 (AHCI = 33%).

Threshold for Congestion Index (CI). In Figure 2, the sensitivity of RBIF (miles-hours of congestion per activation) estimated for each of the three years in the study period to the threshold values of CI is plotted. CI values were gradually increased from 0.1 to 0.9 in 0.1 unit increments, and the resulting RBIF estimated for each direction separately, using a fixed threshold for AHCI (in this case 33%). In addition, the RBIF value corresponding to a fixed speed threshold (45 mph) is shown using the dotted horizontal lines, again for each direction separately. The use of 40 to 45 mph speed threshold is quite a common practice for congestion identification on freeways (*17*, *18*).

The first observation from Figure 2 is that the trend of change in RBIF with different CI thresholds appears to be stable across this three-year period. For the westbound direction in 2014 and for both directions in 2016, RBIF value sharply increased when the threshold for CI exceeds 0.8. In 2015, such increment was observed at a threshold greater than 0.7. Use of the fixed speed threshold (45 mph) yields almost the same RBIF as what a CI threshold value of 0.7 does in all direction-year combinations.

From the above analysis, it is apparent that a CI threshold in the range of 0.7 to 0.8 conforms to a 45-mph fixed speed threshold for all the combinations. Moreover, a CI threshold within this range does not result in any unusual increase in RBIF per activation for the tested routes. Rather, it is the limiting threshold range at which the RBIF rapidly increases.

Additionally, it should be noted that similar explorations on the sensitivity of RBIF using different AHCI thresholds (50% and 60%) were conducted as well. The outcome is that the critical point of the sensitivity curves still lies in the same range of CI threshold (0.7–0.8). Although the RBIF values decrease when a high AHCI threshold value is used, the 45-mph fixed speed threshold conforms to this range of CI threshold. A CI threshold of 0.7 is finally selected for congestion detection in this study.

Threshold for Average Historic Congestion Index (AHCI). Similar to the CI approach, the sensitivity of RBIF per activation to AHCI thresholds is tested in this section. RBIF values for different AHCI thresholds

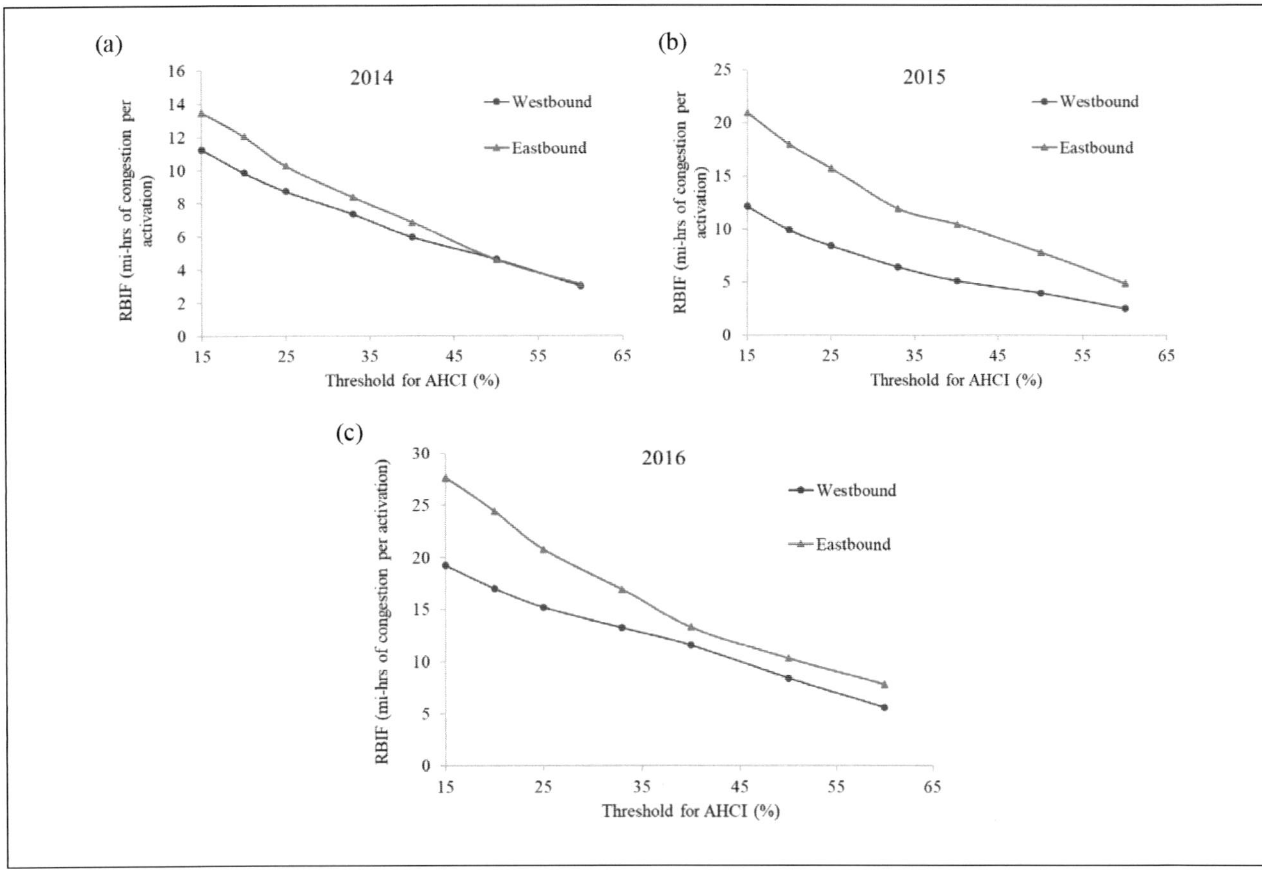

Figure 3. Sensitivity of RBIF to AHCI threshold in (*a*) 2014, (*b*) 2015, and (*c*) 2016.

(varying from 15% to 60%) are plotted for different years in Figure 3. In this analysis, and based on the previous findings, a fixed speed threshold of 45 mph is used for detecting congestion and estimating the RBIF.

It should be noted that raising the AHCI threshold means tightening the constraint on the minimum number of bottleneck activation, resulting in a drop in the magnitude of the RBIF. It is evident from Figure 3 that unlike the threshold for CI, the threshold for AHCI within this range does not exhibit any drastic changes in the RBIF value for any year and direction of I-40. The variation of RBIF per activation with AHCI threshold seems to be almost linear within this range. This makes it difficult to suggest a limiting range for the AHCI threshold. In effect then, the selection of the threshold is left to the operating agency desires on what constitutes recurrent acceptable quality of service. For the purpose of this study, the authors selected an AHCI threshold of 33%. The interpretation of this threshold is that on average, a recurring bottleneck must be activated on at least one out of three weekdays for it to qualify as being "recurring".

Recurring Bottlenecks Identification in Case Study

Once the selection of the CI and AHCI thresholds was made, the process of recurring bottleneck identification and estimation of the performance measures was applied to the study area for all weekdays of each calendar year (2014, 2015, and 2016). The results are summarized in Table 2. In addition, locations and activation periods of the identified bottlenecks were shown earlier in Figure 1.

In Table 2, the bottlenecks are arranged in a decreasing order of the overall RBIF starting from 2014 to 2016. The blank cells indicate that the bottleneck was not present (either shifted or merged) at that location in that respective year. Among the total 13 different locations, only two bottlenecks appeared in all three years (Tag 3 and 5) and total of five bottlenecks appeared in at least two successive years. The remaining eight bottlenecks migrated elsewhere as the corresponding calendar year was over. It should be noted that classical methods of bottleneck identification like RITIS (*7*) generate 52 bottlenecks within the study corridor for the calendar year of 2015. As discussed in a previous study (*19*), RITIS

Table 2. Identified Recurring Bottlenecks and Overall RBIF in 2014 to 2016

				Overall RBIF (mi-hours)		
Tag	Bottleneck TMC	Bottleneck location	Activation peak period	2014	2015	2016
1	125N04861	I-40 EB at Aviation Parkway MM 285	PM	1,389		
2	125 + 04838	I-40 WB at I-440, MM 301	AM	928		
3	125N04836	I-40 EB at US-70, MM 306	PM	451	868	3,278
4	125 + 04965	I-40 WB at Gorman St., MM 295	AM	387	487	
5	125 + 04868	I-40 WB at NC 55, MM 278	PM	246	615	970
6	125 + 04965	I-40 WB at Gorman St, MM 295	PM	35		
7	125-04860	I-40 EB at Harrison Ave., MM 287	PM		2,069	2,758
8	125P17000	I-40 WB at US-70, MM 309	AM		88	
9	125 + 04966	I-40 WB at I-440, MM 293	PM		65	295
10	125-04965	I-40 EB at Gorman St. MM 295	PM		47	
11	125P04966	I-40 WB at I-440, MM 293	AM			2,516
12	125N04869	I-40 EB at Fayetteville Rd., MM 276	AM			179
13	125P04860	I-40 WB at N Harrison Ave., MM 287	AM			166

Note: MM = mile marker.

Table 3. Changes in Case Study Bottlenecks Characteristics Over Time

Category	Years	Bottlenecks emerging (tag)	Bottlenecks disappearing
Emergence and disappearance	2014 vs. 2015	Tag 10	None
	2015 vs. 2016	Tag 11 Tag 12 Tag 13	None None
Merging and separation		Bottlenecks merging	Bottlenecks separating
	2014 vs. 2015	None	None
	2015 vs. 2016	Tag 4 and 8 merged onto Tag 11 Tag 3 and 10 merged onto Tag 3	None None
Shift in bottleneck locations		Bottlenecks migrated upstream	Bottlenecks migrated downstream
	2014 vs. 2015	Tag 2 migrated 7 miles into Tag 8	Tag 6 migrated 2 miles into Tag 9 Tag 1 migrated 2 miles into Tag 7
	2015 vs. 2016	None	None
Change in overall RBIF for spatially stable bottlenecks		Bottleneck tag (% change in Overall RBIF)	
	2014 vs. 2015	Tag 3 (+ 93%) Tag 4 (+ 26%) Tag 5 (+ 150%)	
	2015 vs. 2016	Tag 3 (+ 278%) Tag 5 (+ 58%) Tag 7 (+ 33%) Tag 9 (+ 353%)	

generates such an unusually high number of bottlenecks because it does not distinguish between recurring and non-recurring (i.e., incident created) bottlenecks.

Although Table 2 lists 13 different bottleneck locations, not all of them are independent. Some "new" bottlenecks appeared in a year as a result of a mere shift of location from the previous year, while other bottlenecks got merged by the queue from a downstream bottleneck, and so forth. This identification of recurring bottlenecks is the first step towards investigating the temporal changes in their characteristics. Detailed findings from examining all the changes that occurred in these three years are provided in the following sections.

Tracking the Changes in Recurring Bottlenecks

The changing nature of the identified bottlenecks is examined through the study period and the findings are

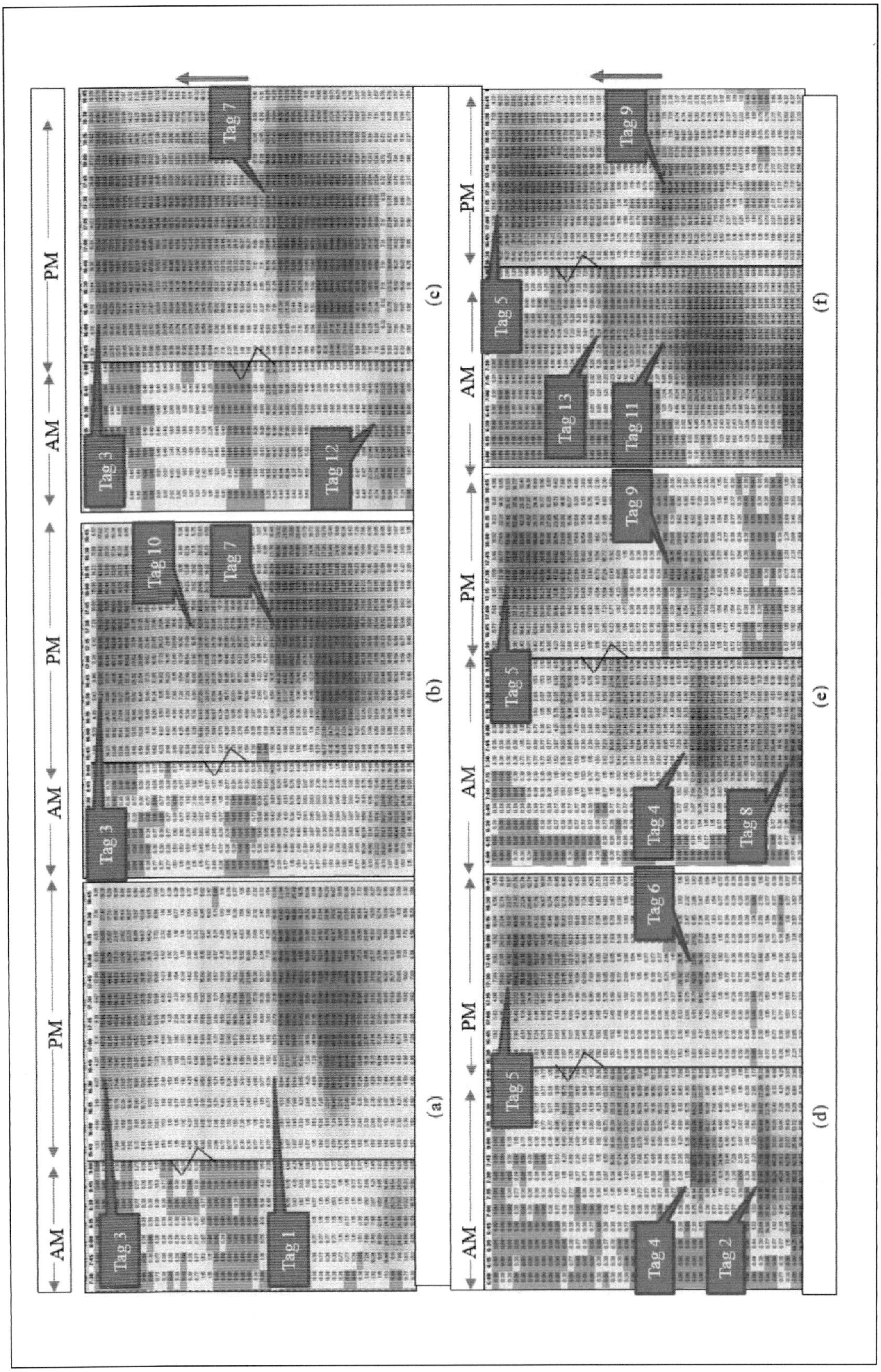

Figure 4. I-40 AHCI on (*a*) EB in 2014; (*b*) EB in 2015; (*c*) EB in 2016; (*d*) WB in 2014; (*e*) WB in 2015; and (*f*) WB in 2016.

summarized in Table 3. Tag numbers refer to those indicated in Table 2.

Figure 4 illustrates all the identified bottlenecks using AHCI contours for each year and route. Cells contain the AHCI value for each space–time combination and are color-coded to visualize the values (red indicates high, and green indicates low). Traffic flows from bottom to top in this figure.

Emergence and Disappearance of Bottlenecks. As shown in Table 3 and Figure 4, none of the recurring bottlenecks disappeared as time passed. Of the four recurring bottlenecks that emerged between 2014–15 and 2015–16, Tag 10 and 11 were likely to be impacted by the I-40 fortify work zone (Figure 1). In 2015, although the AHCI value upstream of the Tag 10 bottleneck did not exceed the threshold (Figure 4*b*), the change in color confirms that it increased significantly compared with 2014 (Figure 4*a*). It reflects the beginning of the impact of the work zone which started in mid-2015. A similar statement can be made about the appearance of Tag 11 bottleneck in 2016. Reduction in roadway capacity due to this long-term fortify work zone may contribute the activations of these two recurring bottlenecks. However, the Overall RBIF for Tag 11 bottleneck is significantly higher than that for Tag 10 bottleneck. The reason is clear from Figure 4*f*, as it appears that the queue from Tag 11 bottleneck merged with upstream bottlenecks. This phenomenon is discussed in a later subsection as well. The other two new AM peak bottlenecks in 2016 (Tag 12 Tag 13) probably emerged because of general increases in traffic volumes in the Triangle area. The gradual increment of AHCI value in these two regions is visible as the AHCI contour color gets darker over time, and crosses the threshold in 2016. To investigate the contribution of land-use changes in the surrounding area, archived AADT values (*15*) within these bottleneck regions were obtained. The recorded AADT in both bottleneck regions showed an increment of about 5% from 2015 to 2016, which suggests that their activations emergence can be contributed to the increase in traffic demand.

Merging and Separation of Bottleneck Queues. Table 3 shows that two cases of bottleneck queues merging occurred in 2015–16, while none got separated. Merging of queue from two or more bottlenecks occurs when one of the bottleneck's impact area expands over time or when a queue from a new downstream bottleneck merges with the old upstream ones. The former situation is observed for Tag 4 and Tag 8 bottlenecks as the new Tag 11 bottleneck in 2016 generated a queue that combined with the other two (Figure 4*e* and *f*). Tag 3 and Tag 10 bottlenecks experienced the latter phenomena as the queue length from Tag 3 became so long that it merged with the Tag 10 bottleneck (Figure 4*b* and *c*). It should be noted that both cases of merging described here are likely to be influenced by the fortify work zone as well. On the other hand, no merging was observed between 2014 and 2015.

Shift in Congestion. As Figure 4 and Table 3 illustrate, no shifting of congestion occurred during 2015–16. The upstream shifting of congestion resulting from Tag 2 bottleneck to Tag 8 represent the longest spatial shift among the three shifts which occurred in 2014–15 (of about 7 miles). This may have been caused by the metering of traffic by Tag 8 bottleneck which caused Tag 2 bottleneck to become a hidden bottleneck in 2015. The intensity of AHCI color in Figure 4*e* indicates that Tag 2 bottleneck is still activated in 2015 but did not cross the threshold.

The two downstream shifting of congestion occurred over a comparatively short distance (2 miles each). The shifting of congestion resulting from Tag 6 bottleneck is likely to be influenced by the fortify work zone since the new shifted congestion (from Tag 9 bottleneck) is located near the starting point of the work zone. The shifting of congestion resulting from Tag 1 bottleneck to Tag 7 indicates a possible increase in demand on the entry ramp at Tag 7 bottleneck (I-40 EB at Harrison Ave., MM 287). Such occurrences are similar to the merging phenomena of bottleneck queues discussed above. From tracking of such a shift, it can be presumed that if the bottleneck at Tag 7 is mitigated, Tag 1 bottleneck may emerge again.

Change in Overall RBIF for Spatially Stable Bottlenecks. Of the 13 different bottlenecks listed in Table 3, five of them maintained their locations at least for two successive years. However, the congestion impact of the sustained bottlenecks increased significantly, as evident from the percent increases in RBIF in Table 3. Regarding the maximum duration of congestion for those bottlenecks, the increase in successive years varied between 15 to 30 minutes for all but Tag 3 bottleneck. The Tag 3 bottleneck duration increased by 45 minutes between 2015 and 2016. Nonetheless, the maximum duration of the recurring congestion was confined to the regular PM peak (16:00 to 18:30). It confirms that those bottlenecks were primarily activated by commuting traffic. To scrutinize the changes in the impact of the sustained bottlenecks, the variation of maximum queue length and number of activations are illustrated in Figure 5.

The most striking observation from Figure 5 is the sharp increase in the queue length for Tag 3 bottlenecks between 2015 and 2016. This can be reasonably attributed to its proximity to the 8-mile long fortify work zone. However, the work zone presence did not influence its number of activations as the increase is almost linear

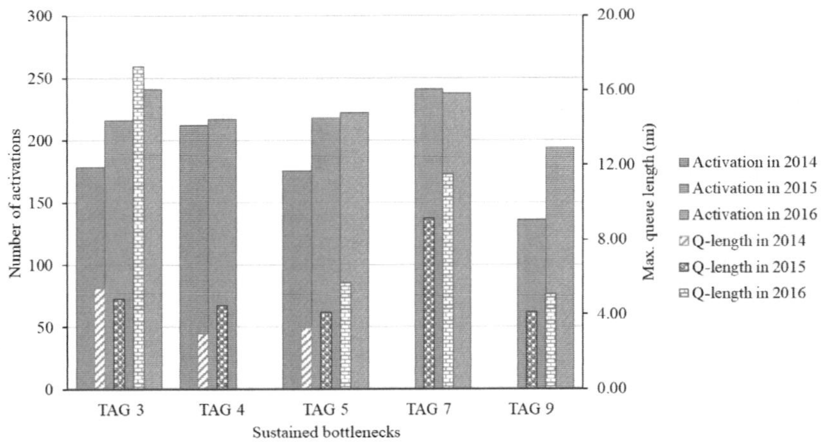

Figure 5. Activations and queue length for spatially stable bottlenecks in study area.

throughout the study period. A noteworthy fact that the maximum queue length of this bottleneck slightly decreased in the 2014–15 period, however that is not reflected in its Overall RBIF as the other two parameters (namely maximum queue length and duration of congestion) also increased.

The increase in the maximum queue length for the other bottlenecks varied between 0.8 to 2.4 miles. Changes in the number of activations did not reveal any abrupt increases, except for Tag 9 bottleneck (increased from 136 to 194 activations between 2015 and 2016). Again, the fortify work zone may have played a significant role in contributing to that increase.

Conclusion

This paper has presented a method for the identification and monitoring of recurring freeway bottlenecks using probe vehicle speed data. Two congestion extent-based performance measures are developed that account for the expected queue length and duration of resulting congestion, along with the number of activations. Upon identification, different phenomena related to the changes in bottleneck characterization and temporal stability are documented and analyzed within a systematic framework.

A case study conducted using the proposed method on I-40 within Wake and Durham counties in North Carolina revealed that the RBIF per activation for each directional route is sensitive to a CI threshold within 0.7 to 0.8. In contrast, the AHCI threshold did not exhibit such sensitivity which led the researcher to select a value of 33% based on perspicacity. The bottleneck identification method revealed 13 different bottleneck locations which existed on the study routes during at least one of the three calendar years from 2014 to 2016.

The changes in the spatiotemporal properties of the recurring bottlenecks over the study years were categorized. Visualization of the AHCI contour revealed that four new bottlenecks emerged during the three-year period while none disappeared from the study routes. Five out of 13 bottlenecks maintained their locations but congestion on those worsened. Two merging cases and three migration cases of congestion were observed. The merging phenomena were associated with queue emanating from a newly emerged or already existing downstream bottleneck. Most of the emergence and merging took place between 2015 and 2016 and were possibly affected by the presence of I-40 Fortify work zone.

Even with the presence of a long-term work zone in the study area, the proposed method identified five bottlenecks that maintained their position for at least two consecutive calendar years. The increasing trend in the Overall RBIF for these bottlenecks reflected the impact of the work zone and rapid growth of traffic in the study area. Exploration of the variables of Overall RBIF revealed that the increment was primarily resulting from the increase in the number of bottleneck activations and queue length.

The identification process of bottlenecks and the framework for monitoring their spatiotemporal characteristics proposed in this study can serve as a useful tool to public agencies. Unlike the classical methods of bottleneck identification, the proposed method offers a flexible way to signal congestion onset while at the same time generating a workable number of high-priority recurring bottlenecks. Additionally, the current practice of bottleneck analysis is to develop performance measures for a single calendar year in a manner which does not allow for analysis of the spatial evolution of bottlenecks. Even high fidelity traffic simulations are unable to capture the heterogeneous dynamics involved in each identified bottleneck. It is expected that the local agencies may use the proposed framework to plan mitigation strategies and to anticipate the possible appearance of currently hidden (or merged) bottlenecks in future. In the future, it is recommended to develop a more rigorous method for distinguishing the phenomena related to bottleneck evolution so as to avoid misinterpretation. Moreover, complex statistical meta-modeling of the multiyear progression should be attempted to identify any correlation of these dynamics with site-specific and planning-level features.

Acknowledgments

This research is based upon the works supported by the North Carolina Department of Transportation (NCDOT) under the Project No. RP 2016-10. We are thankful to NCDOT for their supports.

Author Contributions

The authors confirm contribution to the paper as follows: study conception and design: Ahmed, Rouphail, Tanvir; data collection: Ahmed; analysis and interpretation of results: Ahmed, Rouphail, Tanvir; draft manuscript preparation: Ahmed, Rouphail, Tanvir. All authors reviewed the results and approved the final version of the manuscript.

References

1. FHWA Operations. *Traffic Bottlenecks*. https://ops.fhwa.dot.gov/bn/successful_efforts/. Accessed July 29, 2017.
2. Margiotta, R. A., N. C. Spiller, and J. Halkias. *Traffic Bottlenecks: A Primer Focus on Low-Cost Operational Improvements*. FHWA, U.S. Department of Transportation, Washington, D.C., 2007.
3. Ban, X. J., L. Chu, and H. Benouar. Bottleneck Identification and Calibration for Corridor Management Planning. *Transportation Research Record: Journal of the Transportation Research Board*, 2007. 1999: 40–53.
4. Chen, C., A. Skabardonis, and P. Varaiya. Systematic Identification of Freeway Bottlenecks. *Transportation Research Record: Journal of the Transportation Research Board*, 2004. 1867: 46–52.
5. Florida Department of Transportation. *Bottlenecks on Florida SIS*. Florida Department of Transportation, Tallahassee, FL, 2011.
6. Song, T. J. Recurrent Freeway Bottlenecks Identification and Applications. Raleigh, NC: North Carolina State University, 2016.
7. RITIS, Regional Integrated Transportation Information System. https://vpp.ritis.org/suite/ranking/. Accessed July 27, 2017.
8. INRIX, INRIX. http://inrix.com/. Accessed July 24, 2017.
9. *Highway Capacity Manual 2010*. Transportation Research Board of the National Academies, Washington, D.C., 2010.
10. Huan, L., and R. Bertini. Comparison of Algorithms for Systematic Tracking of Patterns of Traffic Congestion on Freeways in Portland, Oregon. *Transportation Research Record: Journal of the Transportation Research Board*, 2010. 2178: 101–110.
11. Kerner, B. S., H. Rehborn, M. Aleksic, and A. Haug. Recognition and Tracking of Spatial–Temporal Congested Traffic Patterns on Freeways. *Transportation Research Part C: Emerging Technologies*, Vol. 12, No. 5, 2004, pp. 369–400.
12. Kerner, B. S. 2002. Empirical Macroscopic Features of Spatial-Temporal Traffic Patterns at Highway Bottlenecks. *Physical Review E*, Vol. 65, No. 4, 2002, p. 046138.
13. Choe, T., A. Skabardonis, and P. Varaiya. Freeway Performance Measurement System: Operational Analysis Tool. *Transportation Research Record: Journal of the Transportation Research Board*, 2002. 1811: 67–75.
14. Kittelson, W., and M. Vandehey. *SHRP 2 Report S2-L08-RW-1: Incorporation of Travel Time Reliability into the HCM*. Transportation Research Board of the National Academies, Washington D. C., 2013.
15. North Carolina Department of Transportation. *NCDOT AADT Web Map*. https://ncdot.maps.arcgis.com/home/webmap/viewer.html?webmap=b7a26d6d8abd419f8c27-f58a607b25a. Accessed August 1, 2017.
16. North Carolina Department of Transportation. *Fortify I-40/440 Rebuild Project*. https://www.ncdot.gov/fortifync/. Accessed August 1, 2017.
17. Day, C., S. Remias, H. Li, M. Mekker, M. McNamara, E. Cox, D. Horton, and D. Bullock. *2013–2014 Indiana Mobility Report: Summary Version*. Joint Transportation Research Program Indiana Mobility Reports, West Lafayette, IN: Purdue University, 2014.
18. Peterson, L. *WSDOT's Handbook for Corridor Capacity Evaluation*. Olympia, WA: Washington State Department of Transportation, 2014.
19. Ahmed, I., N. M. Rouphail, S. Tanvir, and L. Pan. *Characterizing and Ranking Recurring Freeway Bottlenecks*. Presented at 96th Annual Meeting of the Transportation Research Board, Washington, D.C., 2017.

The Standing Committee on Urban Transportation Data and Information Systems (ABJ30) peer-reviewed this paper (18-06485).

Article

TRR
JOURNAL OF THE TRANSPORTATION RESEARCH BOARD

Transportation Research Record
2018, Vol. 2672(42) 247–256

DOI: 10.1177/0361198118782270
journals.sagepub.com/home/trr
SAGE

Integrated Incident Decision-Support using Traffic Simulation and Data-Driven Models

Tao Wen[1], Adriana-Simona Mihăiţă[1], Hoang Nguyen[1], Chen Cai[1], and Fang Chen[1]

Abstract

This paper introduces the framework of an innovative incident management platform with the main objective of providing decision-support and situation awareness for transport management purposes on a real-time basis. The logic of the platform is to detect and then classify incidents into two types: recurrent and non-recurrent, based on their frequency and characteristics. Under this logic, recurrent incidents trigger the data-driven machine learning module which can predict and analyze the incident impact, in order to facilitate informed decisions for transport management operators. Non-recurrent incidents activate the simulation module, which then evaluates quantitatively the performance of candidate response plans in parallel. The simulation output is used for choosing the most appropriate response plan for incident management. The current platform uses a data processing module to integrate complementary data sets, for the purpose of improving modeling outputs. Two real-world case studies are presented: 1) for recurrent incident management using a data-driven model, and 2) for non-recurrent incident management using traffic simulation with parallel scenario evaluation. The case studies demonstrate the viability of the proposed incident management framework, which provides an integrated approach for real-time incident decision-support on large-scale networks.

An initiative of great value in the era of "Big Data" is to effectively and consistently generate insights and actionable outcomes from multiple data sources. This would typically require several data handling processes, including data forwarding, parsing, and integration, before data becomes usable by dedicated applications, such as machine learning models. These data handling processes were conventionally deployed to enterprise data infrastructure, such as the enterprise data warehouse (EDW). With the advent of cloud-based infrastructure being epitomized by Amazon Web Service (AWS) and Azure, data handling, storage, and computing are increasingly being integrated into a cloud data platform. Such a platform offers agility, scalability, and serviceability with substantial financial cost incentives, when compared with EDW solutions. Generic machine learning analytics are also being integrated with cloud data platforms, for the benefit of generating valuable insights on-demand. Examples of these are IBM Watson and Google Cloud Platform.

Of greater complexity from the generic cloud data platform are the domain platforms. The latter would require a considerable degree of knowledge on a particular domain, for example, urban transportation. Using the incident detection and classification as an example, which is further elaborated in this paper, several data sources would complement each other for improving the reliability and accuracy. Related data sources may include real-time traffic data, real-time incident monitoring data, crowd-sourced data, and so forth. Without a well-developed mathematical modeling framework that implements several fundamental principles, such as the decision tree and support vector machines, it is unlikely the generic analytics would arrive at a deterministic and meaningful outcome.

In this paper, we present a part of the Advanced Data Analytics in Transport (ADAIT) platform (*1*). ADAIT is a cloud data platform hosted on AWS and dedicated to urban transport data analytics. In particular, we address the innovative approach of integrating data analytics with traffic simulation for incident decision-support. This approach exploits the distinction between

[1]DATA61, CSIRO, New South Wales, Australia

Corresponding Author:
Address correspondence to Tao Wen: tao.wen@data61.csiro.au

recurrent and non-recurrent traffic incidents. With an adequate number of observations, information about a recurrent incident can be parsed into several prominent features and represented by a feature space. Data-driven models, that is machine learning models, could then be built to map the feature space into observed impact. The more observed repetitions of the recurrent incident, the better the data-driven model could explain the variance in the incident impact. On the other hand, for a non-recurrent incident, building a viable data-driven model may be prohibitive because of the lack of data. Instead, the detected incident information could be used to populate a simulation model. The latter can then run multiple scenarios in parallel, faster than real-time, to provide the expected impact of the incident. This paper discusses how the two modeling techniques could be integrated via the ADAIT platform to provide incident decision-support capabilities.

Incident Decision-Support Framework

The work presented in this paper focuses on incident detection and impact analysis on the traffic network and is a part of the ADAIT general framework, currently under development in collaboration with the Transport Management Centre (TMC) of Sydney. Figure 1 presents the general flowchart of the incident detection framework, which comprises various types of interconnected modules, from the raw data processing and cleaning to the machine learning predictive module and traffic simulation core.

The occurrence of incidents is one of the main factors that affect urban transport mobility (*2*). To reduce the congestion caused by incidents, an incident decision-support system should have a prompt response time to help operators make timely decisions. Various models and methods have focused on signal priority (*3*), sub-network and tunnel segment (*4*) or incident management system infrastructure (*5*). However, scalability remains an issue. In addition, some simulation-based incident decision-support systems can provide a quantitative estimation of different traffic operations, but struggle to provide automated situation awareness (*6*). The major characteristic of the ADAIT platform is that it offers a continuous situation awareness by real-time monitoring of the traffic condition, and triggers the data-driven pattern reconstruction or traffic simulation modules only when reported incidents appear in the network. If no incident occurs, the platform will continue monitoring and reporting the traffic state in the network. All outputs of each module are stored in a dedicated cloud database and the results are provided for consumption and further

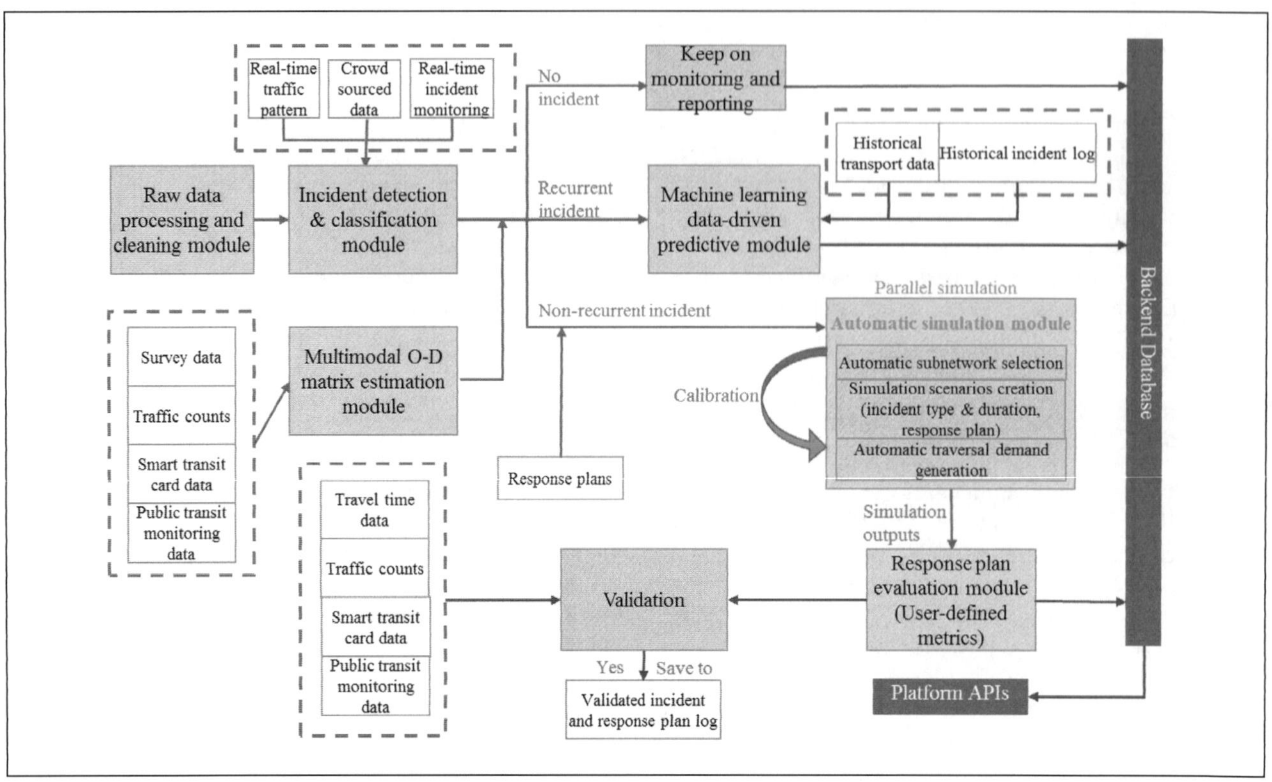

Figure 1. Incident detection and impact analysis framework.

use through public APIs. Amongst the modules of the platform, we detail the following:

1. ***Incident detection and classification module,*** which comprises a) real-time data fusion from various sources for detecting recent incidents, b) incident ranking algorithms based on type and severity, and c) incident duration classification based on available characteristics of the reported accident. The main outcome of this module is to determine if the reported incident follows a recurrent pattern or not to support decision-making for the operator. It also estimates the incident duration and severity to efficiently trigger the simulation module for impact analysis and response plan evaluation.
2. ***Machine learning data-driven predictive module,*** which is based on historical observations of both transport and incident log data, and can predict the severity and duration of the detected incident.
3. ***Automatic simulation module,*** which is triggered only if the incident pattern is identified as non-recurrent. In this case, the incident feature information and candidate response plans are passed to the simulation module. The module then automatically selects the sub-network where the incident took place based on the incident feature information. It further generates a traversal demand matrix for the selected sub-network based on the path assignment obtained from previously running regular macroscopic static traffic assignment on the large-scale network. This will facilitate the application of mesoscopic or microscopic traffic scenarios at the sub-network level for a more detailed simulation of the condition in the affected area while maintaining computation efficiency. Each response plan is simulated in parallel and the corresponding outcome will be evaluated according to user-defined performance metrics. Such parallel simulations can significantly reduce computation time, while providing quantitative insights on a mesoscopic or microscopic level. It is worth mentioning that this type of simulation module requires periodic calibration and validation of the simulation outputs by using existing data sources.
4. ***Multimodal Origin-Destination (O-D) matrix estimation module,*** which estimates the multimodal O-D demand of large-scale network based on historical transport data. The output of this module is a fundamental input for subsequent analysis and is critical to the simulation module's performance.
5. ***Response plan evaluation module,*** which chooses the best plan to mitigate the impact of the incident. The simulated outputs will be compared with the observed traffic data, which will enable the transport operators to make informed decisions by various scenario testing and evaluation.

Combining these various modules with different characteristics and inputs/outputs, from data-driven incident detection/classification/prediction to automatic traffic simulation models, represents a unique and innovative method to evaluate the impact of incidents in a highly affected traffic network. The advantages of the proposed platform in the aforementioned perspectives are:

1. Integrating and fusing multiple data sources to improve the reliability and accuracy of results.
2. Applying machine learning algorithms to learn from previous incidents and make predictions.
3. Detecting and classifying incidents based on real-time data. The innovation is to only trigger the simulation model when the impact of a new incident could not be determined.
4. Automated simulation to provide quantitative evaluation of incident impact and response plan performance.

The logic and mechanism behind the platform are to provide a viable and practical solution to incident management. Such an incident management system needs a detailed decomposition and theoretical analysis in terms of processing flow, traffic modeling, predictive analytics, and response plan selection for mitigating the impact of traffic incidents on both normal and public transport modes. In this paper, we will only focus on detailing the major modules for incident impact analysis and the way the information flow is propagated from one module to another with a major focus on the incident detection, duration prediction, and response plan evaluation.

Module Details

In this section, we describe each module of the decision-support platform, as well as the implications and the afferent data sources needed for obtaining accurate insights.

Data Processing Module

The data sets used in the platform have a wide variety of formats and specifications, are often sparse, and need constant cleaning and monitoring. Amongst them, we cite:

a) *Survey data* such as household travel survey data, census survey data, and so forth, which are very time-consuming and become available every few years; they can provide insights on the O-D demand but may also be obsolete. On this platform, survey data is processed and provided to the multimodal O-D matrix estimation module as prior estimates of O-D matrix.
b) *Traffic counts* are provided by the Sydney Coordinated Adaptive Traffic System (SCATS) system. The integration of SCATS data into the platform is challenging because of a high complexity and duplication of streams, which needs supplementary processing, outlier detection, error elimination, and so forth.
c) *Smart transit card data* provides tap-on and tap-off information of each user traveling in the city by public transport modes.
d) *Public transit monitoring data* includes information such as fixed-route schedules, routes, and bus stop data. The General Transit Feed Specification provides such information on the Sydney network and is used for the validation of simulation results.
e) *Crowd-sourced data* such as Twitter and Waze data are used on the platform as they provide textual information on incidents, and are processed directly by the incident detection and classification module.
f) *Real-time incident monitoring data* provided by the TMC, which reports incidents and actions taken on a real-time basis. This data is used by the incident detection and classification module.
g) *Travel time data* can be provided by many ITS applications; on the current platform Google Travel Time is used for the validation of the simulation outputs.

Multimodal O-D Matrix Estimation Module

The origin-destination demand matrix, which represents the number of trips from one urban centroid to another, is a fundamental element in many transport models. Having a reliable and accurate O-D matrix can significantly enhance the prediction quality (*7*). In this paper, road traffic demand and public transport demand are estimated independently, and data fusion techniques are applied in both estimations. Multimodal O-D matrix estimation algorithms accounting for modal choice require additional information on behavioral analysis and choice modeling (*8*) and is not adopted here because of computation complexity and data availability. Researchers have proposed various methods for data fusion in public transport (*9*, *10*), even the traditional road traffic O-D estimation is an example of data fusion; the survey data are used as prior estimates of O-D matrix and are calibrated based on traffic counts (*11*). However, the integration of the multimodal O-D estimation into the automated incident management platform is a novel approach that offers valuable input for the simulation module.

Incident Detection and Classification Module

Incident Detection. Incident detection is considered an important component of many modern intelligent transport systems. Multiple data sources may provide complementary data, and data fusion can produce a better understanding of the observed situation by decreasing the uncertainty related to the individual (*9*). Traffic control operators can set a threshold to trigger and clear an incident alert using the alert score. The method to estimate the alert score is explained below in Equation 1, while the data used in this paper are explained later.

By analyzing the historical data sources along with confirmed incident logs, the important factor of the individual source is evaluated using attribute ranking algorithms (e.g., information gain, principal component analyses) (*12*).

Besides the reliability of the individual source, the alert score is dependent on spatial and temporal aspects. Furthermore, a recent detection (e.g., within 10 minutes) should receive more attention than a past report (e.g., reported over 30 minutes ago). As a consequence, the spatial-temporal adjusted weight should be considered. The total alert score for an incident over the time is calculated as:

$$score(t) = \sum_{i=1}^{n} R_i W_i^j S(r) T(t) \quad (1)$$

where, i represents data source i, R_i is the reliability score for source i^{th}, W_i^j is the weight for subtype of source i, $S(r)$ is the spatial adjustment function which depends on road type r, and $T(t)$ is the temporal adjustment which depends on time t.

Incident Classification. When an incident is detected and confirmed by the system, a classification process is applied based on the incident description to divide it into different categories including accident, breakdown, delay, and so forth. Historical incident data are also used to train the machine learning incident classification model to further classify incidents based on duration and severity.

For incident duration classification, our system applies the method proposed by Taib et al. (*13*) to predict the duration of an incident that has just happened by using the available characteristics known at the onset, for example, location, time, type of incident, lanes affected, operator in charge, and so forth. Understanding the estimated incident duration is useful to traffic operators when choosing an appropriate response plan. Estimated incident duration is also one of the inputs for the traffic simulation module. Although several machine learning detection methods (*14, 15*) have been applied to detect an incident, few of them combined advanced machine learning, active learning, and outlier detection techniques and achieved approximately 90% accuracy in predicting incident severity (*16*). This severity classification approach is integrated into our proposed system.

Finally, the detailed incident impact on the road network is then predicted by machine learning approaches (for recurrent incidents/congestions) or simulation modules (for unseen incidents).

Machine Learning Data-Driven Predictive Module

When incidents occur in an urban traffic network, they are likely to affect the traffic flows of surrounding areas, especially the traffic leading to the congested roads. Some of the causal congestions follow the same patterns or sequences over the time. Therefore, it is useful to discover frequent patterns (if any) of congestion propagations by reviewing historical data in the traffic networks.

This section reviews an algorithm that finds a congestion propagation pattern by looking at the relationships of congestions from the earliest data record through the latest one (*17*). The main insight is that congestion C_1 is a parent of congestion C_2 if C_1 occurred before C_2 in time and they are spatially connected.

A frequent tree represents an expected congestion propagation pattern when an incident happens on a root segment. A root of a congestion tree is defined as a segment where traffic from other segments is flowing into it and causing congestion.

A Dynamic Bayesian Network (DBN) approach is used to model the spatial–temporal characteristics of a recurrent congestion propagation. Using this method, the probability of a recurrent incident's impact can be estimated. A DBN is usually referred to as a Two-Timeslice Bayesian Network (2TBN) because at any given time T, the value of a variable, is computed from the internal regressors and the immediate prior value (time T-1) (*18*). Therefore, DBN is reasonably close to the real-world phenomenon of traffic congestion in which the status of a segment at a specific time can be determined by its previous condition and previous conditions of connecting segments. However, this assumption depends on the length of the time interval and traffic segment. If either the time interval or traffic segment is too long or too short, the dependency may not be applicable. In our experiment, because of the availability of a data set, the time interval was five minutes and segment lengths were predefined by the data supplier.

To build the DBN traffic network, the road segments are presented by a set of N_h random variables, $O_t^{(i)} \in [0, 1]$, where i represents a congested segment at time t. Snapshot t is a storage of the traffic condition from all segments in the network at time t.

In a DBN, the transition (denoted as $B_{\rightarrow}$) and observation model $P(Z_t|Z_{t-1})$ is then defined as a product of the conditional probability distribution (CPD) in the 2TBN:

$$P(Z_t|Z_{t-1}) = \prod_{i=1}^{N} P(Z_t^{(i)}|Pa(Z_t^{(i)})) \tag{2}$$

where $Z_t^{(i)}$ is the ith node in snapshot t and $Pa\left(Z_t^{(i)}\right)$ are the parents of $Z_t^{(i)}$. The unconditional initial state distribution $P\left(Z_1^{1:N}\right)$ is presented by a standard Bayesian Network, namely B_I. Together, B_I and $B_{\rightarrow}$ define the DBN.

Suppose we have a simple traffic network which comprises three segments: EB and GB are connected to BA. As EB and GB both lead to BA, when BA is congested, it becomes the potential cause for congestions at EB and GB in the next time frame. The corresponding DBN network is presented in Figure 2.

When the propagation pattern is generated, the joint distribution for a known-structure tree which includes T consecutive snapshots (slices) can be obtained by "unrolling" the network until we have T slices, and then multiplying together all of the CPDs:

$$P\left(Z_{1:T}^{(1:N)}\right) = \prod_{i=1}^{N} P_{B_1}(Z_1^{(i)}|Pa(Z_t^{(i)})) \times \prod_{t=2}^{T}\prod_{i=1}^{N} P_{B_{\rightarrow}}(Z_1^{(i)}|Pa(Z_t^{(i)})) \tag{3}$$

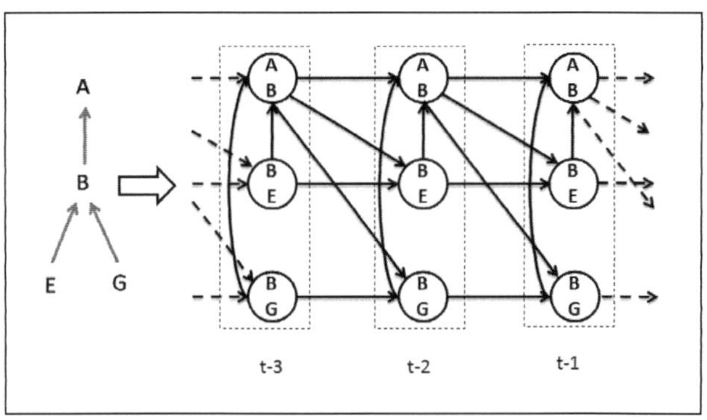

Figure 2. Modeling congestion propagation of a three-segment traffic network by DBN.

In case the detected incident belongs to the root of the congestion propagation tree and the probability to form a propagation pattern is higher than a predefined threshold, the traffic controller may decide to rely on this impact pattern to control the traffic rather than executing simulation, which is more time-consuming. If there is no propagation that exceeded the predefined threshold, the simulation technique is applied to test the impact of an incident on the road network. When the system is implemented, a real-world threshold will be suggested by TMC to decide when it is reliable to use the predicted patterns.

Simulation Module

In the real world, the location, type, and severity of an incident may vary significantly and therefore several incidents may not be accurately and reliably predicted by the machine learning data-driven predictive module. In this case, the simulation module needs to be activated to evaluate the impact of the incident quantitatively. However, simulation of large-scale networks in a reasonable time can hardly be viable because of the computational complexity. Some real-time traffic simulation models have opted for simulation at a corridor/motorway level (*19–21*), but few offer automatic sub-network selection and real-time incident simulation based on duration prediction.

Therefore, we have proposed an automated parallel simulation module to address these issues. In this module, a large-scale traffic simulation model is constructed for the city of Sydney, Australia, and a macroscopic assignment process is implemented every day based on the new incoming data received from the data processing module. Macroscopic assignment models have been well studied by many researchers and many existing models are available (*22*); in the case study, the algorithm by Florian is used (*23*).

To achieve a reasonable computation time, the large-scale network is subdivided into small sub-networks. For example, the Bureau of Transport Statistics in New South Wales, Australia, applies a geometrical zoning configuration which is open to the public (*24*). In addition, some research papers also provide methods for automatic zoning when zoning information is not available (*25*). Based on the incident information (location, severity, lanes affected), the module automatically selects the afferent sub-network, where further analysis and prediction are applied. Then, the traversal demand matrix is generated for the selected sub-network. Here, the traversal demand matrix represents the travelers that travel through, in, and out of the sub-network. To further reduce the computation time, each response plan is simulated simultaneously by the parallel simulation module. This is because transport operators can have multiple candidate response plans, and the simulation module can enable them to choose the most suitable plan. On this platform, the mesoscopic simulation has been adopted because it is computationally more efficient than microscopic simulation, and less data-demanding while capturing the essentials of the traffic dynamics. Note that all the aforementioned simulation processes are automated from the beginning, which is also critical to reducing the computation time to satisfy the strict requirement on response time for incident management purposes.

The reliability of this decision-making process relies on a simulation model that represents the system's behavior closely enough (*26*), and therefore the simulated outputs are validated with several data sources: travel time data (e.g., probe vehicle data, taxi data, or Google Travel Time data), traffic counts (e.g., loop detector data), and public transport data (e.g., smart card data and public transport monitoring data). If the results do not closely approximate the observations, further calibration is conducted to fine-tune the parameters such as sub-network traversal demand, mesoscopic reaction time, and so forth, until the result quality is satisfactory.

Response Plan Evaluation Module

As each city or traffic management center has its own unique characteristics and preferences, it is important that the best response is chosen based on bespoke metrics, so that the ramification of an incident is mitigated in a user-defined way. For example, the average travel time per kilometer can be used to evaluate performance:

$$TT_{network} = \frac{\sum_{i=1}^{N_{network}} TT_i}{N_{network}} \quad (4)$$

where, TT_i is the average travel time per kilometre of vehicle i, $N_{network}$ is the number of vehicles in the network. When travel time reliability or other factors are considered important, they may also be used to choose the most effective response plan. Such a user-defined metric embodies the platform's flexibility and customizability.

Case Study

Here, we present a case study of the Sydney large-scale network consisting of more than 70,000 links and 2,000 centroids. This case study focuses mainly on the incident detection and classification, recurrent incident impact prediction, and non-recurrent incident simulation.

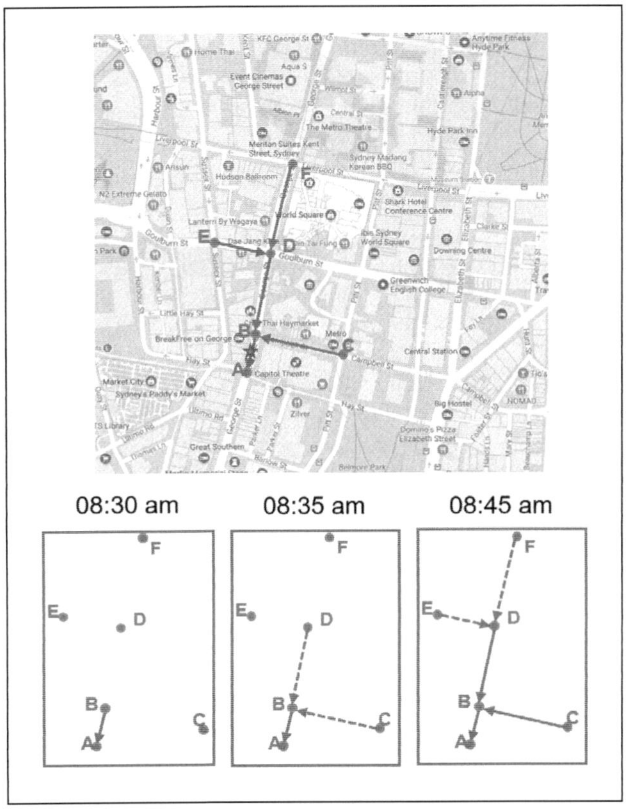

Figure 3. The frequent congestion propagation pattern in the Sydney CBD.

Recurrent Incident: Congestion Propagation and Impact Prediction

When an incident is detected, the incident detection and classification module will also analyze and determine whether the incident is recurrent or not. Here, an example of a recurrent incident is presented, to demonstrate that the machine learning data-driven predictive module can predict the congestion propagation and incident impact for transport operators to make informed decisions.

As illustrated in Figure 3, at 8:30 a.m. on a weekday, an incident was detected at George Street near Campbell Street in the Sydney central business district (CBD). Given that the incident first happened at the root segment BA, using a congestion discovery algorithm and DBN, a five-segment congestion propagation pattern was detected with a joint distribution probability estimated at 74%. This probability is higher than the predefined threshold, therefore this predicted impact can be used by the operator to manage the incident.

The detected congestion pattern was then validated using the real-time traffic data collected from SCATS. The congestion was initially detected at segment BA. Five minutes later, both segments DB and CB also became congested. Until 8:45 a.m., which is 15 minutes after the incident happened, similar congestion propagation patterns were detected on all five segments. The case study shows the efficiency and capability of the machine learning module in decision-support and impact prediction for recurrent incidents.

Non-Recurrent Incident: Parallel Simulation and Performance Evaluation

Although the machine learning data-driven predictive module shows its efficiency in predicting incident impact, several incidents may not be predicted correctly when little historical information is available. The simulation module needs to be triggered to assist transport operators when such non-recurrent incidents are detected. As previously explained, simulation of the whole Sydney network on a mesoscopic or microscopic level is extremely time-consuming, which cannot satisfy the computational time requirements for transport management purposes. Therefore, the Sydney network has been subdivided into many sub-networks beforehand using the Statistical Area definition in (*24*), so that the simulation module can select a sub-network promptly for simulation. The transport simulation model of the Sydney network is implemented in Aimsun, which is regularly calibrated using periodically aggregated SCATS traffic counts and smart card data. The Aimsun network simulation model uses the multimodal O-D matrix previously estimated and runs a macroscopic multimodal traffic assignment for the Sydney network. The output will be saved for later use in generating a traversal demand for the chosen sub-network where the incident happens.

The non-recurrent incident presented here is reported by the incident detection and classification module with the following information:

a. Location (including x and y coordinates): Pyrmont Bridge Road, Pyrmont.
b. Estimated duration: 30 minutes.
c. Severity: major accident affecting all lanes in both directions.
d. Start time: 07:15 a.m.
e. Incident pattern: non-recurrent.

The non-recurrent incident pattern triggers the simulation module. To demonstrate the network state and performance before and after the incident, a two-hour simulation period is chosen, from 7:00 to 9:00 a.m. Using the incident location, a sub-network has been automatically selected from the list of available sub-networks in the city, which is identified as Pyrmont. Pyrmont is a suburb adjacent to Sydney CBD and the

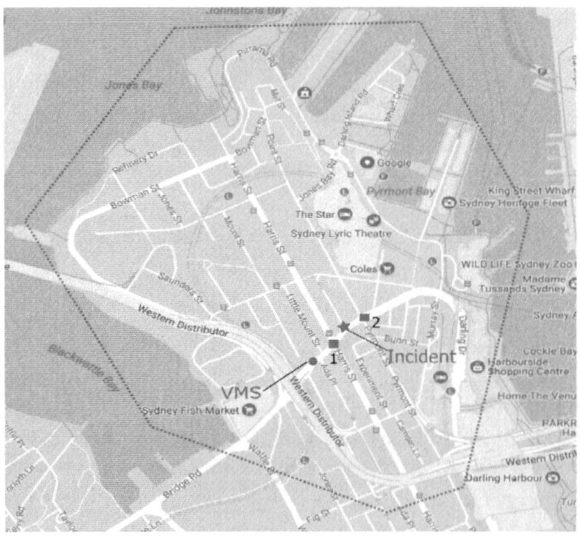

Figure 4. The selected sub-network (Pyrmont, New South Wales).

majority of its area is zoned for commercial purposes (see Figure 4).

After automatically selecting the sub-network area in which the incident has occurred, the traversal demand matrix for Pyrmont is generated and calibrated for the morning peak hour (7:00 to 9:00 a.m.). The model is also validated by comparing the average travel time obtained from simulation (STT) on each road section with the average travel time from Google (GTT) or the average travel time obtained from the SCATS data in Pyrmont (SCATSTT). Figure 5*a* presents an example of comparison between the average STT and GTT on the road section 2839_2840 from Pyrmont, on a Wednesday morning from 7:00 to 9:00 a.m. The plot of travel time every 15 minutes indicates that the simulation provides good results of the TT on this section as it falls between the 5th and 95th percentile of the GTT. This finding is validated once more on a different section (5_2839), where the STT is compared with GTT SCATSTT, which is available for computation (Figure 5*b*).

Based on the received incident information, assume the operators choose the following incident response plans (RPs) for evaluation:

1) RP1: Do nothing.
2) RP2: Redirect all traffic in intersections 1 and 2 (marked as red rectangular in Figure 4) towards adjacent intersections.
3) RP3: Combined actions: activate the variable message sign (VMS) to redirect all off-ramp flow from the bridge toward Little Mount Street (see Figure 4), and redirect all traffic in intersection 1 toward surrounding intersections.
4) RP4: Activate the VMS to redirect all off-ramp flow from the bridge toward Little Mount Street.

RP1 is intended to keep monitoring the network but take no action, in order to evaluate the true impact of the incident if no action is taken. RP2 redirects all traffic in intersections 1 and 2 toward adjacent intersections to prevent vehicles from queuing and eventually blocking the intersection. RP3 has the role of activating the VMS, which will inform drivers to make a left turn before reaching intersection 1. Also, traffic will be redirected at intersection 1 to prevent queuing. RP3 aims to let the major traffic from the bridge (Western distributor) bypass intersection 1. RP4 simplifies RP3 by keeping only the VMS activation action. Note that the parameters for driver behavior in the microsimulation (such as acceleration and reaction time) will remain unchanged because of the difficulty in collecting sufficient data.

These response plans are then simulated in parallel on a microscopic level by the microsimulation engine in Aimsun. Figure 5*c* presents the average travel time per kilometer (including buses and private vehicles) obtained after applying each of the four response plans. RP1 is the baseline and demonstrates that the average travel time reaches a high point at 8:00 a.m., when the incident has already ended. This is due to the accumulation of queue and the increase in traffic and public transport demand. RP4 shows a very marginal improvement over RP1; the average travel time over the two hours is also quite close to RP1. Although RP2 mitigates the congestion between 7:45 a.m. and 8:15 a.m., the travel time increases gradually after 8:15 a.m., making the eventual average travel time over the two hours very similar to RP1. Overall, RP3 performs the best; it smooths the travel time after the incident happens, while having a 7% reduction in average travel time over the two-hour simulation period.

The finding indicates that the best response plan for mitigating congestion produced by a non-recurrent incident is actually a combination of various actions that complement each other and help to reduce the incident clearance time. Therefore, a possible extension of this work is to automatically recommend the best combination of response plans to apply for efficiently easing congestion.

Conclusion

In this paper, we introduced the general framework of the ADAIT platform and explained the main function of each module. The platform can detect and then classify incidents into a recurrent and non-recurrent pattern; the former triggers the machine learning data-driven

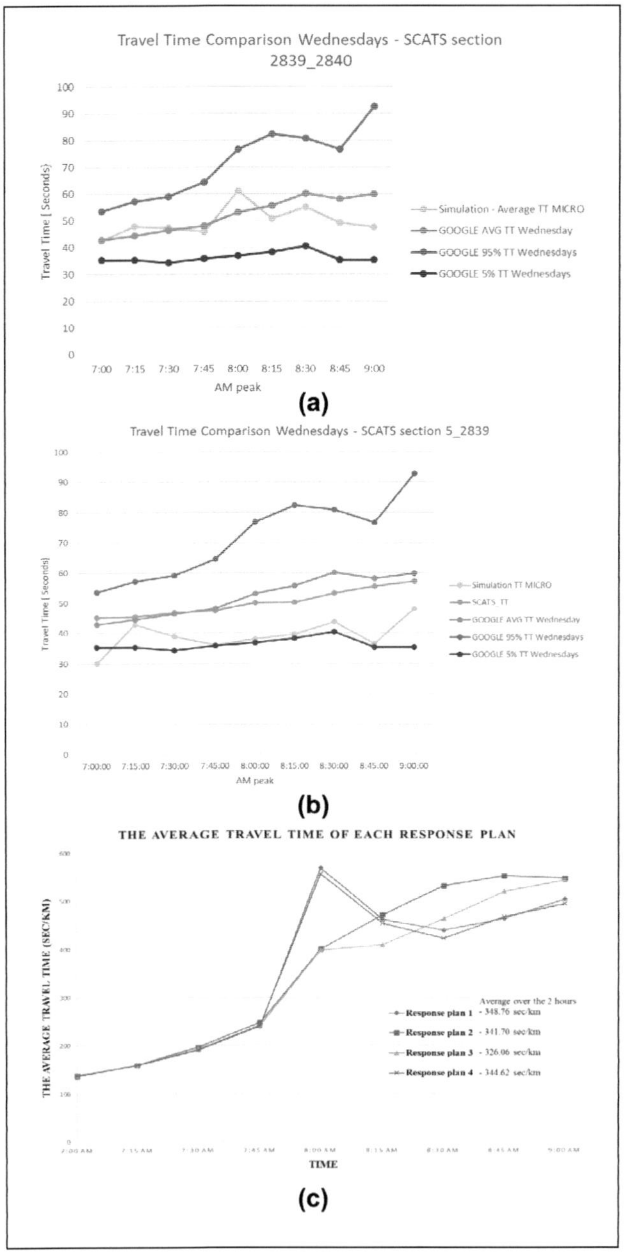

Figure 5. Average travel time comparison between (*a*) simulation and Google Travel Time, (*b*) simulation, Google TT, and SCATS, and (*c*) each response plan (RP1, RP2, RP3, and RP4).

predictive module, which predicts the incident duration and impact, so transport management operators can decide if the simulation module needs to be activated. A non-recurrent incident is directly passed to the simulation module; the performance of candidate response plans is evaluated quantitatively, and then operators can opt for the best plan to mitigate the negative impact of an incident. Case studies demonstrate that the impact of a recurrent incident, such as congestion propagation, can be predicted by the machine learning module, and the simulation module can help choose the best response plan to mitigate the negative incident impact. In short, data-driven incident detection/classification, machine learning analytics for incident prediction, and automatic traffic simulation models are integrated into the cloud-based platform, which represents a unique and innovative method to evaluate the impact of incidents in real-time for large-scale networks.

There are various possible opportunities to further extend the platform: the response plans can be generated automatically by advanced machine learning techniques based on the information of a detected incident (such as location, duration, severity, etc.), and therefore save time on manually input response plans. Also, the platform's modularity allows integration of advanced transport algorithms in each module, which can enhance the platform's applicability. Because the DBN model depends on the length of the time interval and traffic segment, in our future work, the time interval between two continuous snapshots will be considered as an additional parameter. Furthermore, the algorithm will be tested on different traffic network settings with different average segment lengths to evaluate the effect.

Acknowledgments

The work presented in this paper is partially funded by the New South Wales Premier's Innovation Initiative. The authors of this work are grateful for the work and support of the Traffic Management Centre from Transport for New South Wales Australia. Data61 is funded by the Australian Federal Government through Commonwealth Scientific and Industrial Research Organisation.

Author Contributions

The authors confirm contribution to the paper as follows: study conception and design: Tao Wen; data collection: Tao Wen, Hoang Nguyen, Adriana-Simona Mihăiţă; analysis and interpretation of results Tao Wen, Hoang Nguyen, Adriana-Simona Mihăiţă; draft manuscript preparation: all authors. All authors reviewed the results and approved the final version of the manuscript.

References

1. ADAIT. *Advanced Data Analytics in Transportation*. Data61, CSIRO. www.adait.io.
2. Steenbruggen, J., P. Nijkamp, J. M. Smits, and M. Grothe. *Traffic Incident Management: A Common Operational Picture to Support Situational Awareness of Sustainable Mobility*. 2011.
3. Ahmed, F., and Y. Hawas. An Integrated Real-time Traffic Signal System for Transit Signal Priority, Incident Detection and Congestion Management. *Transportation*

Research Part C: Emerging Technologies, Vol. 60, 2015, pp. 52–76.
4. Mitrovich, S., G. Valenti, and M. Mancini. A Decision Support System (DSS) for Traffic Incident Management in Roadway Tunnel Infrastructure. *Proc., 34th European Transport Conference (ETC)*, Strasbourg, France, 2006, p. 240.
5. Li, H., H. Lu, and I. A. Qureshi. Intelligent Decision Support System for Traffic Incident Management. 9th International Conference on *Applications of Advanced Technology in Transportation* (AATT), Chicago, IL, 2006, pp. 292–297.
6. Yurshevich, E., and I. Yatskiv. Decision Support System for Transport System Management and Microscopic Traffic Modeling. *Proc., 7th International Scientific Conference TRANSBALTICA*, 2011, pp. 74–81.
7. Bierlaire, M., and P. L. Toint. Meuse: An Origin-destination Matrix Estimator that Exploits Structure. *Transportation Research Part B: Methodological*, Vol. 29, No. 1, 1995, pp. 47–60.
8. Pattanamekar, P., and D. Park. Estimating Multimodal OD Matrix from Mode Specific OD matrices. Presented at 88th Annual Meeting of the Transportation Research Board, Washington, D.C., 2009.
9. El Faouzi, N.-E., H. Leung, and A. Kurian. Data Fusion in Intelligent Transportation Systems: Progress and Challenges–A Survey. *Information Fusion*, Vol. 12, No. 1, 2011, pp. 4–10.
10. Kusakabe, T., and Y. Asakura. Behavioural Data Mining of Transit Smart Card Data: A Data Fusion Approach. *Transportation Research Part C: Emerging Technologies*, Vol. 46, 2014, pp. 179–191.
11. Wen, T., L. Gardner, V. Dixit, S. T. Waller, C. Cai, and F. Chen. Two Methods to Calibrate the Total Travel Demand and Variability for a Regional Traffic Network. *Computer-Aided Civil and Infrastructure Engineering*, Vol. 33, No. 4, 2018, pp. 288–299.
12. Guyon, I., and A. Elisseeff. An Introduction to Variable and Feature Selection. *Journal of Machine Learning Research*, Vol. 3, No. March, 2003, pp. 1157–1182.
13. Taib, R., D. Yee, F. Chen, and W. Liu. Use of Machine Learning for Improved Incident Management Outcomes. *Proc., 13th ITS Asia Pacific Forum 2014*, Auckland, New-Zealand, 2014, p. 10.
14. Chong, M., A. Abraham, and M. Paprzycki. Traffic accident analysis using machine learning paradigms. Informatica, Vol. 29, No. 1, 2005.
15. Sohn, S. Y., and H. Shin. Pattern Recognition for Road Traffic Accident Severity in Korea. *Ergonomics*, Vol. 44, No. 1, 2001, pp. 107–117.
16. Nguyen, H., C. Cai, and F. Chen. *Automatic Classification of Traffic Incident's Severity Using Machine Learning Approaches*. IET Intelligent Transport Systems, Institution of Engineering and Technology, 2017.
17. Nguyen, H., W. Liu, and F. Chen. Discovering Congestion Propagation Patterns in Spatio-temporal Traffic Data. *IEEE Transactions on Big Data*, Vol. 3, No. 2, 2017, pp. 169–180.
18. Ghahramani, Z. Learning Dynamic Bayesian Networks. *Adaptive Processing of Sequences and Data Structures*, 1998, pp. 168–197.
19. Transport Simulation Systems. *Aimsun in San Diego Integrated Corridor Management Project*, San Diego, California, 2012. https://www.aimsun.com/integrated-corridor-management-project-in-san-diego/.
20. Dia, H., and N. Cottman. Evaluation of Arterial Incident Management Impacts Using Traffic Simulation. *IEEE Proceedings-Intelligent Transport Systems*, No. 153, 2006. pp. 242–252.
21. Hidas, P. Modelling Vehicle Interactions in Microscopic Simulation of Merging and Weaving. *Transportation Research Part C: Emerging Technologies*, Vol. 13, No. 1, 2005, pp. 37–62.
22. Wen, T., C. Cai, L. Gardner, V. Dixit, S. T. Waller, and F. Chen. A Strategic User Equilibrium for Independently Distributed Origin-Destination Demands. *Computer-Aided Civil and Infrastructure Engineering*, Vol. 33, No. 4, 2018, pp. 316–332.
23. Florian, M., and D. Hearn. Network Equilibrium Models and Algorithms. *Handbooks in Operations Research and Management Science*, Vol. 8, 1995, pp. 485–550.
24. Australian Bureau of Statistics. *Main Structure and Greater Capital City Statistical Areas*, 2011. Australian Statistical Geography Standard (ASGS): Volume 1 Institution of Engineering and Technology, 2017.
25. Menon, A. K., C. Cai, W. Wang, T. Wen, and F. Chen. Fine-grained OD Estimation with Automated Zoning and Sparsity Regularisation. *Transportation Research Part B: Methodological*, Vol. 80, 2015, pp. 150–172.
26. Barcelo, J., and J. Casas. Methodological Notes on the Calibration and Validation of Microscopic Traffic Simulation Models. Presented at 83rd Annual Meeting of the Transportation Research Board, Washington, D.C., 2004.

The Standing Committee on Urban Transportation Data and Information Systems (ABJ30) peer-reviewed this paper (18-06575).

Article

TRR
JOURNAL OF THE TRANSPORTATION RESEARCH BOARD

Transportation Research Record
2018, Vol. 2672(42) 257–267

DOI: 10.1177/0361198118796972
journals.sagepub.com/home/trr
SAGE

Using National Performance Management Research Data Set for Corridor Performance Measures: A US-281N Corridor Case Study

Venktesh Pandey[1] and Natalia Ruiz Juri[2]

Abstract

The National Performance Management Research Data Set (NPMRDS), made available by Federal Highway Administration in 2013, provides fine-resolution travel-time data, which have been used in numerous network performance management and operations applications. This article discusses corridor-level performance measures computed using the NPMRDS. Three measures are analyzed on a 20.2-mile long corridor in San Antonio, Texas, including corridor travel time, corridor travel-time reliability, and day-to-day variation in travel time. The primary contributions of this article are the analysis of the impact of using two different approaches for travel-time aggregation across segments—instantaneous and time-dependent approaches—and defining a mean absolute error-based method to identify days when travel times significantly deviate from typical traffic conditions. The findings suggest that the temporal patterns of corridor travel times obtained using instantaneous and time-dependent aggregation approaches are similar; however, instantaneous travel-time estimates lead to inaccuracies that become more apparent during peak hours and for longer segments. In addition, it is found that a *k*-means clustering analysis performed on daily travel-time profiles provides a useful statistic for corridor performance analysis. Using this methodology, 9.23% of weekdays in the 20-month study period are classified as *atypical* for the corridor. The numerical results reinforce the value of the NPMRDS in estimating corridor performance measures and highlight potential limitations of traditional techniques for evaluating corridor performance measures when applied in practice to support enhanced traffic planning and operations.

With the advancements in sensing technologies in the field of intelligent transportation systems, more data are available to support transportation agencies' planning and operation decisions. Available sensors fall in two broad categories: location based and probe based. Location-based sensing includes technologies such as loop detectors and Bluetooth readers, which are fixed to the infrastructure. Probe-based sensing involves tracking the location of a moving vehicle by using data from global positioning system (GPS) or inertial measurement units. Probe-based data have become widely available in the last few decades as a result of an increased penetration of mobile devices equipped with GPS, including cellphones and vehicle onboard devices.

In an effort to inform traffic performance management, operations, and planning using big data, the Federal Highway Administration provides access to fine-resolution travel-time data to state agencies and metropolitan planning organizations (MPOs) (*1*). The National Performance Management Research Data Set (NPMRDS) consists of travel-time information collected at five-minute intervals on traffic message channels (TMCs). The latter are link segments defined throughout the transportation network. The data are reported for both passenger cars and freight vehicles.

Existing literature presents several applications of the NPMRDS, including computing various corridor-level performance metrics (*2–5*), comparing the performance of the data with other datasets (*2*, *6*, *7*), performing before–after analyses (*8*), and using the data for calibration and validation of models.

The objective of this research is to extend the applications of NPMRDS by deriving time-varying mobility

[1]Department of Civil, Architectural and Environmental Engineering, The University of Texas at Austin, Austin, TX
[2]Network Modeling Center, Center for Transportation Research, Cockrell School of Engineering, The University of Texas at Austin, Austin, TX

Corresponding Author:
Address correspondence to Venktesh Pandey: venktesh@utexas.edu

and reliability metrics on corridors. Also exemplified is the potential impact of the methodologies used for data analysis and aggregation, which may be more critical as the use of widespread data becomes central to decision making. As a contribution to the existing literature, this article compares two approaches for computing corridor travel-time metrics and proposes a method to identify deviations of the travel-time on a particular day from the typical travel-time distribution. An extensive analysis is presented for a case study on a congested 20.2-mile long corridor on US-281 going through the city of San Antonio, Texas.

In the next section, existing literature on the use of the NPMRDS and other probe-based data sources is discussed. The methodology is then described and the performance measures introduced, followed by the numerical analyses. The last section summarizes the findings and provides a direction for future research.

Literature Review

The high granularity of the NPMRDS has enabled researchers and agencies to use the dataset for multiple purposes, with corridor performance metrics being the most common use of these data. Venkatanarayana uses NPMRDS in combination with vehicle probe data to determine five performance measures for arterial assessment in Virginia (*2*). The performance metrics include planning-time index, travel-time index, AASHTO reliability indexes, congested hours, and congested miles. The study highlights the lack of data availability on arterials in the dataset. Kaushik et al. compute the travel-time index and planning time index in Maryland to estimate corridor congestion and reliability, respectively (*3*). Bitar analyzes similar metrics on different TMC segments along the I-35 corridor in Oklahoma (*4*), and Eshrask et al. evaluate travel-time index and planning time index for the I-95 corridor through Maryland (*5*).

Studies combining NPMRDS with other data sources include Venkataraman's comparison of NPMRDS data with the Bluetooth sensor and INRIX data (*2*). The latter is another probe-based vehicle travel-time dataset available for purchase. Venkataraman's results suggest that the data from the NPMRDS have fewer observable daily patterns and high day-to-day variability compared with the other two. Margiotta et al. integrate NPMRDS data with freight volume, flow, and demand data to perform freight bottleneck analysis (*6*). Rafferty and Hankley combine the NPMRDS data with the traffic volume counts and determine total vehicle delay and reliability measures as a statistic for the state of Wisconsin evaluated for each month and each corridor (*7*).

Rafferty presents alternative applications of the NPMRDS, including impact assessment of flood closures and lane closures by comparing travel times before and after and traffic analysis during special events (*8*). Sadabadi et al. use the NPMRDS in Maryland to estimate the value of reliability by calibrating appropriate parameters of the stochastic process describing the evolution of travel time (*9*). Hafeez and Kasemsarn use the data to characterize delays at railroad grade crossings and inform possible improvements to the facility (*10*). Other NPMRDS applications include the calibration and validation of existing models including dynamic traffic assignment models (*11*) and four-step planning models.

The direct use of disaggregate probe vehicle data enables additional analysis, including traffic state estimation and travel-time prediction (*12, 13*). Such research typically requires probe vehicle trajectory for model validation purposes, which is not provided by the NPMRDS.

Machine learning algorithms provide additional tools to extract meaningful information from the data. Bitar uses NPMRDS to develop an optimum classifier to identify non-recurring congestion (*4*). Yang et al. use probe vehicle data to identify secondary crashes (*14*). Li and Chen use a data-mining approach to predict travel times under non-recurring congestion by looking at travel-time data collected on small segments between dual loop detectors (*15*).

The findings in the literature suggest that several useful applications of the NPMRDS have been identified and implemented. This article extends the existing applications by considering the impacts of two different aggregation methods for computing corridor travel times, modeling temporal variability of metrics such as reliability, and quantifying the day-to-day variation of travel times.

Methodology

Data Description

The case study analyzed in this paper used 20 months of NPMRDS data for the city of San Antonio (January 2015 to August 2016). Readers may refer to previous work (*3, 4, 8*) for further discussion on the characteristics of the NPMRDS, which includes travel times in 5-minute intervals called epochs for each TMC.

The authors uploaded all the data uploaded to a PostgreSQL database, and used R, an open-source software package for statistical computing and graphics, for statistical analysis. Java, a high-level programming language, was used for converting data into suitable format for the analysis.

Statistical Cleaning

Like any real dataset obtained from the field, the NPMRDS is not free from outliers. Bitar describes four types of outliers in this dataset: discrepancies caused by GPS errors, bias toward lower speeds which are sampled more times in an interval, missing data in low-density areas and off-peak periods, and less data for shorter segments (*4*).

Two levels of statistical cleaning were adopted for the data. The first level of cleaning removed the obvious outliers using the methods presented in Rafferty and Hankley (*7*), which is based on average speeds. Average travel times were converted to average speeds by dividing the length of each TMC by the corresponding travel time. The San Antonio MPO provided the speed limit data, which were complemented by visual inspection using the Google street map feature. Speed values higher than 20 mph over the speed limit, and speeds lower than 2 mph were considered outliers and removed. After the first level of cleaning, 6,673 records (0.31%) were removed from a total of 2,107,725.

The second level of cleaning used statistical methods to remove additional outliers, following a robust methodology from the literature (*16*). The statistical cleaning was implemented on the data filtered based on the time of day, the day type (weekday or weekend), and the TMC id. For each data bin, the median absolute deviation statistic was computed, defined as the median of absolute deviations measured from the median of all data points. A cleaning factor was then selected to determine the range of deviation around the median that was considered acceptable. The cleaning factor was set to 4, based on the authors' previous experience in working with real travel-time datasets. The statistical cleaning removed 21,490 (1.34%) additional observations.

Corridor Travel Time and Reliability Measures

A corridor consists of a sequence of TMCs for which separate travel-time reports are available. There are two broad approaches to computing corridor travel time by aggregating TMCs travel time: instantaneous aggregation and time-dependent aggregation. In this research, the instantaneous travel time for a selected departure time is defined as the sum of mean travel times on all TMC segments at the considered departure time. The time-dependent travel time (also called as experienced travel time) is the travel time on all TMC segments reported at the corresponding time of arrival. Both metrics depend on the length of each time step; the smallest possible time step using NPMRDS data is 5 min, but larger steps may be considered to reduce the computational effort. Chen and Rakha give examples of how using instantaneous travel-time measures may lead to misleading prediction for corridor travel-time (*17*). However, it is included in this analysis as it is an easy-to-compute statistic.

Corridor travel-time reliability measures in the literature include buffer index, planning time index, 95th percentile travel time, standard deviation, coefficient of variation, and misery index (*18*). This study considers the buffer index, planning time index, standard deviation, and coefficient of variation. The buffer index evaluates the amount of time a traveler must add to their average travel time when planning trips to arrive to their destination on time and is defined in Equation 1. Planning time index, on the other hand, is a measure based on system manager's perspective, which evaluates how much worse a system can perform relative to the free flow. It is defined in Equation 2. $TT_{95\%}$ is the 95th percentile travel time, TT_{Mean} is the mean travel time observed during that time interval and $TT_{FreeFlow}$ is the free-flow travel time on the corridor. Standard deviation and coefficient of variation have their usual interpretation and measure the spread of travel-time distribution. All metrics are evaluated for the entire corridor using time-dependent aggregation.

$$\textit{Buffer Index}(BI) = \frac{TT_{95\%} - TT_{\text{Mean}}}{TT_{\text{Mean}}} \tag{1}$$

$$\textit{Planning time Index}(PTI) = \frac{TT_{95\%}}{TT_{\text{FreeFlow}}} \tag{2}$$

In practice, a single value for each measure is reported for a corridor which represents the average performance of the corridor reliability. Though a single measure is useful for a big-picture analysis, the time-of-day variation of these indices can bring meaningful insights for improving corridor operations. This study looks at the time-of-day variation of these four reliability indices, similar to the methods reported in earlier works (*2, 19*).

Measures for Identifying a Typical Day

The analysis of day-to-day variations in travel time provided by the NPMRDS is helpful in understanding whether a day is representative of "typical" traffic conditions, often used as a reference in transportation systems planning and evaluation. To the best of the authors' knowledge, there is not an agreed-upon, data-based definition of "typical" day, even though the concept is used in applications including:

1. Surveys to determine travel choices made by members of the household, which ask about travel patterns on a *typical day*.
2. The four-step planning model, which provides estimates valid for a *typical day* in the future. Calibration data for traffic assignment models is also expected to reflect *an atypical day* in traffic.

From a traffic operations perspective, a typical day is likely the one where the observed traffic is not much higher or lower than the regular traffic. For example, a college football game day might classify as an *atypical* weekend day.

For this analysis, a typical day for a corridor is defined as the one where the observed travel time throughout the day is "similar" to the median travel time. Median travel time is chosen because it is not severely impacted by outliers.

Let $\tau_i(t)$ be the travel-time function observed on the ith day, representing the experienced travel time on the corridor as a function of time of day (t). Similarly, let $\tau_{median}(t)$ represent the median travel-time function obtained by calculating the median of travel times observed on all days for that particular time of day as shown in Equation 3.

$$\tau_{\text{median}}(t) = median(\tau_i(t))\forall i \quad (3)$$

The deviation of travel-time function of an ith day from a typical day was measured by calculating mean absolute error (MAE_i) and root mean square (RMS_i) error as defined in Equations 4 and 5, where n is the number of days being analyzed. For this day-to-day variation analysis, all weekdays from Jan 1, 2015 to Aug 31, 2016 were considered, and thus the value of n is 436. The data were cleaned only using the first level of cleaning to retain the outliers occurring during particular days.

$$MAE_i = \frac{\sum_t |\tau_{\text{median}}(t) - \tau_i(t)|}{n} \quad (4)$$

$$RMS_i = \sqrt{\frac{\sum_t (\tau_{\text{median}}(t) - \tau_i(t))^2}{n}} \quad (5)$$

Calculation of experienced travel time was updated by performing data imputation for missing data for each TMC. The data imputation process assigned the value of travel time the missing epoch as the average of the previous and the next epoch. If the data were also found missing either of previous or the next epoch, the value was replaced by the mean value of travel time for that epoch calculated across all days. The mean statistic was used here to capture the impact of non-recurring congestion while doing data imputation. The experienced corridor travel time calculated for each epoch and day of year was used to determine the MAE and RMS error for a particular day.

In addition, k-means clustering was performed on the travel-time function obtained for each day to identify which days of the year group together to form a *typical* day. k-means clustering is an unsupervised learning approach to identify natural grouping of objects into clusters. Here, the objective was to group all typical days together. The travel time observed in each 5-min interval for a day was transformed to a vector and the Euclidean distance metric was used to compute the similarity between two days. The results are reported in the following section.

Case Study: San Antonio Corridor

The testbed used in this analysis comprises 20 TMC segments in the southbound direction of US-281N corridor in San Antonio. The total corridor length is 20.2 miles. This corridor is the entry point into the city for the travelers from northern regions including Timberwood Park and Springs Branch. This section discusses the results of applying the measures proposed in the previous section on this testbed.

Corridor Travel-Time Measures

Instantaneous and Time-Dependent Aggregation. Figures 1*a* and 1*b*, show the variation of experienced and instantaneous travel times on the corridor averaged over all instances within the analysis period for weekdays and weekends. There are two peak periods on weekdays, a morning peak (6:00 a.m.–8:45 a.m.) and an afternoon peak (4:30 p.m.–7:00 p.m.). A single, mid-day, peak is observed on weekends (9:45 a.m.–2:30 p.m.).

As observed, the instantaneous travel time behaves identically to the experienced travel time on weekends. On weekdays, the estimate for instantaneous travel time lags behind experienced travel time during peak hours. This behavior can be explained by the correlation of travel time in a TMC from one time interval to the next.

The percentage error of instantaneous travel time when compared with time-dependent travel time (Figure 1*c*) is high during the morning and evening peaks on weekdays. On weekends, the percent difference is reasonably close to zero. The corresponding mean absolute percent error (MAPE), evaluated by taking mean of absolute percent difference for all time periods, is 2.04% on weekdays and 1.17% for weekends for the entire day. For weekdays, MAPE value is 6.32% during the morning peak and 6.23% during the evening peak. These findings are consistent with the observations in Chen and Rakha (*17*) for evening peak.

Further analyses were conducted to assess the impact of corridor lengths on the difference between instantaneous and experienced travel times. The northbound direction segments were also included for comparison. Table 1 shows the MAPE values in the morning and evening peak on weekdays for segments of different lengths in both travel directions of the US-281N corridor. As observed, the MAPE values are higher for longer segments indicating that the error in instantaneous

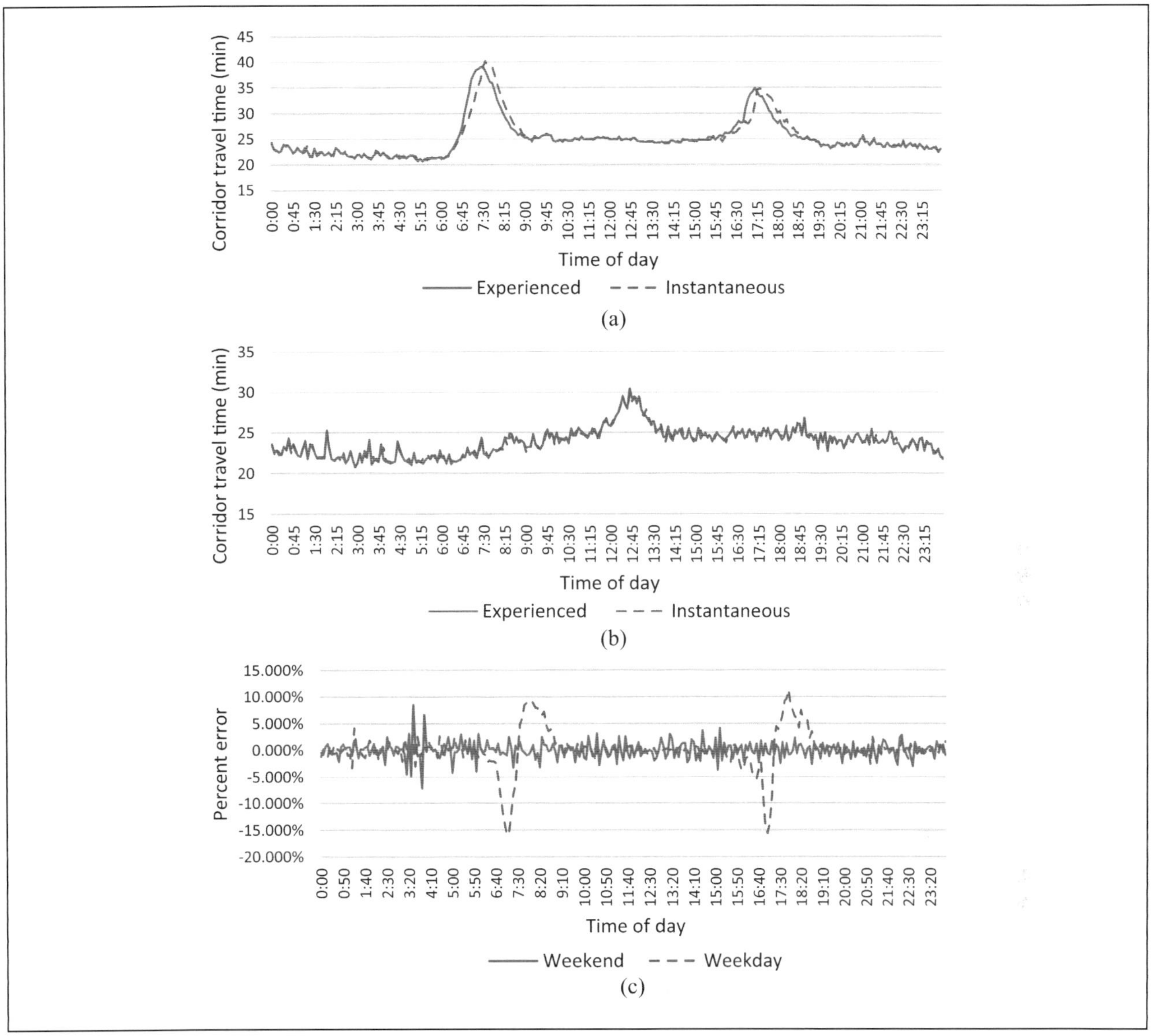

Figure 1. Corridor travel-time analysis including: (*a*) comparison of experienced and instantaneous travel times on weekdays; (*b*) comparison of experienced and instantaneous travel times on weekends; and (*c*) percent error in instantaneous travel time relative to experienced travel time.

Table 1. Mean Absolute Percent Error (MAPE) Value between Instantaneous and Experienced Travel Time on Segments of Different Length for Morning and Evening Peak

Corridor	Segment length	MAPE morning peak (6am–9am)	MAPE evening peak (4pm–7pm)
US-281N southbound	6.4 miles	1.69%	1.34%
	11.9 miles	4.14%	1.12%
	20.2 miles	6.32%	6.23%
US-281N northbound	6.2 miles	0.07%	0.20%
	11.1 miles	0.22%	2.43%
	16.1 miles	1.10%	4.67%

Table 2. Comparison of Travel Times Predicted by Field Counts, NPMRDS Data and the Google Maps Data Over a 7.4-Mile Long Segment of the Corridor

Average travel time (minutes:seconds)			Range of travel time (min)			
	Field counts	NPMRDS		Field counts	NPMRDS	Google maps
6-7 am	14:45	12:30	6-7 am	9–27 min	7–20 min	9–35 min
7-8 am	14:45	18:00	7-8 am	9–27 min	10–25 min	16–40 min
8-9 am	14:45	14:00	8-9 am	9–27 min	9–20 min	14–20 min
Average	14:45	14:49				
3-4 pm	12:56	12:00	3-4 pm	10–17 min	9–13.5 min	10–20 min
4-5 pm	12:56	12:00	4-5 pm	10–17 min	9–14 min	10–22 min
5-6 pm	12:56	12:30	5-6 pm	10–17 min	9–16 min	10–24 min
Average	12:56	12:10				

aggregation increases with the length of the segment. In addition, the MAPE values are lower for the northbound direction, likely as a result of lower congestion (measured as the ratio of peak travel time to the free-flow travel time).

Overall, instantaneous and experienced travel time do not show significant variation for the case study corridor, and thus the former, being easier to compute, can give a reasonably good estimate of true corridor travel time. Further, given that a trip of 20.2-mile commute length is a reasonable upper bound for the typical trips within a city and the MAPE value is less than 7% during peak periods, instantaneous aggregation method may be acceptable for displaying route travel times on variable message signs.

Validation. Field data on observed corridor travel time were made available by TxDOT's consultants on 7.4 miles-long segment of the selected corridor between East Borgfeld Drive and North Loop 1604. Travel times for the same segment estimated using the NPMRDS were compared with the field data. The field data were collected manually using probe vehicles, and the reported values included average travel time and the range of travel time across several days for the month of February, 2016.

Table 2 compares the range and average value of travel times obtained from each dataset and compares them with the historical data provided by Google maps (which does not provide mean travel time values). Travel-time ranges for the NPMRDS were computed by adding and subtracting one standard deviation to the corresponding mean value. The data for all weekdays in the selected 20 months were used for this analysis.

The NPMRDS data match well average travel times from field data, with an average difference of 4 s during the morning peak and 46 s during the evening peak. Similarly, the range observed in the NPMRDS is close to that of the field data. The maximum time reported by Google is typically higher than the one from the NPMRDS and field counts, suggesting that Google maps might be providing conservative estimates to the travelers. This analysis shows the accuracy of prediction of the NPMRDS for the given corridor and thus provides a viable alternative to the field travel-time measurements collected manually.

Corridor Travel-Time Reliability Measures

Figure 2 shows the performance of all four reliability measures and their variation with time of day for weekdays and weekends.

The peak values for buffer time index and planning time index, observed at 7:30 a.m., are 99.4% and 4.11, respectively. Based on these findings, travelers arriving at the corridor during the peak can experience travel times 99.4% higher than the average (37.2 min for this corridor) and should account for such additional time to arrive at their destination on time. Similarly, the planning time index suggests that 95% of travelers may experience travel times four times higher than the average.

The variation of buffer time index and planning time index with the time of day in Figure 2, *a* and *b*, follows a similar trend as the travel-time pattern on the corridor for both weekdays and the weekends. That is, the higher the travel time on the corridor, the higher is the unreliability index. This is because both these measures rely on the 95th-percentile travel time, which shows a similar pattern to the travel time on the corridor because of the log-normal nature of travel-time distribution for a particular time of day (*12*).

On the other hand, the standard deviation and coefficient of variation measures show oscillating but less spread values for the corridor reliability. The mean coefficient of variation across the day for weekdays is 0.19, whereas that for weekends is 0.22, which indicates that travel times on weekends are more variable than on weekdays. Similar behavior is observed for the standard

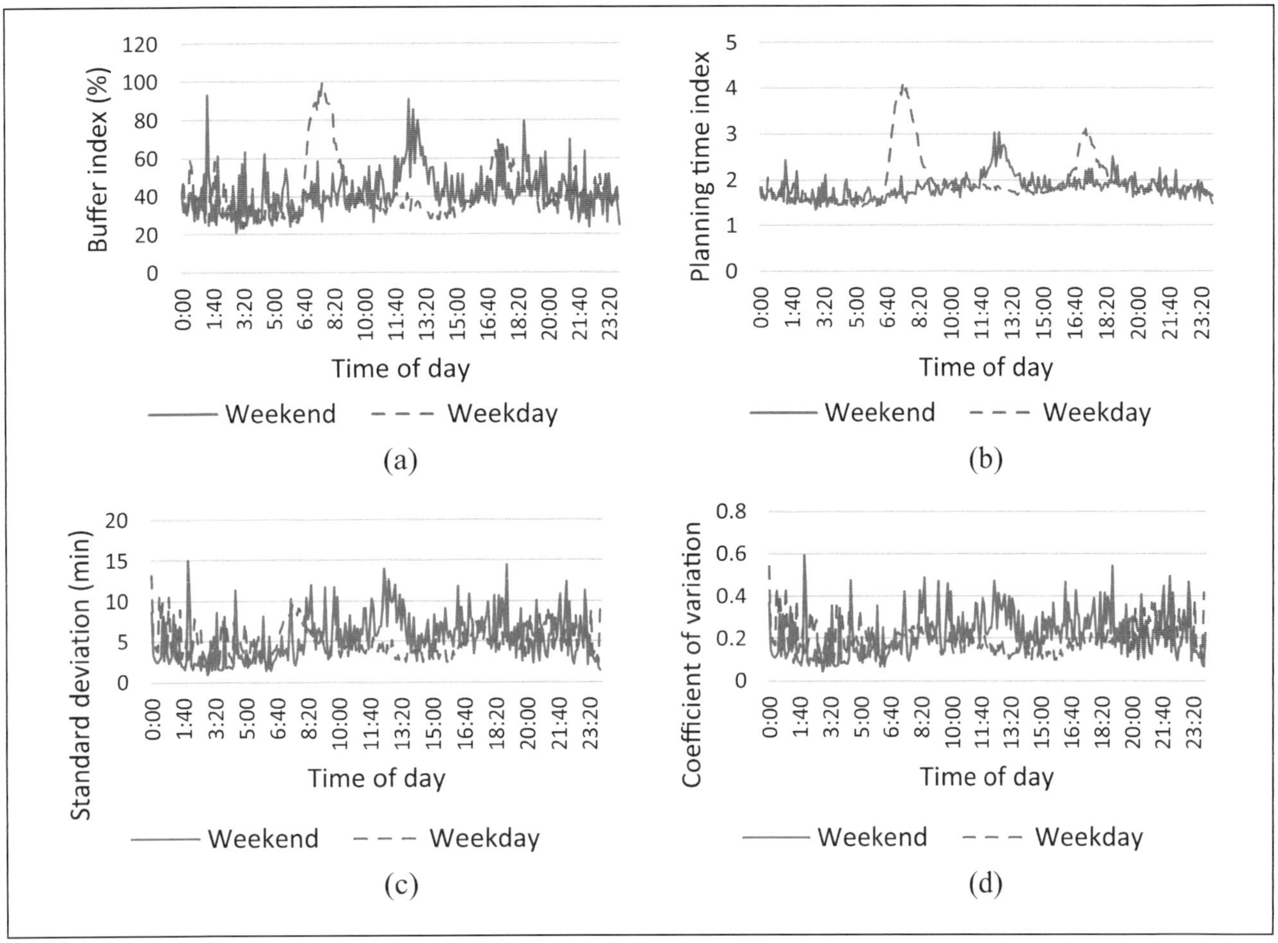

Figure 2. Corridor travel-time reliability measures: (*a*) Buffer time index; (*b*) Planning time index; (*c*) Standard deviation; and (*d*) Coefficient of variation.

deviation measure. Similar to the findings by Pu (*18*), the last two metrics of spread are more stable and robust, which makes them more appropriate to assess reliability over the entire day. This analysis shows that NPMRDS is a useful source for generating time-varying reliability indices for a corridor.

Day-to-Day Variation in Corridor Travel Time

The analysis in this section focuses on weekdays and uses data that have only be cleaned using the first level of cleaning as discussed in the methodology.

Figure 3 shows the plot of MAE and RMS error evaluated for different days of the year plotted in a descending order based on the MAE value. It can be seen that the RMS error and MAE follow the same decreasing trend. However, a lower value of RMS for day *i* compared with another day *j* does not imply that MAE value would be lower for day *i* as well compared with day *j*. This is because the RMS error penalizes the deviation from the median heavily by squaring the term and can cause travel-time function with a high spike at one point to have a high RMS value, whereas the MAE would smooth out the effect of the spike and would not be very high.

MAE was chosen as the preferred statistic to measure the deviation from the median. This is because a travel-time function that has high isolated spikes most likely has recorded an outlier measurement which resulted in an incorrect reading or a spike. Figures 4*a* and 4*b* plot the variation of travel-time function with highest and second highest MAE values and their comparison with the median travel time. As observed, the plots for those travel-time function show high variance of data both in the morning and the evening peak, and thus represent an *atypical* day behavior on the corridor. Figure 4*c* shows the plot of May 5, 2016 which has third highest RMS error because of the spikes but the MAE value is low.

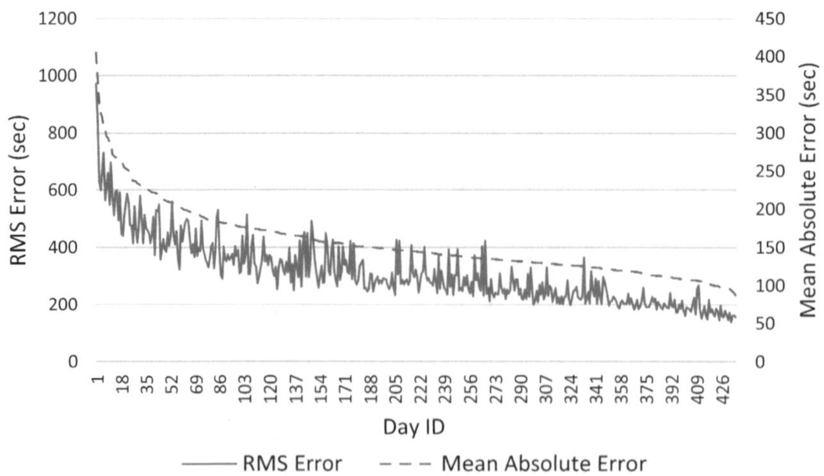

Figure 3. MAE and RMS error on weekdays throughout the analyzed period.

The MAE values can be interpreted as the average error in travel time on a particular day with respect to a typical day. For example, an MAE value of 404.6 s for January 18, 2016 represents that on an average the travel time for each time of day is 6 min 44 s different from the median travel time. This value can thus be used to quantify the *atypical* traffic behavior caused on certain days, useful for applications such as road-user cost calculations for construction projects.

The analysis based on MAE values is useful in identifying groups of particular days on the corridor when the congestion is bad. Figure 5*a* shows the variation in average MAE evaluated for each month analyzed in the dataset. As observed, the month of August is found to have the lowest average MAE for the year 2015. The months of February, November, and December have high MAE values, indicating more prevalence of *atypical* days in those months. The values for the year 2016 show that both the months of August and February have low MAE values.

Figure 5*b* shows the variation in average MAE values with the day of the week. Mondays are found to have the lowest MAE in the year 2015, whereas Wednesdays are found to have the lowest MAE in the year 2016. However, the difference in the average MAE errors is not very significant to make a conclusive argument.

In addition to the analysis based on MAE values, *k*-means clustering was performed using the travel-time values with time of day for each day as a vector. The clustering was performed for two and three clusters. For two clusters, *k*-means classified 52.1% of days in cluster 1 and 47.9% in cluster 2. An even split obtained was inconclusive in deciding whether the days associated to one of the clusters are *atypical* or not. However, for three clusters, *k*-means classified 9.63% of days in cluster 1, 41.9% in cluster 2, and 48.4% in cluster 3. Figures 6*a* and 6*b* show the variation of corridor travel time obtained by each cluster centroid. Figure 6*b* shows that cluster 1 generated for the three-cluster partition has significantly higher travel time during the morning peak, indicating all days classified in that cluster are likely to be *atypical* days. It was thus concluded that 9.63% of the days in the corridor are *atypical* for the analyzed 20 months of data.

Figures 6*a* and 6*b* also show that variation of travel times between cluster centroids is negligible during the times other than morning peak. This suggests that a day can be classified *atypical* only by looking at the traffic during the morning peak. Such a statistic derived from the NPMRDS data can help agencies identify the days which are *atypical* and quantify the number of such days for future planning purposes.

Conclusions

This article looked at the usefulness of NPMRDS in estimating corridor-level performance measures for a case study, a 20.2-mile long US-281 corridor in San Antonio. The focus was on three characteristic performance measures: corridor travel-time measures, corridor travel-time reliability measures, and an MAE measure for quantifying day-to-day variation of travel time on the corridor.

The findings reinforce the usefulness of this dataset in predicting corridor performance measures consistent with other findings in the literature. In addition, it was found that the experienced and instantaneous corridor travel-time curves resemble each other closely, except for the morning and evening peaks where they differ by a MAPE value of 6.32% and 6.23%, respectively. The

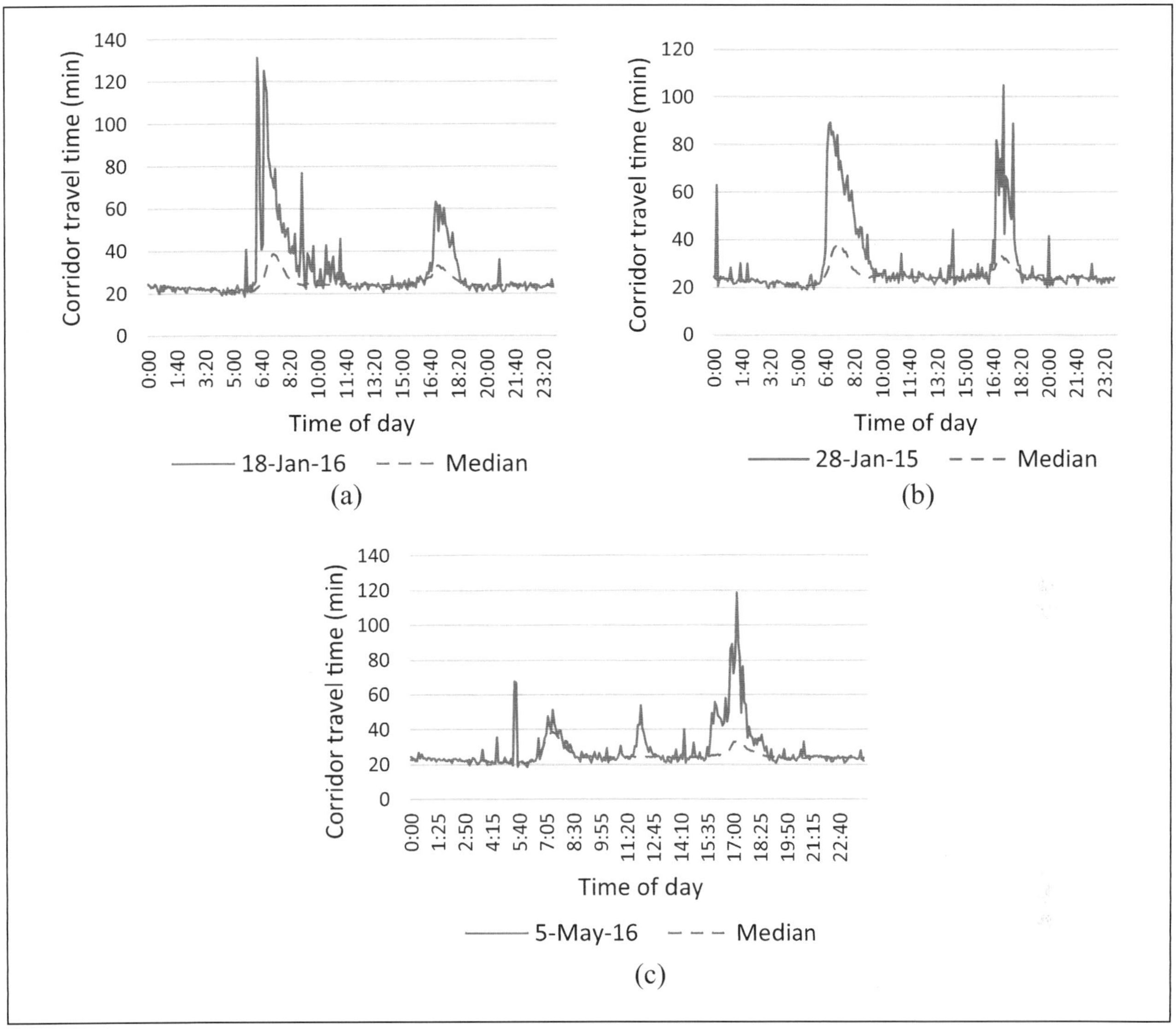

Figure 4. Variation of travel-time function for (*a*) 18 Jan 2016; (*b*) 28 Jan 2015; and (*c*) 5 May 2016.

error values are higher for segments of higher length and during periods of peak congestion. The corridor travel-time data matches well with manually collected field data. The travel-time reliability analysis confirms that the standard deviation and coefficient of variation provide a robust metric of performance throughout the day. MAE values, which were observed to vary by month and day of the week, are found useful in the identification of *atypical* days in a year. The *k*-means clustering analysis suggests that 9.63% of the days on the corridor are *atypical* in the analyzed period. Another interesting finding is that, for the selected corridor, *atypical* days may be identified based only on the morning peak behavior.

The methodologies proposed in this study have important implications. First, the analysis encourages the use of NPMRDS data to obtain travel-time measurements for a particular day, which can help obviate the manual field counts conducted for some of the projects. Second, the analysis suggests that the instantaneous aggregation of corridor travel time provides a reasonably good estimate of the experienced corridor travel time (MAPE value less than 7% during peak hours for a corridor length of 20.2 miles) and can be used when predicting future corridor travel time on variable message signs. Last, the proposed methods can help an agency to identify the number of *atypical* days on the corridor and develop insights such as what time of day contributes most significantly to the *atypical* behavior of a day. An MAE-based measure is another statistic to quantify corridor performance on a particular day that can be useful in applications such as evaluating road-user costs for construction projects.

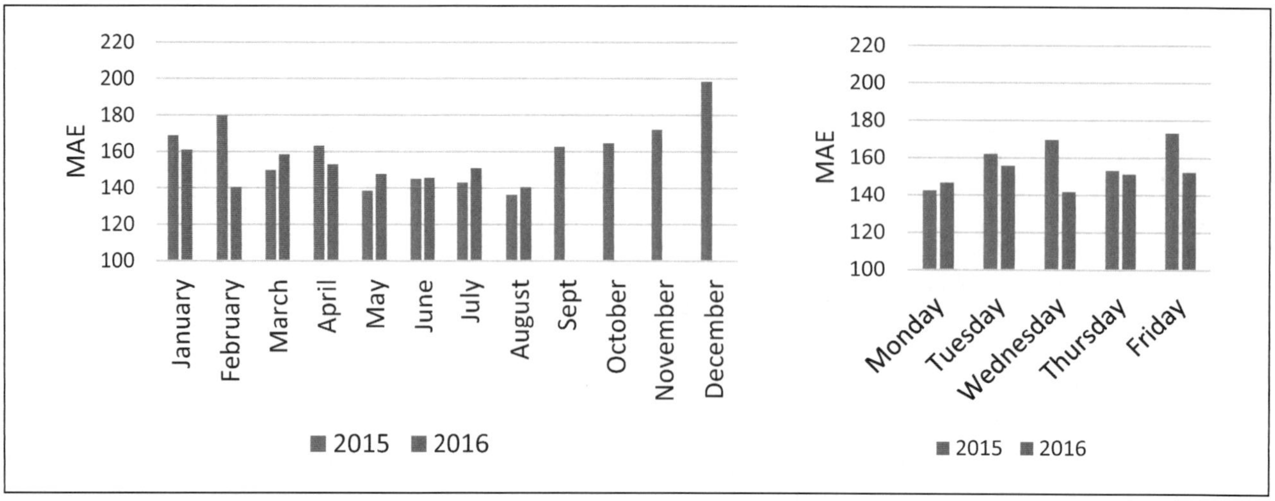

Figure 5. Variation of average MAE values with the month and the day of the week.

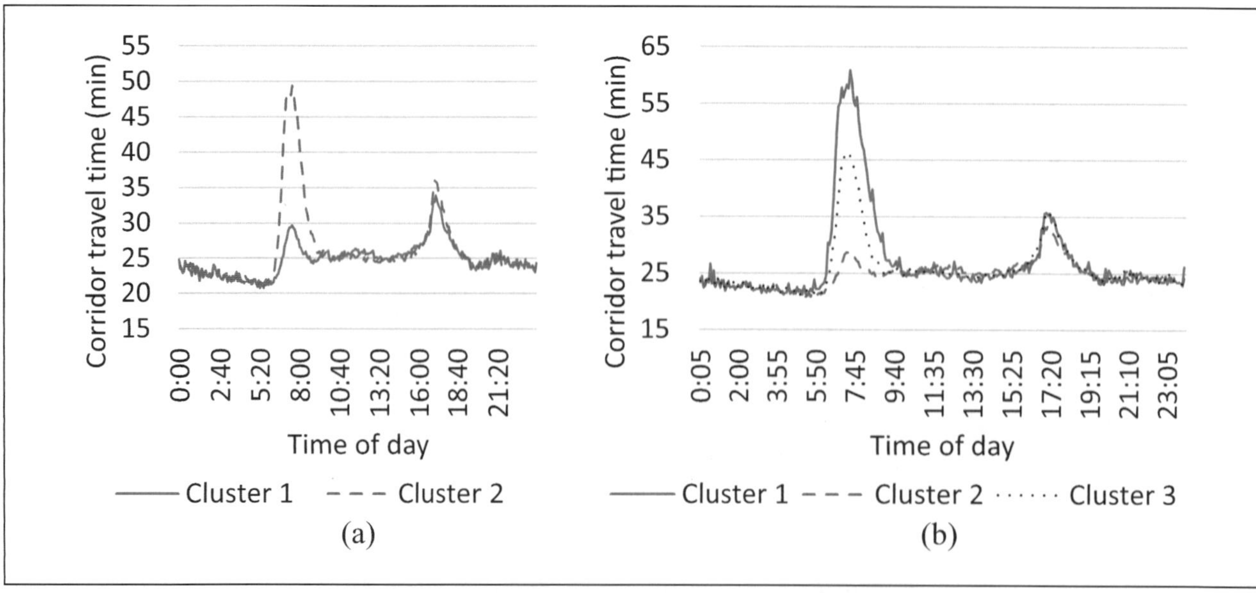

Figure 6. The corridor travel time obtained from cluster centroids by performing *k*-means with (*a*) 2 clusters and (*b*) 3 clusters.

The future work for this research will extend the same analysis for multiple corridors and separate the analysis for passenger cars and freight. The data will also be used to train models that can predict future travel time given the historic records on the corridor. Combining this dataset with other sources such as Bluetooth data and video data will also be a part of the future work.

Acknowledgments

We would like to thank TxDOT San Antonio district for providing us access to the NPMRDS data for the City. The authors would also like to thank Itamar Gal and Amit Gupta for their support in managing the data on the database.

Author Contributions

The authors confirm contribution to the paper as follows: study conception and design: Venktesh and Natalia Ruiz Juri; analysis and interpretation of results: Venktesh Pandey and Natalia Ruiz Juri; draft manuscript preparation: Venktesh Pandey and Natalia Ruiz Juri. All authors reviewed the results and approved the final version of the manuscript.

References

1. Federal Highway Administration. *Operations Performance Measurement*, 2017. https://ops.fhwa.dot.gov/perf_measurement/. Accessed on July 31, 2017

2. Venkatanarayana, R. *Considerations for Calculating Arterial System Performance Measures in Virginia.* No.FHWA/VTRC 17-R2.Virginia Transportation Research Council, Charlottesville, VA, 2017.
3. Kaushik, K., E. Sharifi, and S. E. Young. Computing Performance Measures with National Performance Management Research Data Set. *Transportation Research Record: Journal of the Transportation Research Board*, 2015. 2529: 10–26.
4. Bitar, N. *Big Data Analytics in Transportation Networks using the NPMRDS.* Master thesis. University of Oklahoma, Norman, OK, 2016.
5. Eshragh, S., K. F. Sadabadi, K. Kaushik, and R. M. Juster. Truck and Auto Performance Measurement Using Probe-Based Speed Data: Case Study I-95 Corridor in Maryland. Presented at 94th Annual Meeting of the Transportation Research Board, Washington, D.C., 2015.
6. Margiotta, R., B. Eisele, and J. Short. *Freight Performance Measure Approaches for Bottlenecks, Arterials, and Linking Volumes to Congestion Report.* No. FHWA-HOP-15-033. Washington, D.C., 2015.
7. Rafferty, P., and C. Hankley. Crafting Measures from the National Performance Management Research Data Set. *Proc., 22nd ITS World Congress*, 2015.
8. Rafferty, P. NPMRDS and You: Leveraging Statewide Travel Time Data for Projects, Operations, and Planning. *WisDOT and TOPS lab Webinar Series*, 2016. http://www.glrtoc.org/wp-content/uploads/2015/05/TOPSLab_NPMRDS_Webinar_Slides_Oct2016.pdf . Accessed July 30, 2017
9. Sadabadi, K. F., T. H. Jacobs, S. Erdogan, F. W. Ducca, and L. Zhang L. *SHRP 2 Report S2-L35B-RW-1: Value of Travel Time Reliability in Transportation Decision Making: Proof of Concept—Maryland.* Transportation Research Board of the National Academies, Washington, D.C., 2015.
10. Hafeez, B., and A. Kasemsarn. Using Travel Time Data from the National Performance Management Research Data Set (NPMRDS) to Characterize Hourly Railroad Grade Crossing Delays and Inform Grade Crossing Improvements. Presented at 96th Annual Meeting of the Transportation Research Board, Washington, D.C., 2017.
11. Ashok, K., and M. E. Ben-Akiva. Estimation and Prediction of Time-Dependent Origin-Destination Flows with a Stochastic Mapping to Path Flows and Link Flows. *Transportation Science*, Vol. 36, No. 2, 2002, pp. 184–198.
12. Woodard, D., G. Nogin, P. Koch, D. Racz, M. Goldszmidt, and E. Horvitz. Predicting Travel Time Reliability Using Mobile Phone GPS Data. *Transportation Research Part C: Emerging Technologies*, Vol. 75, 2017, pp. 30–44.
13. Elhenawy, M., H. Chen, and H. A. Rakha. Dynamic Travel Time Prediction using Data Clustering and Genetic Programming. *Transportation Research Part C: Emerging Technologies*, Vol. 42, 2014, pp. 82–98.
14. Yang, H., Z. Wang, K. Xie, and D. Dai. Use of Ubiquitous Probe Vehicle Data for Identifying Secondary Crashes. *Transportation Research Part C: Emerging Technologies*, Vol. 82, 2017, pp. 138–160.
15. Li, C. S., and M. C. Chen. A Data Mining Based Approach for Travel Time Prediction in Freeway with Non-recurrent Congestion. *Neurocomputing*, Vol. 133, 2014, pp. 74–83.
16. Leys, C., C. Ley, O. Klein, P. Bernard, and L. Licata. Detecting Outliers: Do Not Use Standard Deviation Around the Mean, Use Absolute Deviation Around the Median. *Journal of Experimental Social Psychology*, Vol. 49, No. 4, 2013, pp. 764–766.
17. Chen, H., and H. A. Rakha. Multi-step Prediction of Experienced Travel Times Using Agent-Based Modeling. *Transportation Research Part C: Emerging Technologies*, Vol. 71, 2016, pp. 108–121.
18. Pu, W. Analytic Relationships between Travel Time Reliability Measures. *Transportation Research Record: Journal of the Transportation Research Board*, 2011. 2254: 122–130.
19. Mahapatra, S., M. Wolniak, E. Beckett, and K. F. Sadabadi. *Maryland State Highway Mobility Report*, 2015. http://www.roads.maryland.gov/OPPEN/2015%20mobility%20report%20draft_highres_for%20website1.pdf. Accessed July 31, 2017.

The Standing Committee on Urban Transportation Data and Information Systems (ABJ30) peer-reviewed this paper (18-06733).

Article

TRR
JOURNAL OF THE TRANSPORTATION RESEARCH BOARD

Transportation Research Record
2018, Vol. 2672(42) 268–279

DOI: 10.1177/0361198118780204
journals.sagepub.com/home/trr
SAGE

A Comparative Analysis of Tree-Based Ensemble Methods for Detecting Imminent Lane Change Maneuvers in Connected Vehicle Environments

Saleh R. Mousa[1], Peter R. Bakhit[1], Osama A. Osman[1], and Sherif Ishak[2]

Abstract

Lane changing is one of the main contributors to car crashes in the U.S. The complexity of the decision-making process associated with lane changing makes such maneuvers prone to driving errors, and hence, increases the possibility of car crashes. Thus, researchers have been investigating ways to model and predict lane changing maneuvers for optimally designed crash avoidance systems. Such systems rely on the accuracy of detecting the onset of lane-change maneuvers, which requires comprehensive vehicle trajectory data. Connected Vehicles (CV) data provide opportunities for accurate modeling of lane changing maneuvers, especially with the variety of advanced tools available nowadays. The review of the literature indicates that most of the implemented modeling tools do not achieve reliable accuracy for such critical safety application of lane-change prediction. Recently, eXtreme Gradient Boosting (XGB) became a well-recognized algorithm among the computer science community in solving classification problems due to its accuracy, scalability, and speed. This study implements the XGB in predicting the onset of lane changing maneuvers using CV trajectory data. The performance of XGB is compared to three other tree-based algorithms namely, decision trees, gradient boosting, and random forests. The Next Generation SIMulation trajectory data are used to represent the high-resolution CV data. The results indicate that XGB is superior to the other algorithms with a high accuracy value of 99.7%. This outstanding accuracy is achieved when considering vehicle trajectory data two seconds prior to a potential lane change maneuver. The findings of this study are promising for detection of lane change maneuvers in CV environments.

Lane changing is a complex maneuver that involves interaction between vehicles from different lanes, and hence, requires detailed information about the roadway environment. Improper lane changing, when enough gaps are not available or multiple vehicles are concurrently changing lanes at the same location, could significantly affect roadway operability and present a threat to drivers' safety. Past research showed that lane changing has the potential to transform local traffic oscillations to large-scale disturbances that lead to frequent traffic breakdowns (*1–3*).

Lane changing can be defined as a four-phase sequential process that includes planning, preparation, crossover, and adjustment (*4*). The planning phase accounts for the decision-making process in which a yes/no safety assessment-based question is answered as to whether or not to perform the maneuver. Once the lane change decision has been made, the preparation phase starts wherein the vehicle takes some proactive actions to start changing its position towards the target lane. These actions include noticeable variation in the vehicle trajectory along the current lane to align with the available gap in the target lane. Once the vehicle aligns with the gap in the target lane, the crossover phase starts wherein the vehicle significantly changes its acceleration to cross over to the target lane. Finally, the adjustment phase commences wherein the vehicle accelerates/decelerates to adjust its position and speed to create safe gaps with its new lead and following vehicles in the target lane.

[1]Department of Civil and Environmental Engineering, Louisiana State University, Baton Rouge, LA
[2]Department of Civil and Environmental Engineering, The University of Alabama in Huntsville, Huntsville, AL

Corresponding Authors:
Address correspondence to Saleh R. Mousa: smousa6@lsu.edu; Sherif Ishak: sherif.ishak@uah.edu

Modeling such a complex maneuver has been under research for decades to better understand the different phases, and hence, provide the appropriate countermeasures to improve traffic safety and operation. In a recent study by Rahman et al. (*5*), lane changing models were classified into four main categories, namely; discrete choice, rule-based, incentive-based, and artificial intelligence (AI) models. In discrete choice models, lane changing was modeled as a utility maximization process; an example study is presented in (*6*). Rule-based models on the other hand treat lane changing as a binary variable (change or no change) based on predefined conditions in a decision tree process, as in (*7*, *8*). Such models have one main disadvantage, which is the difficulty of calibrating their parameters. In the incentive-based models, lane change decision is based on the gained benefits from remaining in the current lane or crossing over to an adjacent lane (*9*, *10*). This concept is very similar to that used in discrete choice models, except that the gained benefits in incentive-based models are evaluated based on multiple utility functions. Finally, AI models are data-driven and require supervised training. Artificial neural networks and fuzzy logic are example techniques used in the AI category of models (*11–14*). While discrete choice and rule-based models are the most popular, AI models have shown higher accuracy in detecting lane changing, especially that they are based on detailed data from the driving environment.

Recently, tree-based ensemble algorithms have grown in popularity in solving classification problems. These algorithms combine predictions of several decision-tree models to improve the accuracy over a single decision-tree. Unlike most machine learning tools, the possible interpretability of these models makes them a candidate fit for solving classification problems involving data with high dimensionality. eXtreme Gradient Boosting, (XGB) is a recently developed tree-based ensemble algorithm that is a very well recognized algorithm among the computer science community in classification problems for its accuracy, scalability, and speed. The review of the literature indicates that most of the implemented modeling tools do not achieve reliable accuracy for such critical safety application of lane-change detection. Therefore, this study investigates the effectiveness of the advanced XGB algorithm in detecting imminent lane change maneuvers using CV data. This recently developed algorithm has been used (*15*) for short-term travel time prediction and proved to have high prediction accuracy compared to other tree-based ensemble algorithms (*15–17*). In this study, the performance of XGB is compared to other tree-based algorithms namely, decision tree (Tree), random forest (RF), and gradient boosting (GB) as well as the artificial neural network (ANN) lane detection model previously developed in (*18*). Moreover, using the relative importance of the input variables reported by the XGB, a simplified model is developed.

The remainder of the paper presents the methods used in the study, followed by a description of the data used to train and test the models. A discussion of the simplified model's development is then presented, followed by the main findings and conclusions.

Methods

This section provides technical background for the four models used in this study.

Decision Tree (Tree)

This model predicts a response variable by building a set of decision rules from the input variables. The decision rules are presented as nodes, splitting the features' space into regions (subnodes). Each subnode is further split until a specific criterion is met. Each terminal node of these structures is called leaf and is assigned a constant score value (C), which is the average of the response variables values in this node. For a given data set with (n) observations and (m) input variables, the general formulation for this structure can be presented as

$$f(x) = C_{q(x)}, (q : \mathbb{R}^m \rightarrow 1, 2, .., t, C \in \mathbb{R}^m) \quad (1)$$

where $q(x)$ represents the decision rules within a tree that assign a sample of the data to the corresponding leaf index, (t) is the total number of leaves in the tree, and $C_{q(x)}$ represents the score weights assigned to the leaves of the tree.

All the tree-based ensemble algorithms consist of multiple trees that are combined to improve the classification accuracy. As result, a general model ($\hat{y}$) can be written these models as a summation of all classification scores from all trees for a sample (x). The general formulation of these models is presented by the following equations along with Equation 1 as

$$\hat{y}_i(x) = \sum_{T=1}^{T} f_T(x_i), (f_T \in \mathcal{F}) \quad (2)$$

where (T) is the number of trees and ($\mathcal{F}$) is the space of all possible trees. This equation is optimized for the objective function

$$\text{Obj}(\theta) = \sum_{i=1}^{n} l(y_i, \hat{y}_i) + \sum_{T=1}^{T} \Omega(f_T) \quad (3)$$

where the first term is the loss function measuring the difference between prediction ($\hat{y}_i$) and target (y_i). The second term is the regularization term that controls the model complexity and prevents overfitting. A model is overfitting the data, when it starts to model the noise within the data.

Random Forests (RF)

This family of ensemble models implements a bagging technique by generating bootstrap samples to build multiple trees independent of the original training set and then report the average of all trees. RF does not consider all data features at each split and uses a random subset of features during splitting. Such adjustment reduces the correlation among trees and never builds similar trees, which enhances the model accuracy.

Gradient Boosting and eXtreme Gradient Boosting (GB, XGB)

Trees in the GB algorithms are grown sequentially to improve the robustness of the algorithm against overlapping class distributions by optimizing an arbitrary differentiable loss function using gradient descent method. This gradient method gradually minimizes the loss function of the whole system to improve the model accuracy (*16*). XGB, on the other hand, is a more efficient and accurate implementation of GB. Unlike the GB algorithm, which accounts for regularization by only considering a learning rate (L) that shrinks the contribution of each successive tree, XGB implements an additional regularization term that imposes additional control over the model complexity and prevents overfitting. This term Ω is defined as

$$\Omega = \gamma t + \frac{1}{2}\lambda \sum_{j=1}^{t} C_{q(x)j}^{2} \quad (4)$$

where t is the total number of leaves, $C_{q(x)j}^{2}$ is the score on the jth leaf, and γ and λ are two regularization parameters. Depending on the complexity of the model, defined by the number of leaves and sum of squared scores of those leaves, the regularization parameter penalizes the models to minimize that complexity. A detailed analytical overview for the XGB algorithm training can be found in (*17*, *19*).

Data Description

At 100% penetration rate, CV data can be collected from each vehicle and its ambient traffic. In early deployment stages, comprehensive CV data are limited. Thus, vehicle trajectories extracted from Next Generation SIMulation (NGSIM) were used instead. This assumption has been made in previous CV studies such as (*20*, *21*). The study data were collected over a southbound segment of I-395 in Arlington, VA. As shown in Figure 1, the study segment is approximately 3,270 feet long with a posted speed limit of 40 mph. The segment has two off-ramps

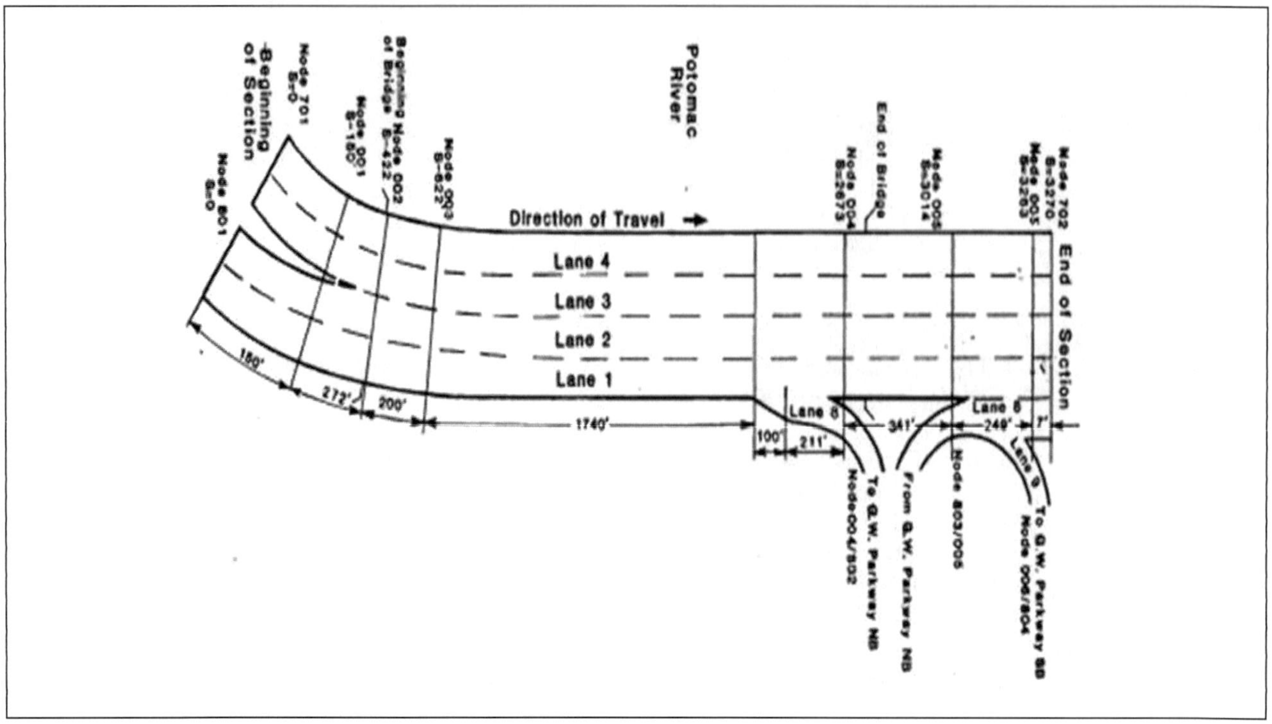

Figure 1. Study segment of I-395.

and an on-ramp with four lanes in the mainline. This geometric setup significantly increases the possibility of lane changes as vehicles merge and diverge to/from the interstate. The data contain 15,223 observations collected from 421 vehicles at a resolution of one second. Each observation has information about vehicle positions, lanes, vehicle dimensions, and lateral distances.

While the one-second resolution is relatively enough to extract useful information, the higher granularity of 0.1 second is obtained to capture minimal changes in the driving pattern of vehicles. The 0.1 second resolution is obtained using the polynomial locally weighted regression procedure (*22*). Further details of the assumptions and the procedures used in data processing can be found in (*6*). This smoothing method has been widely used in practice in previous vehicle-trajectory-related studies such as in (*6*, *21*). Following this step, several variables were extracted from the data to train the lane change detection models. These variables are grouped into two categories: subject-vehicle variables and inter-vehicle performance variables. The former contains the speed, acceleration, and lateral speed of the subject vehicle (the vehicle performing the maneuver), while the latter contains the relative speeds with both the lead and the behind vehicles in the current lane, as well as the gaps with the lead and the behind vehicles in the target lanes (both right and left). Figure 2 shows the four lane change phases along with the features used for building the model. Considering that the last four variables are collected for both the right and left target lanes, this leaves 15 variables from each time step. These variables were selected based on previous studies (*5*, *6*, *8*, *9*).

Detection Models Development

The steps for developing the models are as follows:

Step 1: Determining a Detection Time Window

The first step is to determine the time window size over which lane change maneuvers are to be detected. The size of the time window is defined by the number of time steps over which a change in driving pattern is to be detected before the lane change maneuver is performed. Since those changes in the driving pattern take place during the preparation phase, the time window size should account for that. In this study, a three-second time window is initially implemented.

Step 2: Detection Models Performance Measures

The area under the curve (AUC) metric is a common performance measure for classifiers, where the curve refers to the receiver operating characteristics (ROC) curve. For most of the classifiers, there is a trade-off between the specificity (the ability of the model to correctly classify negative events) and sensitivity (the ability of the model to correctly classify positive events). AUC is a useful tool for evaluating the performance of classifiers on imbalanced data sets since it is a function of both the sensitivity and specificity. Accordingly, the AUC metric is the performance measure used for tuning the hyper-parameters of the developed models using the training/validation data. Moreover, due to the data imbalance, wherein there is a large number of keep lane instances compared to the lane change maneuvers, the performance of the model must be investigated carefully over the minority class observations (lane change instances). Consequently, the sensitivity and accuracy values are used as the performance measures over the testing data and can be measured as shown in Equations 5 and 6 as

$$Sensitivity = \frac{a}{A}$$

$$Accuracy = \frac{a+b}{A+B} \tag{6}$$

where a is the number of correctly classified lane change maneuvers, b is the number of correctly classified lane keeping instances, A is the total number of actual lane changes in the dataset, and B is the total number of lane keeping instances in the dataset.

Step 3: Model Training and Hyper-Parameters Tuning

To perform the full model training for the different algorithms, the data are divided into two parts: 60% for training and validation, and 40% for testing. The previously mentioned 15 variables are extracted from the training dataset for each second of the 3-second time window. The lane change detection models can then be represented as a function of 45 variables as

$$\mathrm{LC}_i = f\begin{pmatrix} V(1), Vlateral(1), \ldots VRLeftFV(1), \\ V(2), Vlateral(2), \ldots VRLeftFV(2), \\ V(3), Vlateral(3), \ldots VRLeftFV(3),, \end{pmatrix} \tag{7}$$

where LC_i is a binary variable equal to 1 when a lane change is about to take place at time step i and 0 otherwise, $V(1)$ is the subject vehicle's speed one second before LC_i, Vlateral(1) is the subject vehicle's lateral speed, $VR\mathrm{Left}FV(1)$ is the relative speed between the subject vehicle and the vehicle ahead in the left lane.

The ensemble tree-based algorithms developed in this study have one or more parameters that need their optimal values to be identified prior to training the model. These parameters are called hyper-parameters and include the maximum tree depth (D), a subset of features (S), a number of trees (T), and a learning rate (L). Depending on the type of the algorithm used, the

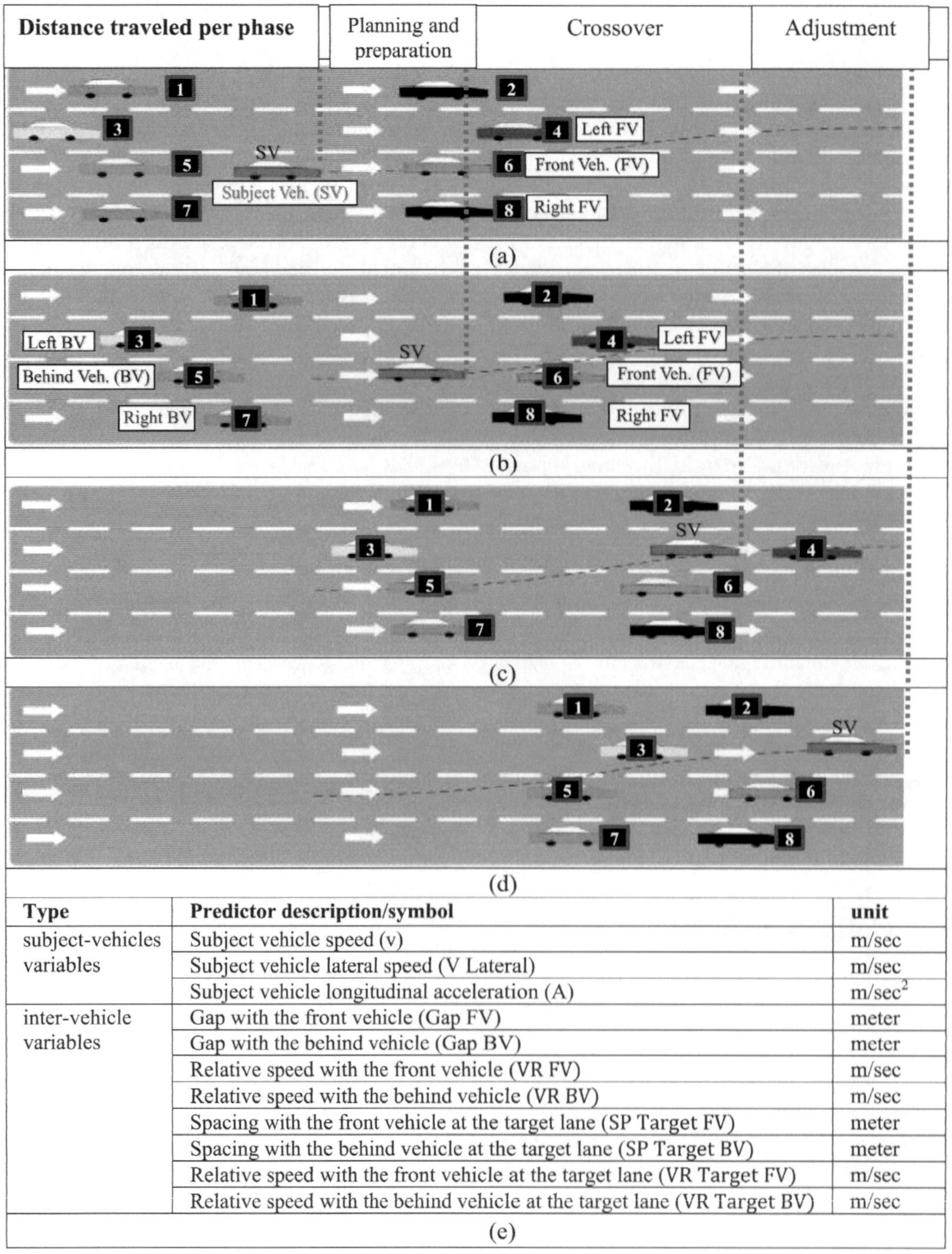

Type	Predictor description/symbol	unit
subject-vehicles variables	Subject vehicle speed (v)	m/sec
	Subject vehicle lateral speed (V Lateral)	m/sec
	Subject vehicle longitudinal acceleration (A)	m/sec^2
inter-vehicle variables	Gap with the front vehicle (Gap FV)	meter
	Gap with the behind vehicle (Gap BV)	meter
	Relative speed with the front vehicle (VR FV)	m/sec
	Relative speed with the behind vehicle (VR BV)	m/sec
	Spacing with the front vehicle at the target lane (SP Target FV)	meter
	Spacing with the behind vehicle at the target lane (SP Target BV)	meter
	Relative speed with the front vehicle at the target lane (VR Target FV)	m/sec
	Relative speed with the behind vehicle at the target lane (VR Target BV)	m/sec

(e)

Figure 2. Lane change phases and the features used for building the model:
a) Start of the planning phase;
b) Start of the crossover phase;
c) Start of the adjustment phase;
d) End of the lane change maneuver; and
e) Features used for building the Model.

prediction model may have one or more of these four parameters. The definition of these parameters, the tuning process of the models, and the impact of these parameters on the accuracy are discussed in detail in the following subsections.

Tree. The Tree model needs to be optimized only for the D parameter, which denotes the maximum number of successive nodes/splits in the tree and is equivalent to the maximum possible depth of the tree. Since D nodes/splits will involve at most D variables, this parameter accounts for modeling the interactions between the input variables by controlling the interaction order of the variables. Increasing D should boost the accuracy of the tree up to a certain value to avoid overfitting the data.

RF. The RF algorithm requires tuning of the parameters S and T only. Unlike the Tree model, the D parameter should not be calibrated in this family of algorithms and each tree in the forest is preferred to grow to the maximum possible depth. Following such practice, the RF is more prone to generate uncorrelated trees. S indicates the size of the variables subset randomly sampled from the original set of variables while developing each tree of the RF. In this study, the value of S ranges from 1 to 45, including all features. T is the total number of trees to be averaged and is a key parameter for all ensemble models. Increasing T should boost the accuracy and reduce the variance of the detection model. However, the computational costs are proportional to the number of trees such that beyond a certain number of trees, the improvement in accuracy becomes negligible. Accordingly, the RF model is optimized for the T and S parameters jointly.

GB and XGB. As an advanced regularized implementation of the GB, XGB shares the same hyper-parameters to be tuned with GB (T, D, and L), in addition to the extra regularization parameters γ and λ. γ and λ are assigned a value of 1 while tuning the hyper-parameters. For both algorithms, L is a regularization parameter that decreases the contribution of each successive tree ($0 < L < 1$) to avoid overfitting the data. While increasing T should boost the accuracy of the detection model, doing so without considering the other parameters (L and D) might cause an overfitting problem.

Grid Search and Ten-Fold Cross-Validation. To tune the hyper-parameters for the different algorithms, a combination of ten-fold and grid search techniques is applied. Grid search is an exhaustive search through all possible combinations of values for the hyper-parameters within a defined space to identify the optimal combination. The different parameters spaces are defined as $D \in [1, 2 \ldots 20]$, $S \in [1, 2 \ldots 45]$, $T \in [1, 2 \ldots 2000]$, and $L \in [0.0001, 0.0005, 0.001, 0.005, 0.01, 0.05, 0.1, 0.5]$. While the learning rate values are commonly assumed to fall between 0.1 and 0.3, this study implemented a wider range of learning rate values due to the large number of trees (1–2000). The searched learning rate values and the varying step size were determined based on a sensitivity analysis and preliminary investigation using different values.

The grid search is guided by a ten-fold cross validation technique in which the data are divided into ten subsets. To perform the ten-fold cross validation, the 60% training/validation dataset is divided into ten subsets. Then, the model training is performed using nine subsets and validation is done using the remaining subset. This is repeated ten times by changing the validation subset. For each trial, the AUC measure is obtained, and the average AUC value is then obtained for the ten trials to evaluate the model performance.

Effect of Hyper-Parameters on the Performance of the Model. For better understanding of the developed models, this section presents the effect of the hyper-parameters on the performance of the RF and XGB models. The GB model demonstrated a similar behavior to the XGB model with relatively less accuracy, and therefore, only the XGB model performance is presented in this paper. By observing the performance of the Tree model for different D values, the optimal AUC value was 0.81 (at $D = 4$). The Tree model experienced no change in accuracy for depths greater than 10. Accordingly, the space of the D hyper-parameter was reduced to $D \in [1, 2 \ldots 10]$ while tuning the GB and XGB models.

For the RF algorithm, Figure 3 illustrates a summary of the model performance results at different T and S values. The figure depicts that increasing T for the same S improves the model prediction accuracy. The figure also shows that lower S values result in improvement in the model accuracy. As T increases beyond 935 and S decreases below 6, the gain in the model accuracy becomes negligible.

For the XGB, a sample of the validation results is presented in Figure 4. The figure shows that the optimal L value is 0.05 at which the model performance significantly improves as T increases. Although an L value of 0.5 results in faster improvement in the model performance, higher AUC values are achieved at a 0.05 L value. This is realized at a tree depth of 4 and 300 trees. As the number of trees exceeds 300, the added accuracy to the model performance becomes unnoticeable. Similarly, D values higher than 4 do not result in any improvement in the model performance.

The four subplots indicate that the AUC value increases as the T value increases until reaching a certain threshold. Each of the subplots reveals that for the same

Table 1. Optimal Hyper-Parameters Combination for the Ensemble Models

	Tree depth (D)	Number trees (T)	Learning rate (L)	Subset of features (S)	AUC
Tree	4	*NA*	*NA*	*NA*	81.4%
RF	*NA*	*935*	*NA*	6	86.6%
GB	*5*	≈ *1500*	L = 0.005	*NA*	93.9%
XGB	*4*	≈ *300*	L = 0.05	*NA*	96.4%

Note: NA = not applicable.

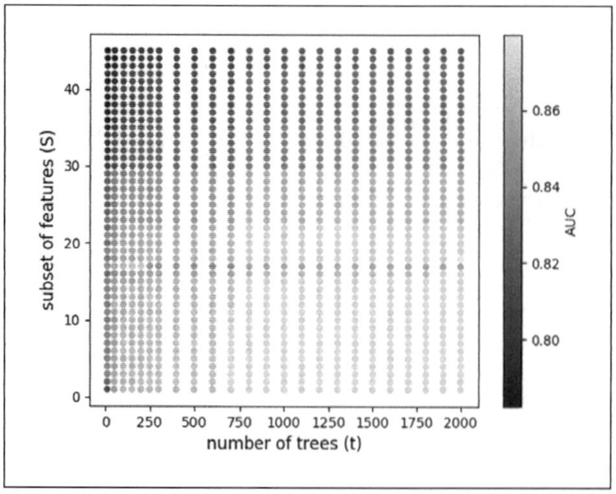

Figure 3. RF validation results.

D value, the slope of the increase in AUC value differs with the L values. The mild slope of the lines at L = 0.0005 indicates that the contribution of each additional tree is much reduced, compared to the slope of the increase in AUC value at L = 0.5. A smaller value of L limits the contribution of each tree in the XGB and requires more trees to be added. Based on the D and T values, the optimal value of L can be different. By inspecting the same L value over subplots *a* and *c*, it is apparent that the XGB model has a higher starting AUC value (AUC at T = 1) at higher D values. This is because a higher D value makes each individual tree more complex and adds more to the model. However, according to subplots *c* and *d*, there is almost no improvement in AUC values when $D > 5$. Overall, it is important to emphasize that the accuracy of the GB and XGB algorithms can be boosted by decreasing L and increasing T, and vice versa. Moreover, higher D values, up to certain threshold, will develop models that can capture minor changes in vehicle performance variable and increase the accuracy.

Table 1 summarizes the optimal hyper-parameters and the associated average AUC values for each algorithm. The table shows that the XGB outperforms the other algorithms with an AUC value of 96.4%, even with less tree depth and number of trees (≈ 300 trees, D = 4) compared to the other algorithms. Additionally, the XGB outperforms the GB in terms of computational costs. For instance, the GB model required 10.02 hours to train all models over the entire grid and 72 seconds for training the best selected model, while the XGB model needed only 4.3 hours and 3 seconds, respectively. All the models were trained using a personal computer with Intel Core i5-4590 CPU 3.3GHz, 8.00 GB RAM, and 64-bit Windows 10.

Step 4: Evaluation of Models' Performance

To evaluate the models' performance, the testing dataset is used to calculate the sensitivity and accuracy measures. Table 2 summarizes the results for the model sensitivity and accuracy as well as the overall confusion matrices obtained for each of the applied algorithms.

The table values are consistent with the validation results as they confirm that the XGB algorithm outperforms all other algorithms with a prediction accuracy of 99.7% and sensitivity of 87.03%. The GB algorithm comes second in the performance followed by the RF and the DT algorithm. The prediction values obtained with the XGB algorithm are also higher than those obtained with the ANN model developed in a previous study for the same dataset (*18*). Thus, the XGB algorithm is selected for the development of a simplified model.

Step 5: Simplified Model Development

Unlike most machine learning models that are difficult to interpret, the XGB reports the relative importance of all variables to the model, which provides better insights and understanding of the model as shown in Figure 5. The relative importance of a variable is the total gain from this variable across all trees (*17*, *19*), normalized such that all variables add up to 100. The higher the value, the more important the variable is. These values provide a technical guidance for developing a simplified model without compromising the detection accuracy.

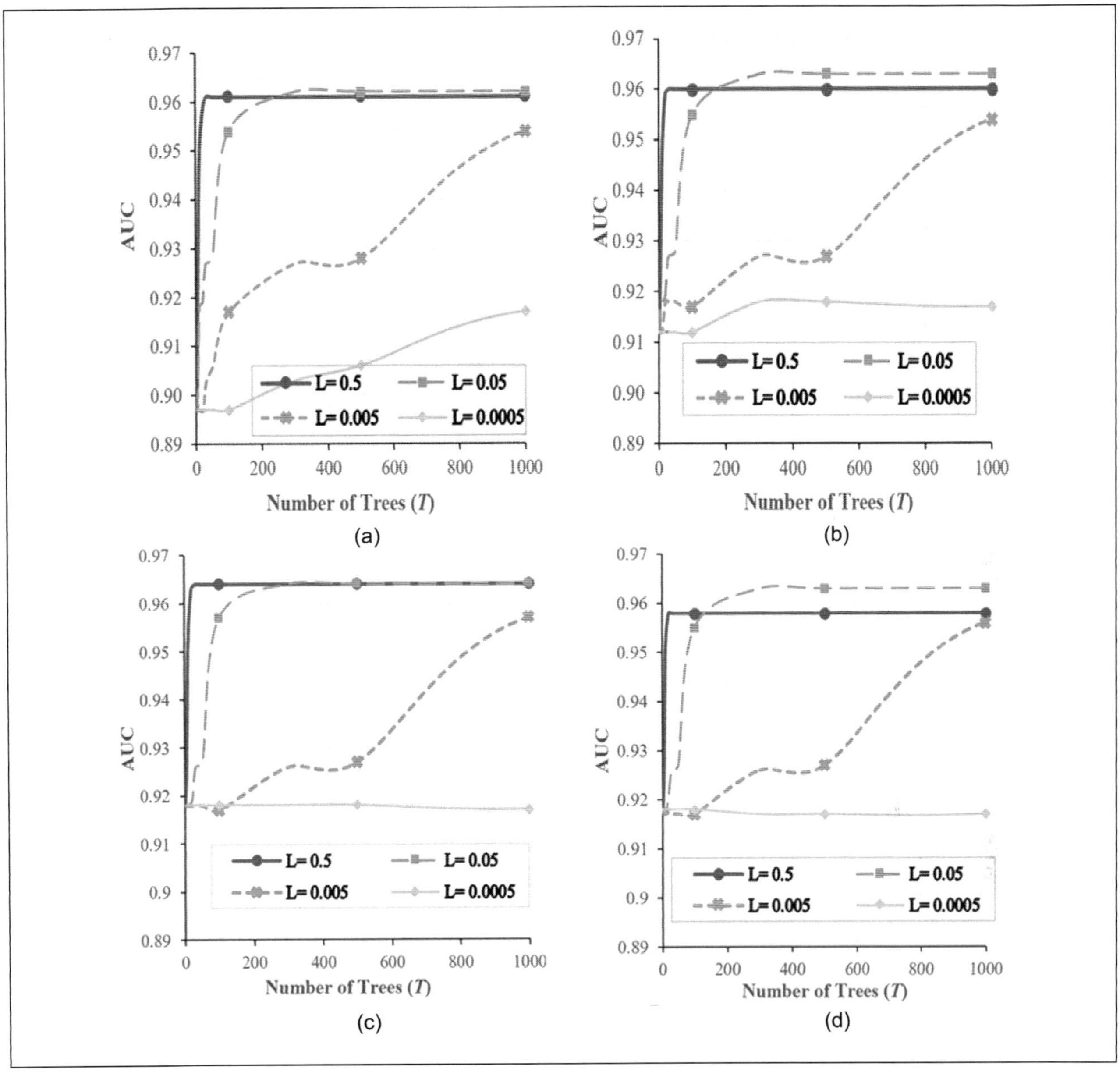

Figure 4. XGB validation results for (*a*) D = 3, (*b*) D = 4, (*c*) D = 5, and (*d*) D = 6.

In any tree, when selecting a certain variable to split a node, the gain of this variable at the splitting node is measured by the change in entropy due to splitting this node into two subnodes. Entropy is a metric to measure the change in purity of the subnodes compared to the splitting node purity (*23*). Figure 5 indicates that the most important variables in the model are V1 (Subject vehicle speed one second before the lane change) followed by the Vlateral1 (Subject vehicle lateral speed one second before the lane change), both contributing by 65% and 13%, respectively, to building the XGB model. The significant contribution (65%) of the V1 was expected since the drivers usually adjust their speed before crossing the lane marking to maintain a speed similar to the vehicles traveling in the target lane. Also, the high importance level (13%) of the Vlateral1 indicates that the variation in the lateral speed is contributed significantly to the detection process. This finding is consistent with the previous studies such as (*24*, *25*).

Aggregating the relative importance values of variables by the time window shows that the first, second, and third seconds account for 86%, 9%, and 3 %,

Table 2. Performance of the Developed Model over the Testing Data

	DT				RF		
True class	Predicted class			True class	Predicted class		
	Lane keep	Lane change			Lane keep	Lane change	
Lane keep	8001	502	*Accuracy* 93.7%	Lane keep	8330	173	*Accuracy* 97.6%
Lane change	45	139	*Sensitivity* 75.6%	Lane change	33	151	*Sensitivity* 82.1%
	GB				XGB		
True class	Predicted class			True class	Predicted class		
	Lane keep	Lane change			Lane keep	Lane change	
Lane keep	8495	8	*Accuracy* 99.6%	Lane keep	8503	0	*Accuracy* 99.7%
Lane change	29	155	*Sensitivity* 84.2%	Lane change	24	160	*Sensitivity* 87.0%

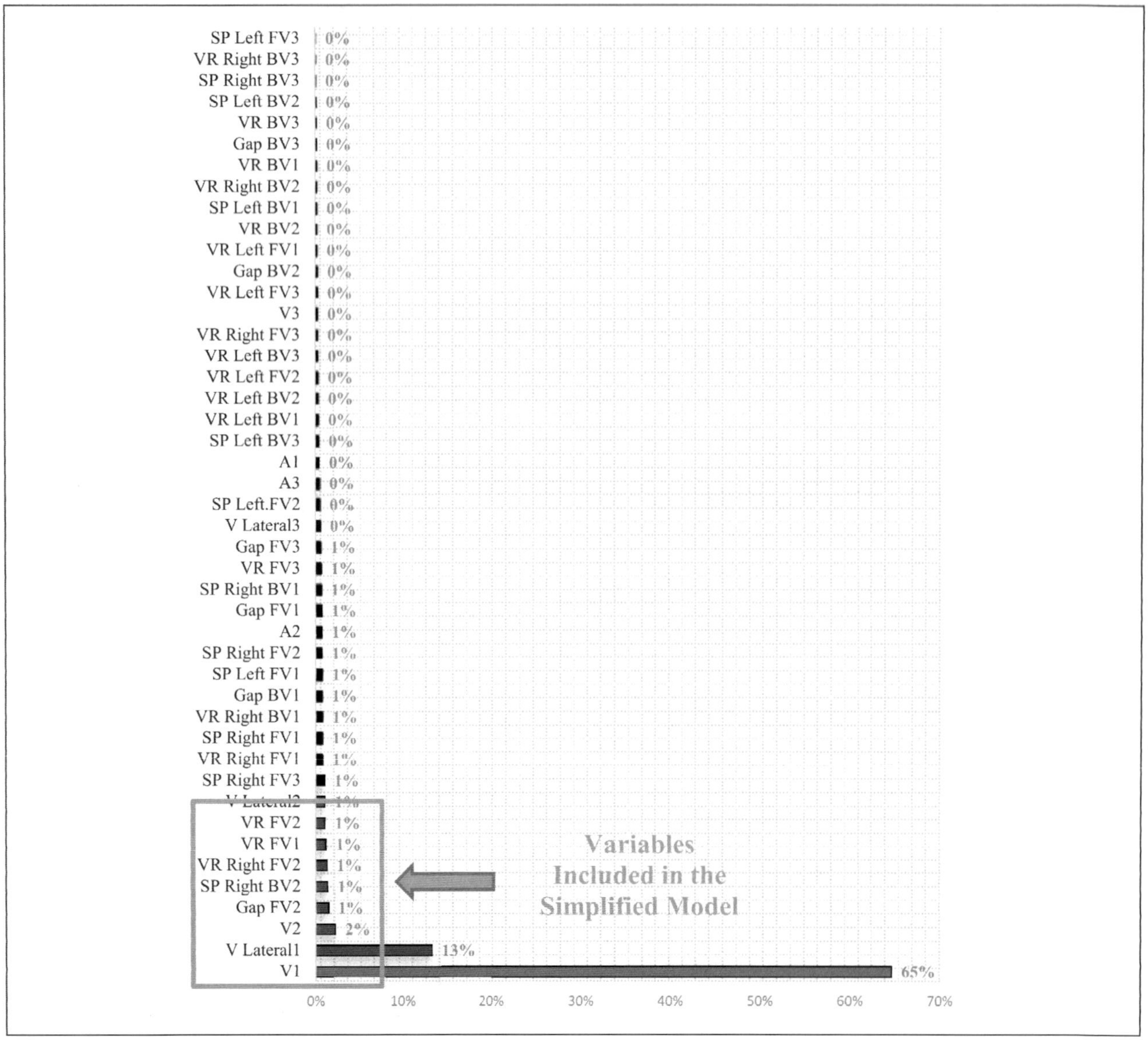

Figure 5. Relative importance values of the variables.

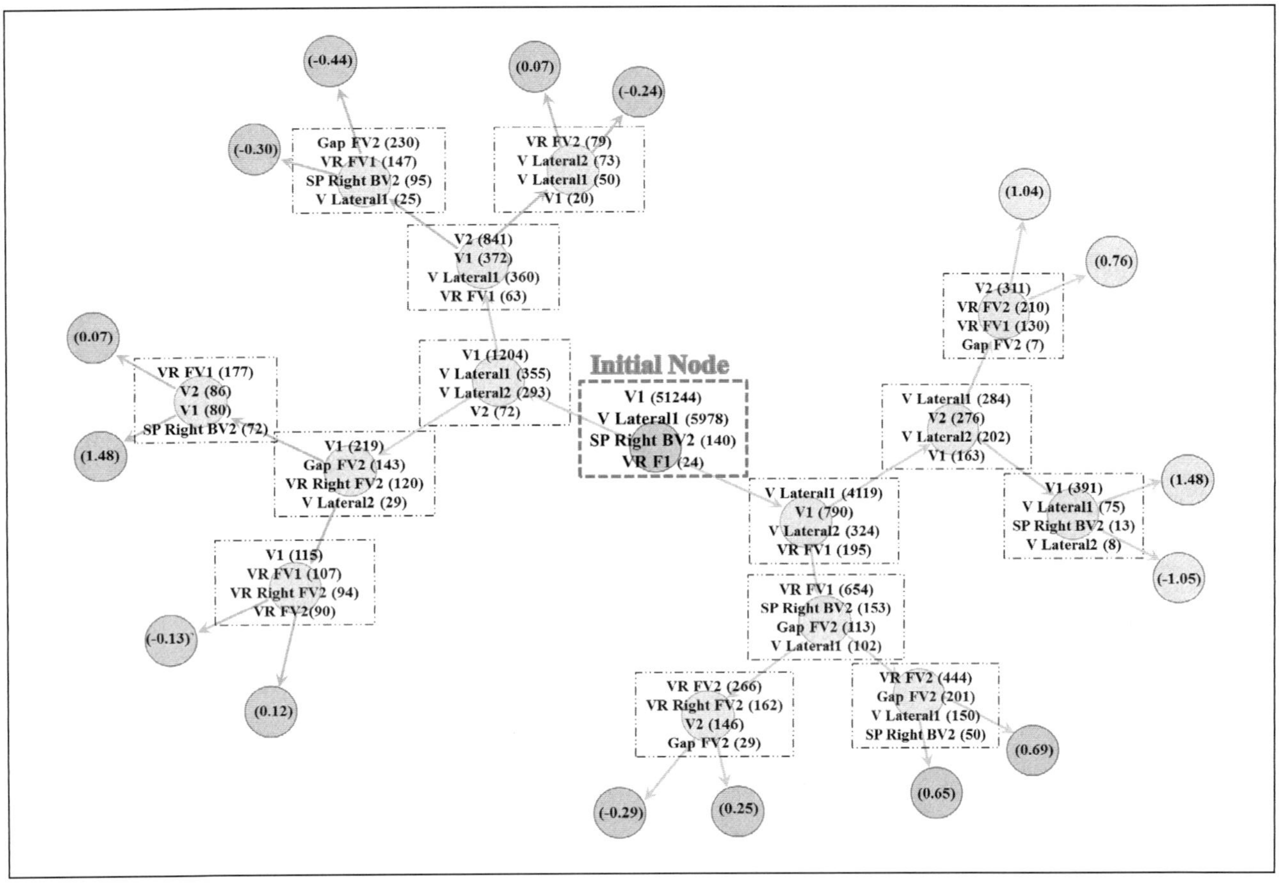

Figure 6. Visualization of the simplified XGB Model.

respectively, in building the model. Due to the very low aggregate importance value of variables in the third second (3%), any variable occurring three seconds prior to the onset of lane changing was excluded from the model. However, not all the variables occurring within two seconds before the onset of lane changing were included in the simplified model since some of these variables had lower importance value than those occurring three seconds prior the onset of the lane changing. Therefore, the following approach was implemented in selecting the variables of the simplified model: (1) arrange all variables in a descending order based on importance value; (2) include variables from the top of the list (most important values) until reaching the first variable occurring in the third second (most important variable in the third second).

Using the training/validation data, the optimal hyperparameter values for the simplified model were $D = 4$, $T \approx 1400$, and $L = 0.01$. The simplified model performed as accurate as the full model over the test data (Accuracy = 99.7%, Sensitivity = 86.4%). Considering that the shape of a tree is only defined by its D, all trees in the XGB model have the same shape and use the same features. Accordingly, all trees for the simplified model can be visualized at once by projecting each tree on the other and adding up at each splitting node the gain per each variable used at this splitting node. Following this procedure, all trees in the simplified XGB model are presented in Figure 6, where for each splitting node the cumulative gain per each variable used at this node and the leaf score is reported. For the terminal nodes (leaves), the leaf scores are the summation of log odd estimates over all trees. These scores can be turned into a probability by performing the logistic transformation. The figure indicates for the initial node that the V1 variable is the most important variable, while other variables become more important at deeper nodes.

Conclusions

This study developed an XGB algorithm for detecting imminent lane change maneuvers. A comparative evaluation was then performed between XGB and three other tree-based algorithms namely; decision tree, random forests, and gradient boosting. The four algorithms were trained using vehicle trajectory data obtained from the NGSIM dataset. NGSIM data was selected to emulate the anticipated high-resolution data that can be obtained from CV environments with 100% market penetration. To tune the multiple models' parameters jointly, a combination of grid search and ten-fold cross validation techniques were applied. Results indicated that XGB is superior to the other algorithms with a high accuracy value of 99.7%. Moreover, the XGB algorithm was used to identify the most important variables in the trained model, and they were the subject vehicle speed and lateral speed one second before the lane change maneuver.

Overall, this study is a pioneer in implementing an XGB lane changing detection model with an accuracy that outperformed all previously developed models. This study demonstrates the effectiveness of XGB in identifying the most important variables in the lane changing detection process and consequently, the optimal time window for modeling lane changing maneuvers accurately can be identified (two seconds in this study). This study provides the models needed for developing more reliable driver assistance systems to alert drivers of imminent lane change maneuvers around them. This can in turn help drivers make optimal decisions and choose the right timings for changing lanes. Similarly, these models can help CV technologies make accurate calculations when performing lane change maneuvers and hence improve the safety of passengers and the entire transportation system. Finally, since the training dataset is limited and can never be exhaustive of all possible real-world scenarios, it is highly recommended that the model be self-adaptive, and that training be continued offline with augmented data to improve its performance.

Author Contributions

The authors confirm contribution to the paper as follows: study conception and design: Saleh R. Mousa, Peter R. Bakhit; data collection: Peter R. Bakhit, Saleh R. Mousa; analysis and interpretation of results: Saleh R. Mousa, Peter R. Bakhit, Osama A. Osman; draft manuscript preparation: Saleh R. Mousa, Peter R. Bakhit, Osama A. Osman, Sherif Ishak. All authors reviewed the results and approved the final version of the manuscript.

References

1. Ahn, S., and M. J. Cassidy. Freeway Traffic Oscillations and Vehicle Lane-change Maneuvers. In *Transportation and Traffic Theory 2007*. Papers Selected for Presentation at ISTTT17, 2007.
2. Sun, J., J. Zhang, and H. M. Zhang. "Investigation of the early-onset breakdown phenomenon at urban expressway bottlenecks in Shanghai." *Transportmetrica B: Transport Dynamics 2*, No. 3, 2014, pp. 215–228.
3. Zheng, Z., S. Ahn, and C. M. Monsere. Impact of Traffic Oscillations on Freeway Crash Occurrences. *Accident Analysis & Prevention*, Vol. 42, No. 2, 2010, pp. 626–636.
4. Van Winsum, W., D. de Waard, and K. A. Brookhuis. Lane Change Manoeuvres and Safety Margins. *Transportation Research Part F: Traffic Psychology and Behaviour*, Vol. 2, No. 3, 1999, pp. 139–149.
5. Rahman, M., M. Chowdhury, Y. Xie, and Y. He. Review of Microscopic Lane-changing Models and Future Research Opportunities. *IEEE Transactions on Intelligent Transportation Systems*, Vol. 14, No. 4, 2013, pp.1942–1956.
6. Ahmed, K., M. Ben-Akiva, H. Koutsopoulos, and R. Mishalani. Models of Freeway Lane Changing and Gap Acceptance Behavior. *Transportation and Traffic Theory*, Vol. 13, 1996, pp. 501–515.
7. Halati, A., H. Lieu, and S. Walker. CORSIM-Corridor Traffic Simulation Model. In *Traffic Congestion and Traffic Safety in the 21st Century: Challenges, Innovations, and Opportunities (Conference)*, Chicago, Ill., 1997, pp. 570–576.
8. Hidas, P. Modelling Vehicle Interactions in Microscopic Simulation of Merging and Weaving. *Transportation Research Part C: Emerging Technologies*, Vol. 13, No. 1, 2005, pp. 37–62.
9. Kesting, A., M. Treiber, and D. Helbing. General Lane-changing Model MOBIL for Car-following Models. *Transportation Research Record: Journal of the Transportation Research Board*, 2007. 1999: 86–94.
10. Schakel, W., V. Knoop, and B. van Arem. Integrated Lane Change Model with Relaxation and Synchronization. *Transportation Research Record: Journal of the Transportation Research Board*, 2012. 2316: 47–57.
11. Brackstone, M., M. McDonald, and J. Wu. *Lane Changing on the Motorway: Factors Affecting its Occurrence, and Their Implications*. Conference Publication No. 454. *Proc., 9th International Conference on Road Transport Information and Control, 1998*. London, UK, IET, 1998, pp. 160–164.
12. Dumbuya, A., A. Booth, N. Reed, A. Kirkham, T. Philpott, J. Zhao, and R. Wood. Complexity of Traffic Interactions: Improving Behavioural Intelligence in Driving Simulation Scenarios. *Complex Systems and Self-organization Modelling*, 2009, pp. 201–209.
13. Hunt, J., and G. Lyons. Modelling Dual Carriageway Lane Changing Using Neural Networks. *Transportation Research Part C: Emerging Technologies*, Vol. 2, No. 4, 1994, pp. 231–245.
14. McDonald, M., J. Wu, and M. Brackstone. Development of a Fuzzy Logic Based Microscopic Motorway Simulation Model. *Proc., IEEE Conference on Intelligent Transportation System, 1997. ITSC'97.*, Boston, USA, IEEE, 1997, pp. 82–87.
15. Mousa, S. R., and S. Ishak. An Extreme Gradient Boosting Algorithm for Freeway Short-Term Travel Time Prediction Using Basic Safety Messages of Connected Vehicles. Presented at 96th Annual Meeting of the Transportation Research Board, Washington, D.C., 2017.
16. Mayr, A., H. Binder, O. Gefeller, and M. Schmid. The Evolution of Boosting Algorithms. *Methods of Information in Medicine*, Vol. 53, No. 6, 2014, pp. 419–427.

17. Chen, T., and C. Guestrin. Xgboost: A Scalable Tree Boosting System. *Proc., 22nd ACM SIGKDD International Conference on Knowledge Discovery and Data Mining*, ACM, 2016. pp. 785–794.
18. Bakhit, Peter R., Osama A. Osman, and Sherif Ishak. Detecting Imminent Lane Change Maneuvers in Connected Vehicle Environments. *Transportation Research Record: Journal of the Transportation Research Board*, 2017. 2645: 168–175.
19. Chen, T., and T. He. Higgs Boson Discovery with Boosted Trees. *NIPS 2014 Workshop on High-energy Physics and Machine Learning*, JMLR: Workshop and Conference Proceedings 42, 2015, pp. 69–80.
20. Talebpour, A., H. Mahmassani, F. Mete, and S. Hamdar. Near-crash Identification in a Connected Vehicle Environment. *Transportation Research Record: Journal of the Transportation Research Board*, 2014. 2424: 20–28.
21. Talebpour, A., H. S. Mahmassani, and S. H. Hamdar. Modeling Lane-changing Behavior in a Connected Environment: A Game Theory Approach. *Transportation Research Procedia*, Vol. 7, 2015, pp. 420–440.
22. Cleveland, W. S., and S. J. Devlin. Locally Weighted Regression: An Approach to Regression Analysis by Local Fitting. *Journal of the American Statistical Association*, Vol. 83, No. 403, 1988, pp. 596–610.
23. Friedman, J. H. Greedy Function Approximation: A Gradient Boosting Machine. *Annals of Statistics* 2001, pp. 1189–1232.
24. Van Leeuwen, C. *Driver Modeling and Lane Change Maneuver Prediction*, 2010.
25. Ding, C., W. Wang, X. Wang, and M. Baumann. A Neural Network Model for Driver's Lane-Changing Trajectory Prediction in Urban Traffic Flow. *Mathematical Problems in Engineering*, Vol. 2013, 2013.

The Data and Information Systems Section (ABJ00) peer-reviewed this paper (18-05191).

Article

TRR
JOURNAL OF THE TRANSPORTATION RESEARCH BOARD

Transportation Research Record
2018, Vol. 2672(42) 280–291

DOI: 10.1177/0361198118798475
journals.sagepub.com/home/trr
SAGE

Enriching Activity-Based Models using Smartphone-Based Travel Surveys

Bat-hen Nahmias-Biran[1], Yafei Han[2], Shlomo Bekhor[3], Fang Zhao[1], Christopher Zegras[4], and Moshe Ben-Akiva[2]

Abstract

Smartphone-based travel surveys have attracted much attention recently, for their potential to improve data quality and response rate. One of the first such survey systems, Future Mobility Sensing (FMS), leverages sensors on smartphones, and machine learning techniques to collect detailed personal travel data. The main purpose of this research is to compare data collected by FMS and traditional methods, and study the implications of using FMS data for travel behavior modeling. Since its initial field test in Singapore, FMS has been used in several large-scale household travel surveys, including one in Tel Aviv, Israel. We present comparative analyses that make use of the rich datasets from Singapore and Tel Aviv, focusing on three main aspects: (1) richness in activity behaviors observed, (2) completeness of travel and activity data, and (3) data accuracy. Results show that FMS has clear advantages over traditional travel surveys: it has higher resolution and better accuracy of times, locations, and paths; FMS represents out-of-work and leisure activities well; and reveals large variability in day-to-day activity pattern, which is inadequately captured in a one-day snapshot in typical traditional surveys. FMS also captures travel and activities that tend to be under-reported in traditional surveys such as multiple stops in a tour and work-based sub-tours. These richer and more complete and accurate data can improve future activity-based modeling.

Transportation planners and travel demand modelers utilize a variety of data sources, the most important source being household travel surveys. Household travel surveys, conducted in many metropolitan areas, are administered to collect detailed information about the characteristics of the household, and each household member's travel and activity for some period of time. High quality data collected from these surveys are essential for estimation of travel and activity models.

Traditionally, household travel surveys have been interview-based, either face-to-face or by telephone. Typically, the interviewers first collect socio-economic and demographic information on the household and individual members. Next, for each participating household member, a detailed travel diary that includes all trips and activities on the previous day(s) is recorded or documented. Due to limitation of memory, such surveys usually collect only one day's data, or at most two days, from each participant. Some known issues for such surveys include inaccuracy in reported times and locations, under-reporting of short trips/activities, and inability to capture variability in user's behavior over a longer period of time.

Several recent household surveys have used GPS and/or other devices to improve data accuracy and reduce non-response (*1–3*). However, passive GPS tracking alone is insufficient to provide the details needed for activity-based models. Therefore, the challenge is to combine both automatic detection of trips and activities, together with reliable user-reported information.

Future Mobility Sensing (FMS) is a smartphone-based platform that can be used for automated household travel surveys. It leverages advanced sensing technologies, machine learning techniques and a user-friendly interface to collect personal travel data (*4*). An initial field test of FMS conducted in Singapore in 2012/2013 showed great potential to provide accurate and rich travel behavior data (*5*). Subsequently, FMS has been employed in large-scale household travel surveys (Phoenix, Arizona, 2016; Tel Aviv and Haifa, Israel, 2016/2017; Singapore 2016/2017) as well as in small

[1]Singapore-MIT Alliance for Research and Technology (SMART), Singapore
[2]Department of Civil & Environmental Engineering, MIT, Cambridge, MA
[3]Department of Transportation and Geo-Information Engineering, Faculty of Civil and Environmental Engineering, Technion, Technion City, Haifa, Israel
[4]Department of Urban Studies & Planning, MIT, Cambridge, MA

Corresponding Author:
Address correspondence to Bat-hen Nahmias-Biran:
bathen@smart.mit.edu

demonstration or research projects (Melbourne, Australia, 2015; Dar es Salaam 2015; London 2016/2017) (*6*).

FMS is designed to overcome key issues faced by traditional travel survey methods such as: oversimplification of trips in a day, under-reporting of short trips, overestimation of travel times, and imprecise reporting of locations and times. It is also more cost-effective than GPS logger-based surveys, since participants use their own smartphones, and it avoids the problems of loggers left uncharged or not carried all the time. In addition, multi-day data can be collected easily at minimal marginal cost.

In this paper, we perform a comparison of data collected by FMS and by traditional surveys in two cities, Singapore and Tel Aviv, and discuss the implications of the enhanced data collection methods for travel behavior modeling. The remainder of this paper is organized as follows. First, a literature review is presented followed by a brief description of the FMS platform. We then describe the four data sets used in this paper, and present the data analysis results. We conclude the paper with a discussion of future directions.

Literature Review

In recent research projects, the feasibility of replacing or complementing the traditional travel diary with a suite of tools that make use of travel data collected smartphone has been studied; for overviews, see Zhao et al. (*5*) and Berger and Platzer (*7*). There have been attempts to make the completion of the whole data collection process on respondents' mobile (cell) phones either manual or semi-automatic (*8–10*). In the past two years, more smartphone-based travel surveys were field evaluated and documented (*11*). SmartMo is a system that is manually completed on respondents' mobile phones and was field-tested in 2013 in Austria (*7*). Another example is MEILI, a system that includes a smartphone application for capturing the movement of users and a web application that allows the users to annotate their movements; it was tested in Stockholm in 2015 (*9*). These systems, although smartphone based, required heavy involvement of participants by providing/verifying the entities and their attributes. An example of a semi-automatic smartphone app is rMove which was used as a multi-day smartphone-based GPS household travel diary survey over a seven-day period in the U.S.A. rMove experiments took place in Indiana, in 2015 (*10*), Ohio in 2016 (*11*), and Washington, D.C., in 2017 (*12*). These studies demonstrated the advantages of technology-based data collection tools; however, their full capabilities and advantages over traditional survey methods, and the implications for travel behavior modeling, are yet to be studied. This paper aims to fill this gap by comparing the travel information collected using traditional surveys and FMS in Singapore and Tel Aviv.

Future Mobility Sensing

FMS is a smartphone-based automated travel survey system (*5, 13*). It consists of three separate but interconnected components: the smartphone app that collects the sensing data; the server that includes the database as well as the data processing and learning algorithms; and the mobile/web interface that users access to view and verify the processed data and answer additional questions to supplement the verified data. The FMS software platform is provided for commercial applications by Mobile Market Monitor (MMM) and has been applied in large-scale surveys in the U.S.A., Singapore and Israel. The Tel Aviv data described in the remainder of this paper was collected using the MMM platform.

The Data Sets

Four data sets were analyzed for the comparison of traditionally collected travel and activity data to data collected via FMS. These data sets include two from Singapore and two from Tel Aviv.

Singapore HITS

The Singapore Land Transport Authority's Household Interview Travel Survey (HITS) is a face-to-face interview-based household travel survey conducted every four years since 2004. The survey collects one day of activity and travel data for each participating household member. It also collects socio-demographic and other information about the households and individuals within the household. The format of the survey follows the standard trip diary-based approach, with travel defined as a one-way journey completed for a purpose. The survey includes walk segments taken as part of a trip (e.g., walking to a bus stop), and walking trips before or after a trip with at least one motorized mode (e.g., walking to work, leaving work using a taxi). Walk-only trips are recorded if they are longer than 10 minutes. The data set used in this paper, HITS 2012, has a sample size of roughly 10,500 households, approximately 1% of Singapore's resident population. The HITS 2012 follows similar format, methods, and objectives to other metropolitan-wide travel surveys.

Singapore FMS

Between October 2012 and September 2013, FMS was field-tested together with LTA's HITS 2012 survey. The FMS recruitment piggybacked on the HITS recruitment

Table 1. Data Sets Information

	Singapore FMS	Singapore HITS	Tel Aviv FMS	Tel Aviv HTS
Year	2012/2013	2012/2013	2016/2017	2014
Valid person-days	3,217	26,209	35,534	8,454
Persons	319	26,209	11,928	8,454
Multiple days?	Yes	No	Yes	No
Weekends included?	Yes	Yes	Yes	Yes

process. Following a HITS interview, the surveyor introduced FMS, and invited the participant to take part. Unlike HITS, which required all members of households to participate, household members could participate in the FMS survey as individuals, to increase the participation rate in the pilot. An FMS participant was considered to have completed the FMS survey after collecting at least 14 days of data and validating at least five of those days. Of the 1,541 recruited users, 793 completed the FMS survey. Cleaning and post-processing of the collected data were performed to eliminate days with data gaps and logical errors. The resulting dataset with 319 participants has been used for the analysis in this paper.

Tel Aviv HTS

A household travel survey (HTS) was carried out in 14 major cities of the Tel Aviv metropolitan area between December 2013 and June 2014. Weekdays and weekend days were surveyed. The survey method was similar to the Jerusalem HTS (*14*), and it is briefly outlined as follows.

The survey was conducted in two phases, both of them with surveyor in-home visits. In the first (recruiting) phase, a surveyor visited the household, collected general information, and provided household members older than 14 with a GPS data logger. In the second phase, the interviewer returned to the household to retrieve the GPS readings and complete the questionnaire about the activities recorded by the GPS logger. The GPS data was retrieved with the use of laptop computers, which assisted the surveyor and the household members to identify their trips and activities.

Tel Aviv FMS

A comprehensive survey of 10,000 households in the Tel Aviv metropolitan area conducted using FMS started in late 2016 (and was on-going at the time of the writing of this paper). The software platform was provided by MMM. In-person recruitment was performed by a local survey research company, which sent interviewers to the sampled households. The interviewers contacted an adult person living in the household and explained the data collection process.

Household members who agreed to participate in the survey were given two options for mobile sensor data collection: they could either install the MMM smartphone app on their phones or carry programmable GPS loggers with them for the duration of the survey. The loggers transmit data to MMM's backend where the data are processed and displayed on the user interface in a similar way to the smartphone-based data. The logger option mitigated the problems of incompatible smartphone models, non-ownership of smartphones, or refusal to install the app on personal phones. Each participant was required to complete at least two days of data collection and verification.

Table 1 summarizes the sample information for each of the data sets; the numbers for FMS data are post-cleaning, after removing days with data gaps. These data were used in the analysis described in the following section.

Data Analysis Results

Comparative analysis was performed on the Singapore and Tel Aviv datasets. FMS data has been shown to overcome several major issues pertaining to traditional surveys: (1) imprecision of locations and times, (2) inability to capture day-to-day variability in individuals' travel patterns, and (3) under-reporting of trips. In this section, we use detailed analysis to demonstrate the richness, completeness, and accuracy of FMS data.

FMS Captures Rich Travel Behavior Data, and Reveals Large Day-to-Day Variability

To group similar daily activity patterns together and identify different types of behavior, we performed clustering analysis on person-day activity records for all four datasets. The unit of analysis is "person-day," a person's activity record during one day. In Singapore HITS and Tel Aviv HTS, each participant has one person-day observation, whereas in FMS, a person can have multiple person-day observations. For each person-day, we divide the 24 hours into 288 five-minute windows. Each window is labeled with its activity type. There are 13 types of activities ("Home," "Work," "Work-related business," "Education," "Shopping," "Personal errands/Prayer," "Medical/Dentist," "Sports/Exercise," "Entertainment/Meals out," "Social visit with friends or family," "Transport/Accompany someone," "Other," and "Missing") coded consistently across the four datasets. Note that "Travel" (change mode/transfer) is a travel activity and was not considered as type of activity for clustering.

The common hierarchical, agglomerative clustering methods share the same algorithmic definition but differ in the way in which inter-cluster distances are updated after each clustering step (*15*). To compute distance between observations, we code each person-day activity pattern into a binary vector. For each activity i, A_i is a vector of length 288, which represents whether a person performed activity i during time window t of the day: $A_i[t] = 1$ if activity i is performed at time window t, and 0 otherwise. We then concatenate A_is over all activity types to have the final binary representation of the person-day activity pattern.

We choose Jaccard distance, a dissimilarity measure for binary data, as the distance metric for clustering. Given two binary activity patterns A and B, Jaccard distance is defined as:

$$d_{AB} = \frac{M_{01} + M_{10}}{M_{01} + M_{10} + M_{11}} \tag{1}$$

where M_{ij} is the total number of time windows where the value of A is i and the value of B is j. Unlike binary distance, Jaccard distance does not include M_{00} in the denominator, meaning the absence of an activity in both sequences at a given time does not contribute to the similarity of two sequences.

Agglomerative hierarchical clustering algorithm is used for clustering the binary activity sequences. It is a well-established clustering algorithm fit for non-Euclidean distance. The algorithm starts by assigning each observation to its own cluster. Until one single cluster is left, repeatedly merge two clusters G and H such that distance between G and H, $d(G,H)$, is the smallest. The distance $d(G,H)$ can be computed in different ways, termed as types of linkage: single linkage, complete linkage, and average linkage. We choose complete linkage (furthest-neighbor linkage), in which the distance between G, H is the largest dissimilarity between two points in opposite groups:

$$d_{\text{complete}}(G,H) = \max_{i \in G,\, j \in H} d_{ij} \tag{2}$$

We chose complete linkage because it generates the most compact and interpretable clusters (*16*). Single linkage and average linkage generate too many small clusters because of the way they compute distance between clusters: closest-neighbor distance (single linkage) or average distance (average linkage).

Agglomerative clustering algorithm generates a dendrogram that shows a hierarchy of where merges happen at what distance. To generate the clusters, we need to select a cutoff distance. The maximum distance within each cluster is below or equal to this cutoff. The smaller the cutoff, the larger the number of clusters. There is no pre-defined optimal value for this cutoff distance. A good choice is the point where there is a large jump between the distances of two successive merges. We noticed a big gap around 0.7, and tried cutoff ranging from 0.6 to 0.8. According to the visualization of each cluster's activity patterns, we found cutoff 0.7 is good because it displays clusters having distinctive interpretations and high similarities within clusters. Most importantly, we chose the same cutoff criteria for both traditional surveys and FMS; within cluster similarity and the number of clusters are comparable across datasets.

Note that cluster membership is not directly comparable between different travel surveys as the sample sizes are very different, that is, a pairwise cluster comparison between survey and FMS is not straightforward as there is no direct correspondence between a survey cluster and an FMS cluster, unless we use some similarity measures. Since this is an exploratory analysis, in this paper we focus on the number of clusters and their diversity that could be discovered from traditional survey versus FMS given the same similarity metrics and similarity cutoffs.

The clustering results show that FMS reveals a much larger set of activity patterns than traditional surveys in both Singapore and Tel Aviv. For example, for full-time workers in Tel Aviv, HTS data lead to a total of 12 clusters, while FMS uncovers 44 clusters overall.

In addition to clusters that represent more common activity behavior and that are captured in both surveys, such as working full days, partial days, or night-shifts, FMS uncovered clear patterns of out-of-work activities, which could not be found in HTS. This analysis supports the hypothesis that traditional surveys under-report out-of-work activities. Figure 1 shows three such clusters from FMS that include sports, social visits, and entertainment/meals out after work, respectively. The x-axis represents time of day, and each user-day is represented by a horizontal stripe. The activities performed in each time window are color-coded, with the y-axis representing the number of user-days belonging to the cluster.

Similarly, for retired people in Tel Aviv, HTS has three activity clusters overall, while FMS uncovers 14 clusters. Figure 2 shows some of the FMS clusters for

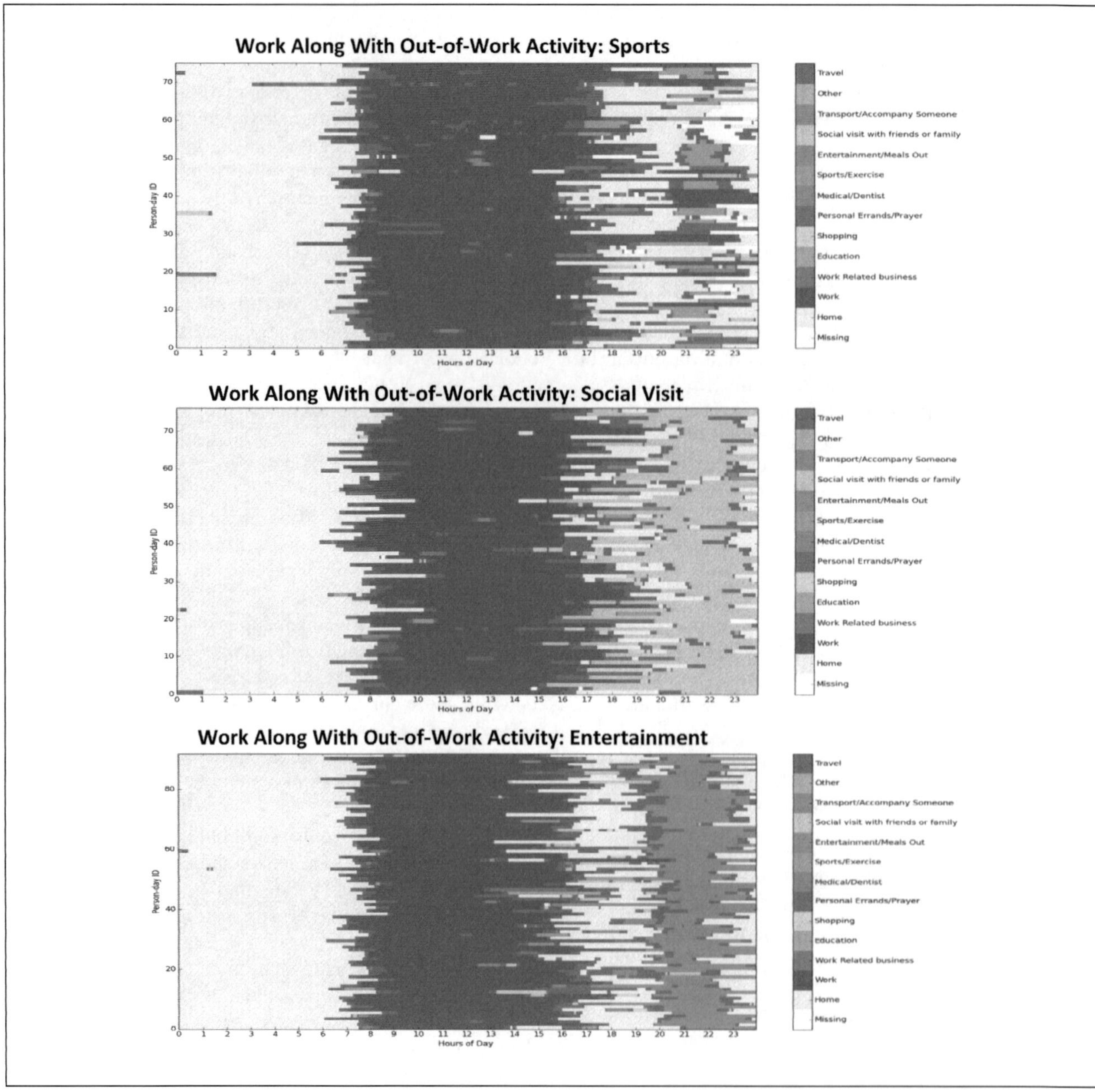

Figure 1. Cluster examples obtained from Tel Aviv FMS: out-of-work activities.

retired individuals in Tel Aviv that are not found in HTS clusters. These activity patterns include sports activities during the day, and social visits or entertainment in the evening.

Besides clustering for each socio-demographic group (employed, retired, etc.), we also performed overall clustering of the whole sample. Figure 3 shows clear activity patterns of leisure activities revealed using FMS. Apart from entertainment activity in the evening, other activities, such as personal errands/prayer in the morning and shopping at noon, are missing in HTS. In activity-based modeling, we usually have work, education, shopping, and "other" as a main purpose of the tour (e.g., *17*, *18*); using detailed data collected in FMS, the "other" can be further divided into well-defined purposes such as prayer, sports, and entertainment.

Clustering done on the Singapore datasets reveals the same trends. Since the traditional Singapore HITS is purely based on user recall, unlike in the Tel Aviv HTS where a GPS logger was used for prompted recall, the

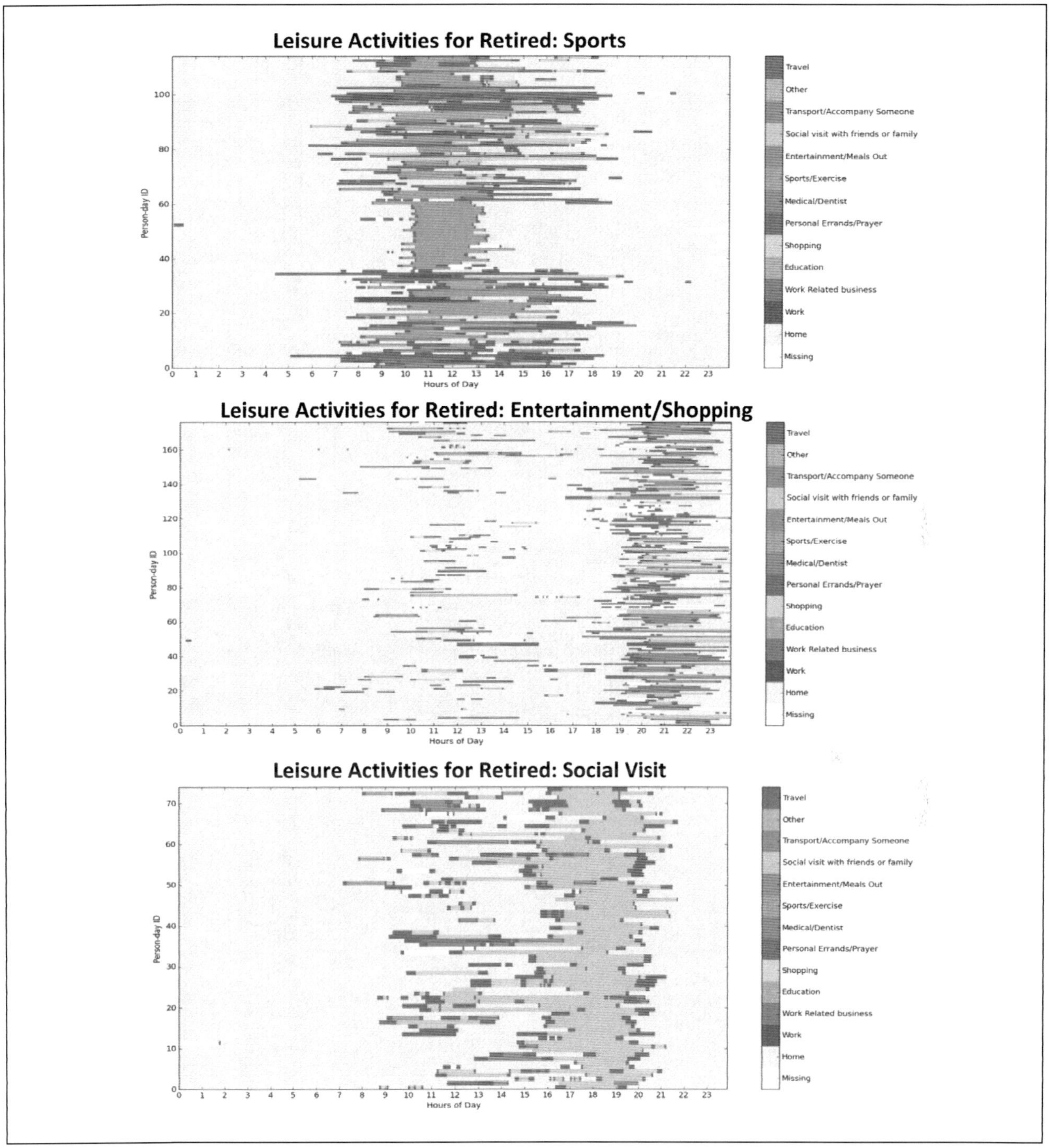

Figure 2. Example of clusters using Tel Aviv FMS data: leisure activities for retired persons.

reported activity patterns are even more uniform and simple. For example, for full-time employed participants, Singapore HITS only has five clusters, as opposed to 12 in the Tel Aviv HTS. The Singapore FMS clustering results can be found in Zhao et al. (*5*); however, due to the small sample size, we cannot perform a comprehensive comparison of clustering results for Singapore.

In many datasets used for model estimation and forecasting, each individual generally represents a single activity pattern. This is because in most traditional

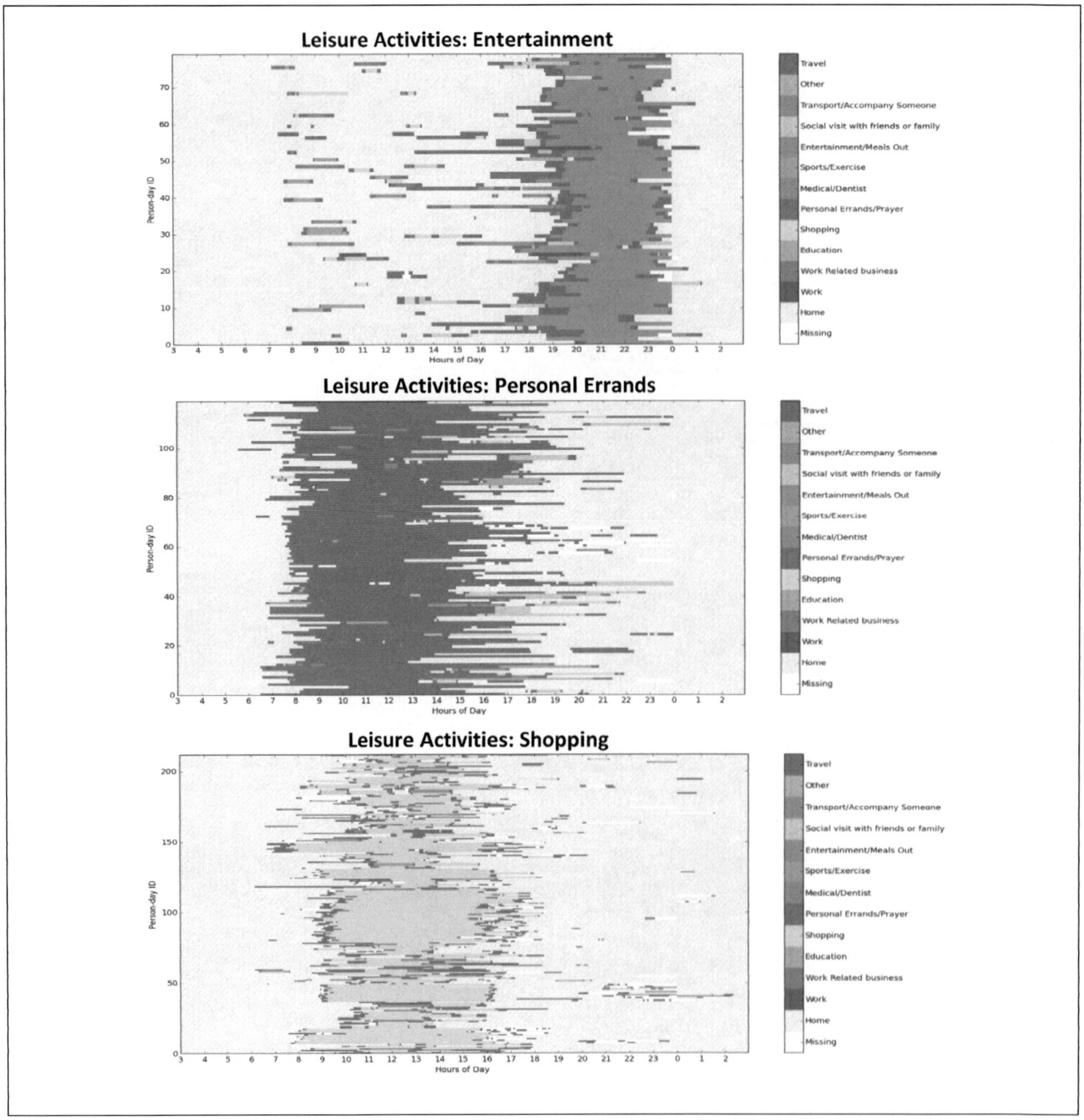

Figure 3. Examples of clusters obtained using Tel Aviv FMS data: leisure activities.

surveys data is collected for a single day, which does not allow day-to-day variability in individuals' activity patterns to be revealed. However, through FMS, we observe significant variability in users' day-to-day activity patterns, which warrants the need for multi-day/long-term data collection. Based on traditional one-day survey data, participants are classified into one main cluster. With multi-day data from FMS, most participants have days that fall into at least two distinct clusters, as shown in Figure 4. Users in Singapore who were tracked for a longer period of time are shown to have a much higher number of clusters. With a single-day sample, the heterogeneity of individuals' daily patterns cannot be identified.

In addition to the clustering analysis, we also identified the activity sequences present in the four data sets. Activity sequence is defined as a sequence of activity

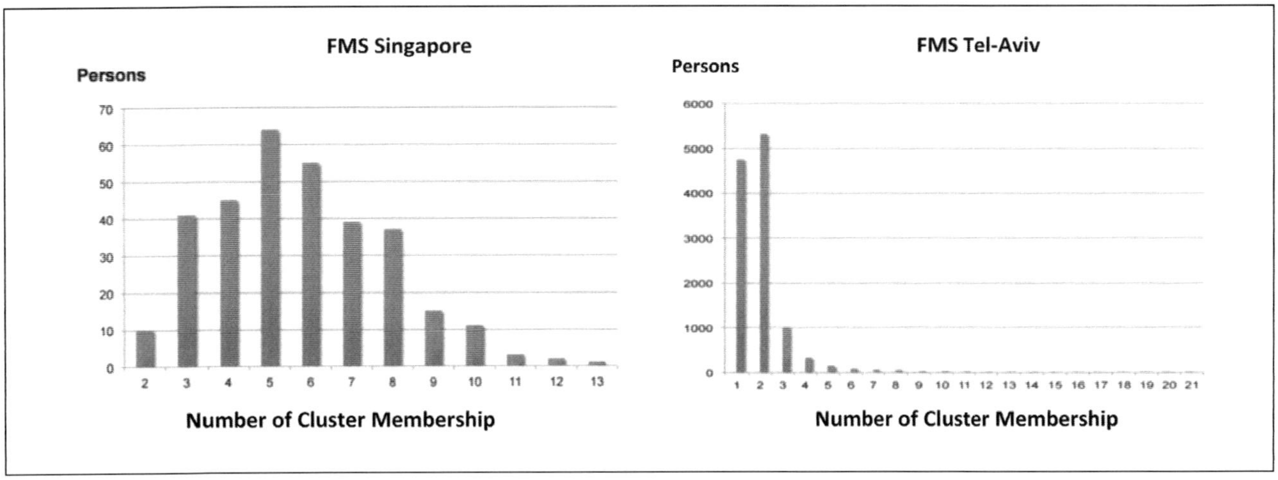

Figure 4. Number of FMS users by counts of distinct cluster membership.

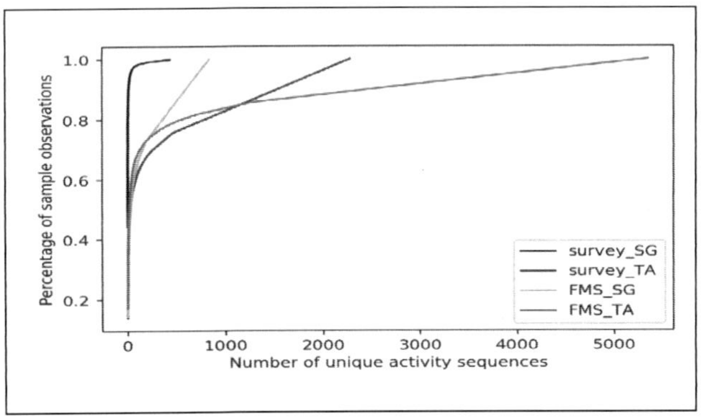

Figure 5. Cumulative percentages of sample covered by different numbers of unique activity sequences.

labels in 24 hours excluding travel (for example, home–work–home, home–work–school–home, and so on). Figure 5 shows the cumulative percentage of observations covered by the unique activity sequences (from most popular to least popular) in the four data sets. In Singapore HITS, more than 90% of the sample is covered by only 10 activity sequences. Despite the much smaller sample size in Singapore FMS, it still covers more unique sequences than HITS. The Tel Aviv HTS data has a large variety of activity sequences, partially due to the GPS-assisted prompted recall, but the data is still not as rich as the FMS data.

FMS Collects More Complete Information

Figure 6 presents the distribution of the number of intermediate stops in a home-based tour in each dataset. In both cities, the proportion of tours with no intermediate stops (that is, with a single destination per tour) is lower in FMS in comparison with previous surveys. For example, in the 2014 Tel Aviv HTS, 70% of tours do not include intermediate stops, while in the FMS dataset over 50% of tours include more than one stop.

The same trend is observed for Singapore. In the HITS dataset, 75% of tours do not include intermediate stops, while in the FMS dataset over 60% of tours include more than one stop. It should be noted that, for both Tel Aviv and Singapore, FMS captures tours with more than four stops, all of which are included in the 4 + stops category. Thus, even with the GPS-assisted data collected in the 2014 Tel Aviv survey, more complex tours were missed.

The completeness of FMS data is also reflected in Table 2, which lists the top three purposes of home-based tours, reported in traditional surveys and FMS, for users of different employment status. In Singapore, for all categories of users (except self-employed), meal/eating-break activity is among their top three purposes in FMS. In contrast, this activity only appears once in the list for HITS. In Israel, for all categories of users (except professional soldiers), personal errands/prayer is among the top three purposes in FMS, while in the traditional survey this activity appears only for unemployed and retired participants. This reflects the under-reporting of short trips/stops in traditional surveys, which is alleviated in FMS.

We also studied the distribution of the number of stops in work-based sub-tours in both Tel Aviv and Singapore (Figure 7). In Singapore, just 2.4% of all work activities had work-based sub-tours in HITS, out of which, 12% of the sub-tours included two stops or more. In contrast, in the FMS dataset, 13% of work activities had work-based sub-tours, and more than 44% of them included two stops or more. Similarly, in Tel Aviv, 7% of work-based sub-tours are detected in the 2014 HTS

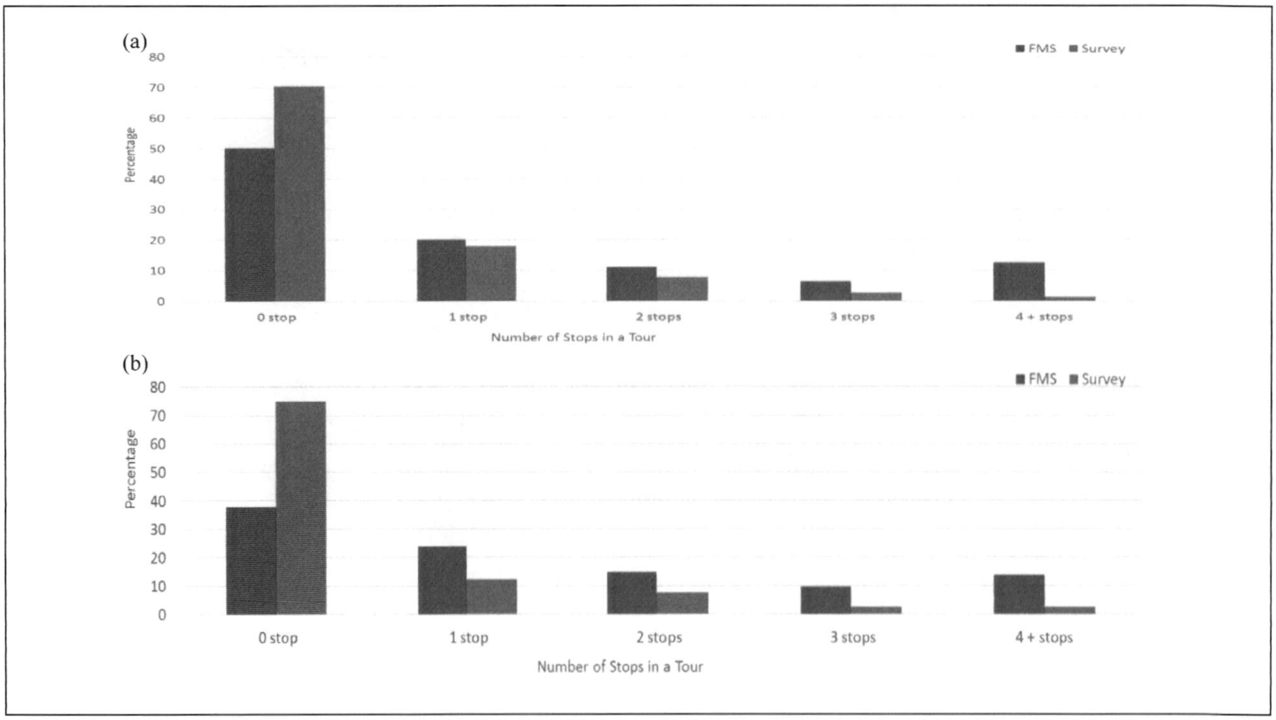

Figure 6. Distribution of number of intermediate stops in a tour: comparison between FMS and traditional survey in (*a*) Tel Aviv and (*b*) Singapore.

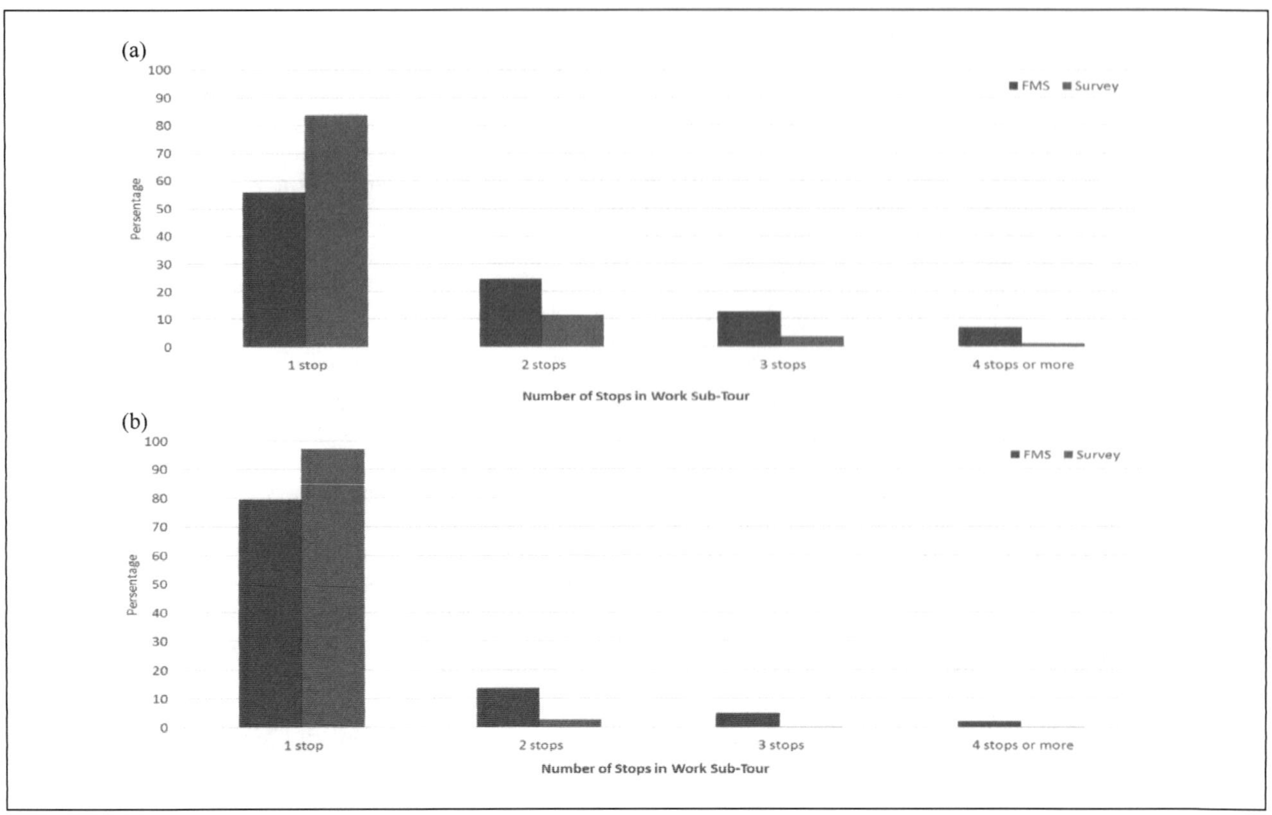

Figure 7. Distribution of number of stops in work-based sub-tours: comparison between FMS and traditional survey in (*a*) Tel Aviv and (*b*) Singapore.

Table 2. Top Three Purposes of Home-Based Tours

Status	HITS	FMS
Singapore		
Full-time worker	Work	Work
	Pick-up/Drop-off	Eating
	Work-related	Personal
Part-time worker	Work	Work
	Pick-up/Drop-off	Eating
	Shopping	Personal
Retired	Others	Eating
	Pick-up/Drop-off	Personal
	Social visit	Recreation
Self-employed	Work-related	Work
	Work	Personal
	Pick-up/Drop-off	Shopping
Homemaker	Pick-up/Drop-off	Eating
	Shopping	Shopping
	Eating	Personal
Full-time student	Education	Education
	Shopping	Work
	Work	Eating
Tel Aviv		
Full-time worker	Work	Work
	Accompany	Accompany
	Shopping	Personal
Part-time worker	Work	Work
	Accompany	Personal
	Shopping	Accompany
Retired	Shopping	Personal
	Personal	Shopping
	Social visit	Social visit
Unemployed (seeking employment)	Shopping	Personal
	Personal	Education
	Accompany	Accompany
Unemployed (not seeking employment)	Shopping	Education
	Social visit	Personal
	Personal	Accompany
Professional soldier	Work	Work
	Shopping	Entertainment
	Education	Social visit
Enlisted soldier	Work	Work
	Other	Social visit
	Social visit	Personal

dataset, compared with 12% of work-based sub-tours in FMS. Out of all work-based sub-tours in 2014 HTS, less than 3% had at least two stops, while in FMS, the number of sub-tours with multiple stops reached almost 20%.

FMS Data Is More Accurate in Location and Time

Location data from FMS are mainly based on GPS sensors, fused with other input data, which are typically more accurate than traditional surveys that are based on user recall and reporting of locations. With regard to time accuracy, Figure 8 compares the time of travel for different activities, plotting the percentage of users traveling for (a) work, (b) home, or (c) meal/eating-break in Singapore (left side) and (a) work, (b) home, or (c) personal errands/prayer in Tel Aviv (right side). For Singapore, HITS data show a narrower travel distribution than FMS, supporting the hypothesis that people tend to report a "typical" day in self-reported surveys, when in reality travel times have a wider spread. In addition, in HITS most users report arriving home in the evening by 8:00 p.m., while, in fact, a significant portion of the users reach home after 10:00 p.m., indicating under-reporting of activities toward the end of the day. Also, three clear peaks emerge in the FMS curve (c), corresponding to trips to breakfast, lunch, and dinner. However, this trend is not clear from HITS data. In fact, the lower values for the HITS curve indicate a much smaller percentage of users reporting meal/eating-break activities.

As already mentioned, unlike Singapore's HITS, the 2014 HTS conducted in Tel Aviv was GPS-assisted and therefore the travel time data are accurate. Note that the curves for both 2014 HTS and 2017 FMS have a very similar shape.

Conclusions

The main purpose of this study was to demonstrate the capabilities of FMS in improving data quality over traditional methods, and the implication of using FMS data for travel behavior modeling. FMS was field-tested in Singapore in 2012/2013 and a large FMS survey is currently being carried out in Tel Aviv metropolitan area. The datasets from these two surveys were used to compare against traditional datasets from HITS conducted in Singapore in 2012 and HTS conducted in the Tel Aviv area in 2014. Results reveal the capabilities of FMS in terms of collecting precise location and time information, uncovering information that was missed using traditional surveys, and revealing day-to-day heterogeneity in individuals' travel patterns that can only be captured by collecting multi-day data. FMS is well suited for this purpose as multi-day data collection can be done with minimal marginal costs and with reduced user burden over time. In addition, it was found that FMS effectively represents out-of-work activities and leisure activities, and alleviates the under-reporting problem in traditional surveys.

The findings reported in this study have major implications for future activity-based modeling. Not only will the information used to estimate the models be more accurate and reliable, but the heterogeneity dimension will add great complexity to the models, and modeling leisure activities will constitute an important part of modelers' work. For example, for Singapore, much richer information can be used for better modeling of work-

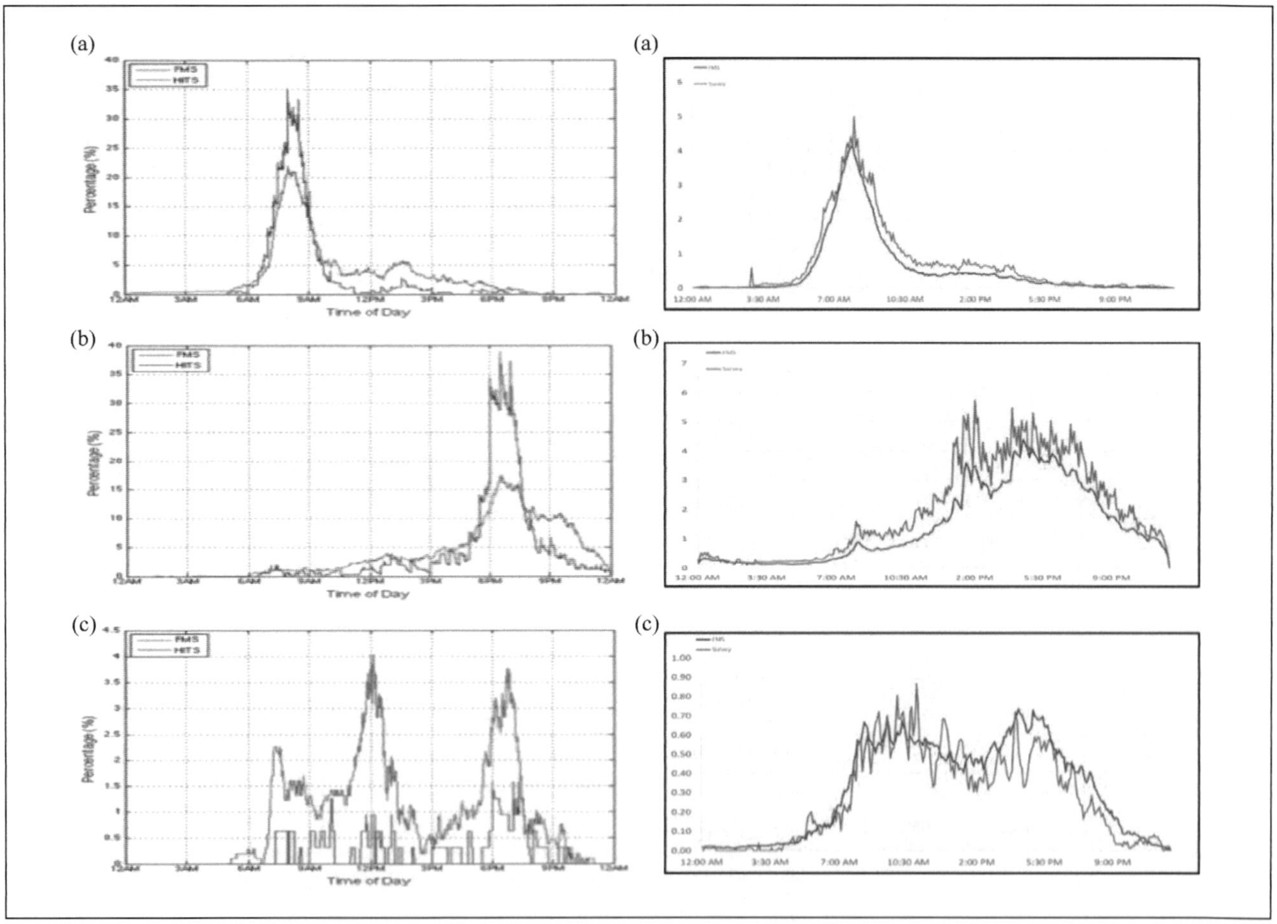

Figure 8. Percentage of participants traveling to (*a*) work; (*b*) home; (*c*) meal/eating-break or personal errands/prayer at different times of day in Singapore (*left*) and Tel Aviv (*right*).

based sub-tours, while in Tel Aviv this dimension, which is missing today, can be added to the Tel Aviv model structure. Likewise, short intra-zone trips could be added to both models using FMS data. Leisure activities can also be added with great detail to both model structures, which are mainly focused on work and education activities. In addition to activity patterns, the enhanced data collection method enables a detailed identification of trip leg modes. Instead of just collecting the main trip mode (e.g., private car, transit, park and ride, and so on), FMS allows the construction of the full mode chain (e.g., walking to the car park, driving, walking to the destination) along with its companions (escorts). Further research will explore the possibilities of including detailed mode chaining and intra-household and social interactions for modeling purposes.

In terms of future improvements for FMS, close examination of FMS data reveals that the quality of the data relies heavily on user verification, and when users are not careful or not familiar with the interface they introduce verification errors. We address this issue by performing cleaning of the collected data. Moving forward, we are continuing to improve the backend inference algorithms, as well as the user interface, in order to minimize user burden and ensure the quality of collected data.

Acknowledgments

The authors gratefully acknowledge the Land Transport Authority of Singapore and Ayalon Highways Co. Ltd for their support and help throughout the field test. This research was supported by the National Research Foundation Singapore through the Singapore MIT Alliance for Research and Technology's Future Urban Mobility IRG research program.

Author Contributions

The authors confirm contribution to the paper as follows: study conception and design: Bat-hen Nahmias-Biran, Shlomo Bekhor, Fang Zhao, Moshe Ben-Akiva; data collection: Shlomo Bekhor, Fang Zhao, Christopher Zegras, Moshe Ben-

Akiva; analysis and interpretation of results: Bat-hen Nahmias-Biran, Yafei Han, Shlomo Bekhor, Fang Zhao, Moshe Ben-Akiva; draft manuscript preparation: Bat-hen Nahmias-Biran, Yafei Han, Shlomo Bekhor, Fang Zhao. All authors reviewed the results and approved the final version of the manuscript.

References

1. *Tel Aviv Regional Travel Survey*. http://www.ayalonhw.co.il/eng/pages/200.aspx
2. Stopher, P., and L. Wargelin, Conducting a Household Travel Survey with GPS: Reports on a Pilot Study. *Proc., 12th World Congress on Transport Research*, Lisbon, Portugal, 2010.
3. Zahabi Seyed A.H., Patterson Zachary, Towards transit trip itinerary from smartphone data: A case study from Montreal, Canada, *CIRRELT*, CIRRELT-2016-07, 2016.
4. Carrion, C., F. C. Pereira, R. Ball, F. Zhao, Y. Kim, K. Nawaranthne, N. Zheng, M. Ben-Akiva, and C. C. Zegras. Evaluating FMS: A Preliminary Comparison with a Traditional Travel Survey. Presented at 93rd Annual Meeting of the Transportation Research Board, Washington, D.C., 2014.
5. Zhao, F., F. C. Pereira, R. Ball, Y. Kim, Y. Han, C. Zegras, and M. Ben-Akiva. Exploratory Analysis of a Smartphone-Based Travel Survey in Singapore. *Transportation Research Record: Journal of the Transportation Research Board*, 2015. 2494: 45–56.
6. Zhao, F., W. Ko, D. Hieu, A. Ghorpade, C. Lui, X. Bing, P. Garbagnoli, J. Ding, B. Atasoy, M. Danaf, I. Viegas de Lima, M. Abou Zeid, and M. Ben-Akiva. *Future Mobility Sensing (FMS) Pilots, Future Urban Mobility Symposium Presentation*, Singapore, July 11, 2017.
7. Berger, M., and M. Platzer. Field Evaluation of the Smartphone-based Travel Behaviour Data Collection App "SmartMo". *Transportation Research Procedia*, Vol. 11, 2015, pp. 263–279.
8. Greaves, S., A. Ellison, R. Ellison, D. Rance, C. Standen, C. Rissel, and M. Crane. A Web-based Diary and Companion Smartphone App for Travel/Activity Surveys. *Transportation Research Procedia, Vol.* 11, 2015, pp. 297–310.
9. Allström, A., I. Kristoffersson, and Y. Susilo. Smartphone based travel diary collection: Experiences from a field trial in Stockholm. *Transportation Research Procedia*, Vol. 26, 2017, pp. 32–38.
10. Greene, E., L. Flake, K. Hathaway, and M. Geilich. A Seven-Day Smartphone-Based GPS Household Travel Survey in Indiana. Presented at 95th Annual Meeting of the Transportation Research Board, Washington, D.C., 2016.
11. Anderson, R. S., P. E. Gregory Giaimo, and L. Flake. *Large-Scale Smartphone-Based Household Travel Survey in Ohio*, working document.
12. *The National Capital Region Travel Survey*. The National Capital Region Transportation Planning Board, Washington metropolitan area. https://regionaltravelsurvey.com/web/pages/about?locale=en-US
13. Zhao, F., A. Ghorpade, F. C. Pereira, C. Zegras, and M. Ben-Akiva. Stop detection in smartphone-based travel surveys. *Transportation Research Procedia*, Vol. 11, 2015, 218–226.
14. Oliveira, M., P. Vovsha, J. Wolf, Y. Birotker, D. Givon, and J. Paasche. Global Positioning System-Assisted Prompted Recall Household Travel Survey To Support Development of Advanced Travel Model in Jerusalem, Israel. *Transportation Research Record: Journal of the Transportation Research Board*, 2011. 2246: 16–23.
15. Anderberg, M. R. *Cluster Analysis for Applications*. Academic Press, New York, 1973.
16. Defays, D. An Efficient Algorithm for a Complete Link Method. *The Computer Journal,* Vol. 20, No. 4, 1977, pp. 364–366.
17. Bekhor, S., C. Dobler, and K. Axhausen. Integration of Activity-based and Agent-based Models: Case of Tel Aviv, Israel. *Transportation Research Record: Journal of the Transportation Research Board*, 2011. 2255: 38–47.
18. Siyu, L. *Activity-based Travel Demand Model: Application and Innovation*. Doctoral dissertation. National University of Singapore, Singapore, 2015.

The Standing Committee on Travel Survey Methods (ABJ40) peer-reviewed this paper (18-05952).

Article

TRR
JOURNAL OF THE TRANSPORTATION RESEARCH BOARD

Transportation Research Record
2018, Vol. 2672(42) 292–301

DOI: 10.1177/0361198118788465
journals.sagepub.com/home/trr
SAGE

Wind Data Collection and Analysis of Topographical Features along a Highway for Traffic Safety Assessment Based on Mobile Mapping Technology

Yatian Pu[1], Feng Chen[1], Peiyan Chen[1], and Xiaodong Pan[1]

Abstract

Strong crosswind is one of the main factors that may cause traffic collisions. Because the wind velocity is influenced by the roadside environment and surrounding terrain, its distribution varies in both the temporal and spatial domains. Therefore, locations with a high probability of strong crosswind should be identified and safety measures should be implemented at these sites. However, geographical data of continuous winds along a highway cannot easily be obtained using existing technology. This prompted the development of a method for geo-location positioning and terrain analysis with elevation data in ArcGIS in combination with mobile mapping technology. The method was applied in a field test conducted on three different highways in China to identify places at which stronger crosswinds occur. The results showed that the proposed method can successfully obtain site-specific wind and crosswind velocity data. It was found that strong winds along the tested highways usually occur at a saddleback or at the border of a riverbank and river, whereas crosswinds are relatively stronger in sections connecting a bridge and tunnel, a bridgehead, or a cross-sea bridge. This information will be useful for future highway projects and traffic safety assessment.

The increasing speed and efficiency of traffic transportation has led to a simultaneous increase in the probability of accidents occurring. According to the *Global Status Report on Road Safety 2015* published by the World Health Organization, 1.25 million people died in traffic accidents every year around the world (*1*). Studies have shown that adverse weather such as rain, snow, pavement icing, fog, wind, and high temperatures are the main objective factors that affect the safety of road traffic (*2, 3*). Among them, Baker and Reynolds found that nearly 50% of traffic accidents in Britain in stormy weather are rollovers, and high trucks face the highest risk (*4*). Moreover, crosswind is particularly dangerous, and Liu et al. found that at a certain speed, the yaw moment and lift force of a vehicle are likely to increase with an increase in the strength of the crosswind, which is a threat to driving safety (*5*). Many studies on vehicle stability under crosswind and traffic safety found that the dynamic response, such as the lateral deviation and lateral acceleration, increases with an increase in the crosswind velocity, and that the critical driving speed is also affected by crosswind (*6–8*).

For traffic safety risks brought by strong crosswinds, actions such as highway closure (e.g., Interstate 80 in Wyoming is closed when the wind gusts over 40 mph) and the installation of traffic safety measures such as wind barriers can be taken. However, because of the prohibitive cost and for reasons of economic efficiency, it is impossible to take measures along the entire length of a highway, especially for highways that stretch over long distances. Thus, it is necessary to identify exposed locations at which topographical features magnify the effect of the wind to enable wind barriers to be installed in these areas. In addition, it is well known that the wind velocity is influenced by the roadside environment and surrounding terrain, and that its distribution varies in time and space. Therefore, the effect on different segments of the same highway can be considerably different at the same time. This means that site-specific wind velocity data along the highway are required for traffic safety assessment purposes.

[1]Key Laboratory of Road & Traffic Engineering of the Ministry of Education, Tongji University, Shanghai, China

Corresponding Author:
Address correspondence to Feng Chen: fengchen@tongji.edu.cn

The problem is that current wind speed data are usually obtained from weather stations, but only some data of places near the basic measured points can be provided, and wind data from locations between two weather stations are usually not available. Furthermore, these weather stations have a scattered distribution; thus, geographical information for continuous wind data along the highway is not readily available through existing technology. A limited number of studies on field wind data collections have been reported in the literature. Snaebjornsson et al. conducted wind velocity measurements by using an anemometer attached to a minivan (*9*). Rutz and Rens designed a data collection system to obtain measurements of the wind speed, wind direction, and strains from selected members at a rapid sampling interval, and a wind direction sensor was used to determine the direction of the wind considering the bridge orientation (*10*). Gonzalo et al. measured the local wind speed profiles by remotely tracking lighter-than-air bubble clusters (*11*). Chen et al. adopted a test truck as a full-size moving "sensor" to obtain wind data at the typical height (*12*). McAuliffe et al. used a sport utility vehicle outfitted with an array of four fast-response pressure probes that could be arranged in vertical or horizontal rake configurations to measure the wind experienced on the road (*13*). These existing studies involve a mobile mapping technique with limited analysis of the effect of the surrounding terrain. This study entailed collecting relevant wind velocity data and identifying places with a stronger crosswind by conducting field tests to obtain real-time measurements (*14–23*).

The paper comprises four parts: an introduction, a description of the field tests, presentation of the wind data analysis, and conclusion. The field tests were carried out by using a mobile wind velocity mapping technique, and a test vehicle equipped with a 3D sonic anemometer and an automobile data recorder was adopted to collect site-specific as well as vehicle-specific wind velocity data in both the time and spatial domains. The field tests were conducted on highways G60, G15, and G92 to apply this technology and determine the characteristics of sites with stronger crosswinds. Ultimately, some landform and geographical features were identified in the analysis using geographic information technology. This information could be applied to future projects involving the selection of locations requiring safety facilities and for traffic safety evaluation.

Field Testing for Wind Velocity Data Collection

Ultrasonic Anemometer

An ultrasonic 3D anemometer manufactured by Osenon's Inc. (A60F-2D2 type) was used to measure the wind velocity data (Figure 1).

It uses the difference in the spread time of ultrasonic waves in the air to measure the wind speed and wind direction. The anemometer uses technology that differs from the general ultrasonic wind-measuring instrument to detect ultrasonic waves in that it adopts an all-digital signal-processing program, thereby improving the accuracy and intelligence of the instrument. The data sampling frequency of the anemometer is 15 Hz, and the instrument can measure wind speed ranging from 0 to 60 m/s with accuracy of 0.2 m/s and a resolution of 0.1

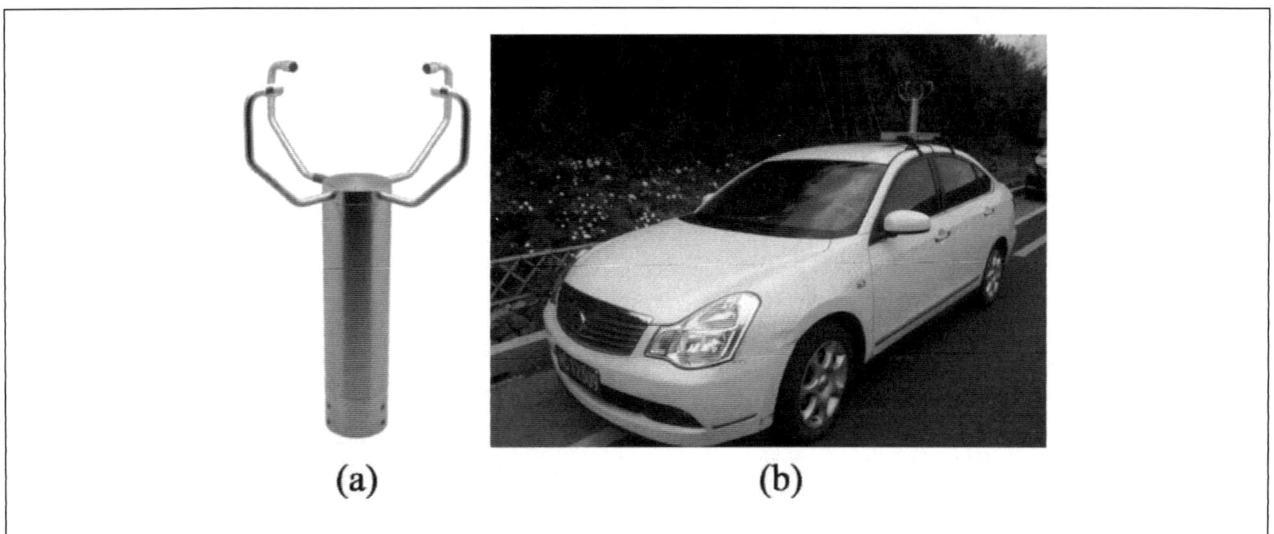

Figure 1. Ultrasonic 3D anemometer (*a*) and the test vehicle fitted with the equipment (*b*).

time	windspeed	windangle	temperature
2016-09-19 09:05:52	3.306804	261.531891	25.5
2016-09-19 09:05:53	2.98875	254.075226	25.5
2016-09-19 09:05:54	1.883197	283.134308	25.5
2016-09-19 09:05:55	1.355442	348.682129	25.5
2016-09-19 09:05:56	1.353295	358.612579	25.5
2016-09-19 09:05:57	2.097133	21.06566	25.5
2016-09-19 09:05:58	2.587826	13.030825	25.5
2016-09-19 09:05:59	2.299384	50.317932	25.5
2016-09-19 09:06:00	2.602708	66.331993	25.5

(a)

Time	Impact (up-down)	Impact (left-right)	Impact (front-back)	speed	X	Y	azimuth angle	frame number	elevation	yaw angle	roll angle	angle of pitch
2016/9/19 13:23:37	-0.108	-0.168	-0.076	80	1423570443	352264315	301	6793	-262.628	-0.076	-0.166	-0.108
2016/9/19 13:23:38	0.04	-0.132	-0.072	80	1423565688	352266818	299	6823	-261.739	-0.072	-0.131	0.04
2016/9/19 13:23:39	-0.084	-0.02	-0.084	82	1423563230	352268013	298	6853	-263.653	-0.084	-0.02	-0.084
2016/9/19 13:23:40	-0.024	-0.088	-0.184	82	1423563230	352268013	298	6883	-264.199	-0.182	-0.088	-0.024
2016/9/19 13:23:41	-0.036	-0.156	-0.084	85	1423560715	352269166	296	6913	-265.029	-0.084	-0.155	-0.036
2016/9/19 13:23:42	-0.048	-0.068	-0.112	84	1423558150	352270285	295	6943	-266.149	-0.112	-0.068	-0.048
2016/9/19 13:23:43	-0.02	-0.056	-0.084	84	1423555561	352271375	295	6973	-266.616	-0.084	-0.056	-0.02

(b)

Figure 2. Example of raw data from anemometer (*a*) and the automobile data recorder (*b*).

(a) (b)

Figure 3. Automobile data recorder (*a*) and video scenes collected by it (*b*).

m/s. As for the wind direction, the range is 0–360°, the accuracy is 3°, and the resolution is 1°. The ultrasonic anemometer monitor software, which was set up on a laptop, can be used to monitor the data recorded by the anemometer. The acquired data include the time, wind speed in intervals of 1 s, wind angle, and temperature (Figure 2).

Considering the convenience of fixation, installation, and possible operation needed during the experimental process, the anemometer was installed on the driving side, 0.5 m above the roof. When the instrument was installed, it was made sure that it was positioned horizontally and the beacon pointed to 0°, namely in the "northerly direction" aligned with the longitudinal axis of the vehicle.

Automobile Data Recorder

The automobile data recorder (Figure 3) can record information such as images and sounds along the way during the driving process, and the recorded information can be provided as evidence for traffic accidents. The automobile data recorder adopted in this experiment is the GDR 35-type supplied by GARMIN Corporation. It can record information about the real-time traffic trajectory, coordinates, time, and speed. The sampling frequency of the GPS receiver is 1 Hz.

During the test, the recorder was mounted behind the wind shield on the front passenger side. The instrument should be debugged and zoomed to ensure that the scenery the camera records is the same as what the driver can see. After the experiment, all the data is recorded in video format and the video data can be converted into Excel format. The contents include the video frame number, vehicle speed, XY coordinates, azimuth angle, elevation, lateral offset, and calculation station. Among these data, it is necessary to change the frame number into the actual time according to the video, with an interval of 1s. Next, the data originating from the GARMIN GDR35 and the anemometer should be synchronized for the following process.

Test Vehicle, Site, and Time

The test vehicle was a NISSAN SYLPHY, a typical sedan style, with the anemometer installed above the top surface of the vehicle (Figure 1). The adoption of the NISSAN SYLPHY as the test vehicle serves two purposes: (1) to acquire general site-specific crosswind velocity data along a highway such that the data are independent of the test vehicle adopted; and (2) to collect vehicle-specific crosswind velocity data with only one anemometer, because the data recorded by the anemometer is relative to the moving car system.

The experimental route covered the G60 Shanghai–Kunming highway, the G15 Shenyang–Haikou highway, the G92 Hangzhou Bay area ring highway, and the Hangzhou Bay Bridge (Figure 4). The length of the main section of the Shanghai–Kunming highway (Xinsong section) is 18.07 km. The G15 Shenyang–Haikou highway is the only one that traverses the southeastern coastal areas, with its Shanghai section starting at the Jiading–Zhuqiao main line (boundaries between Shanghai and the Jiangsu province) and ending at the Qianqiao main line. Its Zhejiang section is mainly composed of three parts, the Ningbo–Taizhou–Wenzhou Expressway, the Hangzhou–Pudong Expressway, and Hangzhou Bay Bridge. This bridge connects Haiyan and Cixi and is the world's third longest cross-sea bridge with a total length of 35 km. The G92 highway is surrounded by Hangzhou Bay, via Hangzhou, ending in Ningbo.

The test was conducted on September 19 and again, separately, on September 29, 2016 because of the existence of windy weather. Typhoon Malakas was located offshore from Zhejiang and moving toward the north on the 18th, landing in Kyushu, Japan on the 20th, and brought heavy rain in Zhejiang and the Shanghai area. On the 29th, the weather was affected by Super Typhoon Megi, which was generated on the surface of the Pacific Ocean near Taiwan's Taitung city on the 23rd and landed in Zhangzhou, Fujian province on the 28th. On the days affected by the typhoon, the average wind speed would be faster and the probability of continuous wind would also be larger compared with the usual days. Thus, the difference between the wind speed at different times along the highway would be easier to distinguish.

Calculation Method for Site-specific Wind and Crosswind Velocity

Following the work of Chen et al. (*12*), the fixed Cartesian coordinate system XYZ was adopted to describe the general wind environment along the

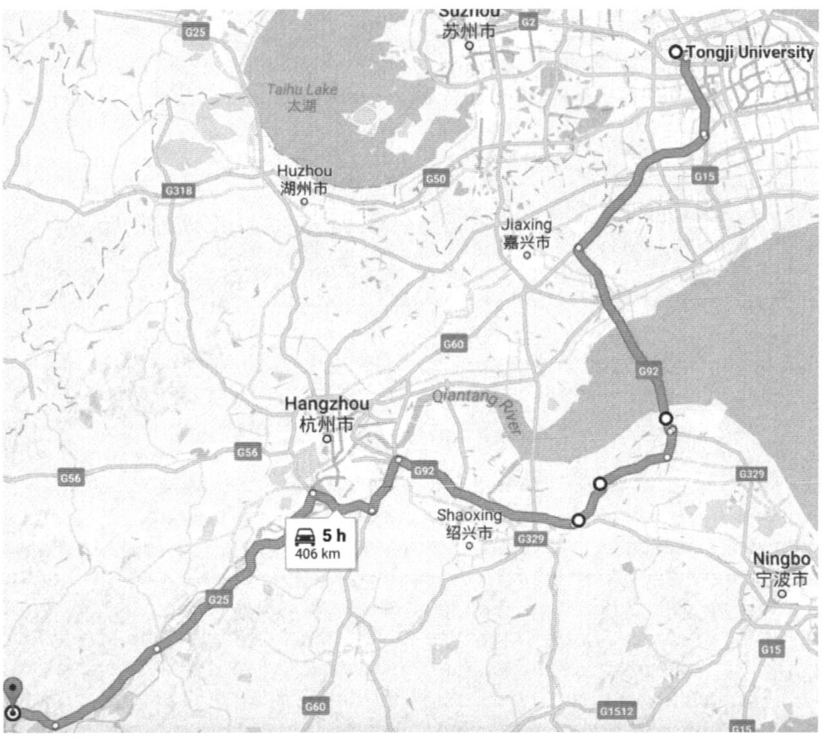

Figure 4. Experimental route.

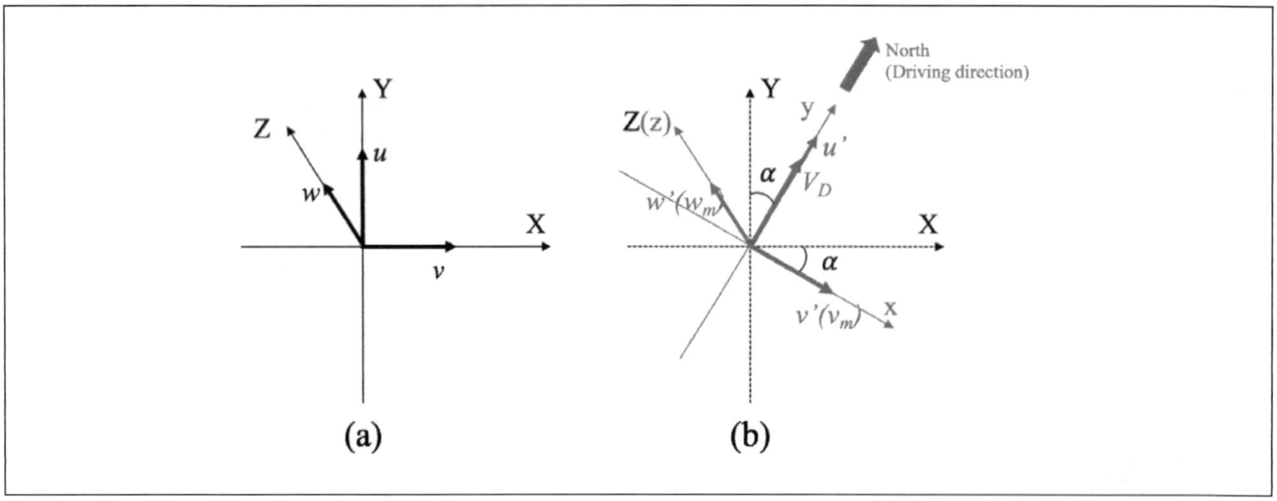

Figure 5. Wind velocity components in the fixed (*a*) and moving (*b*) system.

highway and this reference system is not related to any particular test vehicle. In this fixed coordinate system, the wind velocity can be regarded as a combination of three components *v, u,* and *w*, in the X, Y, and Z directions, respectively. Correspondingly, in the moving XYZ coordinate system attached to the specific test vehicle, the wind velocity can also be regarded as a combination of three components *v', u'* and *w'* (Figure 5). As mentioned before, the "northerly direction" of the anemometer installed on the top of the test vehicle is aligned with the longitudinal axis of the vehicle, thus, the measured wind velocity data of the anemometer can be separated as v_m, u_m, and w_m, representing the wind in the "easterly," "northerly," and "vertical" directions, respectively.

It is worth noting that the majority of highways are not parallel to the horizontal plane; however, because of their relatively small values, their angles do not have a significant impact on the crosswind velocity data. Thus, in this study, the X–Y plane is assumed to be horizontal and the Z-axis is parallel to the z-axis. In this situation, the "northerly" direction data from the anemometer u_m is actually equal to $u' + V_D$, where V_D is the vehicle velocity collected from the driving recorder. The "easterly" direction data of the anemometer $v_m = v'$, represents the crosswind. As for the "vertical" direction, the values of the *w* components of the wind velocities are approximately the same in either the moving or fixed coordinate system, namely $w \approx w' = w_m$.

A study of the impact of the wind velocity on different vehicles driving along the same highway requires the adoption of the site-specific wind velocity, which is dependent on the environment but has little or no dependence on the particular test vehicle. Moreover, the spatially continuous general wind velocity data in the fixed coordinate system can be easily determined according to the following formulas based on the wind velocity measurements by the anemometer:

$$u = (u_m - V_D)\cos\alpha - v_m \sin\alpha, \quad (1)$$

$$v = (u_m - V_D)\sin\alpha + v_m \cos\alpha, \quad (2)$$

$$w = w_m, \quad (3)$$

where α is the angle between the absolute Y-direction and the driving direction, which can be acquired from the GPS data provided by the GARMIN-GDR 35.

Wind Data Analysis

Data Calculation and Collation

The time range of the data recorded by the anemometer and the automobile data recorder is not exactly the same. In addition, the site-specific wind velocity, which can be used for the following analysis, needs to be calculated. Thus, the first step was to match the two kinds of data to achieve a uniform time format. The relevant data from the anemometer and the automobile data recorder were integrated in a data pool, which eventually contained a total of 3350 sets of data.

The site-specific wind velocity is influenced by factors such as the terrain and geographical conditions. The crosswind causes either lateral offset or turnover, and is very likely to cause an accident if the driver cannot undertake corrective action in time. In addition, the instantaneous wind change affects the driver's driving behavior, and thus has an important impact on traffic safety. Therefore, the respective velocities of both the site-specific wind and crosswind need to be determined

Table 1. Examples of Collated Data

Time	X	Y	Vehicle speed (m/s)	Site-specific wind velocity (m/s)	Crosswind velocity (m/s)	Direction of site-specific wind (deg)
2016/9/19 9:07:22	1442853564	366688717	28.611	3.797	1.509	−14.588
2016/9/19 9:07:23	1442856472	366686767	28.889	4.661	1.942	−13.375
2016/9/19 9:07:24	1442859326	366684813	29.167	2.204	1.731	13.769
2016/9/19 9:07:25	1442852213	366682852	29.444	3.047	2.088	5.243
2016/9/19 9:07:26	1442855127	366680884	29.722	4.444	2.269	−6.293
2016/9/19 9:07:27	1442858073	366678912	29.722	4.312	1.422	−17.751
2016/9/19 9:07:28	1442851049	366676932	30	4.195	1.937	9.497

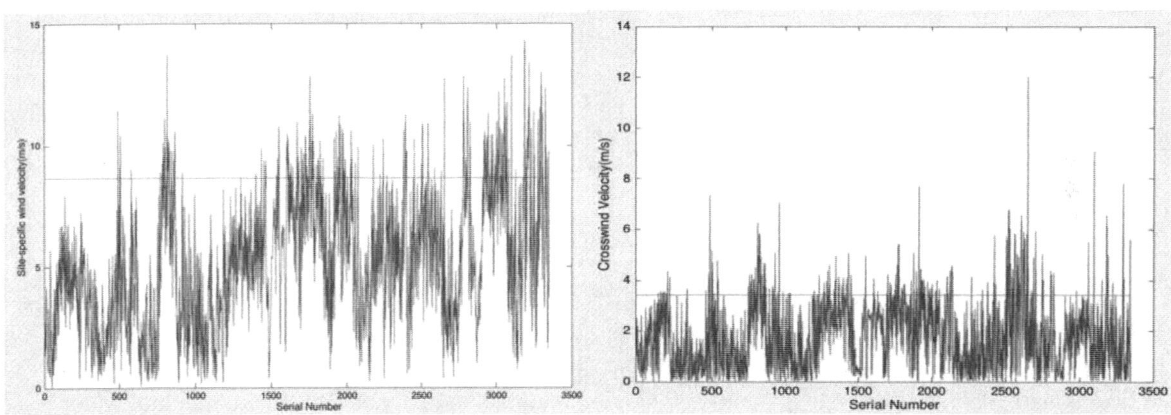

Figure 6. Site-specific wind velocity (*v* in the text) (*a*) and crosswind velocity (*v'* in the text) (*b*) in 1 second.

according to the method mentioned above. Some examples of the collated data are presented (Table 1).

Measurement of Wind Velocity

The wind velocity data fluctuate greatly from each 1-s interval to the next because of the transient feature of wind. However, if the velocity data were processed in time intervals of 5 s or 10 s, a loss of micro-traffic characteristics, such as the short-term peaking speed, might occur. Considering that the average velocity during this period of time would result in a potential loss of explanatory variables and invisible heterogeneity, the original time interval of 1 s was retained in the data analysis. Furthermore, all these data were not completely continuous in time. To observe the change in the wind velocity more easily, all the data were arranged in chronological order and numbered. With the number replaced by the moment, the curves were fitted (Figure 6).

The obvious difference in the wind velocity can be seen from the traces, but the overall velocity level is affected by the general weather conditions prevailing at the time; thus, a relative rather than an absolute value is more reasonable to measure the difference. To find a section with relatively strong wind and crosswind, all the wind velocity data were arranged from small to large when determining the level of the wind, and the 90th percentile value was taken as the evaluation boundary. This approach also enabled the avoidance of possible deviation caused by individual outlier values. The 90th percentile value of the site-specific wind velocity and crosswind velocity is 8.657 m/s and 3.412 m/s, respectively, as represented by the red line (Figure 6).

Geographic Location Positioning in ArcGIS

The corresponding XY coordinates of all the data were imported into ArcGIS for geographical location positioning, represented by green points (Figure 7). In the

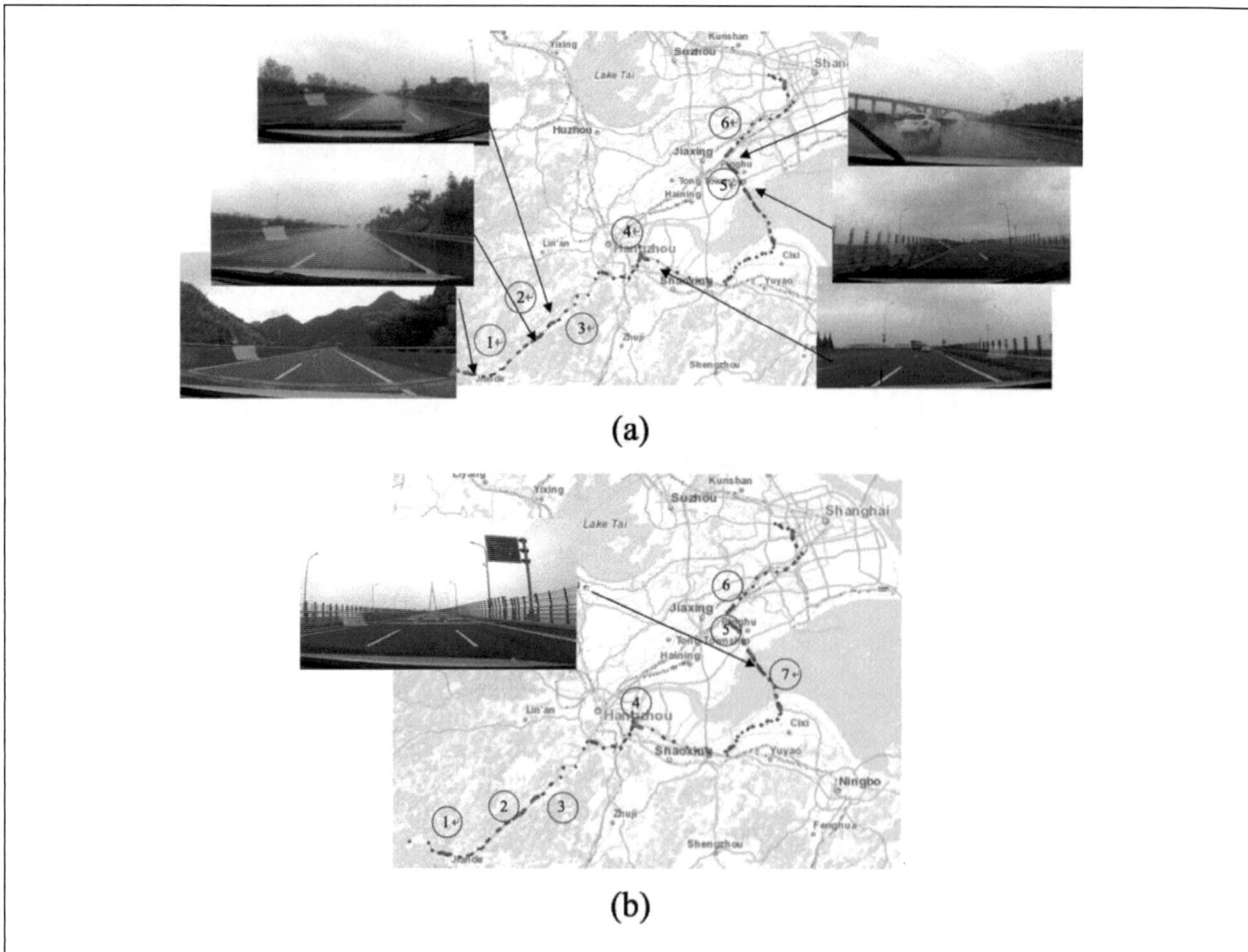

Figure 7. Sites at which the site-specific wind velocity (*a*) and crosswind velocity (*b*) exceed the 90th percentile value.

case of data greater than the 90th percentile value, the corresponding time and site should be identified. Red points are used to mark sites where the site-specific wind or crosswind velocity is greater than 8.657 m/s or 3.412 m/s, respectively. Actual scenes of these sites captured by the video recorder are also shown. Taking these locations as an example, elevation data are used for the following terrain analysis. A comparison of (*a*) and (*b*) shows that the distribution of locations at which the crosswind velocity is greater than the 90th percentile value is similar to that of locations at which the site-specific wind velocity is greater than the 90th percentile value, except for the Hangzhou Bay Bridge segment. The subsequent terrain analysis is therefore based on the crosswind results.

Terrain Analysis with Elevation Image

Because the wind velocity is influenced by the roadside environment and surrounding terrain, site-specific analyses would need to be conducted for areas with a relatively strong wind and crosswind to determine the common characteristics of these places and provide recommendations for future highway safety projects. An elevation image of China (Figure 8) was produced using ArcGIS software based on the SRTMDEM 90M-resolution of the original elevation data. Taking every 5 degrees of latitude and longitude division as a separate file, the deeper colors represent areas of lower elevation, whereas the shallower colors represent higher elevation.

The areas in which the red points are more intensive (for example, the No. 1 to No. 4 sites in Figure 7) are analyzed. Because the terrain in the Shanghai area at sites No. 5 and No. 6 is flat, and the wind environment may be affected by the city buildings, these two sites are not analyzed in detail in this study.

The elevation and contour plots at the four abovementioned sites are shown in Figure 9, with the mountains represented by black triangles. At No. 1 site, points

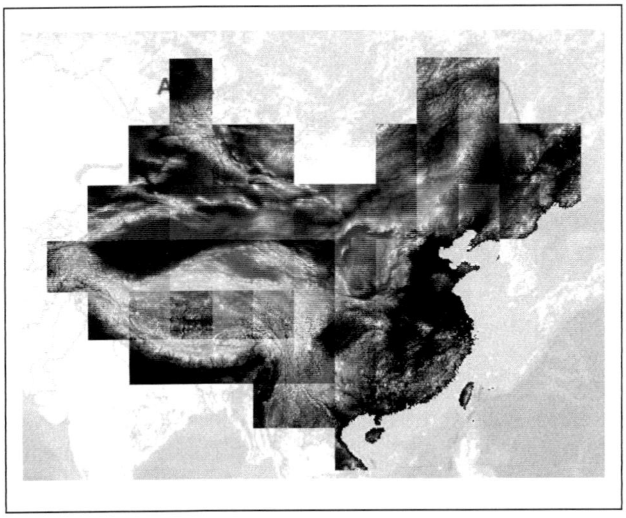

Figure 8. Elevation image of China.

m and p are located between two mountains and their height is lower than the surrounding environments. As the lower part between two peaks is a saddleback, the points indicated by m and p are believed to be on the saddleback. The points denoted n and q are located on the extension line of the saddleback, in the direction vertical to the driving direction at No. 1 site. Thus, it can be shown that the crosswind at n and q mainly comes from the valley. Similarly, No. 2 and No. 3 sites are flanked by mountains on both sides.

As for the No. 4 site, it is located in a large low-terrain area surrounded by three fan-shaped mountains. In addition, it can be seen (Figure 9*d*) that the Qiantang River is located in front of the point at which the roads reach their confluence. There sometimes exists a difference between the airflow over the land and a river. At locations at which the two kinds of airflow merge, atmospheric circulation can be expected to be promoted and a

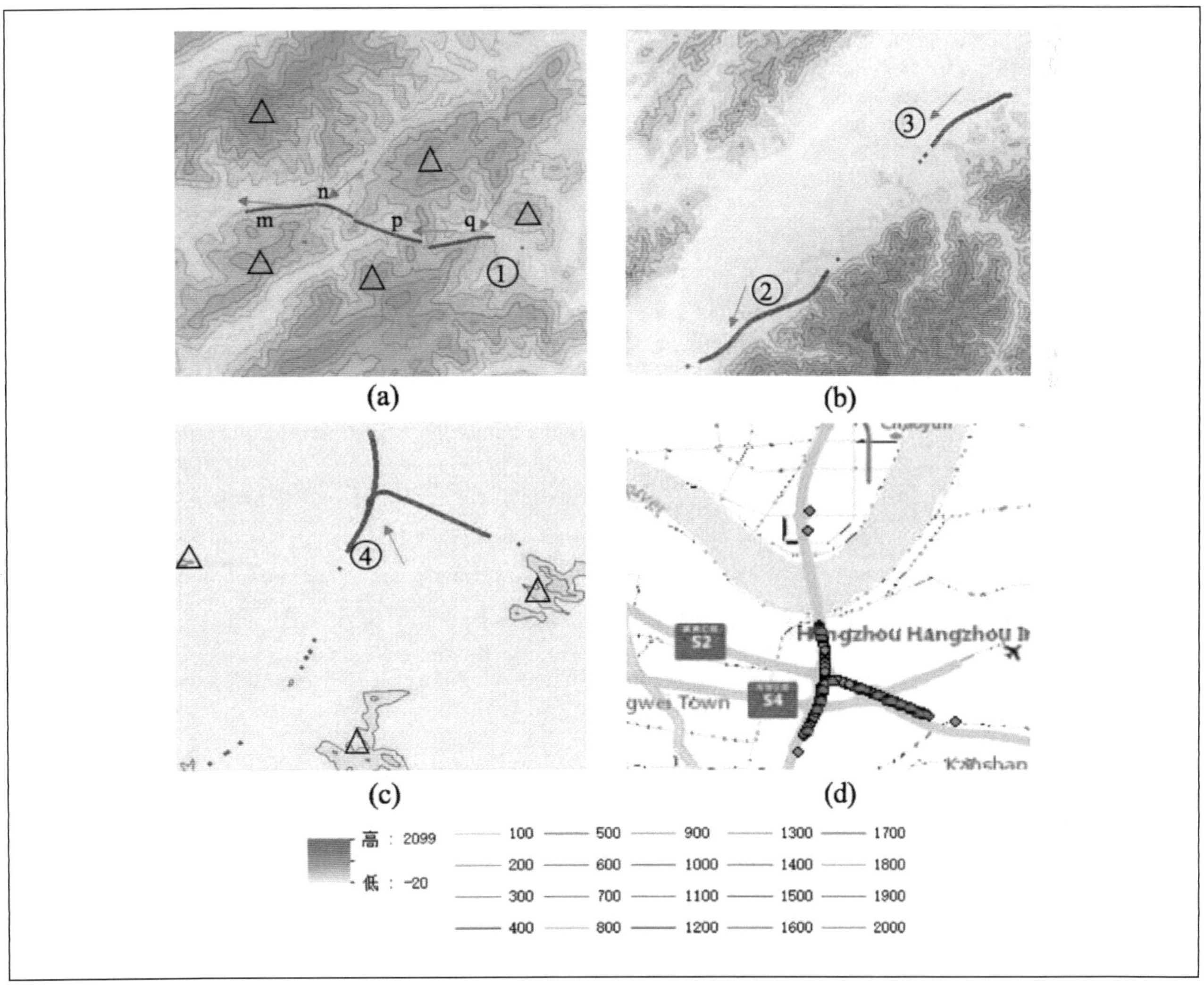

Figure 9. Elevation and contour plots of sites No. 1–4 (*a, b, c*) with a map of No. 4 site (*d*).

specific airflow phenomenon may be produced, which is likely to affect the wind velocity.

Based on the analysis above, all these sites were identified as saddlebacks or as the border between a riverbank and the river. Airflow from an open area into narrow areas such as canyons and valleys accelerates through the narrow area to cause a significant increase in the wind velocity. By observing the direction of the red arrows, which represent the wind direction, it is obvious that the direction of the wind at these locations is very close to and approximately parallel to the direction of the saddlebacks. This illustrates that the wind mainly blows from the saddlebacks or canyons and further validates the preceding explanation. This suggests that strong wind is more prone to occur in saddlebacks and more attention should also be paid to the border between a riverbank and rivers or oceans.

With regard to sites at which the crosswind velocity is relatively large, the video images show that they are usually located at the connecting section of a bridge and tunnel, or the bridgehead in addition to the possible impact of the saddleback terrain mentioned above. Moreover, the reason why the Hangzhou Bay Bridge segment experiences strong crosswind may be that the terrestrial block on the river and the sea is less than that on land; thus, the velocity of the airflow increases without the resistance from buildings and the terrain, allowing the wind to become stronger.

Conclusion

A method combining geo-location positioning and terrain analysis in ArcGIS with mobile mapping technology is introduced in this paper. The method enables wind velocity data in both the time and spatial domains to be obtained along any highway for further terrain analysis at locations with a high probability of strong crosswind. The method was used to conduct a field test on highways G60, G15, and G92 in China, and the results show that two primary types of data can be obtained: (1) site-specific wind velocity data independent of the test vehicle, and (2) crosswind related to specific vehicles.

The wind velocity was calculated on the basis of recorded measurements of the site-specific wind and crosswind. Sites with a wind velocity more than the 90th percentile value, which was taken as the threshold value, were considered to experience strong winds and crosswinds. After the geo-location positioning of these sites in ArcGIS and a terrain analysis based on elevation data, the following conclusions were drawn:

1. The wind velocity is related to the terrain environment, with sites characterized by strong wind usually being a saddleback or an intersection between land and river, and the direction of the wind in these locations is nearly parallel to the direction of the saddleback;
2. The crosswind is affected not only by the terrain environment, but also by traffic scenarios. Vehicles approaching the entrance or exit of a tunnel experience a sudden change in crosswind. That is, the crosswind is relatively significant at the start or endpoint of a bridge or tunnel, a bridgehead, or a cross-sea bridge.

In subsequent studies, field tests could be carried out on additional days along the same section of the highway to obtain data with greater regularity. These geographical conclusions can be further validated by Computational Fluid Dynamics (CFD) and used to provide recommendations for future highway projects. For instance, when selecting sites at which a wind barrier may be necessary, more attention could be paid to the existence of saddlebacks, valleys, bridge and tunnel connections, and intersections between the land and rivers, and so forth. Furthermore, as the basis for traffic safety assessment, the required actual wind data can be measured with this method.

Acknowledgments

This research was jointly sponsored by the Natural Science Foundation of China (51508409 and 71531011), the Fundamental Research Funds for Central Universities, and the project sponsored by SRF for ROCS, SEM.

Author Contributions

The authors confirm contribution to the paper as follows: study conception and design: Yatian Pu, Feng Chen, Xiaodong Pan; data collection: Peiyan Chen; analysis and interpretation of results: Yatian Pu, Feng Chen; draft manuscript preparation: Yatian Pu, Feng Chen. All authors reviewed the results and approved the final version of the manuscript.

References

1. World Health Organization. *Global Status Report on Road Safety 2015*. World Health Organization, Geneva, Switzerland, 2015.
2. Keay, K., and I. Simmonds. The Association of Rainfall and Other Weather Variables with Road Traffic Volume in Melbourne, Australia. *Accident Analysis and Prevention*, Vol. 37, No. 1, 2005, pp. 109–124.
3. Andersson, A., and L. Chapman. The Use of a Temporal Analogue to Predict Future Traffic Accidents and Winter Road Conditions in Sweden. *Meteorological Applications*, Vol. 18, No. 2, 2011, pp. 125–136.
4. Baker, C. J., and S. Reynolds. Wind-induced Accidents of Road Vehicles. *Accident Analysis and Prevention*, Vol. 24, No. 6, 1992, pp. 559–575.
5. Liu, Y., R. Yu, and L. Song. Elementary Study on the Influence of Crosswind on Driving Safety. *Journal of Shandong University of Technology*, Vol. 30, 2016, pp. 70–73.

6. Charuvisita, S., K. Kimurab, and Y. Fujino. Effects of Wind Barrier on a Vehicle Passing in the Wake of a Bridge Tower in Cross Wind and Its Response. *Journal of Wind Engineering and Industrial Aerodynamics*, Vol. 92, No. 7–8, 2004, pp. 609–639.
7. Taukus, A., and D. Miceviciene. Vehicle Stability under Crosswinds. *Intelligent Technologies in Logistics and Mechatronics Systems*, Vol. 1, 2013, pp. 241–246.
8. Batista, M., and M. Perkovič. A Simple Static Analysis of Moving Road Vehicle under Crosswind. *Journal of Wind Engineering and Industrial Aerodynamics*, Vol. 128, 2014, pp. 105–113.
9. Snaebjornsson, J. T., C. J. Baker, and R. Sigbjornsson. Probabilistic Assessment of Road Vehicle Safety in Windy Environments. *Journal of Wind Engineering and Industrial Aerodynamics*, Vol. 95, No. 9–11, 2007, pp. 1445–1462.
10. Rutz, F. R., and K. L. Rens. Wind Pressure and Strain Measurements on Bridges. I: Instrumentation/Data Collection System. *Journal of Performance of Constructed Facilities*, Vol. 22, No. 1, 2008, pp. 1–11.
11. Gonzalo, J., D. Domínguez, D. López, and J. Fernández. Lighter-than-air Particle Velocimetry for Wind Speed Profile Measurement. *Renewable and Sustainable Energy Reviews*, Vol. 33, 2014, pp. 323–332.
12. Chen, S., F. Chen, J. Liu, J. Wu, and B. Bienkiewicz. Mobile Mapping Technology of Wind Velocity Data along Highway for Traffic Safety Evaluation. *Transportation Research Part C: Emerging Technologies*, Vol. 18, No. 4, 2010, pp. 507–518.
13. McAuliffe, B. R., L. Belluz, and M. Belzile. Measurement of the On-Road Turbulence Environment Experienced by Heavy Duty Vehicles. *SAE International Journal of Commercial Vehicles*, Vol. 7, No. 2, 2014, pp. 685–702.
14. Saha, P., M. Ahmed, and R. K. Young. Safety Effectiveness of Variable Speed Limit System in Adverse Weather Conditions on Challenging Roadway Geometry. *Transportation Research Record: Journal of the Transportation Research Board*, 2015. 2521: 45–54.
15. Pinelli, J. P., C. Subramanian, and M. Plamondon. Wind Effects on Emergency Vehicles. *Journal of Wind Engineering and Industrial Aerodynamics*, Vol. 92, No. 7, 2004, pp. 663–685.
16. Snæbjörnsson, J. T., C. J. Baker, and R. Sigbjörnsson. Probabilistic Assessment of Road Vehicle Safety in Windy Environments. *Journal of Wind Engineering and Industrial Aerodynamics*, Vol. 95, No. 9–11, 2007, pp. 1445–1462.
17. Volpe, R., V. Ferrand, A. D. Silva, and L. L. Moyne. Forces and Flow Structures Evolution on a Car Body in a Sudden Crosswind. *Journal of Wind Engineering and Industrial Aerodynamics*, Vol. 128, 2014, pp. 114–125.
18. Maruyama, Y., and F. Yamazaki. Driving Simulator Experiment on the Moving Stability of an Automobile under Strong Crosswind. *Journal of Wind Engineering and Industrial Aerodynamics*, Vol. 94, No. 4, 2006, pp. 191–205.
19. Ren, G., J. Liu, J. Wan, Y. Guo, D. Yu, and J. Liu. Measurement and Statistical Analysis of Wind Speed Intermittency. *Energy*, Vol. 118, 2017, pp. 632–643.
20. Ozay, C., and M. S. Celiktas. Statistical Analysis of Wind Speed Using Two-Parameter Weibull Distribution in Alaçatı Region. *Energy Conversion and Management*, Vol. 121, 2016, pp. 49–54.
21. Li, X., H. Sun, W. Gao, Y. Shi, G. Liu, and Y. Wu. Wind Speed and Direction Measurement Based on Arc Ultrasonic Sensor Array Signal Processing Algorithm. *ISA Transactions*, Vol. 65, 2016, pp. 437–444.
22. Lubitz, W. D., and B. R. White. Wind-tunnel and Field Investigation of the Effect of Local Wind Direction on Speed-up Over Hills. *Journal of Wind Engineering and Industrial Aerodynamics*, Vol. 95, Vol. 8, 2007, pp. 639–661.
23. Stangroom, P. *CFD Modeling of Wind Flow over Terrain.* University of Nottingham, 2004.

The Standing Committee on Statewide Transportation Data and Information Systems (ABJ20) peer-reviewed this paper (18-03594).